Springer Monographs in Mathematics

Springer
Berlin
Heidelberg
New York
Barcelona
Hong Kong
London
Milan
Paris
Singapore
Tokyo

Vladimir Kozlov · Vladimir Maz'ya

Differential Equations with Operator Coefficients

with Applications
to Boundary Value Problems
for Partial Differential Equations

 Springer

Vladimir Kozlov
Vladimir Maz'ya
Linköping University
Department of Mathematics
58183 Linköping
Sweden

Library of Congress Cataloging-in-Publication Data

Kozlov, Vladimir, 1954-
Differential equations with operator coefficients with applications to boundary value problems
for partial differential equations / Vladimir Kozlov, Vladimir Maz'ya.
p. cm.-- (Springer monographs in mathematics)
Includes bibliographical references and indexes.
ISBN 3540651195 (alk. paper)
1. Differential equations. 2. Differential operators. 3. Boundary value problems. I. Maz'ia, V. G.
II. Title. III. Series.
QA372.K853 1999
515'.35--ddc21

Mathematics Subject Classification (1991): Primary 47-02; Secondary 34Gxx, 35Bxx, 35B40, 47F05

ISBN 3-540-65119-5 Springer-Verlag Berlin Heidelberg New York

SPIN 10690912   41/3143-5 4 3 2 1 0 – Printed on acid-free paper

Table of Contents

Part III. Asymptotic Theory of Operator Differential Equations

Introduction

★ The aim of this book is to give a self-contained presentation of a theory of ordinary differential equations with unbounded operator coefficients in a Hilbert or Banach space. This theory has been developed over the last ten years by the authors. We study equations of the form

$$L(t, D_t)u(t) := \sum_{q=0}^{\ell} A_{\ell-q}(t)D_t^q u(t) = f(t) \qquad (0.1)$$

on the real axis $\mathbb{R}$ or semiaxis $t > t_0$, where u and f are vector-valued functions and $D_t = -i\partial_t$.

We deal with the following topics

- conditions of solvability
- classes of uniqueness
- estimates for solutions
- asymptotic representations of solutions as $t \to \infty$

Equations of the form (0.1) have numerous applications, especially to the theory of partial differential equations, and our exposition of abstract results is accompanied by many new applications to this theory.

★ The roots of the theme treated here are the qualitative and asymptotic theories of linear ordinary differential equations; these date back to Liouville, Sturm, Green, Stokes, Poincaré, Lyapunov, to name only a few. In the twentieth century, fundamental contributions to the asymptotic analysis of ordinary differential equations were made by Birkhoff, Perron, Wentzel, Kramers, Brillouin and their numerous successors.

Attempts to extend results of the theory of linear ordinary differential equations to the infinite dimensional case were first made in the forties, and nowadays the theory of abstract differential equations is a boundless domain (see, for example, Fattorini (1984), where the list of references takes more than 100 pages).

An overwhelming majority of the results obtained deal with the class of abstract differential equations for which the Cauchy problem is well-posed. In particular, equations with bounded operator coefficients have been studied in great detail (see Massera, Schäffer (1966), Dalec'kii, Krein, M. (1974)). Even more attention has been paid to those equations with unbounded operator coefficients which can be studied by methods of semigroup theory (see the books Hille, Phillips (1974), Krein, S. (1971), Pazy (1974), Davies (1980) *et al*).

For many interesting applications, the Cauchy problem is not known to be well-posed. Despite this fact, only a few of the papers dealing with (0.1) are free from the assumption of well-posedness (in particular, Evgrafov (1960), Agmon, Nirenberg (1963), Pazy (1967, 1974), Maz'ya, Plamenevskii (1971, 1972), Plamenevskii (1972, 1973) and Bagirov, Kondratiev (1991)). A number of important questions remained open and this has stimulated our work.

★ The book consists of three parts. Part I is devoted to the theory of the equation (0.1) with constant operator coefficients in Hilbert spaces:

$$\mathcal{A}(D_t)u(t) := \sum_{q=0}^{\ell} A_{\ell-q} D_t^q u(t) = f(t). \tag{0.2}$$

Restrictions on the class of equations are dictated by an analogy with elliptic, parabolic and more general quasielliptic equations. We describe zeros, prove solvability and uniqueness results, and obtain power-exponential asymptotic representations of solutions at infinity. The behaviour of a solution u is characterized by the function

$$t \rightarrow \| u \|_{W^\ell(t,t+1)}, \tag{0.3}$$

where $W^\ell(a, b)$ is a vector-valued Sobolev space of order ℓ on the interval (a, b). We show that (0.3) is majorized (modulo a constant factor) by a solution of the scalar ordinary differential equation

$$\left(\partial_t + k_+\right)^{m_+}\left(- \partial_t - k_-\right)^{m_-} w(t) = \|f\|_{L_2(t,t+1;H_0)}.$$

Here $L_2(a, b; H_0)$ is the space of square summable functions on (a, b) with values in a Hilbert space H_0, and the numbers $k_\pm$ and $m_\pm$ depend on the spectral properties of the operator pencil $\mathcal{A}(\lambda)$, where λ is a complex parameter. To be more precise, the numbers $k_\pm$ are chosen in such a way that the strip $k_- < \Im\lambda < k_+$ is free of eigenvalues of $\mathcal{A}$, and $m_\pm$ are the maximum lengths of the Jordan chains corresponding to eigenvalues of $\mathcal{A}$ on the lines $\Im\lambda = k_\pm$. Extensions and variants of this *comparison principle* are systematically exploited in the sequel.

In fact, it is a new feature of our theory, even for the case of constant coefficients, that we study properties of solutions of (0.1) by comparing them with solutions of scalar ordinary differential equations.

★ In Part II we consider the equation (0.1) as a small perturbation (either global or in a neighbourhood of infinity) of the equation (0.2) with constant coefficients, and characterize this perturbation by a majorant $\omega(t)$ of the function

$$t \to \text{const} \, ||L - \mathcal{A}||_{W^\ell(t,t+1) \to L_2(t,t+1;H_0)}.$$

Now the comparison principle takes the form

$$|| \, u \, ||_{W^\ell(t,t+1)} \leq \text{const} \, w(t), \tag{0.4}$$

where w is a solution of the scalar ordinary differential equation

$$\left((\partial_t + k_+)^{m_+} \left(- \partial_t - k_- \right)^{m_-} - \omega(t) \right) w(t) = ||f||_{L_2(t,t+1;H_0)}. \tag{0.5}$$

This equation has been studied in Kozlov, Maz'ya (1997). Information about solutions of this *generalized Sturm-Liouville equation*, combined with the inequality (0.4), implies various solvability and uniqueness theorems as well as estimates for solutions of (0.1). We begin with a mild assumption on smallness of $\omega(t)$ and deduce results for the equation (0.1) formulated in terms of Green's function of the equation (0.5). Then, under special conditions on $\omega(t)$, we derive more explicit statements.

In the last chapter of Part II we discuss some extensions of the previous theory. In particular we generalize our results to abstract hyperbolic equations with dissipation, as well as to abstract differential equations in variational form and in Banach spaces.

★ Part III seems to us the most important. Here we show how the classical asymptotic theory of ordinary differential equations with scalar coefficients can be extended to very general equations with unbounded operator coefficients. We apply a reduction of (0.1) to a first order system and split this system by using a spectral projection of the unperturbed operator. This results in a coupled system of two equations of the first order, one finite- and one infinite-dimensional. The second equation is solved with respect to the infinite-dimensional part of the solution by using our comparison principle. This leads to a certain system of scalar ordinary differential equations of the first order which are perturbed by a weak nonlocal operator. The leading term in the asymptotics of solutions to (0.1) can be found by solving this finite-dimensional system. The main difficulties arising in justifying this formal procedure lie in the correct choice of function spaces for solutions of the split system and in the treatment of the non-local operator.

The asymptotic scheme is new. In comparison with previous work it enables one to obtain asymptotic formulae for solutions of (0.1) under weaker conditions on the coefficients; this has been our ongoing concern.

We demonstrate the power of present approach by finding explicit asymptotic representations of solutions under various assumptions on the spectrum of the pencil $\mathcal{A}(\lambda)$. In fact, dealing with the variable coefficient case in Parts II and III, we successively strengthen these conditions in order to obtain increasingly precise results.

The rich possibilities for applications of the asymptotic technique developed in Part III to nonlinear operator differential equations remain outside the scope of the book.

In an Appendix we give a systematic treatment of the theory of holomorphic operator functions. The basic facts of this theory are used throughout the book.

More detailed information about the contents of the book can be obtained from the Introductions of each chapter.

★ We now give a number of examples from the theory of partial differential equations which illustrate the range of applicability of the methods and results presented in this book.

Consider, for example, the Poisson equation

$$\left(\partial_t^2 + \Delta_x\right) u = f \tag{0.6}$$

in a cylinder $C = \{(x,t) : x \in \Omega, t \in \mathbb{R}\}$, where Ω is a domain in $\mathbb{R}^n$. Clearly, (0.6) is a special case of (0.1) with $A_0 = -1$, $A_1 = 0$ and $A_2 = \Delta_x$.

More generally, the equation (0.2) with constant coefficients is a way of writing the partial differential equation

$$\sum_{q=0}^{\ell} \sum_{|\alpha| \le \ell - q} p_{q,\alpha}(x) \partial_x^\alpha \partial_t^q\, u = f$$

in the same cylinder C, $\partial_x = \{\partial_{x_1}, ..., \partial_{x_n}\}$. A similar equation with $p_{q,\alpha}$ also depending on t can be considered as (0.1) with variable coefficients.

The next example is the equation

$$P(D_x)u = f$$

in a neighbourhood of the origin, where $D_x = -i\partial_x$ and

$$P(D_x) = \sum_{|\alpha|=2m} p_\alpha D_x^\alpha, \quad p_\alpha = \text{const.} \tag{0.7}$$

We introduce spherical coordinates (r, θ), where $r > 0$ and θ is a point on the $(n-1)$-dimensional unit sphere S^{n-1}. After multiplication of (0.7) by r^{2m} and applying the change of variables

$$(\theta, r) \to (\theta, t) \quad \text{with} \quad t = \log r^{-1}, \tag{0.8}$$

we can rewrite (0.7) in the form

$$\sum_{q=0}^{2m} A_{2m-q}(\theta, D_\theta) D_t^q u = e^{-2mt} f, \tag{0.9}$$

where A_j are differential operators on S^{n-1} of order j. One can consider the solution u as a function of the variable t with values in a certain Hilbert or Banach space of functions on S^{n-1}. Then one can treat (0.9) as an ordinary differential equation with operator coefficients independent of t:

$$\mathcal{A}(D_t)u(t) = F(t).$$

For example, the n-dimensional Laplace operator Δ_x generates the abstract ordinary differential operator

$$\mathcal{A}(D_t) = -D_t^2 - i(n-2)D_t + \delta_\theta,$$

where δ_θ is the Laplace-Beltrami operator on S^{n-1}.

The $2m$-th order abstract ordinary differential operator

$$\mathcal{A}(D_t) = \prod_{j=0}^{m-1} \left((iD_t + 2j)(iD_t + 2j - n + 2) + \delta_\theta \right)$$

corresponds to the polyharmonic operator Δ_x^m.

If we start with the general operator with variable coefficients

$$P(x, D_x) = \sum_{|\alpha| \leq 2m} p_\alpha(x) D_x^\alpha,$$

and use the coordinates t, θ, the resulting ordinary differential operator will have t-dependent coefficients.

Another example is the Dirichlet problem

$$\begin{cases} P(x, D_x)u = f & \text{on } K \\ u = \partial_\nu u = \ldots = \partial_\nu^{m-1} u = 0 & \text{on } \partial K \setminus O, \end{cases}$$

where ∂_ν is the outer normal derivative and K is an open n-dimensional cone with vertex at the origin O:

$$K = \{(r, \theta) : r > 0, \ \theta \in \Omega\}, \ \Omega \subset S^{n-1}. \tag{0.10}$$

The corresponding ordinary differential equation has coefficients which are mappings of function spaces on Ω, such as Sobolev and Hölder spaces.

Systems of partial differential equations in a cylinder or a cone can be transformed to ordinary differential equations in a similar way.

Other applications are provided by partial differential equations in domains with more complicated boundary singularities. In fact, sometimes it is convenient to transform such equations to the form (0.1) by a mapping of the domain into a cylinder $\mathcal{C}$.

Consider, for example, the Dirichlet problem with zero boundary conditions for the $2m$-th order elliptic equation

$$P(D_x, D_\tau)u = f \tag{0.11}$$

with constant coefficients in the cuspidal domain

$$G = \{(x, \tau) : |x| < \tau^\alpha, \ \tau \in (0, 1)\}$$

where $\alpha > 1$. By introducing new coordinates $t = (\alpha - 1)^{-1}\, \tau^{1-\alpha}$, $\xi = x\tau^{-\alpha}$ and taking into account that

$$\tau^\alpha(D_x, D_\tau) = \left(D_\xi, -D_t + \frac{\alpha\xi}{(\alpha - 1)t}D_\xi\right),$$

we can transform (0.11) to the equation

$$P(D_\xi, D_t)U + t^{-1}Q(\xi, t, D_\xi, D_t)U = F \tag{0.12}$$

in a semicylinder, where Q is a differential operator with bounded coefficients. The equation (0.12) has the form (0.1), and its coefficients tend to the corresponding coefficients of the original equation as $t \to -\infty$.

The above list of examples does not exhaust the possible applications of the theory of the operator equation (0.1). One could also include higher order parabolic equations both in cylindrical and t-dependent domains, hyperbolic equations with dissipation and integro-differential equations. In the following chapters we turn to some of the examples mentioned above in order to illustrate our theory and obtain new qualitative and asymptotic results for partial differential equations.

★ A prerequisite of this book is an undergraduate university course on real and complex analysis. The reader ought to also be familiar with some basic facts of functional analysis and of the theory of partial differential equations.

Acknowledgements. We are most grateful to C. Schwab and L. I. Hedberg for reading parts of the preliminary version of the manuscript and for their valuable comments. We would like to acknowledge the support of the Swedish Natural Science Research Council (NFR) and the Swedish Research Council for Engineering Sciences (TFR). Our cordial gratitude is due to Springer-Verlag for accepting this book for publication.

Part I

Differential Equations
with Constant Operator Coefficients

1. Power-Exponential Zeros

1.1 Introduction

In order to get an idea of the solutions studied in this chapter consider the equation

$$\left(D_t^2 + A\right) U(t) = 0, \tag{1.1}$$

where A is a non-negative selfadjoint operator in a Hilbert space H with the domain $\mathrm{Dom}A = \mathcal{H}$ (for basic notions and facts of operator theory in a Hilbert space see, for example, Rudin (1973), Reed, Simon (1972) or Gohberg, Goldberg and Kaashoek (1990)).

We are interested in the solutions of (1.1) which have the form

$$U(t) = e^{i\lambda_0 t} \sum_{k=0}^{m} \frac{(it)^k}{k!} \varphi_{m-k} , \tag{1.2}$$

where λ_0 is a complex number, $\varphi_k \in \mathcal{H}$ and $\varphi_0 \neq 0$.

By inserting $U(t)$ in (1.1) we arrive at the equations for $\varphi_0, \ldots, \varphi_m$:

$$(\lambda_0^2 + A)\varphi_0 = 0, \tag{1.3}$$

$$(\lambda_0^2 + A)\varphi_1 + 2\lambda_0\varphi_0 = 0, \tag{1.4}$$

$$(\lambda_0^2 + A)\varphi_{k+2} + 2\lambda_0\varphi_{k+1} + \varphi_k = 0 \quad \text{for} \quad k = 0, \ldots, m-2. \tag{1.5}$$

Non-trivial solutions of (1.3) are called eigenvectors of the quadratic operator pencil

$$\mathbb{C} \ni \lambda \to \mathcal{A}(\lambda) = A + \lambda^2 : \mathcal{H} \to H, \tag{1.6}$$

which correspond to the eigenvalue λ_0 of this pencil. By $\mathbb{C}$ we denote the set of complex numbers and by operator pencils we call polynomials in $\lambda \in \mathbb{C}$ with operator coefficients. (One should distinguish between eigenvalues of pencils and operators!). Solutions $\varphi_1, \varphi_2, \ldots, \varphi_m$ of equations (1.4) and (1.5) are called generalized eigenvectors of the pencil $\mathcal{A}(\lambda)$ corresponding to λ_0.

Equation (1.3) has a non-trivial solution if and only if $\lambda_0 = \pm i\sqrt{\gamma}$, where $\gamma \geq 0$ is an eigenvalue of the operator A. In the case $\gamma > 0$, equation (1.4) has no solutions. This follows from multiplying (1.4) by φ_0. Hence, we obtain the solution

$$U(t) = e^{\pm i\sqrt{\gamma}t}\varphi_0, \tag{1.7}$$

where φ_0 is an arbitrary eigenvector of A corresponding to γ.

In the case $\gamma = 0$, we have $\lambda_0 = 0$ and by (1.3) and (1.4) both φ_0 and φ_1 are zeros of A. The equation for φ_2 becomes $A\varphi_2 = \varphi_0$, which is clearly unsolvable. Hence

$$U(t) = \varphi_1 + t\varphi_0. \tag{1.8}$$

Thus (1.7) and (1.8) exhaust all solutions (1.2) of (1.1).

A similar description of all solutions (1.2) can be given for the general equation

$$\sum_{q=0}^{\ell} A_{\ell-q} D_t^q U(t) = 0 \tag{1.9}$$

with constant operator coefficients acting in a pair of Hilbert spaces: $\mathcal{H} \to H$.

One meets such power-exponential solutions in basic courses on ordinary differential equations with either scalar or matrix coefficients. In the infinite dimensional case these solutions also play an important role. In particular, they determine the asymptotics at infinity of arbitrary solutions, and are used for constructing Green's kernels.

In this chapter we deal with solutions (1.2) of (1.9). We systematically employ basic facts of the theory of Fredholm operator pencils; these are collected without proof in Sect. 1.2. For readers' convenience we give a detailed exposition of these facts in Appendix. This treatment is framed in a more general theory of holomorphic operator functions.

In Sect. 1.3 we consider power-exponential solutions of both (1.9) and the formally adjoint homogeneous equation. We construct bases of these solutions which are biorthogonal with respect to a certain sesquilinear form. These bases will frequently be used in subsequent chapters.

Section 1.4 is devoted to power-exponential solutions of the nonhomogeneous equation with power-exponential right-hand side.

In the last section (section 1.5) we illustrate the above material with examples from the theory of partial differential equations.

1.2 Basics on Operator Pencils

1.2.1 Notation

Let $\mathcal{H}$, H be Hilbert spaces with the norms $\| \cdot \|_{\mathcal{H}}$, $\| \cdot \|_{H}$ and inner products $(\cdot, \cdot)_{\mathcal{H}}$, $(\cdot, \cdot)_{H}$, respectively. Suppose that $\mathcal{H}$ is dense in H and that

$$\| \varphi \|_H \leq C \| \varphi \|_{\mathcal{H}}.$$

We introduce the operator pencil

$$\mathcal{A}(\lambda) = \sum_{q=0}^{\ell} A_{\ell-q}\lambda^q : \mathcal{H} \to H \;, \tag{1.10}$$

where $A_q \in \mathcal{L}(\mathcal{H}, H)$. (By $\mathcal{L}(\mathcal{H}, H)$ we mean the space of linear bounded operators acting from $\mathcal{H}$ into H). We need some notation to describe the behaviour of the resolvent $\mathcal{A}^{-1}(\lambda)$. The following definitions are special cases of those given in Appendix for holomorphic operator functions.

The spectrum of $\mathcal{A}$ is the set of all $\lambda \in \mathbb{C}$ such that $\mathcal{A}(\lambda)$ is not invertible. If the equation $\mathcal{A}(\lambda_0)u = 0$ has a nontrivial solution then λ_0 is called an eigenvalue of the pencil $\mathcal{A}$, and the corresponding nontrivial solutions are called eigenvectors related to λ_0. The maximum number of linear independent eigenvectors is called the geometric multiplicity of λ_0. With an eigenvalue λ_0 of $\mathcal{A}$ we connect the linear set $S(\mathcal{A}, \lambda_0)$ which consists of all elements

$$\Phi(\lambda) = \sum_{k=0}^{m-1} \frac{\varphi_k}{(\lambda - \lambda_0)^{m-k}} \;, \quad \varphi_k \in \mathcal{H} \tag{1.11}$$

such that

$$\mathcal{A}(\lambda)\Phi(\lambda) = o(1) \;\; \text{for small } |\lambda - \lambda_0|.$$

The dimension of $S(\mathcal{A}, \lambda_0)$ is called the algebraic multiplicity of λ_0. If Φ is represented in the form (1.11) with $\varphi_0 \neq 0$ then the number m is called the degree of Φ and is denoted by $\deg \Phi$ and φ_0 is called the leading coefficient of Φ.

The system $\{\Phi_j\}_{j=1}^{J}$ of elements from $S(\mathcal{A}, \lambda_0)$ with $\deg \Phi_{k+1} \leq \deg \Phi_k$, $k = 1, \ldots, J-1$, is called a canonical generating system in $S(\mathcal{A}, \lambda_0)$ if the leading coefficients of these elements are linear independent and every $\Phi \in S(A, \lambda_0)$ can be represented as

$$\Phi(\lambda) = \sum_{j=1}^{J} p_j(\lambda)\Phi_j(\lambda) + r(\lambda),$$

where p_j are polynomials with complex coefficients and $r(\lambda)$ is a polynomial with coefficients in $\mathcal{H}$. As follows from Proposition A.4.6, the numbers $m_j = \deg \Phi_j$, $j = 1, \ldots, J$ are independent of the choice of the canonical generating system. They are called the partial multiplicities of λ_0. By (A.8) the sum of partial multiplicities coincides with the algebraic multiplicity.

A collection of vectors $\varphi_0, \ldots, \varphi_{m-1} \in \mathcal{H}$ is called a Jordan chain corresponding to the eigenvalue λ_0 if φ_0 is an eigenvector and

$$\sum_{k=0}^{n} \frac{1}{k!} \mathcal{A}^{(k)}(\lambda_0)\varphi_{n-k} = 0 \;\; \text{for } n = 1, \ldots, m-1,$$

where $\mathcal{A}^{(k)}(\lambda) = \partial_\lambda^k \mathcal{A}(\lambda)$. According to Proposition A.3.6, the collection $\varphi_0, \ldots, \varphi_{m-1}$ is a Jordan chain if and only if the function (1.11) belongs to $S(\mathcal{A}, \lambda_0)$. A set of Jordan chains

$$\{\varphi_{k,j}\}_{j=0}^{m_k-1}, \ k = 1, \dots, J, \tag{1.12}$$

is called canonical if $\varphi_{1,0}, \dots, \varphi_{J,0}$ is a basis in the set of all eigenvectors corresponding to the eigenvalue λ_0, $m_1 \geq \dots \geq m_J$ and $m_1 + \dots + m_J$ is equal to the algebraic multiplicity of λ_0. In Proposition A.4.4 it is proved that the system (1.12) is canonical if and only if the elements

$$\Phi_k(\lambda) = \sum_{j=0}^{m_k-1} \varphi_{k,j}(\lambda - \lambda_0)^{j-m_k}, \ k = 1, \dots, J,$$

form a canonical generating system in $S(\mathcal{A}, \lambda_0)$.

1.2.2 Decomposition of the Resolvent Near the Pole

As is well-known, for any function $f(\lambda)$ holomorphic at a point $\lambda_0 \in \mathbb{C}$, the function $1/f(\lambda)$ can be represented near λ_0 as the sum of a polynomial in $1/(\lambda - \lambda_0)$ and a holomorphic function. Here we state a similar result for the resolvent $\mathcal{A}^{-1}(\lambda)$ of the operator pencil $\mathcal{A}(\lambda)$.

We suppose that the operator $\mathcal{A}(\lambda)$ is Fredholm for every $\lambda \in \mathbb{C}$ and invertible for at least one value of λ.

By Proposition A.8.4 this assumption implies that the spectrum of the operator pencil $\mathcal{A}(\lambda)$ consists of isolated eigenvalues having finite algebraic multiplicities with the only possible limit point occurring at infinity.

By $\mathcal{H}^*$ we denote the Hilbert space dual to $\mathcal{H}$ with respect to the scalar product in H. We shall use the adjoint operator pencil

$$\mathcal{A}^*(\lambda) = \sum_{q=0}^{\ell} A_{\ell-q}^* \lambda^q \ : H \to \mathcal{H}^*,$$

where the operator A_k^* is adjoint of A_k with respect to the scalar product in H. Clearly, the operator $\mathcal{A}^*(\lambda)$ is also Fredholm for every $\lambda \in \mathbb{C}$ and invertible for at least one value of λ and therefore its spectrum is discrete. Moreover, if λ_* is an eigenvalue of $\mathcal{A}(\lambda)$ then $\overline{\lambda}_*$ is an eigenvalue of $\mathcal{A}^*(\lambda)$ and their geometric, partial and algebraic multiplicities coincide (see Proposition A.9.2).

The local structure of the resolvent $\mathcal{A}^{-1}(\lambda)$ is described in the following fundamental theorem, which is a special case of Theorem A.10.2 and Remark A.10.3.

Theorem 1.2.1. *Let λ_0 be an eigenvalue of $\mathcal{A}$ and let J and $m_1, \dots, m_J$ be its geometric and partial multiplicities. Suppose that*

$$\{\varphi_{k,s}\}, \ s = 0, \dots, m_k - 1, \ k = 1, \dots, J,$$

is a canonical system of Jordan chains of $\mathcal{A}$ corresponding to λ_0. Then

(i) *There exists a unique canonical system of Jordan chains*

$$\{\psi_{k,s}\},\ s = 0,\ldots,m_k - 1,\ k = 1,\ldots,J,$$

of $\mathcal{A}^*$ *corresponding to* $\overline{\lambda}_0$ *such that in a neighbourhood of* λ_0 *the resolvent can be represented as*

$$\mathcal{A}(\lambda)^{-1} = \sum_{k=1}^{J} \sum_{h=0}^{m_k-1} \frac{P_{k,h}}{(\lambda - \lambda_0)^{m_k-h}} + \Gamma(\lambda), \tag{1.13}$$

where

$$P_{k,h} = \sum_{s=0}^{h} (\cdot, \psi_{k,s})_H\ \varphi_{k,h-s} \tag{1.14}$$

and Γ *is holomorphic in a neighbourhood of* λ_0.

(ii) *The canonical system* $\{\psi_{k,s}\}$ *in* (i) *satisfies the biorthogonality condition*

$$\sum_{s=0}^{n} \sum_{\sigma=s+1}^{m_k+s} \frac{1}{\sigma!} (\mathcal{A}^{(\sigma)}(\lambda_0)\varphi_{k,m_k+s-\sigma},\ \psi_{j,n-s})_H = \delta_k^j\ \delta_n^o \tag{1.15}$$

for $k, j = 1,\ldots J$ *and for* $n = 0,\ldots,m_j - 1$.

(iii) *If*

$$\psi_{j,0},\ldots,\psi_{j,m_j-1},\ j = 1,\ldots,J, \tag{1.16}$$

is a collection of Jordan chains of $\mathcal{A}^*$ *corresponding to* $\overline{\lambda}_0$ *which is subject to* (1.15), *then the collection* (1.16) *is a canonical system satisfying* (i).

1.2.3 Two-Term Quadratic Pencils

We illustrate Theorem 1.2.1 by constructing the decomposition (1.13) for the special case

$$\mathcal{A}(\lambda) = A + \lambda^2 : \mathcal{H} \to H, \tag{1.17}$$

where A is a self-adjoint operator in H defined on the Hilbert space $\mathcal{H} \subset H$.

Example 1.2.2. Suppose that A is positive definite and the spectrum of A consists of the eigenvalues $\gamma_k,\ k \geq 1$, satisfying $0 < \gamma_1 < \gamma_2 < \ldots$ and $\gamma_k \to \infty$. Denote by J_k the multiplicity of γ_k and assume that J_k is finite. Let $\alpha_1^{(k)},\ldots,\alpha_{J_k}^{(k)}$ be the eigenvectors of A corresponding to γ_k. We suppose that the set of all eigenvectors forms an orthonormal basis in H.

Clearly, the spectrum of the pencil $\mathcal{A}(\lambda)$ consists of the eigenvalues

$$\lambda_\nu = \begin{cases} i\sqrt{\gamma_\nu} & \text{for } \nu = 1, 2, \ldots \\ -i\sqrt{\gamma_{-\nu}} & \text{for } \nu = -1, -2, \ldots \end{cases} \tag{1.18}$$

and the corresponding eigenvectors are given by

$$\varphi_s^{(\nu)} = \begin{cases} \alpha_s^{(\nu)} & \text{if } \nu \geq 1, \\ \alpha_s^{(-\nu)} & \text{if } \nu \leq -1. \end{cases} \tag{1.19}$$

The equation for the first generalized eigenvector ($A\varphi_1 = -2\lambda\varphi_0$, where φ_0 is an eigenvector corresponding to the eigenvalue λ) is obviously unsolvable. Hence the space $S(\mathcal{A}, \lambda_\nu)$ consists of the elements

$$\Phi(\lambda) = (\lambda - \lambda_\nu)^{-1}\varphi,$$

where φ is either an arbitrary eigenvector of $\mathcal{A}(\lambda)$ corresponding to λ_ν or $\varphi = 0$.

Furthermore, both the algebraic and geometric multiplicities are equal to $J_{|\nu|}$, the partial multiplicities equal 1 and the system $\{\varphi_s^{(\nu)}\}_{s=1}^{J_{|\nu|}}$ is canonical.

The adjoint pencil $\mathcal{A}^*(\lambda)$ has the same form $A + \lambda^2$, except that the new A is a continuous extension of the old one, with domain H and range $\mathcal{H}^*$. The pencil $\mathcal{A}^*(\lambda)$ had the same eigenvalues and eigenvectors as $\mathcal{A}(\lambda)$, and there are no generalized eigenvectors of $\mathcal{A}^*(\lambda)$. The biorthogonality condition (1.15) takes the form

$$\left(\mathcal{A}'(\lambda_\nu)\varphi_k^{(\nu)}, \psi_j^{(\nu)}\right)_H = \delta_k^j,$$

which is the same as

$$\left(\varphi_k^{(\nu)}, \psi_j^{(\nu)}\right)_H = (2\lambda_\nu)^{-1}\delta_k^j.$$

In order to fulfill this equality it is sufficient to define

$$\psi_j^{(\nu)} = (2\bar{\lambda}_\nu)^{-1}\varphi_j^{(\nu)}.$$

We now decompose $\mathcal{A}(\lambda)^{-1}$ near an arbitrary pole λ_ν by referring to Theorem 1.2.1

$$\mathcal{A}(\lambda)^{-1} = \frac{1}{\lambda - \lambda_\nu}\sum_{s=1}^{J_{|\nu|}}(\cdot, \psi_s^{(\nu)})_H\varphi_s^{(\nu)} + \Gamma(\lambda).$$

In fact, we do not need the general Theorem 1.2.1 to obtain this formula for the special case under consideration because

$$(A + \lambda^2)^{-1} = \sum_\nu \sum_{s=0}^{J_{|\nu|}} \frac{(\cdot, \varphi_s^{(\nu)})_H\varphi_s^{(\nu)}}{\lambda^2 - \lambda_\nu^2}.$$

by the spectral decomposition of the operator A.

Example 1.2.3. We shall keep the assumptions of Example 1.2.2 with except that the operator A is non-negative, and $\gamma_0 = 0$ is an eigenvalue of A. Let J_0 be the multiplicity of γ_0 and $\alpha_1^{(0)}, \ldots, \alpha_{J_0}^{(0)}$ be the eigenvectors of A corresponding to γ_0, $(\alpha_p^{(0)}, \alpha_q^{(0)})_H = \delta_p^q$. By turning to the pencil $\mathcal{A}(\lambda) = A + \lambda^2$ we find that $\lambda = 0$ is also its eigenvalue and, as we have seen in the introduction to this chapter, each eigenvector φ_0 of $\mathcal{A}(\lambda)$ corresponding to $\lambda = 0$ generates the first generalized eigenvector φ_1 which belongs to ker A. Moreover, there are no second generalized eigenvectors. Therefore $S(\mathcal{A}, 0)$ consists of the elements

$$\Phi(\lambda) = \lambda^{-2}\varphi_0 + \lambda^{-1}\varphi_1, \quad \varphi_0, \varphi_1 \in \ker A.$$

Furthermore, the geometric multiplicity is J_0, the algebraic multiplicity is $2J_0$, partial multiplicities are equal to 2 and the system

$$\left\{\varphi_{s,0}^{(0)} = \alpha_s^{(0)}, \quad \varphi_{s,1}^{(0)} = 0\right\} \quad, \quad s = 1, \ldots, J_0,$$

is canonical.

The pencil $\mathcal{A}^*(\lambda)$ has the same eigenvectors and generalized eigenvectors corresponding $\lambda = 0$. Let us take a canonical system of $\mathcal{A}^*(\lambda)$ in the form

$$\left\{\psi_{s,0}^{(0)}, \psi_{s,1}^{(0)}\right\}, \quad s = 1, \ldots, J_0,$$

where $\left\{\psi_{s,0}^{(0)}\right\}_{s=1}^{J_0}$ is a basis in ker A. The biorthogonality condition (1.15) takes the form

$$(\varphi_{k,0}, \psi_{j,0})_H = \delta_k^j, \quad (\varphi_{k,0}, \psi_{j,1})_H = 0,$$

where $k, j = 1, \ldots, J_0$. Hence

$$\psi_{j,0} = \varphi_{j,0} \quad \text{and} \quad \psi_{j,1} = 0.$$

Finally,

$$\mathcal{A}(\lambda)^{-1} = \lambda^{-2} \sum_{s=1}^{J_0} (\cdot, \varphi_{s,0})_H \varphi_{s,0} + \Gamma(\lambda)$$

by Theorem 1.2.1. (Incidentally, the last formula follows trivially from the Fourier series for $\mathcal{A}(\lambda)^{-1}$).

We also note that canonical systems of $\mathcal{A}(\lambda)$ and $\mathcal{A}^*(\lambda)$ corresponding to non-zero eigenvalues λ_ν, $\nu = \pm 1, \pm 2, \ldots$, have already been constructed in Example 1.2.2.

1.3 Power-Exponential Solutions of the Homogeneous Equation

1.3.1 Notation. Spaces $Z(\mathcal{A}, \lambda_\nu)$ and $Z(\mathcal{A}^*, \overline{\lambda}_\nu)$

Let the eigenvalues of the operator pencil $\mathcal{A}(\lambda)$ be indexed by integers. With the eigenvalue λ_ν of $\mathcal{A}(\lambda)$ we connect the following notation:

J_ν is the geometric multiplicity of λ_ν;

$m_{\nu k}$, $k = 1, \ldots, J_\nu$, are the partial multiplicities of λ_ν;

$\kappa_\nu = m_{\nu 1} + \ldots + m_{\nu J_\nu}$ is the algebraic multiplicity of λ_ν.

According to the previous section, the number $\overline{\lambda}_\nu$ is an eigenvalue of $\mathcal{A}^*(\lambda)$ and its geometric, partial and algebraic multiplicities coincide with those of the eigenvalue λ_ν of $\mathcal{A}(\lambda)$. By Theorem 1.2.1 there exist canonical systems

$$\{\varphi_{ks}^{(\nu)}\} \quad \text{and} \quad \{\psi_{ks}^{(\nu)}\}, \ k = 1, \ldots, J_\nu, \ s = 0, \ldots, m_{\nu k} - 1,$$

of $\mathcal{A}(\lambda)$ and $\mathcal{A}^*(\lambda)$, respectively, corresponding to the eigenvalues λ_ν and $\overline{\lambda}_\nu$, such that

$$\sum_{s=0}^{n} \sum_{\sigma=s+1}^{m_{\nu k}+s} \frac{1}{\sigma!} (\mathcal{A}^{(\sigma)}(\lambda_\nu) \varphi_{k,m_{\nu k}+s-\sigma}^{(\nu)}, \ \psi_{j,n-s}^{(\nu)})_H = \delta_k^j \, \delta_n^o \tag{1.20}$$

for $k, j = 1, \ldots, J_\nu$, $n = 0, \ldots, m_{\nu j} - 1$.

Denote by $Z(\mathcal{A}, \lambda_\nu)$ the set of solutions of

$$\mathcal{A}(D_t)U = 0 \tag{1.21}$$

having the form

$$U(t) = e^{it\lambda_\nu} \sum_{\sigma=0}^{m} \frac{(it)^\sigma}{\sigma!} u_{m-\sigma}, \ u_{m-\sigma} \in \mathcal{H}. \tag{1.22}$$

By the definition of the Jordan chain, one directly verifies that (1.22) is a solution of (1.21) if and only if λ_ν is an eigenvalue of the pencil $\mathcal{A}(\lambda)$, u_0 is an eigenvector corresponding to λ_ν and $u_1, \ldots, u_m$ are generalized eigenvectors, or equivalently (see Proposition A.3.6), the vector function

$$R(U) = \sum_{\sigma=0}^{m} \frac{u_\sigma}{(\lambda - \lambda_\nu)^{m+1-\sigma}}$$

belongs to $S(\mathcal{A}, \lambda_\nu)$. Thus we get an algebraic isomorphism

$$Z(\mathcal{A}, \lambda_\nu) \ni U \to R(U) \in S(\mathcal{A}, \lambda_\nu).$$

Hence $\dim Z(\mathcal{A}, \lambda_\nu)$ is equal to the algebraic multiplicity κ_ν of λ_ν. It is easily seen that

$$R\big((D_t - \lambda_\nu)^s U\big) = SP\big((\lambda - \lambda_\nu)^s R(U)\big), \tag{1.23}$$

where SP denotes the singular part of a meromorphic function. By Proposition A.4.4, the collection

$$\big\{R_{t\to\lambda}\big(e^{i\lambda_\nu t}\Phi_k^{(\nu)}(it)\big)\big\}_{k=1}^J,$$

with

$$\Phi_k^{(\nu)}(z) = \sum_{\sigma=0}^{m_{\nu k}-1} \frac{z^\sigma}{\sigma!} \varphi_{k,m_{\nu k}-1-\sigma}^{(\nu)}, \tag{1.24}$$

is a canonical generating system in $S(\mathcal{A}, \lambda_\nu)$. Hence (and by (1.23)) the vector functions

$$e^{it\lambda_\nu} D_t^s \Phi_k^{(\nu)}(it), \ k = 1, \ldots J_\nu, \ s = 0, \ldots, m_{\nu k} - 1, \tag{1.25}$$

form a basis in $Z(\mathcal{A}, \lambda_\nu)$.

Analogously, let $Z(\mathcal{A}^*, \overline{\lambda}_\nu)$ be the set of all solutions of

$$\mathcal{A}^*(D_t)V = 0, \tag{1.26}$$

which have the form

$$V(t) = e^{it\overline{\lambda}_\nu} \sum_{\sigma=0}^m \frac{(it)^\sigma}{\sigma!} \ v_{m-\sigma}, \ v_{m-\sigma} \in H.$$

Then the collection of the vector functions

$$e^{it\overline{\lambda}_\nu} D_t^s \Psi_k^{(\nu)}(it), \ k = 1, \ldots, J_\nu, \ s = 0, \ldots, m_{\nu k} - 1, \tag{1.27}$$

where

$$\Psi_k^{(\nu)}(z) = \sum_{\sigma=0}^{m_{\nu k}-1} \frac{z^\sigma}{\sigma!} \psi_{k,m_{\nu k}-1-\sigma}^{(\nu)}, \tag{1.28}$$

is a basis in $Z(\mathcal{A}^*, \overline{\lambda}_\nu)$.

1.3.2 A Biorthogonality Condition

Here we show that the biorthogonality condition (1.20) can be rewritten as a biorthogonality condition for the solutions

$$e^{i\lambda_\nu t} \ D_t^h \Phi_k^{(\nu)}(it) \quad \text{and} \quad e^{i\overline{\lambda}_\mu t} \ D_t^n \Psi_q^{(\mu)}(it).$$

To this end we introduce the sesquilinear form

$$Z(\mathcal{A}, \lambda_\nu) \times Z(\mathcal{A}^*, \overline{\lambda}_\mu) \ni (\mathcal{U}, \ \mathcal{V}) \to \int_{\mathbb{R}} \big(\mathcal{A}(D_t)\eta(t)\mathcal{U}(t), \mathcal{V}(t)\big)_H dt,$$

where η is a smooth function on $\mathbb{R}$, equal to 1 in a neighborhood of $-\infty$ and to 0 in a neighborhood of $+\infty$. This form is well defined since the integration is extended over a finite segment of the real axis. Furthermore, this form is independent of the choice of η. In order to see this, notice that for two different functions η_1 and η_2 we have

$$\int_{\mathbb{R}} \left(\mathcal{A}(D_t)(\eta_1 - \eta_2)(t)\mathcal{U}(t), \mathcal{V}(t) \right)_H dt = 0$$

(by integration by parts).

Proposition 1.3.1. *The biorthogonality condition*

$$\int_{\mathbb{R}} \left(\mathcal{A}(D_t)\big(\eta(t)e^{i\lambda_\nu t} D_t^h \Phi_k^{(\nu)}(it)\big), e^{i\bar{\lambda}_\mu t} D_t^n \Psi_q^{(\mu)}(it) \right)_H dt$$
$$= i\, \delta_\nu^\mu\, \delta_k^q\, \delta_{m_{\nu k}-1-h}^n \tag{1.29}$$

holds for all integers ν, μ; $k = 1, \ldots, J_\nu$; $h = 0, \ldots, m_{\nu k} - 1$; $q = 1, \ldots, J_\mu$; $n = 0, \ldots, m_{\mu q} - 1$.

Remark 1.3.2. In the sequel we shall also use the following modification of (1.29) with $+\infty$ and $-\infty$ interchanged:

$$\int_{\mathbb{R}} \left(\mathcal{A}(D_t)\big(\xi(t)e^{i\lambda_\nu t} D_t^h \Phi_k^{(\nu)}(it)\big), e^{i\bar{\lambda}_\mu t} D_t^n \Psi_q^{(\mu)}(it) \right)_H dt$$
$$= -i\delta_\nu^\mu\, \delta_k^q\, \delta_{m_{\nu k}-1-h}^n. \tag{1.30}$$

Here ξ is a smooth function on $\mathbb{R}$, which is equal to 1 in a neighbourhood of $+\infty$ and to 0 in a neighbourhood of $-\infty$. This follows by putting $\eta = 1 - \xi$ in (1.29).

We give two technical assertions, which will be used in the proof of Proposition 1.3.1.

Lemma 1.3.3. *Let*

$$\eta \in C^\infty(\mathbb{R}) \quad \text{and} \quad \Phi \in C^\infty(\mathbb{R}; \mathcal{H}),$$

where $C^\infty(\mathbb{R}; \mathcal{H})$ *is the set of infinitely differentiable functions on* $\mathbb{R}$ *with values in* $\mathcal{H}$ *Then for complex* λ

$$\mathcal{A}(\lambda + D_t)(\eta\Phi) - \eta\mathcal{A}(\lambda + D_t)\Phi$$
$$= \sum_{j=1}^{\ell} \frac{1}{j!} \mathcal{A}^{(j)}(\lambda) \sum_{\sigma=0}^{j-1} D_t^\sigma \big(D_t \eta D_t^{j-\sigma-1}\Phi \big).$$

Proof. Since

$$\mathcal{A}(\lambda + D_t) = \sum_{j=1}^{\ell} \frac{1}{j!} \mathcal{A}^{(j)}(\lambda) D_t^j$$

it is sufficient to use the identity

$$D_t^j(\eta\Phi) - \eta D_t^j \Phi - \sum_{\sigma=0}^{j-1} D_t^\sigma \left(D_t \eta D_t^{j-\sigma-1}\Phi\right) = 0,$$

which can be verified by induction in j. □

Lemma 1.3.4. *Let*

$$\mathcal{U}(t) = e^{i\lambda_\nu t} P(t) \in Z(\mathcal{A}, \lambda_\nu) \tag{1.31}$$

and

$$\mathcal{V}(t) = e^{i\overline{\lambda}_\mu t} Q(t) \in Z(\mathcal{A}^*, \overline{\lambda}_\mu).$$

Then

$$\int_{\mathbb{R}} \left(\mathcal{A}(D_t)\eta(t)\mathcal{U}(t), \mathcal{V}(t)\right)_H dt \tag{1.32}$$

$$= i e^{i(\lambda_\nu - \lambda_\mu)\tau} \sum_{j=1}^{\ell} \frac{1}{j!} \sum_{\sigma=0}^{j-1} \left(\mathcal{A}^{(j)}(\lambda_\nu) D_\tau^{j-\sigma-1} P(\tau), (D_\tau - \overline{\lambda}_\nu + \overline{\lambda}_\mu)^\sigma Q(\tau)\right)_H.$$

In particular, the right-hand side does not depend on τ.

Proof. The left-hand side of (1.32) is equal to

$$\int_{\mathbb{R}} e^{i(\lambda_\nu - \lambda_\mu)t} \left(\mathcal{A}(D_t + \lambda_\nu)\eta(t) P(t), Q(t)\right)_H dt.$$

Using Lemma 1.3.3 and (1.31) we rewrite this integral as

$$\sum_{j=1}^{\ell} \frac{1}{j!} \int_{\mathbb{R}} e^{i(\lambda_\nu - \lambda_\mu)t} \left(\mathcal{A}^{(j)}(\lambda_\nu) \sum_{\sigma=0}^{j-1} D_t^\sigma (D_t \eta D_t^{j-\sigma-1} P), Q\right)_H dt$$

$$= \sum_{j=1}^{\ell} \frac{1}{j!} \int_{\mathbb{R}} e^{i(\lambda_\nu - \lambda_\mu)t} D_t \eta \left(\mathcal{A}^{(j)}(\lambda_\nu) \sum_{\sigma=0}^{j-1} D_t^{j-\sigma-1} P, (D_t - \overline{\lambda}_\nu + \overline{\lambda}_\mu)^\sigma Q\right)_H dt.$$

Since this expression does not depend on η, we can approximate the distribution $i\delta(t - \tau)$ by the sequence

$$t \to D_t \eta(N(t - \tau)), \quad N = 1, 2, \dots$$

Thus the right-hand sides of the last equality and (1.32) coincide. □

1.3.3 Proof of Proposition 1.3.1

Denote the left-hand side of (1.29) by $I = I(k, h; q, n)$. Let $\nu \neq \mu$. It follows from (1.32) that

$$I = e^{i(\lambda_\nu - \lambda_\mu)\tau} p(\tau),$$

where $p(\tau)$ is a polynomial in τ. Since this expression does not depend on τ we arrive at (1.29) for $\nu \neq \mu$.

Let $\nu = \mu$. First, consider the case $h = 0$. Again, it follows from (1.32) that

$$I = i \sum_{j=1}^{\ell} \sum_{\sigma} \frac{1}{j!} (\mathcal{A}^{(j)}(\lambda_\nu) \varphi_{k,m_{\nu k}-j+\sigma}^{(\nu)}, \psi_{q,m_{\nu q}-n-\sigma-1}^{(\nu)})_H.$$

Here the inner summation is extended over the set

$$\{\sigma : 0 \le \sigma \le j-1, \; j-m_{\nu k} \le \sigma \le m_{\nu q}-n-1\}.$$

Changing the summation order we obtain

$$I = i \sum_{\sigma=0}^{m_{\nu q}-n-1} \sum_{j=\sigma+1}^{m_{\nu k}+\sigma} \frac{1}{j!} (\mathcal{A}^{(j)}(\lambda_\nu) \varphi_{k,m_{\nu k}-j+\sigma}^{(\nu)}, \psi_{q,m_{\nu q}-n-\sigma-1}^{(\nu)})_H. \tag{1.33}$$

By (1.20) the right-hand side is equal to $i\, \delta_k^q\, \delta_{m_{\nu k}-1}^n$.

Now let $h > 0$. Then

$$I(k, h; q, n) =$$
$$\int_{\mathbb{R}} (\mathcal{A}(D_t)(D_t - \lambda_\nu)^h (\eta(t) e^{i\lambda_\nu t} \Phi_k^{(\nu)}(it)), e^{i\overline{\lambda}_\nu t} D_t^n \Psi_q^{(\nu)}(it))_H dt.$$

Integrating by parts we obtain

$$I(k, h; q, n)$$
$$= \int_{\mathbb{R}} (\mathcal{A}(D_t)\eta(t) e^{i\lambda_\nu t} \Phi_k^{(\nu)}(it), e^{i\overline{\lambda}_\nu t} D_t^{n+h} \Psi_q^{(\nu)}(it))_H dt$$
$$= \begin{cases} I(k, 0; q, n+h) & \text{if } n+h \le m_{\nu q} - 1 \\ 0 & \text{if } n+h \ge m_{\nu q}. \end{cases}$$

The proof is complete. $\square$

1.3.4 Two-Term Second Order Equations

We specify the spaces $Z(\mathcal{A}, \lambda_\nu)$ and $Z(\mathcal{A}^*, \overline{\lambda}_\nu)$, as well as the biorthogonality condition (1.29) for equation (1.1).

Example 1.3.5. Consider equation (1.1) corresponding to the operator pencil $\mathcal{A}(\lambda)$ from Example 1.2.2. We have

$$\Phi_s^{(\nu)}(z) = \varphi_s^{(\nu)}, \quad \Psi_s^{\nu}(z) = (2\overline{\lambda}_\nu)^{-1}\varphi_s^{(\nu)}, \quad s = 1, \ldots, J_0,$$

where λ_ν and $\varphi_s^{(\nu)}$ are given by (1.18) and (1.19). The vector-functions

$$e^{it\lambda_\nu}\varphi_s^{(\nu)} \quad \text{and} \quad e^{it\overline{\lambda}_\nu}(2\overline{\lambda}_\nu)^{-1}\varphi_s^{(\nu)}$$

form bases in $Z(\mathcal{A}, \lambda_\nu)$ and $Z(\mathcal{A}^*, \overline{\lambda}_\nu)$ respectively. The biorthogonality condition obtained in Proposition 1.3.1 can easily be checked in the present special case. In fact, for an arbitrary solution U of (1.1), we have

$$\int_{\mathbb{R}} \left((D_t^2 + A)(\eta U), V \right)_H dt$$

$$= -\int_{\mathbb{R}} \left(2\eta'(U', V)_H + \eta''(U, V)_H \right) dt$$

$$= \int_{\mathbb{R}} \eta' \left((U, V')_H - (U', V)_H \right) dt$$

By setting here

$$U(t) = e^{it\lambda_\nu}\varphi_s^{(\nu)}, \quad V(t) = e^{it\overline{\lambda}_\mu}(2\overline{\lambda}_\mu)^{-1}\varphi_j^{(\mu)}$$

we rewrite the last integral in the form

$$-\frac{i}{2}\left(1 + \frac{\lambda_\nu}{\lambda_\mu}\right)\left(\varphi_s^{(\nu)}, \varphi_j^{(\mu)}\right)_H \int_{\mathbb{R}} \eta' e^{it(\lambda_\nu - \lambda_\mu)} dt. \tag{1.34}$$

The scalar product in H fails to vanish only if $|\nu| = |\mu|$ and $s = j$. In the case $\nu = -\mu$, the first bracket vanishes. Finally for $\nu = \mu$, (1.34) equals

$$-i\int_{\mathbb{R}} \eta'(t)dt = i$$

in correspondence with (1.29).

Example 1.3.6. Let us consider the equation (1.1) again and suppose that $\mathcal{A}(\lambda)$ is the same as in Example 1.2.3. In view of Example 1.3.5 we may restrict ourselves to the eigenvalue $\lambda_0 = 0$. Then by Example 1.2.3

$$\Phi_s^{(0)}(z) = z\varphi_s^{(0)}, \quad \Psi_s^{(0)}(z) = z\varphi_s^{(0)}, s = 1, \ldots, J_0$$

and the vector-functions (1.25) and (1.27) (which form the bases in $Z(\mathcal{A}, 0)$ and $Z(\mathcal{A}^*, 0)$) take the form $it\varphi_s^{(0)}, \varphi_s^{(0)}, s = 1, \ldots, J_0$.

1.3.5 A Construction of Canonical Systems of Jordan Chains

The following auxiliary assertion gives a procedure for constructing a canonical system of Jordan chains, starting with a set of special solutions of (1.21).

Lemma 1.3.7. *Let J_ν and $m_{\nu 1}, \ldots m_{\nu J_\nu}$ be the geometric and the partial multiplicities of the eigenvalue λ_ν, and let*

$$e^{i\lambda_\nu t}\Phi_k(it), \ k = 1, \ldots, J_\nu,$$

be elements of $Z(\mathcal{A}, \lambda_\nu)$. Let the vector polynomial Φ_k have the degree $m_{\nu k} - 1$. Suppose that the leading coefficients of Φ_k are linear independent. Then for every τ the collection

$$\left\{ (D_\tau - \lambda_\nu)^{m_{\nu k} - 1 - \sigma} \left(e^{i\lambda_\nu \tau}\Phi_k(i\tau) \right) \right\}_{\sigma=0}^{m_{\nu k} - 1}, \ k = 1, \ldots, J_\nu,$$

is a canonical set of Jordan chains corresponding to λ_ν.

Proof. Our assumptions guarantee that

$$\left\{ R_{t \to \lambda} \left(e^{i\lambda_\nu (t+\tau)}\Phi_k(i(t + \tau)) \right) \right\}_{k=1}^{J_\nu}$$

is a canonical generating system in $S(\mathcal{A}, \lambda_\nu)$. Reference to Proposition A.4.4 completes the proof. □

1.4 Power-Exponential Solutions of the Nonhomogeneous Equation

It is a standard property of ordinary differential operators with constant coefficients that an equation with power-exponential right-hand side has a power-exponential solution. Here we extend this fact to the infinite dimensional case.

Consider the equation

$$\mathcal{A}(D_t)W(t) = e^{i\lambda_0 t} \sum_{0 \le j \le N} \frac{t^j}{j!}\, y_{N-j} \quad \text{on} \quad \mathbb{R}, \tag{1.35}$$

where λ_0 is a complex number and $y_j \in H$, $j = 0, \ldots, N$.

Theorem 1.4.1. *Let $m = 0$ if λ_0 is a regular point of the pencil $\mathcal{A}(\lambda)$ and let m be equal to the maximal partial multiplicity of λ_0 if λ_0 is an eigenvalue of $\mathcal{A}(\lambda)$. There exists a solution of (1.35), which has the form*

$$W(t) = e^{i\lambda_0 t} \sum_{0 \le j \le N+m} \frac{t^j}{j!} x_{N+m-j}, \tag{1.36}$$

where $x_j \in \mathcal{H}$, $j = 0, \ldots, N + m$.

Proof. By inserting (1.36) into (1.35) and by equating coefficients of powers of t, one arrives at

$$\sum_{0 \le k \le h} \frac{\mathcal{A}^{(k)}(\lambda_0)}{k!} x_{h-k} = \begin{cases} y_{h-m} & \text{for } h = m, \ldots, N+m, \\ 0 & \text{for } h = 0, 1, \ldots, m-1. \end{cases} \tag{1.37}$$

We introduce the polynomial

$$y(\lambda) = \sum_{0 \le j \le N} y_j \lambda^j.$$

Consider the equation

$$\mathcal{A}(\lambda_0 + \lambda) x(\lambda) = \lambda^m y(\lambda).$$

By using the formula

$$\mathcal{A}(\lambda_0 + \lambda)^{-1} = \sum_{k=1}^{J} \sum_{j=0}^{m_k-1} P_{kj} \lambda^{j-m_k} + \Gamma(\lambda_0 + \lambda),$$

(which was proved in Theorem 1.2.1), we obtain

$$x(\lambda) = \sum_{k=1}^{J} \sum_{j=0}^{m_k-1} \lambda^{j+m-m_k} P_{kj} y(\lambda) + \lambda^m \Gamma(\lambda_0 + \lambda) y(\lambda).$$

Clearly, $x(\lambda)$ is analytic in a neighbourhood of $\lambda = 0$. To satisfy (1.37) it suffices to put

$$x_j = \frac{x^{(j)}(0)}{j!}, \quad j = 0, \ldots, N+m.$$

The proof is complete. $\square$

Remark 1.4.2. If λ_0 is a regular point of the pencil $\mathcal{A}(\lambda)$ then there is a unique solution of the form (1.36) for equation (1.35). If λ_0 is an eigenvalue of $\mathcal{A}(\lambda)$ such a solution is unique up to an arbitrary zero of $\mathcal{A}(D_t)$, which has the same form (1.36).

1.5 Applications to Elliptic Partial Differential Equations with Constant Coefficients

1.5.1 Neumann Problem in a Cylinder

We start with the Laplace equation

$$\left(D_t^2 - \Delta_x \right) u = 0 \tag{1.38}$$

in the cylinder $\mathcal{C} = \{(x,t) : x \in \Omega, \ t \in \mathbb{R}\}$ where Ω is a domain in $\mathbb{R}^n$ with compact closure and smooth boundary. Let the Neumann boundary condition

$$\partial_\nu u = 0 \quad \text{on} \quad \partial\mathcal{C} \qquad (1.39)$$

be prescribed, where ν is the outward normal.

The function spaces $\mathcal{H}$ and H will be specified by

$$\mathcal{H} = \{\varphi \in W_2^2(\Omega) : \partial_\nu\varphi = 0 \quad \text{on} \quad \partial\Omega\} \quad \text{and} \quad H = L_2(\Omega),$$

where $W_2^\ell(\Omega)$ is the Sobolev space

$$\{\varphi \in L_2(\Omega) : D^\alpha\varphi \in L_2(\Omega) \text{ for all multi-indices } \alpha \text{ with } 0 < |\alpha| \le \ell\}$$

supplied with the norm

$$\|\varphi\|_{W_2^\ell(\Omega)} = \left(\sum_{|\alpha| \le \ell} \int_\Omega |D_x^\alpha\varphi|^2 dx \right)^{1/2}.$$

It is a classical fact that the spectrum of the Laplace operator $-\Delta_x$: $\mathcal{H} \to H$ is a sequence of eigenvalues γ_k of finite multiplicity J_k, $0 = \gamma_0 < \gamma_1 < \dots$. Moreover, the system of eigenfunctions is a basis in $L_2(\Omega)$ (see Ladyzhenskaya (1985), Ch.II, Sect.3).

The operator pencil corresponding to the above Neumann problem is

$$\mathcal{A}(\lambda) = -\Delta_x + \lambda^2 : \mathcal{H} \to H.$$

The Fredholm property of $\mathcal{A}(\lambda)$ is standard, and λ is an eigenvalue of $\mathcal{A}(\lambda)$ if and only if λ^2 is one of the numbers $-\gamma_0, -\gamma_1, \dots$.

We see that $\mathcal{A}(\lambda)$ is a special case of the abstract pencil analyzed in Examples 1.2.3 and 1.3.6. In comparison with these examples the only new information is that $J_0 = 1$ and

$$\varphi_1^{(0)}(x) = \psi_1^{(0)}(x) = (\text{mes}_n\Omega)^{-1/2}.$$

Therefore we shall not duplicate the formulae for the canonical systems and special solutions obtained in the above examples.

1.5.2 The Dirichlet Problem in a Cone

Consider the differential equation with constant coefficients

$$P(D_x)u = 0 \qquad (1.40)$$

in the n-dimensional cone K defined by (0.10). Let Ω be a proper subdomain of S^{n-1} with smooth boundary. Assume that P is of order $2m$ and contains only derivatives of the highest order. Also let P be strongly elliptic, i.e.

$$\Re P(\xi) > 0 \quad \text{for} \quad \xi \ne O. \qquad (1.41)$$

We complement (1.40) with the Dirichlet boundary conditions

$$\partial_\nu^j u = 0 \quad \text{on} \quad \partial K \setminus O \quad \text{for} \quad j = 0, 1, \ldots, m - 1. \tag{1.42}$$

By using spherical coordinates (r, θ) we write the operator $P(D_x)$ in the form

$$P(D_x) = r^{-2m} \sum_{q=0}^{2m} A_{2m-q}(\theta, D_\theta)(-r D_r)^q. \tag{1.43}$$

The coefficients A_q are the same differential operators on S^{n-1} as in (0.9).

By the change of variables (0.8), the above Dirichlet problem is reduced to that in the cylinder

$$\{(\theta, t) : \theta \in \Omega, \ t \in \mathbb{R}\}$$

and the equation (1.40) takes the form

$$\sum_{q=0}^{2m} A_{2m-q}(\theta, D_\theta) D_t^q u = 0.$$

We can now introduce the operator pencil

$$\mathcal{A}(\lambda) = \sum_{q=0}^{2m} A_{2m-q}(\theta, D_\theta) \lambda^q : \mathcal{H} \to H. \tag{1.44}$$

Here

$$H = L_2(\Omega) \quad \text{and} \quad \mathcal{H} = W_2^{2m}(\Omega) \cap \mathring{W}_2^m(\Omega),$$

where $\mathring{W}_2^m(\Omega)$ is the closure of $C_0^\infty(\Omega)$ in $W_2^m(\Omega)$.

1.5.3 Properties of the Operator Pencil (1.44)

Lemma 1.5.1. (i) *The line $\Im \lambda = m - n/2$ on the complex plane $\mathbb{C}$ is free of eigenvalues of $\mathcal{A}(\lambda)$.*

(ii) *The following estimate is valid*

$$\sum_{k=0}^{2m} |\lambda|^{2m-k} \| \varphi \|_{W_2^k(\Omega)} \le c \| \mathcal{A}(\lambda) \varphi \|_{L_2(\Omega)} \tag{1.45}$$

for λ on the same line as in (i) and for all $\varphi \in \mathcal{H}$.

(iii) *Operator (1.44) is Fredholm for all $\lambda \in \mathbb{C}$.*

Proof. First, we check the inequality

$$\int_K |x|^{2(|\beta|-m)} \left| D_x^\beta u(x) \right|^2 dx \le c \int_K \sum_{|\alpha|=m} |D_x^\alpha u(x)|^2 dx \tag{1.46}$$

for $u \in C_0^\infty(K)$ and for $|\beta| < m$. We write $|\nabla v|^2$ in spherical coordinates (r, θ): $|v_r|^2 + r^{-2} |\text{grad}_\theta v|^2$, where $v \in C_0^\infty(K)$, and notice that for all $r > 0$

$$\int_\Omega |v|^2 d\theta \le c \int_\Omega |\mathrm{grad}_\rho v|^2 d\theta.$$

Hence

$$\int_K |x|^{-2k}|v(x)|^2 dx \le c \int_K |x|^{-2(k-1)}|\nabla v(x)|^2 dx. \qquad (1.47)$$

Now (1.46) follows by iterating (1.47).

By (1.41) and (1.46)

$$\Re \int_K P(D_x)u(x) \cdot \overline{u(x)} dx \ge c \int_K \sum_{|\alpha| \le m} |x|^{2(|\alpha|-m)}|D_x^\alpha u|^2 dx. \qquad (1.48)$$

In the coordinates t, θ

$$\Re \int_C e^{(2m-n)t} \mathcal{A}(D_t)u \cdot \overline{u} d\theta dt$$

$$\ge c \int_\mathbb{R} e^{(2m-n)t} \sum_{q=0}^m \| D_t^q u(\,\cdot\,,t) \|^2_{W_2^{m-q}(\Omega)} \, dt.$$

By the Parseval formula (see Lions, Magenes (1972), Ch.I, Sect.2) we have

$$\Re \int_\mathbb{R} \int_\Omega \mathcal{A}(\lambda + i(m-n/2))v(\theta, \lambda) \cdot \overline{v(\theta, \lambda)} d\theta d\lambda$$

$$\ge c \int_\mathbb{R} \sum_{q=0}^m |\lambda|^{2q} \| v(\cdot, \lambda) \|^2_{W_2^{m-q}(\Omega)} \, d\lambda,$$

which implies the inequality

$$\Re \int_\Omega \mathcal{A}(\lambda + i(m-n/2))w(\theta) \cdot \overline{w(\theta)} d\theta$$

$$\ge c \sum_{q=0}^m |\lambda|^{2q} \| w \|^2_{W_2^{m-q}(\Omega)} \qquad (1.49)$$

for all real λ and $w \in C_0^\infty(\Omega)$. By the Lax-Milgram lemma (see Lions, Magenes (1972), Ch.2, Sect.9.1) the operator

$$\mathcal{A}(\lambda + i(m-n/2)) : \mathring{W}_2^m(\Omega) \to \left(\mathring{W}_2^m(\Omega)\right)^*,$$

(where $*$ denotes the dual distribution space), is isomorphic for all real λ. Since $\mathcal{A}(\lambda)$ is an elliptic operator, it follows by a well-known regularity result of Agmon, Douglis and Nirenberg (1959, 1964) that the operator (1.44) is also isomorphic for all λ with $\Im\lambda = m - n/2$.

We turn to assertion (ii). By the Cauchy inequality applied to the left-hand side of (1.49)

$$(1 + |\lambda|)^{2m} \| u \|_{L_2(\Omega)} \le c \| \mathcal{A}(\lambda + i(m-n/2))u \|_{L_2(\Omega)} . \qquad (1.50)$$

Since the operator (1.44) is invertible for $\lambda = i(m - n/2)$, we have

$$\| u \|_{W_2^{2m}(\Omega)} \leq c \| \mathcal{A}(i(m - n/2))u \|_{L_2(\Omega)} .$$

The right-hand side is majorized by

$$c \| \mathcal{A}(\lambda + i(m - n/2))u \|_{L_2(\Omega)} + c \sum_{q=0}^{2m-1} |\lambda|^q \| u \|_{W_2^{2m-q}(\Omega)} .$$

By the well-known interpolation inequality

$$\| u \|_{W_2^{2m-q}(\Omega)} \leq c \| u \|_{W_2^{2m}(\Omega)}^{1-q/2m} \| u \|_{L_2(\Omega)}^{q/2m} \tag{1.51}$$

(see Lions, Magenes (1972), Ch. I, Sect. 9.3) we have

$$\| u \|_{W_2^{2m}(\Omega)} \leq c(\| \mathcal{A}(\lambda + i(m - n/2))u \|_{L_2(\Omega)}$$
$$+ \sum_{q=0}^{2m-1} \| u \|_{W_2^{2m}(\Omega)}^{1-q/2m} (|\lambda|^{2m} \| u \|_{L_2(\Omega)})^{q/2m}).$$

Hence, using (1.50),

$$\| u \|_{W_2^{2m}(\Omega)} \leq c \| \mathcal{A}(\lambda + i(m - n/2))u \|_{L_2(\Omega)} .$$

Estimate (1.45) follows from the last inequality combined with (1.50) and (1.51).

The Fredholm property of (1.44) follows from the invertibility of $\mathcal{A}(i(m - n/2))$ and the compactness of the operator

$$\mathcal{A}(\lambda) - \mathcal{A}(i(m - n/2)) : \mathcal{H} \to H. \quad \square$$

By Lemma 1.5.1, all of the abstract definitions and results given in this chapter apply to the pencil (1.44). In particular, we see that all solutions of (1.40) with zero Dirichlet data which have the form

$$u(x) = r^{-i\lambda} \sum_{\sigma=0}^{M} \frac{(-i \log r)^\sigma}{\sigma!} u_{M-\sigma}, \quad u_{M-\sigma} \in \mathcal{H}, \tag{1.52}$$

are described in terms of eigenvalues, eigenfunctions and generalized eigenfunctions of $\mathcal{A}(\lambda)$.

1.5.4 The Adjoint Pencil

In order to describe eigenvalues, eigenfunctions and generalized eigenfunctions of $\mathcal{A}^*(\lambda)$, it is convenient to express $\mathcal{A}^*(\lambda)$ by the operator pencil $\mathcal{B}(\lambda)$ which corresponds to the Dirichlet problem for the formally adjoint operator $P^*(D_x)$ in K. On one hand,

$$P^*(D_x) = r^{-2m}\mathcal{B}(rD_r).$$

On the other hand, by (1.43)

$$P^*(D_x)u = \sum_{q=0}^{2m} A_{2m-q}^*(\theta, D_\theta)(rD_r - in)^q(r^{-2m}u)$$

$$= r^{-2m}\sum_{q=0}^{m} A_{2m-q}^*(\theta, D_\theta)(rD_r + i(2m - n)).$$

Hence

$$\mathcal{A}^*(\lambda) = \mathcal{B}(\lambda - i(2m - n)). \tag{1.53}$$

Although $\mathcal{A}^*$ acts from H to $\mathcal{H}^*$, its eigenfunctions and generalized eigenfunctions belong to $\mathcal{H}$ by a well-known regularity result (see Agmon, Douglas and Nirenberg (1959, 1964)). Therefore the eigenvalues, eigenfunctions and generalized eigenfunctions of $\mathcal{A}^*(\lambda)$ are the same as those of the pencil

$$\mathcal{B}(\lambda - i(2m - n)) : \mathcal{H} \to H.$$

In particular, if $P(D_x) = P^*(D_x)$ (i.e. the coefficients p_α are real), then

$$\mathcal{A}^*(\lambda) = \mathcal{A}(\lambda - i(2m - n)). \tag{1.54}$$

This implies that the number λ_0 together with $\overline{\lambda}_0 + i(2m - n)$ are eigenvalues of $\mathcal{A}(\lambda)$, and their geometric, algebraic and partial multiplicities are the same.

1.5.5 The Dirichlet Problem in a Half-Space

In the case

$$K = \mathbb{R}_+^n = \{x = (x', x_n) : x_n > 0\}$$

all solutions of (1.40), (1.42), which have the form (1.52) can be written down explicitly. In this case, any function in the domain of the pencil $\mathcal{A}(\lambda)$ is subject to zero Dirichlet conditions at the boundary of the upper hemisphere S_+^{n-1}.

In order to describe special solutions (1.52) one needs the Poisson kernels $\mathcal{P}_k$, $k = 0, \ldots, m - 1$, i.e positive homogeneous solutions of degree $k + 1 - n$ of the Dirichlet problem

$$\begin{cases} P(D_x)\mathcal{P}_k(x) = 0 & \text{on } \mathbb{R}_+^n \\ \partial_{x_n}^j \mathcal{P}_k(x) = \delta(x')\delta_k^j & \text{for } x_n = 0, \ j = 0, \ldots, m - 1, \end{cases}$$

where $\delta(x')$ is the Dirac function at O with respect to the variable $x' = (x_1, \ldots, x_{n-1})$.

Proposition 1.5.2. *The function* (1.52) *is a solution of* (1.40), (1.42) *if and only if*

(i) *the exponent* λ *in* (1.52) *is either equal to* im, $i(m+1),\ldots$ *or to* $i(m-n)$, $i(m-n-1),\ldots$;

(ii) *if* $-i\lambda = N$ *then*

$$u(x) = x_n^m \sum_{|\alpha|=N-m} a_\alpha x^\alpha + \sum_{k=0}^{m-1} \sum_{|\beta|=k+1-n-N} b_\beta \partial_{x'}^\beta \mathcal{P}_k(x), \qquad (1.55)$$

where a_α, b_β *are constant and the first term in the right-hand side satisfies* (1.40). *If* $N < m$ *or* $N > k+1-n$ *for some* $k = 0,\ldots,m-1$, *then the corresponding terms in* (1.55) *are omitted.*

Proof. Since the function (1.52) belongs to $W_2^{2m}(Q)$ for every subdomain Q of $\mathbb{R}_+^n$ such that the closure of Q does not contain 0, it follows that (1.52) is in $C^\infty(\overline{\mathbb{R}_+^n} \setminus O)$. Hence the coefficients $u_{M-\sigma}$ are smooth on $\overline{S}_+^{n-1}$. Using the partial hypoellipticity on the boundary (see Hörmander (1976), Theorem 4.3.1) one can show that the function (1.52) belongs to $\mathcal{H}_{(2m,-\infty)}^{loc}(\mathbb{R}_+^n)$ (this space is introduced in Hörmander (1976), Sect.10.4). Therefore the boundary value problem (1.40), (1.42) can be treated in the distributional sense. Thus we can replace the Dirichlet boundary conditions (1.42) by

$$\partial_\nu^j u = \sum c_\alpha D_{x'}^\alpha \delta(x') \quad \text{for } x_n = 0 \text{ and for } j = 0, 1, \ldots, m-1, \qquad (1.56)$$

where the sum is finite and c_α is constant. The formula (1.55) gives all power-logarithmic solutions of the problem (1.40), (1.42), with prescribed homogeneity. $\square$

1.5.6 Elliptic Equations in $\mathbb{R}^n \setminus O$

All the solutions (1.52) can be explicitly described in the case of the partial differential equation

$$P(D_x)u = 0 \quad \text{in} \quad \mathbb{R}^n \setminus O. \qquad (1.57)$$

Here, the operator P is the same as in Sect. 1.5.2 except that it need only satisfy the ellipticity condition $P(\xi) \neq 0$ for $\xi \neq O$ (rather than being strongly elliptic). The operator pencil $\mathcal{A}(\lambda)$ has the same form as in Sect. 1.5.2 and $\Omega = S^{n-1}$. We have $\mathcal{H} = W_2^{2m}(S^{n-1})$, $H = L_2(S^{n-1})$. As before we omit the proofs of the Fredholm property for $\mathcal{A}(\lambda)$, and of the invertibility of $\mathcal{A}(\lambda)$ for sufficiently large real λ.

According to Sections 1.2 and 1.3, the set of solutions (1.52) contains all information about eigenvalues, eigenfunctions and generalized eigenfunctions of $\mathcal{A}(\lambda)$. In the case under consideration this set can be completely characterized as follows.

Denote by $G(x)$ a fundamental solution of the operator $P(D_x)$, i.e. a solution of the equation

$$P(D_x)G(x) = \delta(x) \quad \text{in} \quad \mathbb{R}^n,$$

where δ is the Dirac function in $\mathbb{R}^n$. If either $2m \geq n$, n is odd, or $2m < n$ we require G to have the form

$$G(x) = r^{2m-n}Q(\theta).$$

In the case $2m \geq n$, n is even, one can choose G to satisfy

$$G(x) = R(x)\log r + r^{2m-n}S(\theta).$$

Here Q and S are smooth functions on S^{n-1} and R is a homogeneous polynomial of degree $2m - n$ (see John (1955)).

Proposition 1.5.3. *The function u of the form* (1.52) *is a solution of* (1.57) *if and only if the following conditions are satisfied:*
(i) $\lambda = ik$ or $\lambda = i(2m - n - k)$ with any nonnegative integer k,
(ii) the function u is given by

$$u(x) = P_k(x)$$

if $-i\lambda = k \geq 0$ for $2m < n$ or $-i\lambda = k > 2m - n$ for $2m \geq n$, by

$$u(x) = \sum_{|\alpha|=k} c_\alpha D_x^\alpha G(x) \tag{1.58}$$

if $-i\lambda = 2m - n - k < 0$, and by

$$u(x) = P_k(x) + \sum_{|\alpha|=2m-n-k} c_\alpha D_x^\alpha G(x)$$

if $-i\lambda = k$ and $0 \leq k \leq 2m - n$. Here $c_\alpha \in \mathbb{C}$ and P_k is a homogeneous polynomial of degree k.

Proof. First, let $\Im\lambda > -n$. Then (1.52) is a distribution. Since the support of the distribution $P(D_x)u$ is O, equation (1.57) can be rewritten as

$$P(D_x)u = \sum_\alpha c_\alpha D_x^\alpha \delta(x) \quad \text{on} \quad \mathbb{R}^n,$$

where the sum is finite and c_α is constant. We set

$$u(x) = \sum_\alpha c_\alpha D_x^\alpha G(x) + v(x).$$

The function v satisfies the homogeneous equation $P(D_x)v(x) = 0$ on $\mathbb{R}^n$, therefore v is a polynomial. Thus we arrive at (i) and (ii) for $\Im\lambda > -n$.

Let $\Im\lambda \leq -n$. Consider solutions (1.52) of the equation

$$P^*(D_x)u = 0 \quad \text{in} \quad \mathbb{R}^n \setminus O, \tag{1.59}$$

where $P^*(D_x)$ is the formal adjoint operator of $P(D_x)$. According to Sect. 1.5.4 the complex number λ is an eigenvalue of the pencil $\mathcal{A}(\lambda)$ if and only if $\overline{\lambda} - i(2m - n)$ is an eigenvalue of the operator pencil $\mathcal{A}^*(\lambda)$ associated with the problem (1.59), and the algebraic multiplicities of these eigenvalues coincide. Since $\Im(\overline{\lambda} - i(2m - n)) \geq 2m$, we can apply the theorem to equation (1.59). We thus obtain the description of the eigenvalues of the pencil $\mathcal{A}(\lambda)$ and their multiplicities in the half-plane $\Im\lambda \leq -n$. This gives (i) and (ii), since the dimension of solutions (1.58) to (1.57) is equal to the algebraic multiplicity of $\lambda = i(2m - n - k)$. $\square$

In particular, we obtain from this result that eigenvalues of $\mathcal{A}(\lambda)$ are either $0, i, 2i, \ldots$ or $(2m - n)i, (2m - n - 1)i, \ldots$ The partial multiplicities of these eigenvalues do not exceed 2. Generalized eigenfunctions appear if and only if n is even and $2m \geq n$.

In the following chapters we will repeatedly meet the operator pencils mentioned in this section, as well as other examples of operator pencils from the theory of partial differential equations.

1.6 Comments

1.2.2 Properties of general polynomial operator pencils were first studied in Keldysh (1951, 1971), where, in particular, Theorem 1.2.1 can be found (see also Gohberg, Sigal (1971) and Gohberg, Goldberg and Kaashoek (1990)).

1.5.2 Change of variables (0.8) is a starting point of the theory of general elliptic boundary value problems in domains with conic singularities (Kondratiev (1967), Maz'ya, Plamenevskii (1978), Dauge (1996), Nazarov, Plamenevskii (1994) and Kozlov, Maz'ya and Rossmann (1997)).

1.5.3 Spectral properties of the operator pencils of the Dirichlet problem for higher order elliptic equations in a cone were studied in Kozlov, Maz'ya (1988) and Kozlov, Maz'ya (1991).

2. Differential Operator Equations in Weighted Sobolev Spaces

2.1 Introduction

In this chapter we consider the ordinary differential equation with constant operator coefficients

$$\mathcal{A}(D_t)u = f \quad \text{on} \quad \mathbb{R}. \tag{2.1}$$

We assume that the coefficients are bounded operators acting in pairs of Hilbert spaces.

In Sect.2.2 we define the class of pencils $\mathcal{A}(\lambda)$ to be studied in what follows.

Next, we introduce some spaces of vector valued functions of the variable t and describe their properties (Sect.2.3). In particular we consider a local Sobolev space $W_{\text{loc}}^\ell(\mathbb{R})$, where ℓ is the order of $\mathcal{A}(D_t)$, as well as a Sobolev space $W_\beta^\ell(\mathbb{R})$ with the weight $\exp(2\beta t)$.

In Sect.2.4 a theorem on the unique solvability of (2.1) in $W_\beta^\ell(\mathbb{R})$ is proved.

Sect.2.6 is devoted to estimates of Green's operator function of the equation (2.1). Asymptotic formulae for Green's function at infinity based upon Theorem 1.2.1 are given in Sect. 2.7. In Sect. 2.8 we use these results to obtain asymptotic representations as $t \to \pm\infty$ of solutions of (2.1) in the space $W_\beta^\ell(\mathbb{R})$. An auxiliary local estimate for solutions of (2.1) is found in Sect.2.9.

We illustrate general results by applications to the Dirichlet problem in a cylinder and a cone (Sections 2.5 and 2.10).

2.2 The Operator Pencil $\mathcal{A}(\lambda)$

2.2.1 Conditions on $\mathcal{A}(\lambda)$

Let $\mathcal{H}$ and H be the same Hilbert spaces as in Sect. 1.2.1. We consider the polynomial operator pencil (1.10) satisfying

Condition I *The operator*

$$\mathcal{A}(\lambda) : \mathcal{H} \to H$$

is Fredholm for every $\lambda \in \mathbb{C}$ and invertible at least for one value of λ.

We introduce a collection of Hilbert spaces

$$\{H_j\}_{j=0}^{\ell}$$

with norms $\| \cdot \|_j$ such that

$$\mathcal{H} = H_\ell \subset H_{\ell-1} \subset \ldots \subset H_0 = H$$

and

$$\| u \|_j \leq \| u \|_{j+1} \quad \text{for} \quad j = 0, 1, \ldots, \ell - 1. \tag{2.2}$$

The above embeddings are dense since $\mathcal{H}$ is dense in H. We suppose that the operator A_q maps H_q continuously into H_0. The operator pencil we are going to study is also subject to the following

Condition II *There exists $r > 0$ such that*

$$\sum_{q=0}^{\ell} |\lambda|^q \| \varphi \|_{\ell-q} \leq c \| \mathcal{A}(\lambda)\varphi \|_0 \tag{2.3}$$

for all $\varphi \in H_\ell$ and for all real λ with $|\lambda| \geq r$.

This condition is often used in the theory of operator pencils and corresponding ordinary differential equations with operator coefficients (see, for example, Agmon, Nirenberg (1963), Agranovich, Vishik (1964), Maz'ya, Plamenevskii (1972), Bagirov, Kondratiev (1991)).

Proposition 2.2.1. *There exists $\theta \in (0, \pi/2)$ such that for all*

$$\lambda \in S_{r,\theta} = \{ z \in \mathbb{C} : \ |\arg(\pm z)| \leq \theta, \ |\Re z| \geq r \},$$

inequality (2.3) holds (r is as in Condition II).

Proof. Let $\lambda = \tau + ia$, where $|\tau| \geq r$ and $|a| \leq \varepsilon|\tau|$. Then by (2.3)

$$\sum_{q=0}^{\ell} |\lambda|^q \| \varphi \|_{\ell-q} \leq c_1 \| \mathcal{A}(\tau)\varphi \|_0$$

$$\leq c_1 \Big(\| \mathcal{A}(\tau)\varphi \|_0 + \| (\mathcal{A}(\lambda) - \mathcal{A}(\tau))\varphi \|_0 \Big). \tag{2.4}$$

Furthermore, we have

$$\| (\mathcal{A}(\lambda) - \mathcal{A}(\tau))\varphi \|_0 \leq c \varepsilon \sum_{q=1}^{\ell} |\lambda|^q \| \varphi \|_{\ell-q} .$$

This together with (2.4) implies (2.3) for $\lambda \in S_{r,\theta}$ with sufficiently small θ. $\square$

2.2.2 Examples of Pencils Satisfying Conditions I and II

Example 2.2.2. An example of the pencil satisfying Conditions I and II is provided by the two-term operator pencil

$$\mathcal{A}(\lambda) = A + \lambda^2, \tag{2.5}$$

where A is a positive definite selfadjoint operator in $H = H_0$ with the domain $\mathcal{H} = H_2$. Suppose that H_2 is compactly embedded into H_0 and put $H_1 = \mathrm{Dom}(A^{1/2})$. Then A has a discrete spectrum consisting of positive eigenvalues γ_j, $j = 1, 2, \ldots$. The spectrum of the operator pencil $\mathcal{A}(\lambda)$ is formed by the eigenvalues $\lambda_j^{\pm} = \pm i\, \gamma_j^{1/2}$ and Condition II above follows from

$$\| \mathcal{A}(\lambda)\varphi \|_H^2 = \| A\varphi \|_H^2 + 2\Re\lambda^2 (A\varphi, \varphi)_H + |\lambda|^4 \| \varphi \|_H^2 .$$

The pencil (2.5) is a generalization of the Laplace operator with zero Dirichlet data in a cylinder.

Example 2.2.3. The heat operator $iD_t - \Delta$ in a cylinder is a special case of the pencil

$$\mathcal{A}(\lambda) = A + i\lambda$$

where A is the same as before. Here $\mathcal{H} = H_1 = \mathrm{Dom}A$, the spectrum of $\mathcal{A}$ consists of the eigenvalues $i\gamma_j$, $j = 1, \ldots$, and Condition II is a consequence of

$$\| \mathcal{A}(\lambda)\varphi \|^2 = \| A\varphi \|^2 - 2\Im\lambda(A\varphi, \varphi) + |\lambda|^2 \| \varphi \|^2 .$$

Example 2.2.4. In what follows we will frequently meet the ordinary differential operator

$$\mathcal{M}(\partial_t) = (\partial_t + k_+)^{m_+} (-\partial_t - k_-)^{m_-}, \tag{2.6}$$

where $k_+ > k_-$ and m_+, m_- are positive integers. In this case, $\ell = m_+ + m_-$ and $H_0 = \ldots = H_\ell = \mathbb{C}$. The operator pencil $\mathcal{A}(\lambda)$ is the operator of multiplication by the polynomial

$$\mathcal{M}(i\lambda) = (i\lambda + k_+)^{m_+} (-i\lambda - k_-)^{m_-}.$$

The validity of both conditions is obvious. The pencil $\mathcal{M}(i\lambda)$ has two eigenvalues $\lambda_\pm = ik_\pm$ with the geometric multiplicity 1, the partial multiplicities $m_\pm$ and the algebraic multiplicities $m_\pm$. The eigenvectors and generalized eigenvectors are

$$\varphi_0^{(\pm)} = 1, \quad \varphi_j^{(\pm)} = 0, \; j = 1, \ldots, m_\pm - 1.$$

2.2.3 Notation

We identify H^* with H by the scalar product $(\cdot, \cdot)_H$. This scalar product leads to the following realization of the dual space H_j^* of H_j.

Let $h \in H$. We introduce the norm

$$\|h\|_{-j} = \sup\{|(g, h)_H| : g \in H_j, \|g\|_j = 1\}.$$

The completion of H with respect to this norm is a subspace of H_j^*, which, in fact, coincides with H_j^* since H_j is dense in H. By continuity the product $(g, h)_H$ can be defined for all $g \in H_j$ and $H \in H_j^*$. Clearly,

$$H_0 = H \subset H_1^* \subset \cdots \subset H_\ell^* = \mathcal{H}^*.$$

The operator pencil $\mathcal{A}^*(\lambda)$ maps H into $\mathcal{H}^*$ and

$$\left(\mathcal{A}^*(\lambda)h, x\right)_H = \left(h, \mathcal{A}(\overline{\lambda})x\right)_H.$$

As in Sect. 1.4, we denote by $\{\varphi_{kj}^{(\nu)}\}$ and $\{\psi_{kj}^{(\nu)}\}$ (respectively) the canonical systems for $\mathcal{A}(\lambda)$ and $\mathcal{A}^*(\lambda)$ corresponding to the eigenvalues λ_ν and $\overline{\lambda}_\nu$, and subject to the biorthogonality condition (1.20).

According to Theorem 1.2.1, for λ in some neighbourhood of the eigenvalue λ_ν we have

$$\mathcal{A}^{-1}(\lambda) = \sum_{1 \le k \le J_\nu} \sum_{j=0}^{m_{\nu k}-1} \frac{P_{kj}^{(\nu)}}{(\lambda - \lambda_\nu)^{m_{\nu k}-j}} + \Gamma_\nu(\lambda), \tag{2.7}$$

where

$$P_{kj}^{(\nu)} = \sum_{s=0}^{j} (\cdot, \psi_{ks}^{(\nu)})_H \phi_{k,j-s}^{(\nu)}. \tag{2.8}$$

We shall use the solutions (1.25) and (1.27) of (1.21) and (1.26) respectively, and put

$$U_{kh}^{(\nu)}(t) = e^{i\lambda_\nu t} D_t^{m_{\nu k}-1-h} \Phi_k^{(\nu)}(it), \tag{2.9}$$

$$V_{kh}^{(\nu)}(t) = i\, e^{i\overline{\lambda}_\nu t} D_t^h \Psi_k^{(\nu)}(it), \tag{2.10}$$

where $\Phi_k^{(\nu)}$ and $\Psi_k^{(\nu)}$ are given by (1.24) and (1.28). To simplify the notation we sometimes use a unique numeration for the pairs (k, h), i.e. we denote

$$U_\sigma^{(\nu)}(t) = U_{kh}^{(\nu)}(t), \tag{2.11}$$

$$V_\sigma^{(\nu)}(t) = V_{kh}^{(\nu)}(t), \tag{2.12}$$

where $\sigma = 1, \ldots, \kappa_\nu$.

By Proposition 1.3.1, the vector functions $U_\sigma^{(\nu)}$, $V_\sigma^{(\nu)}$ satisfy the biorthogonality condition

$$\int_{\mathbb{R}} \left(\mathcal{A}(D_t)(\eta(t)U_\sigma^{(\nu)}(t)), V_\rho^{(\mu)}(t)\right)_H dt = \delta_\nu^\mu \, \delta_\sigma^\rho, \tag{2.13}$$

where $\nu, \mu \in Z$, $\sigma = 1, \ldots, \kappa_\nu$, $\rho = 1, \ldots, \kappa_\mu$.

2.3 Some Spaces of Vector Valued Functions

Here we give definitions and study simple properties of some spaces of vector valued functions of the real variable t.

2.3.1 Sobolev Spaces

Definition 2.3.1. By $L_2(a,b;H_q)$ we denote the space of vector valued functions $u : (a,b) \to H_q$ with the finite norm

$$\| u \|_{L_2(a,b;H_q)} = \left(\int_a^b \| u(t) \|_q^2 \, dt \right)^{1/2}.$$

Definition 2.3.2. Let $W^\ell(a,b)$ be the space of distributions u on (a,b) with values in H_ℓ (see Lions, Magenes (1972), Sect.1.3) such that

$$D_t^j u \in L_2(a,b;H_{\ell-j}), \quad j = 0,1,\ldots,\ell.$$

We equip $W^\ell(a,b)$ with the norm

$$\| u \|_{W^\ell(a,b)} = \left(\int_a^b \sum_{0 \le q \le \ell} \| D_t^q u(t) \|_{\ell-q}^2 \, dt \right)^{1/2}.$$

Definition 2.3.3. Let $L_{2,\mathrm{loc}}(\mathbb{R};H_q)$ be the locally convex space of functions $u : \mathbb{R} \to H_q$ equipped with the seminorms

$$\| u \|_{L_2(\tau,\tau+1,H_q)}, \quad \tau \in \mathbb{R}.$$

Definition 2.3.4. The locally convex space of distributions on $\mathbb{R}$ with values in H_ℓ such that

$$D_t^j u \in L_{2,\mathrm{loc}}(\mathbb{R};H_{\ell-j}), \quad j = 0,1,\ldots,\ell,$$

equipped with the seminorms

$$\| u \|_{W^\ell(\tau,\tau+1)}, \quad \tau \in \mathbb{R},$$

will be denoted by $W_{\mathrm{loc}}^\ell(\mathbb{R})$.

Functions in W^ℓ satisfy the following "pointwise" estimate.

Proposition 2.3.5. *Let* $u \in W^\ell(a,a+\delta)$, *where* $\delta > 0$. *Then for all* $t \in (a,a+\delta)$ *the estimate*

$$\| D_t^q u(t) \|_{\ell-q-1} \le c_\delta \| u \|_{W^\ell(a,a+\delta)}, \tag{2.14}$$

holds (where $q = 0,1,\ldots,\ell-1$ *and* c_δ *does not depend on* a).

Proof. By the Newton-Leibniz formula

$$\| v(t) \|_j^2 \le 2 \int_a^{a+\delta} \left(\delta \| D_\tau v(\tau) \|_j^2 + \delta^{-1} \| v(\tau) \|_j^2 \right) d\tau,$$

which together with (2.2) leads to (2.14). $\square$

2.3.2 Weighted Sobolev Spaces

Definition 2.3.6. We introduce the spaces

$$L_{2,\beta}(\mathbb{R}; H_0) \quad \text{and} \quad W_\beta^\ell(\mathbb{R})$$

of functions in $L_{2,\mathrm{loc}}(\mathbb{R}; H_0)$ and $W_{\mathrm{loc}}^\ell(\mathbb{R})$ with finite norms

$$\| u \|_{L_{2,\beta}(\mathbb{R};H_0)} = \left(\int_{\mathbb{R}} e^{2\beta t} \| u(t) \|_0^2 \, dt \right)^{1/2},$$

$$\| u \|_{W_\beta^\ell(\mathbb{R})} = \left(\int_{\mathbb{R}} e^{2\beta t} \sum_{q=0}^\ell \| D_t^q u(t) \|_{\ell-q}^2 \, dt \right)^{1/2},$$

where $\beta \in \mathbb{R}$.

By Parseval's equality (see Lions, Magenes (1972), Ch. I, Sect. 2) we obtain

$$\| f \|_{L_{2,\beta}(\mathbb{R};H_0)}^2 = 2\pi \int_{\Im\lambda=\beta} \| \tilde{f}(\lambda) \|_0^2 \, d\lambda, \tag{2.15}$$

where $\tilde{f}$ is the Fourier transform defined by

$$\tilde{f}(\lambda) = \int_{\mathbb{R}} e^{-i\lambda t} f(t) dt$$

and, analogously,

$$\| u \|_{W_\beta^\ell(\mathbb{R})}^2 = 2\pi \int_{\Im\lambda=\beta} \sum_{0 \le q \le \ell} |\lambda|^{2q} \| \tilde{u}(\lambda) \|_{\ell-q}^2 \, d\lambda. \tag{2.16}$$

Note that by Definition 2.3.6 the norms

$$\| u \|_{W_\beta^\ell(\mathbb{R})} \quad \text{and} \quad \| e^{\beta t} u \|_{W_0^\ell(\mathbb{R})}$$

are equivalent.

We shall also need two-weighted spaces containing functions with different behaviour at $+\infty$ and $-\infty$.

Definition 2.3.7. Let $L_{2,\beta_-,\beta_+}(\mathbb{R}; H_0)$ and $W_{\beta_-,\beta_+}^\ell(\mathbb{R})$ be the spaces of functions in $L_{2,\mathrm{loc}}(\mathbb{R}; H_0)$ and $W_{\mathrm{loc}}^\ell(\mathbb{R})$ with finite norms

$$\| u \|_{L_{2,\beta_-,\beta_+}(\mathbb{R};H_0)} = \left(\sum_\pm \int_{\mathbb{R}_\pm} e^{2\beta_\pm t} \| u(t) \|_0^2 \, dt \right)^{1/2},$$

$$\| u \|_{W_{\beta_-,\beta_+}^\ell(\mathbb{R};H_0)} = \left(\sum_\pm \int_{\mathbb{R}_\pm} e^{2\beta_\pm t} \sum_{q=0}^\ell \| D_t^q u(t) \|_{\ell-q}^2 \, dt \right)^{1/2},$$

where

$$\mathbb{R}_\pm = \{ t \in \mathbb{R} : \ t \gtrless 0 \}.$$

Proposition 2.3.8. *If $u \in W_{\beta_-,\beta_+}^\ell(\mathbb{R})$ then*

$$\sum_{q=0}^{\ell-1} \| D_t^\ell u(t) \|_{\ell-q-1} = o\big(e^{-\beta_\pm t}\big) \quad as \quad t \to \pm\infty. \qquad (2.17)$$

Proof. By Definition 2.3.7

$$e^{\beta_\pm t} u \in W^\ell(\mathbb{R}_\pm),$$

and the result follows from Proposition 2.3.5. $\square$

2.4 Solvability in $W_\beta^\ell(\mathbb{R})$

It is clear that the operator

$$\mathcal{A}(D_t): \ W_\beta^\ell(\mathbb{R}) \to L_{2,\beta}(\mathbb{R}; H_0) \qquad (2.18)$$

is bounded for every $\beta \in \mathbb{R}$. In the next theorem we prove the unique solvability of (2.1) in $W_\beta^\ell(\mathbb{R})$.

Theorem 2.4.1. *Let $\beta \in \mathbb{R}$. Suppose that there are no eigenvalues of $\mathcal{A}(\lambda)$ on the line $\Im\lambda = \beta$. Then the operator (2.18) is invertible and the inverse operator is given by*

$$u(t) = \frac{1}{2\pi} \int_{\Im\lambda=\beta} e^{i\lambda t} \mathcal{A}^{-1}(\lambda)\tilde{f}(\lambda)d\lambda \qquad (2.19)$$

and the following estimate holds:

$$\| u \|_{W_\beta^\ell(\mathbb{R})} \leq C \| f \|_{L_{2,\beta}(\mathbb{R};H_0)}. \qquad (2.20)$$

Proof. Let $u \in W_\beta^\ell(\mathbb{R})$ be a solution of (2.1). By the Fourier transform

$$\mathcal{A}(\lambda)\tilde{u}(\lambda) = \tilde{f}(\lambda)$$

for almost every λ with $\Im\lambda = \beta$. Clearly, $\tilde{f}(\lambda) \in H_0$ and $\tilde{u}(\lambda) \in H_\ell$. Since the line $\Im\lambda = \beta$ consists of regular points of the pencil $\mathcal{A}(\lambda)$, it follows that

$$\tilde{u}(\lambda) = \mathcal{A}^{-1}(\lambda)\tilde{f}(\lambda). \qquad (2.21)$$

By (2.3),

$$\int_{\Im\lambda=\beta} \sum_{q=0}^{\ell} |\lambda|^{2q} \| \tilde{u}(\lambda) \|_{\ell-q}^2 \, d\lambda < C \int_{\Im\lambda=\beta} \| \tilde{f}(\lambda) \|_0^2 \, d\lambda,$$

which leads to (2.20) by (2.15) and (2.16). Applying the inverse Fourier transform to (2.21) we arrive at (2.19).

To prove the solvability we note that the right-hand side in (2.19) belongs to $W_\beta^\ell(\mathbb{R})$ by Parseval's theorem and that it satisfies (2.1). $\square$

Remark 2.4.2. If there is at least one eigenvalue λ_0 of $\mathcal{A}(\lambda)$ with $\Im\lambda_0 = \beta$, then the operator (2.18) is not isomorphic. In fact, let φ_0 be an eigenvector corresponding to λ_0 normalized by $\| \varphi_0 \|_0 = 1$. We set

$$u_\varepsilon(t) = \left(\frac{2\varepsilon}{\pi}\right)^{1/4} e^{-\varepsilon t^2 + i\lambda_0 t}\varphi_0.$$

Then

$$\| u_\varepsilon \|_{W^\ell_\beta(\mathbb{R})} \geq \| u_\varepsilon \|_{L_{2,\beta}(\mathbb{R};H_0)} = 1.$$

On the other hand, since

$$\mathcal{A}(D_t)(e^{i\lambda_0 t}\varphi_0) = 0,$$

it follows that

$$\| \mathcal{A}u_\varepsilon \|_{L_{2,\beta}(\mathbb{R};H_0)} = O(\varepsilon^{1/2})$$

and the estimate (2.20) fails.

2.5 Application to the Dirichlet Problem in a Cylinder

Consider the Dirichlet problem

$$\begin{cases} P(x, D_x, D_t)u = f & \text{on } \mathcal{C} \\ \partial_\nu^j u = 0 & \text{on } \partial\mathcal{C},\ j = 0,\ldots,m-1, \end{cases} \tag{2.22}$$

in the cylinder $\mathcal{C} = \{(x,t) : x \in \Omega,\ t \in \mathbb{R}\}$, where Ω is a domain in $\mathbb{R}^n$ with compact closure and smooth boundary. (The case of smooth compact manifold $\overline{\Omega}$ with or without boundary can be treated in exactly the same way).

The operator P in (2.22) is given by

$$P(x, D_x, D_t) = \sum_{q=0}^{\ell} A_{\ell-q}(x, D_x)D_t^q.$$

We assume that ℓ and $2m/\ell$ are integers and that A_k is a differential operator with smooth coefficients. Let s_k be the order of A_k. The operator P is said to be of order type $(2m, \ell)$ if the order of A_k does not exceed $2mk/\ell$. The sum of terms in A_k of order $2mk/\ell$ will be denoted by A_k^0.

We say that $\mathcal{A}(x, D_x, D_t)$ is $(2m, \ell)$-elliptic if

$$\sum_{q=0}^{\ell} A_{\ell-q}^0(x,\xi)\tau^q \neq 0$$

for all non-zero $(\xi, \tau) \in \mathbb{R}^n \times \mathbb{R}$ and all $x \in \overline{\Omega}$. In the case $n = 1$, we require additionally that there are precisely m roots of the polynomial

$$z \to \sum_{q=0}^{\ell} A_{\ell-q}^0(x, z)\tau^q$$

in the half-plane $\Im z > 0$ for all $\tau \neq 0$, $x \in \partial\Omega$.

We note that $(2m, 2m)$-elliptic operators are standard elliptic operators of order $2m$. Furthermore, the class of $(2m, 1)$-elliptic operators includes parabolic operators.

We introduce the operator pencil

$$\mathcal{A}(\lambda) = P(x, D_x, \lambda) : \mathcal{H} \to H \qquad (2.23)$$

where $H = L_2(\Omega)$ and

$$\mathcal{H} = \left\{ \varphi \in W_2^{2m}(\Omega) : \partial_\nu^j \varphi = 0 \quad \text{on} \quad \partial\Omega, \; j = 0, \ldots, m-1 \right\}.$$

The coefficients $A_k(x, D_x)$ will be considered as operators from

$$H_k = W_2^{2mk/\ell}(\Omega) \cap \mathcal{H} \quad \text{to} \quad H_0 = H.$$

The following result shows that the pencil $\mathcal{A}(\lambda)$ satisfies Conditions I and II of Sect. 2.2.1.

Theorem 2.5.1. (see, for example, Agmon, Nirenberg (1963), Th. 5.2, 5.3). *The mapping (2.23) is Fredholm for all $\lambda \in \mathbb{C}$ and one-to-one for all real λ of sufficiently large absolute value. For these λ the estimate (2.3) holds for all $\varphi \in \mathcal{H}$.*

The space $W_\beta^\ell(\mathbb{R})$ introduced in Sect. 2.3.2 takes the form

$$\Big\{ u : e^{\beta t} D_x^\alpha D_t^k u \in L_2(\mathcal{C}) \text{ for } |\alpha| + \frac{2mk}{\ell} \leq 2m$$
$$\text{and } \partial_\nu^j u = 0 \text{ on } \partial\mathcal{C} \text{ for } j = 0, \ldots, m-1 \Big\}.$$

We shall denote this space by $\dot{W}_{2,\beta}^{2m,\ell}(\mathcal{C})$ and supply it with the norm

$$\left(\int e^{2\beta t} \sum_{|\alpha| + \frac{2mk}{\ell} \leq 2m} \left| D_x^\alpha D_t^k u(x, t) \right|^2 dx dt \right)^{1/2}.$$

In this case, the space $L_{2,\beta}(\mathbb{R}; H_0)$ is realized as

$$L_{2,\beta}(\mathcal{C}) = \left\{ u : e^{\beta t} u \in L_2(\mathcal{C}) \right\}.$$

By Theorem 2.4.1 and Remark 2.4.2, the operator

$$\mathcal{A}(D_t) : \dot{W}_{2,\beta}^{2m,\ell}(\mathcal{C}) \to L_{2,\beta}(\mathcal{C})$$

is isomorphic if and only if there are no eigenvalues of $\mathcal{A}(\lambda)$ on the line $\Im \lambda = \beta$.

2.6 Green's Kernel

2.6.1 Definition of Green's Kernel

The following assertion will be used to define Green's kernel $G(t)$ of the operator inverse to (2.18).

Lemma 2.6.1. *Suppose that the line $\Im\lambda = \beta$ does not contain eigenvalues of $\mathcal{A}(\lambda)$. Then for $t \neq 0$, the limit*

$$\lim_{N\to+\infty} \int_{-N}^{N} e^{it\tau} \mathcal{A}^{-1}(\tau + i\beta) d\tau \qquad (2.24)$$

exists in the space $\mathcal{L}(H_0, H_{\ell-1})$.

Proof. By differentiating $\mathcal{A}^{-1}(\lambda)$ and using (2.3) we verify that

$$\| D_\lambda^m \mathcal{A}^{-1}(\lambda) \|_{H_0 \to H_q} \leq c_m \left(1 + |\lambda|\right)^{q-\ell-m} \qquad (2.25)$$

for $\lambda \in S_{r,\theta}$, $q = 0, \ldots, \ell$ and $m = 0, 1, \ldots$. By using the fact that $D_\tau e^{it\tau} = t e^{it\tau}$ and integrating by parts, the integral in (2.24) can be transformed as

$$-\frac{1}{t} \int_{-N}^{N} e^{it\tau} D_\tau \mathcal{A}^{-1}(\tau + i\beta) d\tau - \frac{1}{it} \left(e^{itN} \mathcal{A}^{-1}(N + i\beta) \right.$$
$$\left. - e^{-itN} \mathcal{A}^{-1}(-N + i\beta) \right).$$

For $m = 1$ and $q = \ell - 1$, (2.25) implies that this operator sequence tends to the integral

$$-\frac{1}{t} \int_{-\infty}^{\infty} e^{it\tau} D_\tau \mathcal{A}^{-1}(\tau + i\beta) d\tau$$

in the space $\mathcal{L}(H_0, H_{\ell-1})$. Thus, for $t \neq 0$,

$$\lim_{N\to+\infty} \int_{-N}^{N} e^{it\tau} \mathcal{A}^{-1}(\tau + i\beta) d\tau = -\frac{1}{t} \int_{-\infty}^{\infty} e^{it\tau} D_\tau \mathcal{A}^{-1}(\tau + i\beta) d\tau, \quad (2.26)$$

where the integral in the right-hand side is absolutely convergent in the norm of $\mathcal{L}(H_0, H_{\ell-1})$. $\square$

Now we introduce the operator

$$G(t) = \frac{1}{2\pi} \int_{\Im\lambda=\beta} e^{it\lambda} \mathcal{A}^{-1}(\lambda) d\lambda, \qquad (2.27)$$

interpreted in the sense of the Cauchy principal value integral. By (2.26)

$$G(t) = -\frac{1}{2\pi t} \int_{\Im\lambda=\beta} e^{it\lambda} D_\lambda \mathcal{A}^{-1}(\lambda) d\lambda \qquad (2.28)$$

with absolutely convergent integral in $\mathcal{L}(H_0, H_{\ell-1})$ on the right.

2.6.2 Properties of $G(t)$

Proposition 2.6.2. *Let the strip* $k_- < \Im\lambda < k_+$ *be free of eigenvalues of the operator pencil* $\mathcal{A}(\lambda)$. *Then the operator function* (2.27) *has the following properties.*

a) *It does not depend on* $\beta \in (k_-, k_+)$.

b) *For all* $|t| \geq 1$, $k = 0, 1, \ldots$,

$$\| D_t^k G(t) \|_{H_0 \to H_\ell} \leq c_{k,\gamma} \, e^{-\gamma t}, \tag{2.29}$$

where γ *is an arbitrary number from* (k_-, k_+).

c) *For all* t *such that* $|t| \leq 1$, $k = 0, 1, \ldots, \ell - 1$,

$$\| D_t^k G(t) \|_{H_0 \to H_{\ell-k-1}} \leq c \left(\log \frac{1}{|t|} + 1 \right). \tag{2.30}$$

Proof. a). Since the operator function $\mathcal{A}^{-1}(\lambda)$ is analytic in the strip $k_- < \Im\lambda < k_+$, the independence of β follows from (2.28).

b) By (2.25) and (2.28) we have

$$t^{m+N} D_t^m G(t) = \frac{1}{2\pi} \int_{\Im\lambda=\beta} e^{it\lambda} (-D_\lambda)^{m+N} \left(\lambda^m \mathcal{A}^{-1}(\lambda) \right) d\lambda$$

for $m = 0, 1, \ldots$ and $N = 2, 3, \ldots$ Direct estimation of the integrand leads to (2.29).

c) Let r and θ be the same numbers as in Proposition 2.2.1 and let ρ be a sufficiently large positive number depending on β, r and θ. We replace the line of integration in (2.28) by the broken line

$$L_\beta^\pm = \left\{ z : \Im z = \beta, \ |\Re z| \leq \rho \right\}$$
$$\cup \left\{ z = \rho + i\beta + se^{\pm i\theta/2} : s > 0 \right\}$$
$$\cup \left\{ z = -\rho + i\beta - se^{\pm i\theta/2} : s > 0 \right\},$$

choosing $\pm$ in accordance with $t \gtrless 0$. This is possible due to (2.25). Thus we get the following representation for G:

$$G(t) = \frac{1}{2\pi} \int_{L_\beta^\pm} e^{it\lambda} \mathcal{A}^{-1}(\lambda) d\lambda \quad \text{for} \quad \pm t > 0. \tag{2.31}$$

For λ on the broken lines $L_\beta^\pm$ we have

$$\| \mathcal{A}^{-1}(\lambda) \|_{H_0 \to H_{\ell-k-1}} \leq c \left(1 + |\lambda| \right)^{-1-k}, \quad k = 0, \ldots, \ell - 1.$$

Therefore for λ on the horizontal part of $L_\beta^\pm$ we obtain

$$\| e^{it\lambda} \lambda^k \mathcal{A}^{-1}(\lambda) \|_{H_0 \to H_{\ell-k-1}} \leq c \, e^{-\beta t} \left(1 + |\lambda| \right)^{-1}.$$

Furthermore for $t > 0$ and

$$\lambda = \pm \rho + i\beta \pm s \, e^{\pm i\theta/2}, \quad s > 0,$$

we have

$$\| \, e^{it\lambda} \lambda^k \mathcal{A}^{-1}(\lambda) \, \|_{H_0 \to H_{\ell-k-1}} \leq c \, e^{-\beta - st \sin(\theta/2)} (1+s)^{-1}. \qquad (2.32)$$

We conclude that for $t > 0$

$$\| \, D_t^k G(t) \, \|_{H_0 \to H_{\ell-k-1}}$$

$$\leq c \, e^{-\beta t} \Big(\int_{-\rho}^{\rho} \frac{ds}{1+|s|} + \int_0^\infty e^{-ts \sin(\theta/2)} \frac{ds}{1+s} \Big). \qquad (2.33)$$

The estimate (2.30) follows for $t > 0$. The argument is the same for $t < 0$. □

2.6.3 Integral Representation of Solutions

Theorem 2.6.3. *The inverse operator defined in* Theorem 2.4.1 *is given by*

$$u(t) = \int_{\mathbb{R}} G(t-\tau) f(\tau) d\tau. \qquad (2.34)$$

(Due to $f \in L_{2,\beta}(\mathbb{R}; H_0)$ and the estimates (2.29) and (2.30), the last integral converges absolutely with respect to the norm in $H_{\ell-1}$).

Proof. By (2.31) the integral in the right-hand side of (2.34) is equal to

$$\frac{1}{2\pi} \Big(\int_{t>\tau} \int_{L_\beta^+} e^{i(t-\tau)\lambda} \mathcal{A}^{-1}(\lambda) d\lambda f(\tau) d\tau$$

$$+ \int_{t<\tau} \int_{L_\beta^-} e^{i(t-\tau)\lambda} \mathcal{A}^{-1}(\lambda) d\lambda f(\tau) d\tau \Big). \qquad (2.35)$$

For $t > \tau$ and $\lambda \in L_\beta^+$ we have by (2.32), (2.33) and the fact that $f \in L_{2,\beta}(\mathbb{R}; H_0)$, that the first double integral is absolutely convergent in the norm of $H_{\ell-1}$. The same is obviously true for the second integral.

We note that for $f \in L_{2,\beta}(\mathbb{R}; H_0)$ the integral

$$\int_{\Im\lambda=\beta} \| \, \mathcal{A}^{-1}(\lambda) \int_{-\infty}^t e^{i\lambda(t-\tau)} f(\tau) d\tau \, \|_{\ell-1} |d\lambda| \qquad (2.36)$$

is finite. In fact, it does not exceed

$$C \Big(\int_{\Im\lambda=\beta} \frac{|d\lambda|}{(1+|\lambda|)^2} \Big)^{1/2} \Big(\int_{\Im\lambda=\beta} \| \int_{-\infty}^t e^{i\lambda(t-\tau)} f(\tau) d\tau \, \|_0^2 \, |d\lambda| \Big)^{1/2},$$

which is majorized by

$$C e^{-\beta t} \Big(\int_{\mathbb{R}} \| \int_{\mathbb{R}} e^{-is\tau + \beta\tau} \chi_t(\tau) f(\tau) d\tau \|_0^2 ds \Big)^{1/2}.$$

By Parseval's theorem this function is estimated by

$$c\, e^{-\beta t} \parallel f \parallel_{L_{2,\beta}(\mathbb{R};H_0)} \cdot$$

The same argument gives the convergence of the integral

$$\int_{\Im\lambda=\beta} \parallel \mathcal{A}^{-1}(\lambda) \int_t^\infty e^{i\lambda(t-\tau)} f(\tau)d\tau \parallel_{\ell-1} |d\lambda|.$$

Our aim is to replace the integration paths $L_\beta^\pm$ in (2.35) by the line $\Im\lambda = \beta$. To this end we introduce the intervals

$$\sigma_N^\pm = \{\lambda = \pm N + iy : 0 < y - \beta < (N - \rho)\tan(\theta/2)\}$$

with sufficiently large N, and show that the integrals

$$\int_{\sigma_N^\pm} e^{i\lambda t} \mathcal{A}^{-1}(\lambda) \int_{-\infty}^t e^{-i\lambda\tau} f(\tau)d\tau d\lambda$$

tend to zero in the norm in $H_{\ell-1}$ provided $f \in L_{2,\beta}(\mathbb{R};H_0)$. In fact, by (2.32) the $H_{\ell-1}$-norms of both integrals do not exceed

$$\frac{c}{N}\int_0^{c_0 N} \int_{-\infty}^t e^{-(x+\beta)(t-\tau)} \parallel f(\tau)\parallel_0 d\tau dx$$

$$= \frac{c}{N}\int_{-\infty}^t \frac{1 - e^{-c_0 N(t-\tau)}}{t-\tau} e^{-\beta(t-\tau)} \parallel f(\tau)\parallel_0 d\tau$$

$$\leq c_1 N^{-1/2} e^{-\beta t} \parallel f \parallel_{L_{2,\beta}(\mathbb{R};H_0)} \cdot$$

Therefore, L_β^+ in (2.35) may be replaced by the line $\Im\lambda = \beta$. The same argument can be applied to L_β^-, which shows that (2.35) is equal to

$$\frac{1}{2\pi}\int_{\Im\lambda=\beta} e^{i\lambda t} \mathcal{A}^{-1}(\lambda) \int_\mathbb{R} e^{-i\lambda\tau} f(\tau)d\tau d\lambda.$$

The result follows from (2.19). $\square$

2.7 Asymptotic Decompositions of Green's Kernel

2.7.1 Representations for $G(t)$

According to Proposition 2.6.2, a), Green's kernel $G(t)$ depends on $k_\pm$ and not on $\beta \in (k_-, k_+)$. Let the line $\Im\lambda = d$ be free of eigenvalues of $\mathcal{A}(\lambda)$. We can define a new Green's kernel

$$G^{(d)}(t) = \frac{1}{2\pi}\int_{\Im\lambda=d} e^{it\lambda} \mathcal{A}^{-1}(\lambda)d\lambda. \qquad (2.37)$$

By (2.29)

$$\| D_t^k G^{(d)}(t) \|_{H_0 \to H_\ell} \leq c_k e^{-dt} \quad \text{for} \quad |t| \geq 1. \tag{2.38}$$

Here we find a relation between G and $G^{(d)}$.

Let us introduce the operator function

$$P_\nu(t) = \frac{1}{2\pi} \int_{S_\nu} e^{it(\lambda-\lambda_\nu)} \mathcal{A}^{-1}(\lambda) d\lambda, \tag{2.39}$$

where S_ν is a small circle centered at the eigenvalue λ_ν. Making use of (2.7) together with Cauchy's residue theorem we have

$$P_\nu(t) = i \sum_{1 \leq k \leq J_\nu} \sum_{s=0}^{m_{\nu k}-1} \frac{(it)^s}{s!} P^{(\nu)}_{k,m_{\nu k}-1-s}, \tag{2.40}$$

where $P^{(\nu)}_{k,s}$ is defined by (2.8). In particular, this representation yields that P_ν is the operator polynomial of degree $m_\nu - 1$, where

$$m_\nu = \max_{1 \leq k \leq J_\nu} m_{\nu k}, \tag{2.41}$$

and the coefficients of P_ν are finite dimensional operators acting from H_0 into H_ℓ.

Theorem 2.7.1. *Let d_-, d_+ be real numbers such that $d_- < k_-$, $d_+ > k_+$. Suppose that there are no eigenvalues of the operator pencil $\mathcal{A}(\lambda)$ on the lines $\Im \lambda = d_\pm$. Then*

$$G(t) = \sum_{k_+ \leq \Im \lambda_\nu < d_+} e^{i\lambda_\nu t} P_\nu(t) + G^{(d_+)}(t), \tag{2.42}$$

$$G(t) = - \sum_{d_- < \Im \lambda_\nu \leq k_-} e^{i\lambda_\nu t} P_\nu(t) + G^{(d_-)}(t). \tag{2.43}$$

Proof. Due to (2.25) the difference $G(t) - G^{(d_+)}(t)$ is equal to the sum of integrals in (2.39), multiplied by $e^{i\lambda_\nu t}$, with the summation being extended over the eigenvalues situated between the lines $\Im \lambda = k_+$ and $\Im \lambda = d_+$. Reference to (2.39) gives (2.42). The proof of (2.43) is analogous. $\square$

Due to (2.38), the formulae (2.42) and (2.43) can be interpreted as asymptotic representations of $G(t)$ as $t \to \pm\infty$.

2.7.2 Representations for $G(t - \tau)$

Lemma 2.7.2. *The following formula holds:*

$$\sum_{s=0}^{m_{\nu k}-1} \frac{(it - i\tau)^s}{s!} P_{k,m_{\nu k}-1-s}^{(\nu)}$$

$$= \sum_{s=0}^{m_{\nu k}-1} \left(\cdot \, , D_\tau^s \Psi_k^{(\nu)}(i\,\tau) \right)_{H_0} D_t^{m_{\nu k}-1-s} \Phi_k^{(\nu)}(it), \qquad (2.44)$$

where the vector-functions $\Psi_k^{(\nu)}$ and $\Phi_k^{(\nu)}$ are defined by (1.28) and (1.24).

Proof. The right-hand side of (2.44) is equal to

$$\sum_{s=0}^{m_{\nu k}-1} \left(\cdot \, , \sum_{j=0}^{m_{\nu k}-1-s} \frac{(i\tau)^j}{j!} \psi_{k,m_{\nu k}-1-s-j}^{(\nu)} \right)_{H_0} \sum_{q=0}^{s} \frac{(it)^q}{q!} \phi_{k,s-q}^{(\nu)}$$

$$= \sum_{j=0}^{m_{\nu k}-1} \sum_{q=0}^{m_{\nu k}-1-j} \sum_{s=q}^{m_{\nu k}-1-j} \frac{(-i\,\tau)^j}{j!} \frac{(it)^q}{q!} \left(\cdot \, , \psi_{k,m_{\nu k}-1-s-j}^{(\nu)} \right)_{H_0} \phi_{k,s-q}^{(\nu)}$$

$$= \sum_{j=0}^{m_{\nu k}-1} \sum_{q=0}^{m_{\nu k}-1-j} \sum_{s=0}^{m_{\nu k}-1-j-q} \frac{(-i\tau)^j}{j!} \frac{(it)^q}{q!} \left(\cdot \, , \psi_{k,m_{\nu k}-1-s-j-q}^{(\nu)} \right)_{H_0} \phi_{ks}^{(\nu)}.$$

Introducing the new index $n = q + j$ we obtain that the right-hand side of the last relation equals

$$\sum_{n=0}^{m_{\nu k}-1} \sum_{q=0}^{n} \sum_{s=0}^{m_{\nu k}-1-n} \frac{(it)^q}{q!} \frac{(-i\tau)^{n-q}}{(n-q)!} \left(\cdot \, , \psi_{k,m_{\nu k}-1-n-s}^{(\nu)} \right)_{H_0} \phi_{ks}^{(\nu)}$$

$$= \sum_{n=0}^{m_{\nu k}-1} \frac{(it - i\tau)^n}{n!} \sum_{s=0}^{m_{\nu k}-1-n} \left(\cdot \, , \psi_{k,m_{\nu k}-1-n-s}^{(\nu)} \right)_{H_0} \phi_{ks}^{(\nu)}.$$

Making use of (2.8) we see that the last expression coincides with the left-hand side of (2.44). $\square$

Theorem 2.7.3. *Green's kernel admits the representations*

$$G(t - \tau) = - \sum_{k_+ \leq \Im\lambda_\nu < d_+} \sum_{\sigma=1}^{\kappa_\nu} \left(\cdot \, , V_\sigma^{(\nu)}(\tau) \right)_{H_0} U_\sigma^{(\nu)}(t)$$

$$+ G^{(d_+)}(t - \tau), \qquad (2.45)$$

and

$$G(t - \tau) = \sum_{d_- < \Im\lambda_\nu \leq k_-} \sum_{\sigma=1}^{\kappa_\nu} \left(\cdot \, , V_\sigma^{(\nu)}(\tau) \right)_{H_0} U_\sigma^{(\nu)}(t)$$

$$+ G^{(d_-)}(t - \tau), \qquad (2.46)$$

where $d_\pm$, $G^{(d_\pm)}$ are as in Theorem 2.7.1 and $U_\sigma^{(\nu)}$, $V_\sigma^{(\nu)}$ were introduced in Sect.2.2.3.

Proof. By (2.39) and (2.44)

$$P_\nu(t-\tau) = i \sum_{k=0}^{J_\nu} \sum_{s=0}^{m_{\nu k}-1} \left(\cdot\, , D_\tau^s \Psi_k^{(\nu)}(i\,\tau) \right)_{H_0} D_t^{m_{\nu s}-1-s} \Phi_k^{(\nu)}(it). \qquad (2.47)$$

The result now follows from Theorem 2.7.1 and (2.11), (2.12). □

Since by (2.38),

$$\| D_t^k G^{(d_\pm)}(t-\tau) \|_{H_0 \to H_\ell} \le c_k e^{-d_\pm(t-\tau)} \quad \text{for} \quad |t-\tau| \ge 1, \qquad (2.48)$$

the representations (2.45) and (2.46) can be considered as asymptotic formulae for $G(t-\tau)$ as $t-\tau \to \pm\infty$.

Remark 2.7.4. By combining (2.47) and (2.9)–(2.12) we arrive at the representation

$$\sum_{\sigma=1}^{\kappa_\nu} \left(\cdot\, , V_\sigma^{(\nu)}(\tau) \right)_{H_0} U_\sigma^{(\nu)}(t) = e^{i\lambda_\nu(t-\tau)} P_\nu(t-\tau). \qquad (2.49)$$

This will be used in the sequel.

Remark 2.7.5. It is worth noting that the bases $\{U_\sigma^{(\nu)}\}$ and $\{V_\sigma^{(\nu)}\}$ can be chosen rather arbitrarily in order that all the subsequent theory be valid. In fact, it is enough to take bases $\{u_j\}_{j=1}^{\kappa_\nu}$ and $\{v_k\}_{k=1}^{\kappa_\nu}$ in the spaces $Z(\mathcal{A}, \lambda_\nu)$ and $Z(\mathcal{A}^*, \overline{\lambda}_\nu)$ subject to the biortogonality condition

$$\int_{\mathbb{R}} \left(\mathcal{A}(D_t)\eta(t)u_j(t), v_k(t) \right)_{H_0} dt = \delta_j^k,$$

where η is the same function as in (1.29). Then, in particular,

$$\sum_{\sigma=1}^{\kappa_\nu} \left(\cdot\, , V_\sigma^{(\nu)}(\tau) \right)_{H_0} U_\sigma^{(\nu)}(t) = \sum_{\sigma=1}^{\kappa_\nu} \left(\cdot\, , v_j(\tau) \right)_{H_0} u_j(t),$$

which gives another representation for $e^{i\lambda_\nu(t-\tau)} P_\nu(t-\tau)$.

2.8 Asymptotics of Solutions in $W_\beta^\ell(\mathbb{R})$

2.8.1 Asymptotic Representations

Theorem 2.7.3 leads to a formula for the difference of two solutions of equation (2.1).

Proposition 2.8.1. *Let* $\beta, \gamma \in \mathbb{R}$ *and* $\beta < \gamma$. *Suppose that there are no eigenvalues of* $\mathcal{A}(\lambda)$ *on the lines* $\Im \lambda = \beta$, $\Im \lambda = \gamma$. *Let*

$$f \in L_{2,\beta}(\mathbb{R}, H_0) \cap L_{2,\gamma}(\mathbb{R}; H_0)$$

and u_1, u_2 *be the solutions of* (2.1) *from the spaces* $W_\beta^\ell(\mathbb{R})$, $W_\gamma^\ell(\mathbb{R})$ *respectively. Then*

$$u_2(t) - u_1(t) = \sum_{\beta < \Im \lambda_\nu < \gamma} \int_\mathbb{R} e^{i\lambda_\nu(t-\tau)} P_\nu(t-\tau) f(\tau) d\tau \qquad (2.50)$$

or equivalently,

$$u_2(t) - u_1(t) = \sum_{\beta < \Im \lambda_\nu < \gamma} \sum_{\sigma=1}^{\kappa_\nu} C_{\nu\sigma} U_\sigma^{(\nu)}(t), \qquad (2.51)$$

where

$$C_{\nu\sigma} = \int_\mathbb{R} \left(f(\tau), V_\sigma^{(\nu)}(\tau) \right)_{H_0} d\tau. \qquad (2.52)$$

Proof. Let k_-, k_+ be real numbers such that $k_- < \beta < k_+$ and the strip $k_- < \Im \lambda < k_+$ is free of eigenvalues.

First, let $\gamma \in (\beta, k_+)$. By Proposition 2.6.2 a) and Theorem 2.6.3, $u_1 = u_2$.

Now, let $\gamma > k_+$. By using Lemma 2.7.1 with $d_+ = \gamma$ together with Theorem 2.6.3 we arrive at (2.50). Assertion (2.51) follows from Theorem 2.7.3 in the same way. $\square$

Representations (2.50) and (2.51) can be considered as asymptotic formulae for u_1 as $t \to +\infty$ and for u_2 as $t \to -\infty$. For example, the following assertion which follows directly from Proposition 2.8.1.

Corollary 2.8.2. *Let* $\beta, \gamma \in \mathbb{R}$ *and* $\beta < \gamma$. *Suppose that there are no eigenvalues of* $\mathcal{A}(\lambda)$ *on the lines* $\Im \lambda = \beta$, $\Im \lambda = \gamma$. *Let* $\xi f \in L_{2,\gamma}(\mathbb{R})$, *where* ξ *is a smooth function equal to* 1 *for* $t > 2$ *and to* 0 *for* $t < 1$. *If* u *satisfies* $\mathcal{A}(D_t)u = f$ *for* $t > 0$ *and* $\zeta u \in W_\beta^\ell(\mathbb{R})$, *where* $\zeta \in C^\infty(\mathbb{R})$, $\xi \zeta = \zeta$, *then*

$$\xi \left(u - \sum_{\beta < \Im \lambda_\nu < \gamma} \sum_{\sigma=1}^{\kappa_\nu} C_{\nu\sigma} U_\sigma^{(\nu)} \right) \in W_\gamma^\ell(\mathbb{R})$$

where $C_{\nu\sigma} = const$ *and* $U_\sigma^{(\nu)}$ *are the power-exponential solutions defined by* (2.11).

2.8.2 Solutions of the Homogeneous Equation

We give a corollary of Proposition 2.8.1 and Theorem 2.4.1 which shows that zeros of $\mathcal{A}(D_t)$ with exponential growth at infinity are linear combinations of $U_\sigma^{(\nu)}$.

Proposition 2.8.3. *Let β, $\gamma \in \mathbb{R}$, $\beta < \gamma$, and let $u \in W_{\gamma,\beta}^\ell(\mathbb{R})$ satisfy the equation*

$$\mathcal{A}(D_t)u(t) = 0 \quad on \quad \mathbb{R}. \tag{2.53}$$

Then

$$u(t) = \sum_{\beta < \Im\lambda_\nu < \gamma} \sum_{\sigma=1}^{\kappa_\nu} C_{\nu\sigma} U_\sigma^\nu(t). \tag{2.54}$$

In particular, if the strip $\beta < \Im\lambda < \gamma$ is free of the spectrum of $\mathcal{A}(\lambda)$ then $u = 0$.

Proof. First suppose that the lines $\Im\lambda = \beta$, $\Im\lambda = \gamma$ are free of eigenvalues of $\mathcal{A}(\lambda)$.

Let ξ be a smooth function as in Corollary 2.8.2. We have

$$(1 - \xi)u \in W_\gamma^\ell(\mathbb{R}) \quad \text{and} \quad \mathcal{A}(D_t)\big((1 - \xi)u\big) \in W_\beta^\ell(\mathbb{R}) \cap W_\gamma^\ell(\mathbb{R}).$$

By Proposition 2.8.1,

$$(1 - \xi)u = \sum_{\beta < \Im\lambda_\nu < \gamma} \sum_{\sigma=1}^{\kappa_\nu} C_{\nu\sigma} U_\sigma^{(\nu)}(t) + v,$$

where $v \in W_\beta^\ell(\mathbb{R})$. By rewriting this equality in the form

$$u = \sum_{\beta \leq \Im\lambda_\nu < \gamma} \sum_{\sigma=1}^{\kappa_\nu} C_{\nu\sigma} U_\sigma^{(\nu)}(t) + v + \xi u$$

we obtain

$$\mathcal{A}(D_t)(v + \xi u) = 0 \quad on \quad \mathbb{R}.$$

Since $\xi u \in W_\beta^\ell(\mathbb{R})$, we have $v + \xi u \in W_\beta^\ell(\mathbb{R})$. By Theorem 2.4.1 this implies that $v + \xi u = 0$.

Now consider the general case. Choose a positive number ε such that the lines $\Im\lambda = \beta - \varepsilon$ and $\Im\lambda = \gamma + \varepsilon$ are free of eigenvalues of $\mathcal{A}(\lambda)$. Since $u \in W_{\gamma+\varepsilon,\beta-\varepsilon}^\ell(\mathbb{R})$ we obtain

$$u(t) = \sum_{\beta-\varepsilon < \Im\lambda_\nu < \gamma+\varepsilon} \sum_{\sigma=1}^{\kappa_\nu} C_{\nu\sigma} U_\sigma^{(\nu)}(t).$$

By taking into account the fact that the functions $U_\sigma^{(\nu)}$ do not belong to $W_{\gamma,\beta}^\ell(\mathbb{R})$ for $\Im\lambda_\nu \leq \beta$ and for $\Im\lambda_\nu \geq \gamma$, we arrive at (2.54). $\square$

2.9 A Local Estimate for Solutions

We shall use the following local estimate.

Lemma 2.9.1. *Let $\delta > 0$ and let $u \in W^\ell(a - 3\delta, \, a + 3\delta)$ be a solution of*

$$\mathcal{A}(D_t)u = f \quad on \quad (a - 3\delta, \, a + 3\delta).$$

Then

$$\| \, u \, \|_{W^\ell(a,a+\delta)}$$
$$\leq C_\delta \Big(\, \| \, f \, \|_{L_2(a-3\delta, \, a+3\delta; H_0)} + \int_{a-3\delta}^{a+3\delta} \| \, u(\tau) \, \|_{\ell-1} \, d\tau \Big), \qquad (2.55)$$

where C_δ does not depend on a.

Proof. It is sufficient to prove the assertion for $a = 0$. Let $\eta \in C^\infty(\mathbb{R}), \eta(t) = 1$ for $|t| < 2\delta$ and $\eta(t) = 0$ for $|t| > 3\delta$. Clearly,

$$\mathcal{A}(D_t)(\eta u) = \eta f + \big[\mathcal{A}(D_t), \eta\big]u, \qquad (2.56)$$

where $[B, C] = BC - CB$.

Let β be a real number such that the line $\Im\lambda = \beta$ does not intersect the spectrum of $\mathcal{A}(\lambda)$. Since $\eta u \in W_\beta^\ell(\mathbb{R})$, we have

$$(\eta u)(t) = \int_\mathbb{R} G(t - \tau)(\eta f)(\tau)d\tau$$
$$+ \int_\mathbb{R} G(t - \tau)\big[\mathcal{A}(D_\tau), \eta\big]u(\tau)d\tau = v_1(t) + v_2(t).$$

By Theorem 2.4.1 we obtain

$$\| \, v_1 \, \|_{W_\beta^\ell(\mathbb{R})} \leq c \, \| \, \eta f \, \|_{L_{2,\beta}(\mathbb{R};H_0)} \, .$$

Hence

$$\| \, v_1 \, \|_{W^\ell(0,\delta)} \leq c_\delta \, \| \, f \, \|_{L_2(-3\delta, \, 3\delta; \, H_0)} \, . \qquad (2.57)$$

One can verify the identity

$$\Big[\mathcal{A}(D_t)\eta(t) - \eta(t)\mathcal{A}(D_t)\Big]u(t)$$
$$= \sum_{0 \leq m \leq \ell-1} \frac{1}{m!} \, D_t^m \Big(\sum_{1 \leq k \leq \ell-m} \frac{(m + k)!}{k!} \, (D_t^k\eta(t)) A_{\ell-m-k}u(t) \Big).$$

Thus

$$v_2(t) = \sum_{0 \leq m \leq \ell-1} \frac{1}{m!} \sum_{1 \leq k \leq \ell-m} \frac{(m + k)!}{k!}$$
$$\times \int_\mathbb{R} D_t^m G(t - \tau) \, (D_\tau^k\eta(\tau)) A_{\ell-m-k}u(\tau)d\tau.$$

If $t \in (0, \delta)$ then $|t - \tau| \geq 2\delta$ for τ in the support of the function $(d^k/d\tau^k)\eta(\tau)$ for $k \geq 1$. Using (2.29) we obtain

$$\| v_2 \|_{W^\ell(0,\delta)} \leq \int_{-3\delta}^{3\delta} \| u(\tau) \|_{\ell-1} \, d\tau. \tag{2.58}$$

The result follows from (2.57) and (2.58). □

2.10 Application to the Dirichlet Problem in a Cone

This is a continuation of Sect. 1.5.2–1.5.4, where the Dirichlet problem for the $2m$-th order elliptic operator P in the cone $K \subset \mathbb{R}^n$ was considered. We introduce the weighted Sobolev space

$$\dot{V}_{2,\delta}^{2m}(K) = \{ u : r^{\delta - 2m + |\alpha|} D_x^\alpha u \in L_2(K), |\alpha| \leq 2m,$$
$$\text{and} \quad \partial_\nu^j u = 0 \quad \text{on} \quad \partial K, \quad j = 0, \ldots, m - 1 \}$$

and the space

$$L_{2,\delta}(K) = \{ u : r^\delta u \in L_2(K) \}.$$

The mapping (0.8) transforms the Dirichlet problem in the cone to the problem in the cylinder $\mathcal{C}$ treated in Sect. 2.5. By (0.8) the spaces $W_{2,\beta}^{2m,2m}(\mathcal{C})$ and $L_{2,\beta}(\mathcal{C})$ are respectively isometric to $\dot{V}_{2,\delta}^{2m}(K)$ and $L_{2,\delta-2m}(K)$ with $\delta = -\beta + 2m - n/2$. Hence, by Sect. 2.5, the operator

$$P(D_x) : \dot{V}_{2,\delta}^{2m}(K) \to L_{2,\delta}(K)$$

is an isomorphism if and only if there are no eigenvalues of the pencil $\mathcal{A}(\lambda)$ on the line $\Im\lambda = -\delta + 2m - n/2$.

By setting $t = -\log r$ in (2.9) and (2.10) we obtain the bases of special "power-logarithmic" solutions to $P(D_x)\mathcal{U}(x) = 0$ and $P^*(D_x)\mathcal{V}(x) = 0$ in K with zero Dirichlet data on $\partial K \backslash \{0\}$:

$$\mathcal{U}_{k,h}^{(\nu)}(x) = r^{-i\lambda_\nu}(-rD_r)^{m_{\nu k}-1-h}\Phi_k^{(\nu)}(-i\log r),$$

$$\mathcal{V}_{k,h}^{(\nu)}(x) = ir^{2m-n-i\bar{\lambda}_\nu}(-rD_r)^h\Psi_k^{(\nu)}(-i\log r),$$

where $k = 1, \ldots, J_\nu$, $h = 0, \ldots, m_{\nu k}$ and

$$\Phi_k^{(\nu)}(z) = \sum_{\sigma=0}^{m_{\nu k}-1} \frac{z^\sigma}{\sigma!} \varphi_{k,m_{\nu k}-1-\sigma}^{(\nu)}(\theta),$$

$$\Psi_k^{(\nu)}(z) = \sum_{\sigma=0}^{m_{\nu k}-1} \frac{z^\sigma}{\sigma!} \psi_{k,m_{\nu k}-1-\sigma}^{(\nu)}(\theta).$$

We have made use of the fact obtained in Sect. 1.5.4 that $P^*(D_x)$ generates the pencil $\mathcal{A}^*(\lambda + i(2m - n))$, (see (1.53)). We use the same notation as in Sect.

1.2: J_ν is the geometric multiplicity of λ_ν, $m_{\nu k}$ are the partial multiplicities of λ_ν and

$$\left\{\varphi_{k,h}^{(\nu)}\right\} \quad \text{and} \quad \left\{\psi_{k,h}^{(\nu)}\right\} \tag{2.59}$$

are canonical systems of $\mathcal{A}(\lambda)$ and $\mathcal{A}^*(\lambda)$ corresponding to the eigenvalues λ_ν and $\overline{\lambda}_\nu$. The systems (2.59) are subject to the biorthogonality condition (1.20). Following Sect. 2.2.3, we introduce unique numeration for the pairs (k,h), i.e. we denote

$$\mathcal{U}_\sigma^{(\nu)}(x) = \mathcal{U}_{k,h}^{(\nu)}(x) \, ,$$
$$\mathcal{V}_\sigma^{(\nu)}(x) = \mathcal{V}_{k,h}^{(\nu)}(x) \, , \ 1 \leq \sigma \leq \kappa_\nu \, ,$$

where κ_ν is the algebraic multiplicity of λ_ν. The biorthogonality property (2.13) takes the form

$$\int_K P(D_x)(\zeta(|x|)\mathcal{U}_\sigma^{(\nu)}(x)) \cdot \overline{\mathcal{V}_\rho^{(\mu)}}(x)dx = \delta_\nu^\mu \delta_\sigma^\rho \, ,$$

where $\nu, \mu \in Z$, $\sigma, \rho = 1, \ldots, \kappa_\nu$ and $\zeta \in C^\infty(\mathbb{R}_+)$, $\zeta = 1$ in a neighbourhood of $+\infty$ and $\zeta = 0$ in a neighbourhood of 0.

With this notation we can restate Proposition 2.8.1 on the asymptotics of solutions to the problem

$$\begin{cases} P(D_x)u = f & \text{on } K \\ \partial_\nu^j u = 0 & \text{on } \partial K \backslash \{O\} \text{ for } j = 0, \ldots, m-1. \end{cases} \tag{2.60}$$

Theorem 2.10.1. *Let $\beta, \delta \in \mathbb{R}$ and $\delta < \beta$. Suppose that there are no eigenvalues of $\mathcal{A}(\lambda)$ on the lines $\Im\lambda = -\beta + 2m - n/2$ and $\Im\lambda = -\delta + 2m - n/2$. Also let $f \in L_{2,\beta}(K) \cap L_{2,\delta}(K)$ and u_1, u_2 be the solutions of (2.60) from the spaces $V_{2,\beta}^{2m}(K)$, $\dot{V}_{2,\delta}^{2m}(K)$ respectively. Then*

$$u_2(x) - u_1(x) = \sum_\nu \sum_{\sigma=1}^{\kappa_\nu} C_{\nu\sigma}\mathcal{U}_\sigma^{(\nu)}(x) \, ,$$

where the summation is extended over the set

$$\{\nu \in \mathbb{Z} : -\beta + 2m - n/2 < \Im\lambda_\nu < -\delta + 2m - n/2\}$$

and the coefficients $C_{\nu\sigma}$ are given by

$$C_{\nu\sigma} = \int_K f(x)\overline{\mathcal{V}_\sigma^{(\nu)}}(x)dx. \tag{2.61}$$

In conclusion, we note that by Theorem 2.7.3, Green's function $G(x,y)$ of the problem (2.60) has immediate asymptotic representations near 0 and ∞ which are stated in terms of solutions $\mathcal{U}_\sigma^{(\nu)}$ and $\mathcal{V}_\sigma^{(\nu)}$ (cf Maz'ya, Plamenevskii (1979)).

2.11 Comments

Condition I in Sect. 2.2.1 is a direct generalization of the ellipticity of parameter dependent boundary value problems (see Agranovich, Vishik (1964) and Agmon, Nirenberg (1963)). Equation (2.1) (under the Conditions I and II in Sect. 2.2.1) was considered by Maz'ya, Plamenevskii (1972); Theorem 2.4.1 on the unique solvability can be found therein.

The class of $(2m, l)$-elliptic equations considered in Sect. 2.5 was introduced by Agmon, Nirenberg (1963). Green's kernel estimates similar to (2.29) and (2.30) were obtained by Evgrafov (1960) for the case of the first order operator $D_t + A$ with constant A (see also Plamenevskii (1972)). For general boundary elliptic problems in a cone, representations of Green's kernels of the type (2.46) and (2.47) were obtained by Maz'ya, Plamenevskii (1979). In an earlier paper (Maz'ya, Plamenevskii (1975)) similar formulae to (2.52) were found for coefficients in the asymptotics of solutions of such problems.

The power-exponential asymptotic expansion for solutions with exponential growth

$$u(t) \sim \sum_{\Im \lambda_j \geq \gamma} e^{i\lambda_j t} P_j(t) \quad \text{as } t \to +\infty \tag{2.62}$$

dates to Agmon, Nirenberg (1963). Here, γ is a number characterizing the growth of the solution and P_j are polynomials in t whose coefficients are eigenvectors and generalized eigenvectors corresponding to the eigenvalue λ_j of the operator pencil A. The estimate of the remainder term in $W_\gamma^\ell(\mathbb{R})$ corresponds to the exposition given by Kondratiev (1967) for the elliptic boundary value problem in a cone.

3. Solutions in a Local Sobolev Space

3.1 Introduction

In this chapter, we connect with any strip $k_- < \Im\lambda < k_+$ free of eigenvalues of $\mathcal{A}(\lambda)$ a class of solutions of equation (2.1) which have the order $o(e^{-k_- t})$ as $t \to +\infty$ and $o(e^{-k_+ t})$ as $t \to -\infty$. We show the uniqueness of solutions in this class and give conditions on f ensuring existence (Theorem 3.3.2). In Proposition 3.8.1, we derive a representation of the difference of two solutions corresponding to different strips; this can be understood as an asymptotic formula for each of these solutions at infinity (see Proposition 3.8.2). The power exponential terms in the asymptotics are the same as in Sect. 2.8.1, but the class of admissible right-hand sides of the equation is larger and the estimate for the remainder term is different.

We characterize the behaviour of the solution u and of the right-hand side f by the functions

$$\mathcal{U}(t) = \Big(\int_t^{t+1} \sum_{q=0}^{\ell} \| D_t^q u(\tau) \|_{\ell-q}^2 \, d\tau \Big)^{1/2},$$

$$\mathcal{F}(t) = \Big(\int_t^{t+1} \| f(\tau) \|_0^2 \, d\tau \Big)^{1/2}.$$

In Theorem 3.3.2 we obtain the pointwise estimate for $\mathcal{U}(t)$:

$$\mathcal{U}(t) \leq c\Big\{ \int_t^{\infty} e^{k_-(\tau-t)}(1 + \tau - t)^{m_- - 1}\mathcal{F}(\tau)d\tau$$

$$+ \int_{-\infty}^{t} e^{k_+(\tau-t)}(1 + t - \tau)^{m_+ - 1}\mathcal{F}(\tau)d\tau \Big\},$$

where $m_\pm$ are the maximum lengths of the Jordan chains corresponding to eigenvalues of $\mathcal{A}(\lambda)$ on the lines $\Im\lambda = k_\pm$.

In Sect. 3.5 we show that this estimate is equivalent to the following comparison principle between $\mathcal{U}(t)$ and the solution $w(t)$ of the ordinary differential equation

$$(-1)^{m_-}(\partial_t + k_+)^{m_+}(\partial_t + k_-)^{m_-} w(t) = \mathcal{F}(t) \quad \text{on} \quad \mathbb{R}, \tag{3.1}$$

satisfying $w(t) = o(e^{-k_\mp t})$ as $t \to \pm\infty$:

$$\mathcal{U}(t) \le c\, w(t).$$

This result is based upon an explicit formula for Green's function of (3.1) which reveals its positivity.

We conclude the chapter with applications to boundary value problems in a cylinder and a cone.

3.2 Zeros of $\mathcal{A}(D_t)$

3.2.1 Uniqueness for Homogeneous Equation

We continue our study of the equation (2.53). Let k_- and k_+ be real numbers, $k_- < k_+$. Suppose the strip $k_- < \Im\lambda < k_+$ is free of the spectrum of $\mathcal{A}(\lambda)$.

According to Proposition 2.8.3, a solution u of (2.53) belonging to the space $W^\ell_{k_+,k_-}(\mathbb{R})$ is zero. Clearly, any function in this space satisfies

$$\lim_{t\to\pm\infty} e^{k_\mp t} \parallel u \parallel_{W^\ell(t,t+\delta)} = 0 \tag{3.2}$$

with an arbitrary $\delta > 0$. We now show that (3.2) can be replaced by a weaker condition.

Proposition 3.2.1. *Let $u \in W^\ell_{\mathrm{loc}}(\mathbb{R})$ be a solution of (2.53) satisfying*

$$\liminf_{t\to\pm\infty} e^{k_\mp t} \parallel u \parallel_{W^\ell(t,t+\delta)} = 0 \tag{3.3}$$

with some $\delta > 0$. Then $u = 0$.

Proof. Choose $\zeta \in C^\infty(\mathbb{R})$ such that $\zeta(t) = 1$ for $t > 1$ and $\zeta(t) = 0$ for $t < 0$. Denote by $\{T_j\}_{j\ge 1}$ a sequence of positive numbers, $T_j \to +\infty$ and

$$e^{k_- T_j} \parallel u \parallel_{W^\ell(T_j, T_j+\delta)} \to 0 \quad \text{as} \quad j \to \infty. \tag{3.4}$$

Introduce the cut-off function $\eta_j \in C^\infty(\mathbb{R})$ such that $\eta_j(t) = 1$ for $t < T_j$, $\eta_j(t) = 0$ for $t > T_j + \delta$ and the derivatives $\partial_t^n \eta_j$, $n = 0, \dots, \ell$, are uniformly bounded with respect to j. Clearly

$$\mathcal{A}(D_t)(\eta_j \zeta u) = f + g_j \quad \text{on} \quad \mathbb{R},$$

where $f = [\mathcal{A}(D_t), \zeta]u$ and $g_j = [\mathcal{A}(D_t), \eta_j]u$.

Let $\beta < k_-$ and let the line $\Im\lambda = \beta$ be free of eigenvalues of $\mathcal{A}(\lambda)$. Then by Theorem 2.4.1

$$\parallel \eta_j \zeta u \parallel_{W^\ell_\beta(\mathbb{R})} \le c\big(\parallel f \parallel_{L_{2,\beta}(\mathbb{R};H_0)} + \parallel g_j \parallel_{L_{2,\beta}(\mathbb{R};H_0)} \big).$$

By (3.4) the second norm on the right tends to zero as $j \to \infty$. Thus $\zeta u \in W_\beta^\ell(\mathbb{R})$ for all $\beta < k_-$. Analogously, one shows that $(1 - \zeta)u \in W_\beta^\ell(\mathbb{R})$ for all $\beta > k_+$. Now we can use Proposition 2.8.3 to obtain the equality

$$u(t) = \sum_\pm \sum_{\Im \lambda_\nu = k_\pm} \sum_{\sigma=1}^{\kappa_\nu} c_{\nu\sigma} U_\sigma^{(\nu)}(t),$$

where $U_\sigma^{(\nu)}$ are special solutions of $\mathcal{A}u = 0$ defined by (2.11) and $c_{\nu\sigma} = \text{const}$. By (3.3), $c_{\nu\sigma} = 0$. $\square$

Remark 3.2.2. Condition (3.3) is equivalent (up to the value of δ) to a formally weaker requirement:

$$\liminf_{t \to \pm\infty} e^{k_\mp t} \int_t^{t+\delta} \| u(\tau) \|_{\ell-1} \, d\tau = 0. \tag{3.5}$$

This is a consequence of the local estimate (2.55).

Remark 3.2.3. Condition (3.3) is best possible in the following sense. Suppose that the line $\Im \lambda = k_+$ contains an eigenvalue λ_j. Then $e^{i\lambda_j t}\varphi$ is a non-trivial solution of (2.53) (φ is an eigenfunction of $\mathcal{A}(\lambda)$ corresponding to λ_j). This solution violates (3.3) as $t \to -\infty$.

3.2.2 Behaviour of Zeros at Infinity

In the next proposition we assume that only one of the relations (3.3) is valid, and give some quantitative information about the solution of (2.53). We use the notation

$$m_\pm = \max_\nu m_\nu \tag{3.6}$$

with the maximum being taken over the set

$$\{\nu : \Im \lambda_\nu = k_\pm\}.$$

Here m_ν is the maximum partial multiplicity of λ_ν (see (2.41)). If one of the lines $\Im \lambda = k_+$, $\Im \lambda = k_-$ contains no eigenvalues then we put $m_+ = 1$ or $m_- = 1$ respectively.

Proposition 3.2.4. (i) *Let $u \in W_{\text{loc}}^\ell(\mathbb{R})$ be a non-trivial solution of (2.53) such that*

$$\liminf_{t \to -\infty} e^{k_+ t} \| u \|_{W^\ell(t,t+\delta)} = 0 \tag{3.7}$$

for some $\delta > 0$. Then

$$\limsup_{t \to -\infty} (1 + |t|)^{1-m_-} e^{k_- t} \| u \|_{W^\ell(t,t+\delta)} < \infty \tag{3.8}$$

and

$$\liminf_{t \to +\infty} e^{k_- t} \| u \|_{W^\ell(t,t+\delta)} > 0. \tag{3.9}$$

(ii) *The same assertion holds with $k_\pm$, m_-, $\pm\infty$ replaced by $k_\mp$, m_+, $\mp\infty$.*

Proof. We restrict ourselves to (i), since (ii) can be verified in a similar way. If (3.9) were false then $u = 0$ (by Proposition 3.2.1), contradicting the assumption of Proposition 3.2.4.

Let ζ be the same cut-off function as in Proposition 3.2.1. Then

$$\mathcal{A}(D_t)((1 - \zeta)u) = F ,$$

where F is a function with compact support. Since $(1 - \zeta)u$ satisfies (3.3) and since (3.2) holds for an arbitrary function in $W_\beta^\ell(\mathbb{R})$, $\beta \in (k_-, k_+)$, Proposition 3.2.1 and Theorem 2.4.1 together imply

$$(1 - \zeta)u \in W_\beta^\ell(\mathbb{R}).$$

Hence Proposition 2.8.1 gives

$$(1 - \zeta)\left(u - \sum_{\Im\lambda_\nu=k_-} \sum_{\sigma=1}^{\kappa_\nu} c_{\nu\sigma}U_\sigma^{(\nu)}(t)\right) \in W_\gamma^\ell(\mathbb{R})$$

with some $\gamma < k_-$. So (3.8) holds. $\square$

3.3 Unique Solvability of the Nonhomogeneous Equation

3.3.1 An Auxiliary Existence Result

As before, let the strip $k_- < \Im\lambda < k_+$ be free of the spectrum of $\mathcal{A}(\lambda)$. We introduce the function

$$\mu(t) = \begin{cases} e^{k_-t}(1+t)^{m_--1} & \text{for } t \geq 0 \\ e^{k_+t}(1+|t|)^{m_++1} & \text{for } t \leq 0 \end{cases} \tag{3.10}$$

with $m_\pm$ defined by (3.6).

We start with the following existence result.

Lemma 3.3.1. *Let $f \in L_{2,\text{loc}}(\mathbb{R}; H_0)$ and*

$$\int_\mathbb{R} \mu(\tau) \| f(\tau) \|_0 \, d\tau < \infty. \tag{3.11}$$

Then the equation (2.1) has a solution satisfying

$$\| u \|_{W^\ell(t,t+1)} \leq c\Bigg\{ \left(\int_{t-3}^{t+3} \| f(\tau) \|_0^2 \, d\tau \right)^{1/2}$$

$$+ \int_\mathbb{R} \mu(\tau - t) \| f(\tau) \|_0 \, d\tau \Bigg\} \quad \text{for all } t \in \mathbb{R} . \tag{3.12}$$

Proof. Let η be a smooth function such that $\eta(t) = 1$ for $|t| < 2$ and $\eta(t) = 0$ for $|t| > 3$. Suppose f has a compact support. According to Theorem 2.4.1 the equation (2.1) has a solution $u \in W_\beta^\ell(\mathbb{R})$, $\beta \in (k_-, k_+)$. We represent it in the form $u = u_1 + u_2$, where

$$A(D_t)u_1 = \eta(t - a)f \,,$$
$$A(D_t)u_2 = \big(1 - \eta(t - a)\big)f = f_2.$$

Applying Theorem 2.4.1 to u_1 we get

$$e^{\beta a} \parallel u_1 \parallel_{W^\ell(a,a+1)} \leq c \parallel u_1 \parallel_{W_\beta^\ell(\mathbb{R})} \leq c \parallel \eta(\cdot - a)f \parallel_{L_{2,\beta}(\mathbb{R};H_0)} \,.$$

Hence

$$\parallel u_1 \parallel_{W^\ell(a,a+1)} \leq c\Big(\int_{a-3}^{a+3} \parallel f(\tau) \parallel_0^2 d\tau \Big)^{1/2}. \tag{3.13}$$

We turn to the estimate of u_2. Let $d_\pm$ be the same numbers as in Theorem 2.7.1. By Theorems 2.7.3 and 2.7.1 we have

$$u_2(t) = \int_\mathbb{R} \zeta(t - \tau)\big(G^{(d_+)}(t - \tau) + \sum_{k_+ \leq \Im\lambda_\nu < d_+} e^{i\lambda_\nu(t-\tau)} P_\nu(t - \tau)\big) f_2(\tau) d\tau$$

$$+ \int_\mathbb{R} (1 - \zeta(t - \tau))\big(G^{(d_-)}(t - \tau) - \sum_{d_- < \Im\lambda_\nu \leq k_-} e^{i\lambda_\nu(t-\tau)} P_\nu(t - \tau)\big) f_2(\tau) d\tau \,,$$

where the notation is the same as in Theorem 2.7.1 and ζ is a smooth function on $\mathbb{R}$ such that $\zeta(t) = 1$ for $t > 0$ and $\zeta(t) = 0$ for $t < -1$.

Since $f_2(\tau) = 0$ for $t \in (a - 2, a + 2)$, we obtain by (2.48):

$$\parallel u_2 \parallel_{W^\ell(a,a+1)}^2 \leq c\Big\{ \int_a^{a+1} \Big(\int_\mathbb{R} \varphi(t - \tau) \parallel f_2(\tau) \parallel_0 d\tau\Big)^2 dt$$

$$+ \int_a^{a+1} \Big(\int_{-1}^\infty e^{-k_+(t-\tau)}(1 + |t - \tau|)^{m_+ - 1} \parallel f_2(\tau) \parallel_0 d\tau\Big)^2 dt$$

$$+ \int_a^{a+1} \Big(\int_{-\infty}^1 e^{-k_-(t-\tau)}(1 + |t - \tau|)^{m_- - 1} \parallel f_2(\tau) \parallel_0 d\tau\Big)^2 dt\Big\},$$

where

$$\varphi(t) = \begin{cases} e^{d_- t} & \text{for } t \geq 0 \\ e^{d_+ t} & \text{for } t < 0. \end{cases}$$

Using the Minkowski inequality we find

$$\parallel u_2 \parallel_{W^\ell(a,a+1)} \leq c\Big\{ \int_\mathbb{R} \Big(\int_a^{a+1} \varphi^2(\tau - t)dt\Big)^{1/2} \parallel f_2(\tau) \parallel_0 d\tau$$

$$+ \int_\mathbb{R} \Big(\int_a^{a+1} \mu^2(\tau - t)dt\Big)^{1/2} \parallel f_2(\tau) \parallel_0 d\tau\Big\} \leq$$

$$\leq c \int_\mathbb{R} \mu(\tau - t) \parallel f_2(\tau) \parallel_0 d\tau. \tag{3.14}$$

From (3.13) and (3.14) we deduce (3.12) for f with compact support.

It remains to remove the assumption on the compact support of f. To this end, let f be a function in $L_{2,\mathrm{loc}}(\mathbb{R}; H_0)$ satisfying (3.11). Choose a sequence of functions $f_1, \ldots, f_k, \ldots$ such that

a) $f_k \in L_{2,\mathrm{loc}}(\mathbb{R}; H_0)$ and the support of f_k is compact,

b) the following relation holds:

$$\int_{\mathbb{R}} \mu(t) \, \| f(t) - f_k(t) \|_0 \, dt \to 0 \quad \text{as} \quad k \to \infty,$$

c) for every positive N

$$\int_{-N}^{N} \| f(t) - f_k(t) \|_0^2 \, dt \to 0 \quad \text{as} \quad k \to \infty.$$

Let u_k be the solution of the equation $\mathcal{A}(D_t) u_k = f_k$ satisfying (3.12). In view of (3.12),

$$\| u_k - u_m \|_{W^\ell(t,t+1)} \le c \Big\{ \Big(\int_{t-3}^{t+3} \| f_k(\tau) - f_m(\tau) \|_0^2 \, d\tau \Big)^{1/2}$$
$$+ \int_{\mathbb{R}} \mu(\tau - t) \, \| f_k(\tau) - f_m(\tau) \|_0 \, d\tau \Big\}.$$

By b) and c) the right-hand side tends to 0 for every t. Hence the sequence $\{u_k\}$ converges in $W_{\mathrm{loc}}^\ell(\mathbb{R})$. Thus $u = \lim u_k$ satisfies (2.1) and (3.12). $\square$

3.3.2 Unique Solvability

Now we state the principal result of this section.

Theorem 3.3.2. (i) (Existence) *Let* $f \in L_{2,\mathrm{loc}}(\mathbb{R}; H_0)$ *and*

$$\int_{\mathbb{R}} \mu(\tau) \, \| f \|_{L_2(\tau,\tau+1;H_0)} \, d\tau < \infty. \tag{3.15}$$

There exists a solution $u \in W_{\mathrm{loc}}^\ell(\mathbb{R})$ *of (2.1) satisfying*

$$\| u \|_{W^\ell(t,t+1)} \le c \int_{\mathbb{R}} \mu(\tau - t) \, \| f \|_{L_2(\tau,\tau+1;H_0)} \, d\tau. \tag{3.16}$$

This solution also satisfies

$$\| u \|_{W^\ell(t,t+1)} = \begin{cases} o(e^{-k_- t}) & \text{as } t \to +\infty \\ o(e^{-k_+ t}) & \text{as } t \to -\infty. \end{cases} \tag{3.17}$$

(ii) (Uniqueness) a) *If* u_1, u_2 *are solutions of (2.1) in* $W_{\mathrm{loc}}^\ell(\mathbb{R})$ *subject to (3.17) then* $u_1 = u_2$.

b) *Let* $f \in L_{2,\mathrm{loc}}(\mathbb{R}; H_0)$ *satisfy (3.15). Solution* $u \in W_{\mathrm{loc}}^\ell(\mathbb{R})$ *of (2.1) subject to (3.3) is unique.*

Proof. (i). First we show that the right-hand side in (3.12) does not exceed (up to a constant factor) the right-hand side in (3.16). Indeed,

$$
\int_{\mathbb{R}} \mu(\tau - t) \, \| \, f(\tau) \, \|_0 \, d\tau \leq c \int_{\mathbb{R}} \| \, f(\tau) \, \|_0 \int_{\tau-1}^{\tau} \mu(x - t) dx \, d\tau
$$

$$
= c \int_{\mathbb{R}} \mu(x - t) \int_{x}^{x+1} \| \, f(\tau) \, \|_0 \, d\tau \, dx
$$

$$
\leq c \int_{\mathbb{R}} \mu(x - t) \, \| \, f \, \|_{L_2(x,x+1;H_0)} \, dx. \tag{3.18}
$$

Moreover, the first term on the right in (3.12) is obviously majorized by the last integral in (3.18). Therefore the solution constructed in Lemma 3.3.1 satisfies (3.16).

We turn to (3.17). The right-hand side of (3.16) is equal to

$$
c \int_{t}^{\infty} e^{k_-(\tau-t)}(1 + \tau - t)^{m_- - 1} \, \| \, f \, \|_{L_2(\tau,\tau+1;H_0)} \, d\tau
$$

$$
+ c \int_{-\infty}^{t} e^{k_+(\tau-t)}(1 + t - \tau)^{m_+ - 1} \, \| \, f \, \|_{L_2(\tau,\tau+1;H_0)} \, d\tau \, . \tag{3.19}
$$

Consider the case $t \to +\infty$. The first term in (3.19) can be estimated by

$$
c \, e^{-k_- t} \int_{t}^{\infty} e^{k_- \tau}(1 + \tau)^{m_- - 1} \, \| \, f \, \|_{L_2(\tau,\tau+1;H_0)} \, d\tau.
$$

Due to (3.15) the last expression is $o(e^{-k_- t})$ as $t \to +\infty$. The second term in (3.19) can be rewritten in the form

$$
c \, e^{-k_+ t} \int_{-\infty}^{N} e^{k_+ \tau}(1 + t - \tau)^{m_+ - 1} \, \| \, f \, \|_{L_2(\tau,\tau+1;H_0)} \, d\tau
$$

$$
+ c \int_{N}^{t} e^{k_+(\tau-t)}(1 + t - \tau)^{m_+ - 1} e^{k_- \tau} \, \| \, f \, \|_{L_2(\tau,\tau+1;H_0)} \, d\tau, \tag{3.20}
$$

where N is a positive number. Since $k_+ > k_-$, the first term in (3.20) is $o(e^{-k_- t})$. Using the boundedness of the function $e^{(k_- - k_+)x}(1 + x)^{m_+ - 1}$ for $x \geq 0$ we can estimate the second term in (3.20) by

$$
c \, e^{-k_- t} \int_{N}^{t} e^{k_- \tau} \, \| \, f \, \|_{L_2(\tau,\tau+1;H_0)} \, d\tau.
$$

Taking N sufficiently large we complete the proof for $t \to +\infty$. The case $t \to -\infty$ is analogous.

(ii) a) The uniqueness of the solution satisfying (3.17) follows from Proposition 3.2.1.

b) Let u_0 be the solution constructed in (i). Then $u - u_0$ satisfies the homogeneous equation (2.53). Since u_0 fulfills (3.17), the difference $u - u_0$ is subject to (3.3). Hence $u = u_0$ by Proposition 3.2.1. $\square$

3.4 Solutions Corresponding to a Strip

We choose a strip $k_- < \Im \lambda < k_+$ which does not contain eigenvalues of $\mathcal{A}(\lambda)$. Theorem 3.3.2 on the unique solvability of (2.1) suggests the following:

Definition 3.4.1. A solution $u \in W_{\text{loc}}^\ell(\mathbb{R})$ of the equation (2.1) subject to (3.17) is called the (k_-, k_+)-solution (or the solution corresponding to the strip $k_- < \Im \lambda < k_+$).

By Theorem 3.3.2, the (k_-, k_+)-solution is unique and it exists provided f is subject to (3.15). In particular, let $f \in L_{2,\beta}(\mathbb{R}; H_0)$ where $\beta \in (k_-, k_+)$. Then f satisfies (3.15) by the inequality

$$
\int_{\mathbb{R}} \mu(\tau) \parallel f \parallel_{L_2(\tau,\tau+1;H_0)} d\tau
$$

$$
\leq \left(\int_{\mathbb{R}} \mu(\tau)^2 e^{-2\beta\tau} d\tau \right)^{1/2} \left(\int_{\mathbb{R}} e^{2\beta\tau} \parallel f \parallel_{L_2(\tau,\tau+1;H_0)}^2 d\tau \right)^{1/2}.
$$

Therefore, since (3.17) follows from $u \in W_\beta^\ell(\mathbb{R})$, Theorem 3.3.2 (ii) implies that the (k_-, k_+)-solution belongs to $W_\beta^\ell(\mathbb{R})$.

Consider the case when one of the lines $\Im \lambda = k_\pm$ is free of eigenvalues of $\mathcal{A}(\lambda)$. Let k'_-, k'_+ be real numbers such that $k_- \geq k'_-$, $k'_+ \geq k_+$ and the strip $k'_- < \Im \lambda < k'_+$ still contains no eigenvalues of $\mathcal{A}(\lambda)$. Then by the uniqueness assertion in Theorem 3.3.2 any solution of (2.1) corresponding to the strip $k_- < \Im \lambda < k_+$ corresponds to the strip $k'_- < \Im \lambda < k'_+$ as well.

3.5 Comparison Principle

3.5.1 Comparison Equation and Its Green Function

Here we show that one may consider estimate (3.16) as a comparison principle for solutions of the operator differential equation (2.1) on one side and a solution of the ordinary differential equation

$$
(-1)^{m_-} (\partial_t + k_+)^{m_+} (\partial_t + k_-)^{m_-} w(t) = \parallel f \parallel_{L_2(t,t+1;H_0)} \quad \text{on} \quad \mathbb{R} \quad (3.21)
$$

(comparison equation) on the other side. The operator on the left has been already met in Example 2.2.4, where it was denoted by $\mathcal{M}(\partial_t)$. We shall constantly use this notation in the sequel.

Let g be Green's function of the operator on the left in (3.21), i.e. the solution of the equation

$$
\mathcal{M}(\partial_t)g(t) = \delta(t) \quad \text{on} \quad \mathbb{R}, \qquad (3.22)
$$

subject to

$$
g(t) = o\big(e^{-k_\mp t}\big) \quad \text{as} \quad t \to \pm\infty.
$$

Lemma 3.5.1. *The following formulae hold:*

$$g(t) = e^{-k_+ t} \sum_{s=0}^{m_+ - 1} \frac{t^{m_+ - 1 - s}}{(m_+ - 1 - s)!} \frac{(m_- + s - 1)!}{(m_- - 1)! s!} (k_+ - k_-)^{-m_- - s} \qquad (3.23)$$

for $t > 0$ and

$$g(t) = e^{-k_- t} \sum_{s=0}^{m_- - 1} \frac{(-t)^{m_- - 1 - s}}{(m_- - 1 - s)!} \frac{(m_+ + s - 1)!}{(m_+ - 1)! s!} (k_+ - k_-)^{-m_+ - s} \qquad (3.24)$$

for $t < 0$.

Proof. By using the Fourier transform we obtain

$$g(t) = i^{m_- - m_+} \frac{1}{2\pi} \int_{\Im \lambda = \beta} \frac{e^{i\lambda t} d\lambda}{(\lambda - ik_+)^{m_+} (\lambda - ik_-)^{m_-}},$$

where $\beta \in (k_-, k_+)$. For $t > 0$ we have

$$g(t) = i e^{-k_+ t} i^{m_- - m_+} \sum_{s=0}^{m_+ - 1} \frac{(it)^{m_+ - 1 - s}}{(m_+ - 1 - s)!}$$

$$\times \frac{(-m_-)(-m_- - 1) \ldots (-m_- - s + 1)}{s!} (ik_+ - ik_-)^{-m_- - s},$$

which yields (3.23).

If $t < 0$ then

$$g(t) = -i e^{-k_- t} i^{m_- - m_+} \sum_{s=0}^{m_- - 1} \frac{(it)^{m_- - 1 - s}}{(m_- - 1 - s)!}$$

$$\times \frac{(-m_+)(-m_+ - 1) \ldots (-m_+ - s + 1)}{s!} (ik_- - ik_+)^{-m_+ - s}$$

and we arrive at (3.24). $\square$

Lemma 3.5.2. (i) *The following inequalities hold:*

$$c_1 \mu(-t) \le g(t) \le c_2 \mu(-t), \quad t \in \mathbb{R}, \qquad (3.25)$$

where c_1, c_2 are positive constants which depend only on $k_\pm, m_\pm$, and μ is defined by (3.10).

(ii) *Let $j = 0, \ldots, m_+ - 1$ and $k = 0, \ldots, m_- - 1$. Then*

$$(\partial_t + k_+)^j (-\partial_t - k_-)^k g(t) > 0. \qquad (3.26)$$

Proof. (i) The result follows by comparison of (3.10) with (3.23), (3.24).
(ii) The left-hand side of (3.26) is Green's function of the operator

$$(\partial_t + k_+)^{m_+ - j} (-\partial_t - k_-)^{m_- - k},$$

implying its positivity. $\square$

3.5.2 Solvability Criterion for the Comparison Equation

The following basic fact on the solvability of the equation

$$\mathcal{M}(\partial_t)w(t) = h(t), \quad t \in \mathbb{R} \tag{3.27}$$

can be found in Kozlov, Maz'ya (1997), Proposition 1.3.1. Here h is a given function in the space of locally summable functions $L_{1,\text{loc}}(\mathbb{R})$. The solution w is sought in the space $W_{1,\text{loc}}^{m_+ + m_-}(\mathbb{R})$ of functions, absolutely continuous with their derivatives up to order $m_+ + m_- - 1$.

Proposition 3.5.3. *The condition*

$$\int_{-\infty}^{\infty} g(-\tau) \mid h(\tau) \mid d\tau < \infty$$

or equivalently,

$$\int_0^{\infty} e^{k-\tau}(1+\tau)^{m_- - 1}|h(\tau)|d\tau < \infty, \tag{3.28}$$

and

$$\int_{-\infty}^0 e^{k+\tau}(1+|\tau|)^{m_+ - 1}|h(\tau)|d\tau < \infty \tag{3.29}$$

is sufficient (and in the case $h \geq 0$ necessary as well) for the existence of the solution w of (3.27) *subject to*

$$\liminf_{t \to \pm\infty} e^{k \mp t}|w(t)| = 0. \tag{3.30}$$

This solution is represented in the form

$$w(t) = \int_{-\infty}^{\infty} g(t-\tau)h(\tau)d\tau \tag{3.31}$$

and satisfies

$$\partial_t^j w(t) = \begin{cases} o(e^{-k_- t}) & as \quad t \to +\infty, \\ o(e^{-k_+ t}) & as \quad t \to -\infty, \end{cases} \tag{3.32}$$

where $j = 0, 1, \ldots, m_+ + m_- - 1$.

Clearly, all solutions of the homogeneous equation

$$\mathcal{M}(\partial_t)\xi(t) = 0 \quad \text{on} \quad \mathbb{R}$$

are

$$\xi(t) = e^{-k_+ t}\Pi_+(t) + e^{-k_- t}\Pi_-(t),$$

where $\Pi_\pm$ are arbitrary polynomials of degrees $m_\pm - 1$. Therefore

$$w(t) = o(e^{-k_\mp t}) \quad \text{as} \quad t \to \pm\infty \tag{3.33}$$

describes a uniqueness class of solutions to equation (3.27).

Remark 3.5.4. The equation (3.27) is a simple model of the general operator equation (2.1) (see Example 2.2.4). Condition (3.15) for this special case takes the form

$$\int_{\mathbb{R}} \mu(\tau)|h(\tau)|d\tau < \infty. \tag{3.34}$$

According to Proposition 3.5.3, the inequality (3.34) is not only sufficient but for $h \geq 0$ is also necessary for the solvability of (3.27). Moreover if $h \geq 0$ the solution satisfies the two-sided estimates

$$c_1 w(t) \leq \int_{\mathbb{R}} \mu(\tau - t)h(\tau)d\tau \leq c_2 w(t).$$

Thus the condition (3.15) and the estimate (3.16) are best possible for the whole class of operator differential equations under consideration.

3.5.3 Comparison Principle

We turn to the operator differential equation $\mathcal{A}(D_t)u = f$. By Proposition 3.5.3 the ordinary differential equation (3.21) is solvable if and only if

$$\int_{\mathbb{R}} g(-\tau) \parallel f \parallel_{L_2(\tau,\tau+1;H_0)} d\tau < \infty. \tag{3.35}$$

In the same proposition we have proved that there exists a solution w represented in the form

$$w(t) = \int_{\mathbb{R}} g(t - \tau) \parallel f \parallel_{L_2(\tau,\tau+1;H_0)} d\tau. \tag{3.36}$$

This solution satisfies (3.33).

Now we are in a position to state the above mentioned comparison principle.

Theorem 3.5.5. *Let $f \in L_{2,\mathrm{loc}}(\mathbb{R}; H_0)$ satisfy (3.15) and let w be the unique solution of (3.21) subject to (3.33). Then there exists a (k_-, k_+)–solution of (2.1) which satisfies*

$$\parallel u \parallel_{W^\ell(t,t+1)} \leq b\, w(t) \tag{3.37}$$

with $b = \mathrm{const}$ (according to Theorem 3.3.2(ii) this solution is unique).

Proof. The result follows from Theorem 3.3.2 together with (3.36) and (3.25). $\square$

3.5.4 A General Asymptotic Representation
of the (k_-, k_+)-Solution

Here we obtain a certain representation of the (k_-, k_+)-solution which, under some circumstances, can be interpreted as an asymptotic formula at $\pm\infty$. This interpretation is valid under the same assumption (see (3.35)) on the right-hand side f which guarantees the existence of a solution.

We introduce real numbers d_- and d_+ such that $d_- < k_-$, $d_+ > k_+$ and the lines $\Im\lambda = d_\pm$ contain no eigenvalues of $\mathcal{A}(\lambda)$.

Let $\chi(t)$ be a smooth function on $\mathbb{R}$ equal to 1 for $t > 1$ and 0 for $t < 0$. We introduce the operator function

$$G_{d_-,d_+}(t-\tau) = G(t-\tau)$$

$$-\chi(t-\tau) \sum_{d_-<\Im\lambda_\nu<k_-} \sum_{\sigma=1}^{\kappa_\nu} (\cdot, V_\sigma^{(\nu)}(\tau))_{H_0} U_\sigma^{(\nu)}(t)$$

$$+(1-\chi(t-\tau)) \sum_{k_+<\Im\lambda_\nu<d_+} \sum_{\sigma=1}^{\kappa_\nu} (\cdot, V_\sigma^{(\nu)}(\tau))_{H_0} U_\sigma^{(\nu)}(t), \qquad (3.38)$$

where G is Green's kernel corresponding to the strip $k_- < \Im\lambda < k_+$ and $U_\sigma^{(\nu)}$ and $V_\sigma^{(\nu)}$ were introduced in Sect. 2.2.3.

Using (2.49) we can rewrite (3.38) as

$$G_{d_-,d_+}(t-\tau) = G(t-\tau)$$

$$-\chi(t-\tau) \sum_{d_-<\Im\lambda_\nu<k_-} e^{i\lambda_\nu(t-\tau)} P_\nu(t-\tau)$$

$$+(1-\chi(t-\tau)) \sum_{k_+<\Im\lambda_\nu<d_+} e^{i\lambda_\nu(t-\tau)} P_\nu(t-\tau). \qquad (3.39)$$

Lemma 3.5.6. *Let*

$$\int_{-\infty}^0 e^{d_+\tau} \|f\|_{L_2(\tau,\tau+1;H_0)} d\tau +$$

$$+ \int_0^\infty e^{d_-\tau} \|f\|_{L_2(\tau,\tau+1;H_0)} d\tau < \infty$$

and let

$$v(t) = \int_{\mathbb{R}} G_{d_-,d_+}(t-\tau) f(\tau) d\tau.$$

Then

$$\|v\|_{W^\ell(t,t+1)} \le c \Big(\int_{-\infty}^t e^{-d_+(t-\tau)} \|f\|_{L_2(\tau,\tau+1;H_0)}$$

$$+ \int_t^\infty e^{-d_-(t-\tau)} \|f\|_{L_2(\tau,\tau+1;H_0)} d\tau \Big). \qquad (3.40)$$

Proof. By using Proposition 2.6.2 and Theorem 2.7.3 we obtain for $k = 0, 1, ..., \ell - 1$:

$$\|D_t^k v(t)\|_{H_{\ell-k-1}} \leq c\Big(\int_{|t-\tau|\leq 1} (1 + |\log |t - \tau||) \, \|f(\tau)\|_0 d\tau$$
$$+ \int_{-\infty}^{t-1} e^{-d_+(t-\tau)} \|f(\tau)\|_0 d\tau + \int_{t+1}^{\infty} e^{-d_-(t-\tau)} \|f(\tau)\|_0 d\tau \Big).$$

Hence

$$\|D_\tau^k v\|_{L_2(t,t+1;H_{\ell-k-1})} \leq c\Big(\int_{-\infty}^{t} e^{-d_+(t-\tau)} \|f\|_0 d\tau$$
$$+ \int_{t}^{\infty} e^{-d_-(t-\tau)} \|f(\tau)\|_0 d\tau \Big). \tag{3.41}$$

By (3.18) the right-hand side of this inequality can be replaced by the right-hand side of (3.40).

Let $h(t) = \mathcal{A}(D_t)v(t)$. It follows from (3.39) that

$$h(t) = f(t) - \int_{t-1}^{t} Q(t - \tau)f(\tau)d\tau, \tag{3.42}$$

where

$$Q(t) = [\mathcal{A}(D_t), \chi(t)] \sum e^{i\lambda_\nu t} P_\nu(t)$$

and the summation is extended over the set

$$\{\nu : d_- < \Im\lambda_\nu < k_- \quad \text{and} \quad k_+ < \Im\lambda_\nu < d_+\}.$$

By (2.55)

$$\|v\|_{W^\ell(t,t+1)} \leq c\Big(\|f\|_{L_2(t-4,t+4)} + \int_{t-4}^{t+4} \|V(\tau)\|_{\ell-1} d\tau\Big).$$

Hence (by (3.41)) we arrive at (3.40). $\square$

The following theorem is the principal result of this section.

Theorem 3.5.7. *Let $f \in L_{2,\text{loc}}(\mathbb{R})$ satisfy (3.35). Then the (k_-, k_+)-solution of (2.1) admits the representation*

$$u(t) = \sum_{d_- < \Im\lambda_\nu < k_-} c_{\sigma\nu}^-(t)U_\sigma^{(\nu)}(t) + \sum_{k_+ < \Im\lambda_\nu < k_+} c_{\sigma\nu}^+(t)U_\sigma^{(\nu)}(t) + v(t),$$

where the coefficients $c_{\sigma\nu}^\pm(t)$ are defined by

$$c_{\sigma\nu}^-(t) = \int_{\mathbb{R}} \chi(t - \tau)\big(f(\tau), V_\sigma^{(\nu)}(\tau)\big)_{H_0} d\tau$$

and

$$c_{\sigma\nu}^+(t) = \int_{\mathbb{R}} (\chi(t - \tau) - 1)\big(f(\tau), V_\sigma^{(\nu)}(\tau)\big)_{H_0} d\tau.$$

The function v satisfies (3.40).

Proof. Follows directly from Lemma 3.5.6 and (3.38). $\square$

We show that for equation (2.1) with a second order operator $\mathcal{A}(D_t)$ the assertion of Theorem 3.5.7 takes especially transparent form.

Example 3.5.8. Consider the equation

$$(D_t^2 + A)u(t) = f(t) \quad \text{on} \quad \mathbb{R}, \tag{3.43}$$

where A is the same positive definite selfadjoint operator in $H = H_0$ as in Ex. 2.2.2. We shall use all the notations introduced in Ex. 2.2.2 and Ex. 1.2.2. Let $k_\pm = \pm\gamma_1^{1/2}$ and let

$$\int_{\mathbb{R}} e^{-\gamma_1^{1/2}|\tau|} \|f\|_{L_2(\tau,\tau+1;H_0)} d\tau < \infty.$$

By Theorem 3.5.5 there exists a unique (k_-, k_+)-solution of (3.43) subject to

$$\|u\|_{W^2(t,t+1)} \le c \int_{\mathbb{R}} e^{-\gamma_1^{1/2}|t-\tau|} \|f\|_{L_2(\tau,\tau+1;H_0)} d\tau.$$

Moreover, by Theorem 3.5.7 with arbitrary d_- and d_+ in the intervals

$$\left(-\gamma_2^{1/2}, -\gamma_1^{1/2}\right) \quad \text{and} \quad \left(\gamma_1^{1/2}, \gamma_2^{1/2}\right)$$

respectively, the solution u satisfies

$$u(t) = \sum_{\pm}\sum_{j=1}^{J_1} c_j^\pm(t) e^{\pm\gamma_1^{1/2}t} \alpha_j^{(1)} + v(t),$$

where

$$c_j^-(t) = \frac{1}{2\gamma^{1/2}} \int_{\mathbb{R}} \chi(t-\tau)\left(f(\tau), \alpha_j^{(1)}\right)_{H_0} e^{\gamma_1^{1/2}\tau} d\tau,$$

$$c_j^+(t) = \frac{1}{2\gamma^{1/2}} \int_{\mathbb{R}} (1-\chi(t-\tau))\left(f(\tau), \alpha_j^{(1)}\right)_{H_0} e^{-\gamma_1^{1/2}\tau} d\tau$$

and v is subject to (3.40) with $\ell = 2$.

In particular, let

$$f(t)) = \begin{cases} t^\sigma e^{-\gamma_1^{1/2}t}q & \text{for } t > 1 \\ 0 & \text{otherwise,} \end{cases}$$

where $q \in H_0$ and $\sigma > 0$. Then

$$c_j^-(t) = \frac{(q, \alpha_j^{(1)})_{H_0}}{2\gamma_1^{1/2}(\sigma+1)} t^{\sigma+1} + O(t^\sigma),$$

$$c_j^+(t) = O\left(t^\sigma e^{-2\gamma_1^{1/2}t}\right)$$

and

$$\|v\|_{W^2(t,t+1)} = O\left(t^\sigma e^{-\gamma_1^{1/2}t}\right) \tag{3.44}$$

as $t \to +\infty$. Hence we arrive at the asymptotic formula

$$u(t) = \frac{t^{\sigma+1}}{2\gamma_1^{1/2}(\sigma+1)} \sum_{j=1}^{J_1} (q, \alpha_j^{(1)})_{H_0} \alpha_j^{(1)} + w(t),$$

where w satisfies the same estimate as v in (3.44).

3.6 Estimates for Solutions on a Semiaxis

In the next lemma we obtain an estimate for solutions of the equation

$$\mathcal{A}(D_t)u = f \quad \text{for} \quad t > t_0. \tag{3.45}$$

The meaning of this result is that a solution of (3.45) which is subject to a rough upper estimate admits a better majorant; this majorant depends on f.

We assume that $f \in L_{2,\text{loc}}(t_0, \infty; H_0)$, i.e. $f \in L_2(t_0, t_1; H_0)$ for any finite $t_1 > t_0$. Analogously, the solution u will be sought in the space $W_{\text{loc}}^\ell(t_0, \infty)$, i.e. $u \in W^\ell(t_0, t_1)$ for any finite $t_1 > t_0$.

Lemma 3.6.1. *Let $u \in W_{\text{loc}}^\ell(t_0, \infty)$ be a solution of (3.45) with*

$$\int_{t_0}^\infty (|t|+1)^{m_- - 1} e^{k_- t} \| f \|_{L_2(t,t+1;H_0)} \, dt < \infty . \tag{3.46}$$

If

$$\liminf_{t \to +\infty} e^{k_- t} \| u \|_{W^\ell(t,t+1)} = 0 \tag{3.47}$$

then for $t > t_0 + 1$

$$\| u \|_{W^\ell(t,t+1)} \le c \left\{ \int_{t_0}^\infty \mu(\tau - t) \| f \|_{L_2(\tau,\tau+1;H_0)} \, d\tau \right.$$

$$\left. + \mu(t_0 - t) \| u \|_{W^{\ell-1}(t_0,t_0+1)} \right\} , \tag{3.48}$$

where the constant c is independent of t_0, u, f and μ is defined by (3.10).

Proof. By ζ we denote a smooth function on $\mathbb{R}$ equal to 1 for $t > t_0 + 1$ and to 0 for $t < t_0 + 1/2$. Then

$$\mathcal{A}(D_t)(\zeta u) = F , \tag{3.49}$$

where $F = \zeta f + [\mathcal{A}(D_t), \zeta]u$. By (3.46), Theorem 3.3.2 is applicable to the equation (3.49). Hence ζu is the (k_-, k_+)-solution. By (3.16) $\| \zeta u \|_{W^\ell(t,t+1)}$ is estimated by

$$c \left\{ \int_{t_0-1/2}^{\infty} \mu(\tau - t) \parallel \zeta f \parallel_{L_2(\tau,\tau+1;H_0)} d\tau + \mu(t_0 - t) \parallel u \parallel_{W^{\ell-1}(t_0,t_0+1)} \right\}.$$

Since $\int_{t_0-1/2}^{t_0}$ is majorized by the integral in (3.48) we arrive at (3.48). $\square$

In what follows we shall use both Lemma 3.6.1 and its analog for the equation

$$\mathcal{A}(D_t)u = f \quad \text{for} \quad t < t_0, \tag{3.50}$$

the proof of which is quite similar.

Lemma 3.6.2. *Let* $u \in W_{\mathrm{loc}}^{\ell}(-\infty, t_0)$ *be a solution of* (3.50) *with*

$$\int_{-\infty}^{t_0} (|t| + 1)^{m_+ - 1} e^{k+t} \parallel f \parallel_{L_2(t,t+1;H_0)} dt < \infty. \tag{3.51}$$

If

$$\liminf_{t \to -\infty} e^{k+t} \parallel u \parallel_{W^{\ell}(t,t+1)} = 0 \tag{3.52}$$

then for $t < t_0 - 1$

$$\parallel u \parallel_{W^{\ell}(t,t+1)} \leq c \left\{ \int_{-\infty}^{t_0} \mu(\tau - t) \parallel f \parallel_{L_2(\tau,\tau+1;H_0)} d\tau \right.$$

$$\left. + \mu(t_0 - t) \parallel u \parallel_{W^{\ell-1}(t_0-1,t_0)} \right\}. \tag{3.53}$$

3.7 The Phragmén-Lindelöf Principle

It is well known (and easily checked) that one of the following two alternatives is valid for solutions of the Dirichlet problem

$$\begin{cases} (\partial_t^2 + \partial_x^2)u(t,x) = 0 & \text{for } x \in (0,h) \text{ and } t > t_0, \\ u(t,0) = u(t,h) = 0 & \text{for } t > 0. \end{cases}$$

The function u either grows at least as $\exp(\pi t/h)$ or admits the estimate $O(\exp(-\pi t/h))$. This fact is a variant of the classical Phragmén-Lindelöf principle. Here we show that it has a direct generalization to solutions of the operator equation (3.45). As before we assume that there are no eigenvalues in the strip $k_- < \Im\lambda < k_+$.

Theorem 3.7.1. *Let* $u \in W_{\mathrm{loc}}^{\ell}(t_0, \infty)$ *be a solution of* (3.45) *with*

$$\int_{t_0}^{\infty} e^{k+t} \parallel f \parallel_{L_2(t,t+1;H_0)} dt < \infty. \tag{3.54}$$

Then either

$$\liminf_{t \to +\infty} e^{k-t} \parallel u \parallel_{W^{\ell}(t,t+1)} > 0 \tag{3.55}$$

or

$$\limsup_{t \to +\infty} t^{1-m_+} e^{k+t} \parallel u \parallel_{W^{\ell}(t,t+1)} < \infty. \tag{3.56}$$

Proof. We note first that (3.54) implies (3.46). If (3.55) fails then the estimate (3.48) is valid by Lemma 3.6.1. Now it is sufficient to estimate the integral in (3.48) by

$$ct^{m_+-1}e^{-k_+t}\int_{t_0}^{\infty}e^{k_+\tau}\|f\|_{L_2(\tau,\tau+1;H_0)}d\tau$$

for large positive t. This follows from the estimate

$$\mu(\tau-t)\le ce^{k_+(\tau-t)}t^{m_+-1},\quad t,\tau\ge 1,$$

(which can be verified directly). $\square$

3.8 Asymptotics of Solutions Corresponding to a Strip

3.8.1 A Representation for the Difference of Two Solutions

Here we extend the results of Sect. 2.8 to (k_-,k_+)-solutions. We start with a generalization of Proposition 2.8.1 and give a representation for the difference of solutions corresponding to two strips.

Let the strips

$$k_-^{(i)}<\Im\lambda<k_+^{(i)},\ i=1,2,$$

where $k_+^{(1)}\le k_-^{(2)}$, be free of eigenvalues of $\mathcal{A}(\lambda)$, and let $m_\pm^{(i)}$ be defined by (3.6) with $k_\pm$ replaced by $k_\pm^{(i)}$.

Proposition 3.8.1. *Let* $f\in L_{2,\mathrm{loc}}(\mathbb{R};H_0)$ *satisfy*

$$\int_{-\infty}^{0}e^{k_+^{(1)}\tau}(1+|\tau|)^{m_+^{(1)}-1}\|f\|_{L_2(\tau,\tau+1;H_0)}\,d\tau<\infty,\tag{3.57}$$

$$\int_{0}^{\infty}e^{k_-^{(2)}\tau}(1+\tau)^{m_-^{(2)}-1}\|f\|_{L_2(\tau,\tau+1;H_0)}\,d\tau<\infty\tag{3.58}$$

Then the solutions $u_i,\ i=1,2,$ *corresponding to the strips* $k_-^{(i)}<\Im\lambda<k_+^{(i)}$ *satisfy*

$$u_2(t)-u_1(t)=\sum_{k_+^{(1)}\le\Im\lambda_\nu\le k_-^{(2)}}\sum_{\sigma=1}^{\kappa_\nu}c_{\nu\sigma}U_\sigma^{(\nu)}(t),\tag{3.59}$$

where

$$c_{\nu\sigma}=\int_{\mathbb{R}}\big(f(\tau),V_\sigma^{(\nu)}(\tau)\big)_{H_0}d\tau.\tag{3.60}$$

The functions $U_\sigma^{(\nu)}$ *and* $V_\sigma^{(\nu)}$ *are defined by* (2.11) *and* (2.12).

Proof. Let f have a compact support. Then $u_2 \in W_\gamma^\ell(\mathbb{R})$ with $\gamma \in (k_-^{(2)}, k_+^{(2)})$ and $u_1 \in W_\beta^\ell(\mathbb{R})$ with $\beta \in (k_-^{(1)}, k_+^{(1)})$ (see Sect.3.4). Hence the result follows from Proposition 2.8.1.

Now consider the general case. Choose a sequence $\{f_j\}_{j\geq 1}$ such that

a) $f_j \in L_{2,\mathrm{loc}}(\mathbb{R}; H_0)$ and f_j has compact support,

b) $\int_0^\infty e^{k_-^{(2)}\tau}(1+\tau)^{m_--1} \| f(\tau) - f_j(\tau) \|_{L_2(\tau,\tau+1;H_0)} \, d\tau$
$+ \int_{-\infty}^0 e^{k_+^{(1)}\tau}(1+|\tau|)^{m_++1} \| f(\tau) - f_j(\tau) \|_{L_2(\tau,\tau+1;H_0)} \, d\tau \to 0$
as $j \to \infty$.

Denote by u_{1j}, u_{2j} the solutions of equation (2.1) for $f = f_j$ corresponding to the strips

$$k_-^{(1)} < \Im \lambda < k_+^{(1)}, \quad k_-^{(2)} < \Im \lambda < k_+^{(2)}$$

respectively. Due to the estimate (3.16) for u_{1j} and to the analogous estimate for u_{2j}, the sequence $\{u_{1j}\}$ converges to u_1 in $W_{\mathrm{loc}}^\ell(\mathbb{R})$ and the sequence $\{u_{2j}\}$ converges to u_2 in the same space. Assumption b) implies

$$\int_{\mathbb{R}} \left(f_j(\tau), V_\sigma^{(\nu)}(\tau)\right)_{H_0} d\tau \to c_{\nu\sigma} \quad \text{as} \quad j \to \infty.$$

Hence we arrive at (3.59). $\square$

3.8.2 An Asymptotic Formula

Due to Theorem 3.3.2

$$\| u_1 \|_{W^\ell(t,t+1)} = o\left(e^{-k_-^{(1)}t}\right) \quad \text{for} \quad t \to +\infty, \tag{3.61}$$

$$\| u_2 \|_{W^\ell(t,t+1)} = o\left(e^{-k_+^{(2)}t}\right) \quad \text{for} \quad t \to -\infty. \tag{3.62}$$

So (3.59) is an asymptotic formula for u_2 as $t \to +\infty$ and for u_1 as $t \to -\infty$ with remainder terms characterized by (3.61) and (3.62). We also give a variant of the asymptotic formula for solutions of equation (3.45) on the semiaxis $t > t_0$.

Proposition 3.8.2. *Let* $u \in W_{\mathrm{loc}}^\ell(t_0, \infty)$ *be a solution of* (3.45) *and let*

$$\int_{t_0}^\infty e^{k_-^{(2)}\tau}(1+|\tau|)^{m_-^{(2)}-1} \| f \|_{L_2(\tau,\tau+1;H_0)} \, d\tau < \infty. \tag{3.63}$$

Let

$$\liminf_{t\to\infty} e^{k_-^{(1)}t} \| u \|_{W^\ell(t,t+1)} = 0. \tag{3.64}$$

Then for $t \geq t_0 + 1$

$$u(t) = \sum_{k_+^{(1)}\leq\Im\lambda_\nu\leq k_-^{(2)}} \sum_{\sigma=0}^{\kappa_\nu} c_{\nu\sigma} U_\sigma^{(\nu)}(t) + v(t) , \tag{3.65}$$

where $c_{\nu\sigma}$ are constants and

$$\| v \|_{W^\ell(t,t+1)} \leq c \left\{ \int_{t_0}^{\infty} \mu^{(2)}(\tau - t) \| f \|_{L_2(\tau,\tau+1;H_0)} \, d\tau \right.$$

$$\left. + \mu^{(2)}(t_0 - t) \| u \|_{W^{\ell-1}(t_0,t_0+1)} \right\} \tag{3.66}$$

with a constant c independent of u, f and t_0. The function $\mu^{(2)}$ is defined by (3.10) with $k_\pm$, $m_\pm$ replaced by $k_\pm^{(2)}$, $m_\pm^{(2)}$.

Proof. Let ζ be a smooth function which is equal to 1 for $t > t_0 + 1$ and 0 for $t < t_0 + 1/2$. We rewrite (3.45) as

$$A(D_t)\zeta u = \zeta f + [A(D_t), \zeta]u \quad \text{on} \quad \mathbb{R}. \tag{3.67}$$

By (3.64) and Theorem 3.3.2, the function ζu is the solution of (3.57) corresponding to the strip $(k_-^{(1)}, k_+^{(1)})$. From Proposition 3.8.1 it follows that

$$\zeta u = \sum_{k_+^{(1)} \leq \Im \lambda_\nu \leq k_-^{(2)}} \sum_{\sigma=1}^{\kappa_\nu} c_{\nu\sigma} U_\sigma^{(\nu)} + v, \tag{3.68}$$

where v is the $(k_-^{(2)}, k_+^{(2)})$-solution of

$$A(D_t)v = \zeta f + [A(D_t), \zeta]u \quad \text{on} \quad \mathbb{R}.$$

The use of (3.48) for v completes the proof. $\square$

3.8.3 Description of Solutions to Homogeneous Equation

The following result is a refinement of Proposition 2.8.3, where we showed that the zeros of $A(D_t)$ from the two-weight space $W_{\gamma,\beta}^\ell(\mathbb{R})$ are linear combinations of power-logarithmic zeros. Here, we arrive at the same representation for solutions in the local space $W_{\text{loc}}^\ell(\mathbb{R})$.

Proposition 3.8.3. *Let $u \in W_{\text{loc}}^\ell(\mathbb{R})$ satisfy (2.53). Assume furthermore that*

$$\liminf_{t \to +\infty} e^{k_-^{(1)}t} \| u \|_{L_2(t,t+1;H_0)} = 0 \tag{3.69}$$

and

$$\liminf_{t \to -\infty} e^{k_+^{(2)}t} \| u \|_{L_2(t,t+1;H_0)} = 0. \tag{3.70}$$

Then

$$u(t) = \sum_{k_+^{(1)} \leq \Im \lambda_\nu \leq k_-^{(2)}} \sum_{\sigma=1}^{\kappa_\nu} c_{\nu\sigma} U_\sigma^{(\nu)}(t), \tag{3.71}$$

where $c_{\nu\sigma}$ are constants.

Proof. By Lemma 3.6.1 with $k_\pm$ replaced by $k_\pm^{(1)}$, we have for large $t > 0$:

$$\| u \|_{W^\ell(t,t+1)} \le c\, e^{-k_+^{(1)} t} t^{m_+^{(1)} - 1}.$$

We now apply Lemma 3.6.2 to the function u to obtain the bound

$$\| u \|_{W^\ell(t,t+1)} \le c\, e^{-k_-^{(2)} t} |t|^{m_-^{(2)} - 1}$$

for large negative t. Hence

$$u \in W^\ell_{k_-^{(2)} + \varepsilon,\; k_+^{(1)} - \varepsilon}(\mathbb{R}),$$

where ε is an arbitrary positive number. Reference to Proposition 2.8.3 completes the proof. $\square$

3.9 Applications to Boundary Value Problems

3.9.1 The Dirichlet Problem in a Cylinder

Consider the Dirichlet problem for a $(2m, \ell)$-elliptic operator in the cylinder $\mathcal{C}$ (as in Sect. 2.5). We shall retain all the notation used in Sect. 2.5. Here we reformulate the abstract comparison principle (3.37) for the Dirichlet problem in question. The role of the function $t \to \| u \|_{W^\ell(t,t+1)}$ will be played by

$$t \to \| u \|_{W_2^{2m,\ell}(\mathcal{C}_t)}$$
$$= \left(\int_t^{t+1} \int_\Omega \sum_{\frac{|\alpha|}{2m} + \frac{k}{\ell} \le 1} \left| D_x^\alpha D_\tau^k u(x,\tau) \right|^2 dx d\tau \right)^{1/2}, \qquad (3.72)$$

where

$$\mathcal{C}_t = \Omega \times (t, t+1).$$

The right-hand side f in (2.22) will be characterized by the function

$$\mathbb{R} \ni t \to \| f \|_{L_2(\mathcal{C}_t)}. \qquad (3.73)$$

The space of functions u with

$$\| u \|_{W_2^{2m,\ell}(\Omega \times (t,t+1))} < \infty \quad \text{for all} \quad t \in \mathbb{R}$$

and satisfying $\partial_\nu^j u = 0$ on $\partial\mathcal{C}$, $j = 0, \dots, m-1$, will be denoted by $\mathring{W}_{2,\mathrm{loc}}^{2m,\ell}(\mathcal{C})$. This space corresponds to the space $W_{\mathrm{loc}}^\ell(\mathbb{R})$ (see Definition 2.3.4). Analogously, the space $L_{2,\mathrm{loc}}(\mathbb{R}; H_0)$ becomes

$$L_{2,\mathrm{loc}}(\mathcal{C}) = \left\{ f : \| f \|_{L_2(\mathcal{C}_t)} < \infty \quad \text{for all} \quad t \in \mathbb{R} \right\}.$$

With this notation we can rewrite all the results of the present chapter for the Dirichlet problem under consideration. In particular, by (3.37) and (3.36), we have the estimate

$$\| u \|_{W_2^{2m,\ell}(\mathcal{C}_t)} \leq b(P) \int_{\mathbb{R}} g(t - \tau) \, \| \, Pu \, \|_{L_2(\mathcal{C}_\tau)} \, d\tau, \qquad (3.74)$$

where $u \in \dot{W}_{2,\text{loc}}^{2m,\ell}(\mathcal{C})$ is subject to

$$\| u \|_{W_2^{2m,\ell}(\mathcal{C}_t)} = o\left(e^{-k_{\mp} t}\right) \quad \text{as} \quad t \to \pm\infty,$$

and g is Green's function of the operator $\mathcal{M}(\partial_t)$ given by (3.23) and (3.24).

As the simplest example we take the Dirichlet problem for the heat equation:

$$\begin{cases} (\partial_t - \Delta_x) \, u = f & \text{on } \mathcal{C} \\ u = 0 & \text{on } \partial\mathcal{C}. \end{cases}$$

Let

$$\dot{W}_2^2(\Omega) = \{ u \in W_2^2(\Omega) : \; u = 0 \; \text{on} \; \; \partial\Omega \}.$$

Eigenvalues of the operator pencil

$$\mathcal{A}(\lambda) = i\lambda - \Delta : \dot{W}_2^2(\Omega) \to L_2(\Omega)$$

form the sequence $\lambda_j = i\gamma_j$, where $0 < \gamma_1 < \gamma_2 < \ldots$ is the sequence of eigenvalues of the operator $-\Delta$ with zero Dirichlet conditions on $\partial\Omega$.

By choosing $k_- = \gamma_j$ and $k_+ = \gamma_{j+1}$ we obtain the existence and uniqueness result stated in Theorem 3.3.2 with

$$\mu(t) = \begin{cases} e^{\gamma_j t} & \text{for } t \geq 0 \\ e^{\gamma_{j+1} t} & \text{for } t < 0. \end{cases}$$

Analogously, the Dirichlet problem for the Laplace equation

$$\begin{cases} (\partial_t^2 + \Delta_x) \, u = f & \text{on } \mathcal{C} \\ u = 0 & \text{on } \partial\mathcal{C} \end{cases}$$

can be treated. In particular, by taking $k_\pm = \pm\sqrt{\gamma_1}$ we must set

$$\mu(t) = e^{-\sqrt{\gamma_1}|t|}$$

in the formulation of Theorem 3.3.2. $\square$

3.9.2 The Neumann Problem in a Cylinder

Consider the Neumann problem for the Laplace equation:

$$\begin{cases} \left(D_t^2 - \Delta_x\right) u = f & \text{on } \mathcal{C} \\ \partial_\nu u = 0 & \text{on } \partial\mathcal{C}. \end{cases} \tag{3.75}$$

The pencil $\mathcal{A}(\lambda) = -\lambda^2 + \Delta_x$ will be defined on the space

$$\left\{ u \in W_2^2(\Omega) : \partial_\nu u = 0 \quad \text{on} \quad \partial\Omega \right\}. \tag{3.76}$$

It is a classical result that $\mathcal{A}(\lambda)$ is Fredholm and that for all $\lambda \in \mathbb{R}$

$$\sum_{|\alpha| \leq 2} |\lambda|^{2-|\alpha|} \, \| D_x^\alpha u \|_{L_2(\Omega)} \leq c \, \| \mathcal{A}(\lambda) u \|_{L_2(\Omega)} \, .$$

Since the operator $-\Delta_x$ with the domain (3.76) is selfadjoint, the last estimate implies the invertibility of $\mathcal{A}(\lambda)$ for all $\lambda \in \mathbb{R}\setminus\{0\}$. Therefore, (3.75) is a special case of equation (2.1) analyzed in this chapter.

We set

$$\| u \|_{W^2(t,t+1)} = \left(\int_t^{t+1} \int_\Omega \sum_{|\alpha|+j \leq 2} |D_x^\alpha D_\tau^j u(x,\tau)|^2 \, dx d\tau \right)^{1/2}, \tag{3.77}$$

$$\| f \|_{L_2(t,t+1;H_0)} = \left(\int_t^{t+1} \int_\Omega |f(x,\tau)|^2 \, dx d\tau \right)^{1/2} \tag{3.78}$$

and define the spaces $W_{\text{loc}}^2(\mathbb{R})$ and $L_{2,\text{loc}}(\mathbb{R}; H_0)$ by these seminorms (see Definitions 2.3.4 and 2.3.3).

By using the notation introduced in Sect. 1.5.1, one can reformulate Theorem 3.3.2 for the problem (3.75). By taking, for instance, $k_- = 0, k_+ = \sqrt{\gamma_1}$ we have Theorem 3.3.2 with

$$\mu(t) = \begin{cases} 1 + t & \text{for } t \geq 0 \\ e^{\sqrt{\gamma_1}t} & \text{for } t \leq 0. \end{cases}$$

Choosing $k_- = -\sqrt{\gamma_1}$ and $k_+ = 0$ leads to Theorem 3.3.2 with

$$\mu(t) = \begin{cases} e^{-\sqrt{\gamma_1}t} & \text{for } t \geq 0 \\ 1 + |t| & \text{for } t \leq 0. \end{cases}$$

As an application of our abstract results on the asymptotics of solutions to (2.1), we shall state a corollary of Proposition 3.8.1 for the Neumann problem (3.75). We set

$$k_-^{(1)} = -\sqrt{\gamma_1} \quad , \quad k_+^{(1)} = k_-^{(2)} = 0 \quad \text{and} \quad k_+^{(2)} = \sqrt{\gamma_1} \, .$$

Then

$$m_-^{(1)} = m_+^{(2)} = 1 \quad \text{and} \quad m_+^{(1)} = m_-^{(2)} = 2$$

(see Example 1.2.3). We subject f to the requirement

$$\int_{-\infty}^{+\infty} (1+|t|)\, \|\, f\, \|_{L_2(\Omega \times (t,t+1))}\, dt < \infty$$

and consider solutions u_1 and u_2 of (3.75) satisfying

$$\|\, u_1\, \|_{W^2(t,t+1)} = \begin{cases} o\left(e^{\sqrt{\gamma_1}t}\right) & \text{as } t \to +\infty \\ o(1) & \text{as } t \to -\infty. \end{cases}$$

and

$$\|\, u_2\, \|_{W^2(t,t+1)} = \begin{cases} o(1) & \text{as } t \to +\infty \\ o\left(e^{-\sqrt{\gamma_1}t}\right) & \text{as } t \to -\infty. \end{cases}$$

Then by Proposition 3.8.1 and Sect. 1.2.3

$$u_2(x,t) - u_1(x,t) = c_1 + c_2 t\ ,$$

where

$$c_1 = \frac{-1}{\operatorname{mes}_n \Omega} \int_C f(x,\tau)\tau\, dx d\tau,$$

$$c_2 = \frac{1}{\operatorname{mes}_n \Omega} \int_C f(x,\tau)\, dx d\tau.$$

3.9.3 The Dirichlet Problem in a Cone

Let us consider the Dirichlet problem for a $2m$-th order elliptic operator in the cone K

$$\begin{cases} P(D_x)u = f & \text{on } K \\ \partial_\nu^j u = 0 & \text{on } \partial K \backslash \{0\},\ j = 0, \dots, m-1. \end{cases} \tag{3.79}$$

The operator of this problem has already been treated in Sections 1.5.2 and 2.10. We preserve the same notation.

Since the multiplication by r^{2m} and the change of variables (0.8) transform the Dirichlet problem in K to the Dirichlet problem for the equation (0.9) in the cylinder C, we are within the scope of Sect. 3.9.1 with $\ell = 2m$.

Therefore, all results of the present chapter have their counterparts for the Dirichlet problem in the cone. In order to write these results explicitly we need only to make the inverse change of variables $(\theta, t) \to (\theta, r),\ r = \exp(-t)$. Then the solution of (3.79) is characterized by the function

$$\mathbb{R}_+ \ni r \to \|\, u\, \|_{W_2^{2m}(K_r)}\ , \tag{3.80}$$

where

$$K_r = \{x \in K : e^{-1}r < |x| < r\}$$

and

$$\| u \|_{W_2^{2m}(K_r)} = \left(\sum_{|\alpha| \leq 2m} r^{2|\alpha|-n} \| D_x^\alpha u \|_{L_2(K_r)}^2 \right)^{1/2}.$$

The behaviour of the right-hand side is described by the function

$$\mathbb{R}_+ \ni r \to r^{-n/2} \| f \|_{L_2(K_r)} . \tag{3.81}$$

These two functions correspond to (3.77) and (3.78).

In what follows, the notations $u \in V_{2,\mathrm{loc}}^{2m}(K)$ and $f \in L_{2,\mathrm{loc}}(K)$ denote that the functions (3.80) and (3.81) are finite for every $r > 0$.

Let $\Upsilon(\rho) = \mu(\log \rho^{-1})$; i.e.

$$\Upsilon(\rho) = \begin{cases} \rho^{-k_-} (1 + |\log \rho|)^{m_- - 1} & \text{for } \rho \leq 1 \\ \rho^{-k_+} (1 + \log \rho)^{m_+ - 1} & \text{for } \rho \geq 1. \end{cases}$$

Here the numbers k_- and k_+ are chosen in such a way that the strip $k_- < \Im\lambda < k_+$ does not contain eigenvalues of the pencil (1.44) and $m_\mp$ are maximum lengths of the Jordan chains corresponding to eigenvalues of this pencil with $\Im\lambda = k_\mp$. We also put $m_+ = 1$ ($m_- = 1$) if there are no eigenvalues on the line $\Im\lambda = k_+$ ($\Im\lambda = k_-$).

We can now state all results of this chapter for the Dirichlet problem in K. For instance, Theorem 3.3.2 can be reformulated as the following statement on unique solvability of (3.79).

Theorem 3.9.1. *Let $f \in L_{2,\mathrm{loc}}(K)$ and*

$$\int_0^\infty \Upsilon(\rho)\rho^{2m-n/2} \| f \|_{L_2(K_\rho)} \frac{d\rho}{\rho} < \infty.$$

There exists a solution $u \in V_{2,\mathrm{loc}}^{2m}(K)$ of (3.79) satisfying

$$\| u \|_{W_2^{2m}(K_r)} \leq c \int_0^\infty \Upsilon\left(\frac{\rho}{r}\right) \rho^{2m-n/2} \| f \|_{L_2(K_\rho)} \frac{d\rho}{\rho}. \tag{3.82}$$

This solution also satisfies

$$\| u \|_{W_2^{2m}(K_r)} = \begin{cases} o(r^{k_-}) & \text{as } r \to 0 \\ o(r^{k_+}) & \text{as } r \to +\infty. \end{cases} \tag{3.83}$$

Additionally, condition (3.83) describes a class of uniqueness for solutions in $V_{2,\mathrm{loc}}^{2m}(K)$ of the problem (3.79).

By (3.25), the inequality (3.82) can be interpreted as the comparison principle

$$\| u \|_{W_2^{2m}(K_r)} \leq b(P)w(\log r^{-1}), \tag{3.84}$$

where $w(t)$ is a solution of

$$\mathcal{M}(\partial_t)w(t) = e^{(n/2-2m)t} \| f \|_{L_2(K_{e^{-t}})}$$

satisfying (3.30).

Remark 3.9.2. Suppose that $P(D_x)$ has real coefficients. It was shown in Lemma 1.5.1(i) that the line $\Im\lambda = m - n/2$ is free of eigenvalues of $\mathcal{A}(\lambda)$. Let κ be the supremum of $k > 0$ such that the strip

$$m - n/2 \le \Im\lambda < m - n/2 + k$$

does not contain eigenvalues of $\mathcal{A}(\lambda)$. Then, according to Sect. 1.5.4 (see (1.54)), the strip $|\Im\lambda - m + n/2| < \kappa$ is free of eigenvalues of $\mathcal{A}(\lambda)$. Thus, we can set $k_\pm = m - n/2 \pm \kappa$ in the last theorem.

Under the complementary assumption that the cone K can be described by the inequality $x_n > F(x')$, $x' \in \mathbb{R}^{n-1}$, it was shown in Kozlov, Maz'ya (1988, 1991) that $\kappa > 1/2$, and that this estimate cannot, in general, be improved. If $K = \mathbb{R}^n_+$ then $\kappa = n/2$ (see Sect. 1.5.5). For $n = 2$ the better estimate $\kappa > 1$ is valid provided the opening α of the angle K is $< \pi$. Moreover, κ is a decreasing function of α on the interval $[\pi, 2\pi]$ and equal to 1 for $\alpha = \pi$ and $1/2$ for $\alpha = 2\pi$ (see Kozlov (1989, 1991)). Some other information on κ for special differential operators can be found in the survey Kozlov, Maz'ya (1996).

We finish this section by formulating a direct corollary of Proposition 3.8.1 on the asymptotics of solutions to the problem (3.79) as $r \to 0$ and $r \to \infty$.

Let $k_\pm^{(j)}$ and $m_\pm^{(j)}$, $j = 1, 2$, have the same meaning as in Sect. 3.8 for the operator pencil (1.44). We shall use the bases

$$\left\{\mathcal{U}_\sigma^{(\nu)}\right\} \quad \text{and} \quad \left\{\mathcal{V}_\sigma^{(\nu)}\right\}$$

of power logarithmic zeros of $P(D_x)$ and $P^*(D_x)$ introduced in Sect. 2.10.

Proposition 3.9.3. *Let $f \in L_{2,\mathrm{loc}}(K)$ and let*

$$\int_1^\infty \rho^{-k_+^{(1)} + 2m - n/2} (1 + \log\rho)^{m_+^{(1)} - 1} \| f \|_{L_2(K_\rho)} \frac{d\rho}{\rho} < \infty$$

$$\int_0^1 \rho^{-k_-^{(2)} + 2m - n/2} (1 + |\log\rho|)^{m_-^{(2)} - 1} \| f \|_{L_2(K_\rho)} \frac{d\rho}{\rho} < \infty.$$

Then the solutions u_i, $i = 1, 2$, corresponding to the strips $k_-^{(i)} < \Im\lambda < k_+^{(i)}$ satisfy

$$u_2(x) - u_1(x) = \sum_{k_+^{(1)} \le \Im\lambda_\nu \le k_-^{(2)}} \sum_{\sigma=1}^{\kappa_\nu} C_{\nu\sigma} \mathcal{U}_\sigma^{(\nu)}(x),$$

where the coefficients $C_{\nu\sigma}$ are given by (2.61).

3.10 Comments

Exponential estimates at infinity (especially theorems of the Phragmén-Lindelöf type) which develop the earlier work by Lax (1957) on the same subject were obtained in the papers of Evgrafov (1960) and Agmon, Nirenberg (1963).

The starting point of this chapter was the paper by Kozlov, Maz'ya (1985), where elliptic boundary value problems in a cone are studied in local Sobolev spaces. The comparison principle (3.37) appeared in Kozlov, Maz'ya (1991-1996), report 1, where most of results of the present chapter can be found.

4. Two-Weight L_2-Estimates

4.1 Introduction

In this chapter we study equation (2.1) with the right-hand side f in the space $L_2(\mathbb{R}; H_0; \Gamma)$ of abstract functions $f : \mathbb{R} \to H_0$ with the norm

$$\| f \|_{L_2(\mathbb{R};H_0;\Gamma)} = \left(\int_{\mathbb{R}} \Gamma^2(\tau) \, \| f(\tau) \|_0^2 \, d\tau \right)^{1/2}.$$

We are interested in solutions in the space $W^\ell(\mathbb{R}; \gamma)$ of functions $u : \mathbb{R} \to H_\ell$ endowed with the norm

$$\| u \|_{W^\ell(\mathbb{R};\gamma)} = \left(\int_{\mathbb{R}} \gamma^2(\tau) \sum_{q=0}^{\ell} \| D_\tau^q u(\tau) \|_{\ell-q}^2 \, d\tau \right)^{1/2}.$$

Here Γ and γ are weight functions from a certain class (defined in Sect.4.2).

In Sect. 4.3 we take real numbers k_- and k_+ such that $k_- < k_+$ and the strip $k_- < \Im\lambda < k_+$ is free of eigenvalues of the pencil $\mathcal{A}(\lambda)$. Then we show in Theorem 4.3.1 that $W^\ell(\mathbb{R}; \gamma)$ is a class of uniqueness for (2.1) provided that both integrals

$$\int_{-\infty}^{0} e^{-2k_+\tau} \gamma^2(\tau) d\tau \quad \text{and} \quad \int_{0}^{\infty} e^{-2k_-\tau} \gamma^2(\tau) d\tau \tag{4.1}$$

diverge. According to Theorem 4.3.3, this result is the best possible one under the additional condition that the spectrum of $\mathcal{A}(\lambda)$ contains eigenvalues with arbitrarily large positive and negative imaginary parts. To be more precise, we show that the uniqueness in $W^\ell(\mathbb{R}; \gamma)$ implies the existence of a strip $k_- < \Im\lambda < k_+$ free of the spectrum of $\mathcal{A}(\lambda)$ such that both integrals (4.1) diverge.

In Theorem 4.4.1 we prove the solvability in $W^\ell(\mathbb{R}; \gamma)$ of the equation (2.1) with the right-hand side from $L_2(\mathbb{R}; H_0; \gamma)$. We obtain conditions on Γ and γ for the validity of the two-weight estimate

$$\| u \|_{W^\ell(\mathbb{R};\gamma)} \leq c \, \| f \|_{L_2(\mathbb{R};H_0;\Gamma)} . \tag{4.2}$$

Our starting point is the estimate (3.16) for the function

$$t \to \| u \|_{W^\ell(t,t+1)} \cdot$$

To obtain (4.2) we apply two-weight L_2-estimates for Riemann-Liouville integral operators (as proved by Stepanov (1989)) to the right-hand side of (3.16).

A special case of (4.2) is the estimate

$$\| u \|_{W^\ell_\beta(\mathbb{R})} \leq c \, \| f \|_{L_{2,\beta}(\mathbb{R};H_0)}$$

considered in Sect. 2.4. This estimate is valid if and only if there are no eigenvalues of $\mathcal{A}(\lambda)$ on the line $\Im \lambda = \beta$. In particular, it follows from our general theorem (Theorem 4.4.1) that in the presence of eigenvalues on $\Im \lambda = \beta$, the two-weight estimate (4.2) holds for

$$\Gamma(t) = \big(1 + |t|\big)^m \gamma(t)$$

if

$$\gamma(t) = e^{\beta t} \begin{cases} (1 + t)^{\sigma^{(+)}} & \text{for } t \geq 0 \\ (1 + |t|)^{\sigma^{(-)}} & \text{for } t \leq 0, \end{cases} \tag{4.3}$$

where $\sigma(+) + m < 1/2$ and $\sigma^{(-)} > -1/2$, and m is the maximum length of Jordan chains corresponding to eigenvalues of $\mathcal{A}(\lambda)$ on the line $\Im \lambda = \beta$. (see Sect. 4.4.4).

The chapter is completed with a direct application of these results to the Dirichlet problem in a cone.

4.2 Weighted Sobolev Spaces

Before introducing the weighted Sobolev spaces of vector valued functions which will be used, we define a class of weight functions γ.

We say that a positive measurable function γ on $\mathbb{R}$ belongs to the class A if there exist positive constants c_1, c_2 such that

$$c_1 \gamma(t) \leq \gamma(t + h) \leq c_2 \gamma(t) \tag{4.4}$$

for all $t \in \mathbb{R}$ and $h \in (0, 1]$.

From (4.4) it follows immediately that

$$a_1 e^{-b_1 |t|} \leq \gamma(t) \leq a_2 e^{b_2 |t|} \quad \text{for all} \quad t \in \mathbb{R}, \tag{4.5}$$

where a_1, a_2, b_1, b_2 are positive numbers.

Let $W^\ell(\mathbb{R}; \gamma)$ be the space of functions $u : \mathbb{R} \to H_\ell$ endowed with the norm

$$\|u\|_{W^\ell(\mathbb{R};\gamma)} = \Big(\int_{\mathbb{R}} \gamma^2(\tau) \sum_{q=0}^{\ell} \|D_\tau^q u(\tau)\|_{\ell-q}^2 d\tau \Big)^{1/2},$$

and let $L_2(\mathbb{R}; H_q; \gamma)$ consist of functions $u : \mathbb{R} \to H_q$ with finite norm

$$\| u \|_{L_2(\mathbb{R}; H_q; \gamma)} = \left(\int_{\mathbb{R}} \gamma^2(\tau) \| u(\tau) \|_q^2 \, d\tau \right)^{1/2}.$$

Clearly the function $\gamma(t) = e^{\beta t}$, $\beta \in \mathbb{R}$, belongs to $\mathbb{A}$ and

$$W^\ell(\mathbb{R}; e^{\beta t}) = W_\beta^\ell(\mathbb{R}), \quad L_{2,\beta}(\mathbb{R}; H_0) = L_2(\mathbb{R}; H_0; e^{\beta t}).$$

Proposition 4.2.1. *Let $\gamma \in \mathbb{A}$. Then*

a) *The norm*

$$\left(\int_{\mathbb{R}} \gamma^2(t) \| u \|_{W^\ell(t, t+1)}^2 \, dt \right)^{1/2}$$

is equivalent to the norm $\| u \|_{W^\ell(\mathbb{R}; \gamma)}$.

b) *The norm*

$$\left(\int_{\mathbb{R}} \gamma^2(t) \| u \|_{L_2(t, t+1; H_q)}^2 \, dt \right)^{1/2}$$

is equivalent to the norm $\| u \|_{L_2(\mathbb{R}; H_q; \gamma)}$.

Proof. It is sufficient to verify b). We have

$$\int_a^b \gamma^2(t) \int_t^{t+1} \| u(\tau) \|_q^2 \, d\tau \, dt$$
$$= \int_a^{b+1} \| u(\tau) \|_q^2 \int_{\max(\tau-1, a)}^{\min(\tau, b)} \gamma^2(t) dt.$$

In view of (4.4) this implies

$$c \int_{a+1}^b \gamma^2(\tau) \| u(\tau) \|_q^2 \, d\tau \leq \int_a^b \gamma^2(\tau) \| u \|_{L_2(t, t+1; H_q)}^2 \, d\tau$$
$$\leq c' \int_a^{b+1} \gamma^2(\tau) \| u(\tau) \|_q^2 \, d\tau , \tag{4.6}$$

where $c = \min(c_1, c_2^{-1})$, $c' = \max(c_2, c_1^{-1})$. The result follows. $\square$

4.3 Uniqueness of Solutions in $W^\ell(\mathbb{R}; \gamma)$

Let k_+ and k_- be two numbers satisfying $k_- < k_+$, and let the strip $k_- < \Im \lambda < k_+$ be free of eigenvalues of the pencil $\mathcal{A}(\lambda)$.

Theorem 4.3.1. *Let $\gamma \in \mathbb{A}$ and let*

$$\int_{-\infty}^0 e^{-2k_+ \tau} \gamma^2(\tau) d\tau = \infty \tag{4.7}$$

$$\int_0^\infty e^{-2k_- \tau} \gamma^2(\tau) d\tau = \infty. \tag{4.8}$$

If $u \in W^\ell(\mathbb{R}; \gamma)$ satisfies the equation $\mathcal{A}(D_t)u = 0$ on $\mathbb{R}$ then $u = 0$.

Proof. Suppose $u \in W^\ell(\mathbb{R}; \gamma)$ and $\mathcal{A}u = 0$ on $\mathbb{R}$. Due to (4.5) we have $u \in W_{\beta_2, \beta_1}(\mathbb{R})$ for some real numbers $\beta_1 < \beta_2$. Making use of Proposition 2.8.3 we obtain

$$u = \sum_{\beta_1 < \Im \lambda_\nu < \beta_2} \sum_{\sigma=1}^{J_\nu} c_{\nu\sigma} U_\sigma^{(\nu)}(t).$$

By (4.7) we have

$$\int_{-\infty}^0 \| U_\sigma^{(\nu)}(t) \|_\ell^2 \, \gamma^2(\tau) d\tau = \infty \quad \text{for} \quad \Im \lambda_\nu \geq k_+,$$

and by (4.8)

$$\int_0^\infty \| U_\sigma^{(\nu)}(t) \|_\ell^2 \, \gamma^2(\tau) d\tau = \infty \quad \text{for} \quad \Im \lambda_\nu \leq k_-.$$

Hence $c_{\nu\sigma} = 0$, which implies $u = 0$. $\square$

We give an example of a special family of weight functions γ satisfying the conditions of the previous theorem.

Example 4.3.2. Let $\gamma \in \mathbb{A}$ and

$$\gamma(t) = \begin{cases} e^{\alpha_0 t} t^{\alpha_1} & \text{for } t > N, \\ e^{\beta_0 t} |t|^{\beta_1} & \text{for } t < -N, \end{cases} \tag{4.9}$$

where N is a sufficiently large positive number. Clearly, the divergence of the integrals in (4.7) and (4.8) is equivalent to the pair of the relations:

a) either $\alpha_0 > k_-$ or $\alpha_0 = k_-$ and $\alpha_1 \geq -1/2$;

b) either $\beta_0 < k_+$ or $\beta = k_+$ and $\beta_1 \geq -1/2$.

The last theorem admits an inversion.

Theorem 4.3.3. *Let there exist eigenvalues of $\mathcal{A}(\lambda)$ with arbitrary large positive and negative imaginary parts. If $W^\ell(\mathbb{R}; \gamma)$ is a class of uniqueness for (2.1) then there exists a strip $k_- < \Im \lambda < k_+$, free of eigenvalues, such that (4.7) and (4.8) are valid.*

Proof. By the assumption of uniqueness, no solution U_ν^σ of $\mathcal{A}(D_t)U = 0$ belongs to $W^\ell(\mathbb{R}; \gamma)$. Hence

$$\int_{\mathbb{R}} e^{-2\Im \lambda_\nu \tau} \gamma^2(\tau) d\tau = \infty \tag{4.10}$$

for all $\nu \in Z$. We set

$$s_+ = \sup \left\{ \Im \lambda_\nu : \int_{-\infty}^0 e^{-2\Im \lambda_\nu \tau} \gamma^2(\tau) d\tau < \infty \right\}$$

and

$$s_- = \inf \left\{ \Im\lambda_\nu : \int_0^\infty e^{-2\Im\lambda_\nu\tau}\gamma^2(\tau)d\tau < \infty \right\}.$$

Since by (4.5)

$$C_1 e^{-N|t|} < \gamma(t) < C_2 e^{N|t|}$$

for some positive C_1, C_2 and N, it follows that $s_\pm$ are finite. In other words, we can replace inf and sup by min and max. By (4.10), $s_+ > s_-$. Clearly,

$$\int_0^\infty e^{-2\Im\lambda_\nu\tau}\gamma^2(\tau)d\tau = \infty$$

if $\Im\lambda_\nu < s_+$ and

$$\int_{-\infty}^0 e^{-2\Im\lambda_\nu\tau}\gamma^2(\tau)d\tau = \infty$$

if $\Im\lambda_\nu > s_-$. Therefore, the required conditions are satisfied by any strip $k_- < \Im\lambda < k_+$ which is free of the spectrum and such that $k_- \geq s_-$ and $k_+ \leq s_+$. $\square$

4.4 Existence of Solutions in $W^\ell(\mathbb{R}; \gamma)$

4.4.1 Principal Result

Theorem 4.4.1. *Let the strip $k_- < \Im\lambda < k_+$ be free of eigenvalues of $\mathcal{A}(\lambda)$ and let the weight functions γ and Γ from $\mathbb{A}$ satisfy*

$$\sup_{t\in\mathbb{R}} \int_{Q_\pm(t)} e^{2k_\pm(\sigma-\tau)}|\sigma - \tau|^{2(m_\pm-1)}\frac{\gamma^2(\tau)}{\Gamma^2(\sigma)}d\sigma d\tau < \infty. \qquad (4.11)$$

Here $Q_\pm$ are quadrants $\{(\sigma, \tau) \in \mathbb{R}^2 : \tau \gtrless t \gtrless \sigma\}$. (This condition actually consists of two inequalities, corresponding to plus and minus.) Then for every $f \in L_2(\mathbb{R}; H_0; \Gamma)$ there exists a solution $u \in W^\ell(\mathbb{R}; \gamma)$ of $\mathcal{A}(D_t)u = f$ on $\mathbb{R}$. Moreover, the estimate

$$\| u \|_{W^\ell(\mathbb{R};\gamma)} \leq c \| f \|_{L_2(\mathbb{R};H_0;\Gamma)} \qquad (4.12)$$

holds.

Before proving this theorem we formulate some well-known criteria for boundedness of special one dimensional integral operators in pairs of weighted L_2-spaces.

4.4.2 Auxiliary Results on Operators of Multiple Integration

The following assertion, with a slight modification of the original proof, is due to Stepanov (1989).

We shall need the operators $\mathcal{I}_k$ and $\mathcal{J}_k$, $k = 0, 1, ...$, given by

$$(\mathcal{I}_k h)(t) = \int_{-\infty}^{t} (t - \tau)^k h(\tau) d\tau \tag{4.13}$$

and

$$(\mathcal{J}_k h)(t) = \int_{t}^{\infty} (t - \tau)^k h(\tau) d\tau. \tag{4.14}$$

Proposition 4.4.2. *Let φ, Φ be positive measurable functions on $\mathbb{R}$. The inequality*

$$\int_{\mathbb{R}} |\varphi(t)(\mathcal{I}_k h)(t)|^2 dt \le c \int_{\mathbb{R}} |\Phi(t)h(t)|^2 dt \tag{4.15}$$

holds if and only if

$$A := \sup_{t \in \mathbb{R}} \int_{t}^{\infty} \int_{-\infty}^{t} (\tau - \sigma)^{2k} \frac{\varphi^2(\tau)}{\Phi^2(\sigma)} d\sigma d\tau < \infty. \tag{4.16}$$

Moreover, the best constant c in (4.15) satisfies

$$2^{-1-2k} A \le c \le 2^{2(k+2)} A. \tag{4.17}$$

Proof. The case $k = 0$ is well known (see, for example, Muckenhoupt (1972) or Maz'ya (1985), Theorem 1.3.1/4) and moreover, (4.17) can be improved:

$$A \le c \le 4A. \tag{4.18}$$

Let $k \ge 1$. For the "only if" part we assume that (4.15) holds with a finite constant $c > 0$ and

$$\int_{-\infty}^{t} \frac{d\sigma}{\Phi^2(\sigma)} < \infty \quad \text{for all } -\infty < t < \infty. \tag{4.19}$$

If we reduce inequality (4.15) to the functions h such that supp $h \subset (-\infty, t]$ with a fixed $t \in \mathbb{R}$, we find

$$c \int_{-\infty}^{t} |h(\sigma)\Phi(\sigma)|^2 d\sigma \ge \int_{t}^{\infty} \left| \varphi(\sigma) \int_{-\infty}^{t} (\tau - \sigma)^k h(\sigma) d\sigma \right|^2 d\tau$$

$$\ge \int_{t}^{\infty} (\tau - t)^{2k} \varphi^2(\tau) d\tau \left| \int_{-\infty}^{t} h(\sigma) d\sigma \right|^2$$

and by the reverse Hölder inequality

$$c \ge \int_{t}^{\infty} \int_{-\infty}^{t} (\tau - t)^{2k} \frac{\varphi^2(\tau)}{\Phi^2(\sigma)} d\sigma d\tau. \tag{4.20}$$

Arguing analogously for the functions h of the form $h(\sigma) = (t - \sigma)^k g(\sigma)$, where supp $g \subset (-\infty, t]$, $g \geq 0$, and assuming that

$$\int_{-\infty}^t \frac{(t - \tau)^{2k} d\sigma}{\varPhi^2(\sigma)} < \infty \quad \text{for all } -\infty < t < \infty, \tag{4.21}$$

we obtain

$$c \int_{-\infty}^t (t - \sigma)^{2k} |g(\sigma)\varPhi(\sigma)|^2 d\sigma$$

$$\geq \int_t^\infty |\varphi(\tau) \int_{-\infty}^t (t - \sigma)^k (\tau - \sigma)^k g(\sigma) d\sigma|^2 d\tau$$

$$\geq \int_t^\infty |\varphi(\tau)|^2 d\tau \left| \int_{-\infty}^t (t - \sigma)^{2k} g(\sigma) d\sigma \right|^2.$$

Hence, the reverse Hölder inequality implies

$$c \geq \int_t^\infty \int_{-\infty}^t (t - \sigma)^{2k} \frac{\varphi^2(\tau)}{\varPhi^2(\sigma)} d\sigma d\tau. \tag{4.22}$$

The estimates (4.20) and (4.22) give the lower bound (4.17). The assumptions (4.19) and (4.21) can be removed by applying the above argument to (4.15) with $\varPhi(\sigma)$ replaced by $\varPhi(\sigma) + \varepsilon|\sigma|^{2k} + \varepsilon$, $\varepsilon > 0$, and then letting $\varepsilon \to 0$.

For the "if" part we write

$$\varLambda := \int_{-\infty}^\infty \left[\varphi(\sigma) \int_{-\infty}^\tau (\tau - \sigma)^k h(\sigma) d\sigma \right]^2 d\tau$$

$$= \int_{-\infty}^\infty \varphi^2(\tau) \int_{-\infty}^\tau (\tau - \sigma)^k h(\sigma) d\sigma \int_{-\infty}^\tau (\tau - s)^k h(s) ds.$$

By changing the order of integration we arrive at

$$\varLambda = \int_{-\infty}^\infty h(\sigma) \int_\sigma^\infty (\tau - \sigma)^k \varphi^2(\tau) \int_{-\infty}^\tau (\tau - s)^k h(s) ds d\tau d\sigma$$

$$= \int_{-\infty}^\infty h(\sigma) \int_\sigma^\infty (\tau - \sigma)^k \varphi^2(\tau) \left[\int_{-\infty}^\sigma + \int_\sigma^\tau \right] (\tau - s)^k h(s) ds d\tau d\sigma$$

$$= \int_{-\infty}^\infty h(\sigma) d\sigma \int_\sigma^\infty (\tau - \sigma)^k \varphi^2(\tau) d\tau \int_{-\infty}^\sigma (\tau - \sigma + \sigma - s)^k h(s) ds$$

$$+ \int_{-\infty}^\infty h(\sigma) d\sigma \int_\sigma^\infty h(s) ds \int_s^\infty (\tau - s + s - \sigma)^k (\tau - s)^k \varphi^2(\tau) d\tau.$$

Hence

$$\Lambda \le 2^{k-1} \int_{-\infty}^{\infty} h(\sigma)d\sigma \Big[\int_{\sigma}^{\infty} (\tau - \sigma)^{2k}\varphi^2(\tau)d\tau \int_{-\infty}^{\sigma} h(s)ds$$

$$+ \int_{\sigma}^{\infty} (\tau - \sigma)^{k}\varphi^2(\tau)d\tau \int_{-\infty}^{\sigma} (\sigma - s)^{k}h(s)ds$$

$$+ \int_{\sigma}^{\infty} h(s)ds \int_{s}^{\infty} (\tau - s)^{2k}\varphi^2(\tau)d\tau$$

$$+ \int_{\sigma}^{\infty} (s - \sigma)^{k}h(s)ds \int_{s}^{\infty} (\tau - s)^{k}\varphi^2(\tau)d\tau \Big]$$

$$= 2^{k} \Big[\int_{-\infty}^{\infty} h(\sigma) \int_{-\infty}^{\sigma} h(s)ds \int_{\sigma}^{\infty} (\tau - \sigma)^{2k}\varphi^2(\tau)d\tau d\sigma$$

$$+ \int_{-\infty}^{\infty} h(\sigma) \int_{-\infty}^{\sigma} (\sigma - s)^{k}h(s)ds \int_{\sigma}^{\infty} (\tau - \sigma)^{k}\varphi^2(\tau)d\tau d\sigma \Big]$$

$$:= 2^{k}(\Lambda_1 + \Lambda_2).$$

Applying Hölder's inequality we arrive at

$$\Lambda_1 \le ||h\Phi||_{L_2(\mathbb{R})} \, H_1^{1/2},$$

where

$$H_1 = \int_{-\infty}^{\infty} \Phi^{-2}(\sigma) \Big(\int_{-\infty}^{\sigma} h(s)ds \Big)^2 \Big(\int_{\sigma}^{\infty} (\tau - \sigma)^{2k}\varphi^2(\tau)d\tau \Big)^2 d\sigma$$

and using the upper bound (4.18) (corresponding to $k = 0$) we obtain

$$H_1^{1/2} \le 2a||h\Phi||_{L_2(\mathbb{R})}, \tag{4.23}$$

where

$$a^2 = \sup_{t \in \mathbb{R}} \int_{t}^{\infty} \Phi^{-2}(\sigma) \Big(\int_{\sigma}^{\infty} (\tau - \sigma)^{2k}\varphi^2(\tau)d\tau \Big)^2 d\sigma \int_{-\infty}^{t} \Phi^{-2}(s)ds.$$

From definition (4.16) it follows that

$$\int_{t}^{\infty} \Phi^{-2}(\sigma) \Big(\int_{\sigma}^{\infty} (\tau - \sigma)^{2k}\varphi^2(\tau)d\tau \Big)^2 d\sigma$$

$$\le A^2 \int_{t}^{\infty} \Phi^{-2}(\sigma) \Big(\int_{-\infty}^{\sigma} \Phi^{-2}(s)ds \Big)^{-2} d\sigma \le A^2 \Big(\int_{-\infty}^{t} \Phi^{-2}(s)ds \Big)^{-1}.$$

Thus, $a \le A$ and (4.23) leads to the estimate

$$\Lambda_1 \le 2A||h\Phi||_{L_2(\mathbb{R})}. \tag{4.24}$$

Analogously,

$$\Lambda_2 \le ||h\Phi||_{L_2(\mathbb{R})}H_2^{1/2}, \tag{4.25}$$

where

$$H_2 = \int_{-\infty}^{\infty} \Phi^{-2}(\sigma) \left(\int_{\sigma}^{\infty} (\tau - \sigma)^k \varphi^2(\tau) d\tau \right)^2 (\mathcal{I}_k h)^2(\sigma) d\sigma.$$

Without loss of generality we may assume that h has a compact support. Then integrating by parts we find

$$H_2 = \int_{-\infty}^{\infty} \int_t^{\infty} \Phi^{-2}(\sigma) \left(\int_{\sigma}^{\infty} (\tau - \sigma)^k \varphi^2(\tau) d\tau \right)^2 d(\mathcal{I}_k h)^2(\sigma)$$

and the Minkowski inequality yields

$$H_2 \leq \int_{-\infty}^{\infty} \left(\int_t^{\infty} \varphi^2(\tau) \left(\int_t^{\tau} (\tau - \sigma)^{2k} \Phi^{-2}(\sigma) d\sigma \right)^{1/2} \right)^2 d(\mathcal{I}_k h)^2(t)$$

$$\leq A \int_{-\infty}^{\infty} \left(\int_t^{\infty} \varphi^2(\tau) \left(\int_\tau^{\infty} \varphi^2(s) ds \right)^{-1/2} d\tau \right)^2 d(\mathcal{I}_k h)^2(t)$$

$$\leq 4A \int_{-\infty}^{\infty} \left(\int_t^{\infty} \varphi^2(\tau) d\tau \right) d(\mathcal{I}_k h)^2(t)$$

$$= 4A \int_{-\infty}^{\infty} \varphi^2(\tau) \left(\int_{-\infty}^{\tau} (\tau - s)^k h(s) ds \right)^2 d\tau.$$

Summing up the estimates obtained for Λ_1 and Λ_2 we arrive at

$$\|\varphi \mathcal{I}_k h\|_{L_2(\mathbb{R})}^2 \leq 2^{k+1} \left(A \|h\Phi\|_{L_2(\mathbb{R})}^2 + A^{1/2} \|h\Phi\|_{L_2(\mathbb{R})} \|\varphi \mathcal{I}_k h\|_{L_2(\mathbb{R})} \right),$$

and denoting

$$x = \frac{\|\varphi \mathcal{I}_k h\|_{L_2(\mathbb{R})}}{A^{1/2} \|h\Phi\|_{L_2(\mathbb{R})}}$$

we find that

$$x^2 \leq 2^{k+1}(1 + x).$$

Hence

$$x \leq 2^k + (2^{2k} + 2^{k+1})^{1/2}$$

and

$$\|\varphi \mathcal{I}_k h\|_{L_2(\mathbb{R})} \leq 2^{k+2} A^{1/2} \|h\Phi\|_{L_2(\mathbb{R})}.$$

The upper bound (4.17) follows. $\square$

We turn to the operator $\mathcal{J}_k$.

Corollary 4.4.3. *The inequality*

$$\int_{\mathbb{R}} |\varphi(t)(\mathcal{J}_k h)(t)|^2 dt \leq c \int_{\mathbb{R}} |\Phi(t) h(t)|^2 dt \tag{4.26}$$

holds if and only if

$$\sup_{t \in \mathbb{R}} \int_{-\infty}^t \int_t^{\infty} (\sigma - \tau)^{2k} \frac{\varphi^2(\tau)}{\Phi^2(\sigma)} d\sigma d\tau < \infty. \tag{4.27}$$

The best constant c in (4.26) satisfies (4.17).

Proof. The operator adjoint of $\mathcal{J}_k$ in $L_2(\mathbb{R})$ coincides with $\mathcal{I}_k$. Hence (4.26) holds if and only if the estimate

$$\int_{\mathbb{R}} |\Phi^{-1}(t)(\mathcal{I}_k h)(t)|^2 dt \le c \int_{\mathbb{R}} |\varphi^{-1}(t)h(t)|^2 dt \tag{4.28}$$

is valid. By Proposition 4.4.2 we obtain that (4.28) holds if and only if (4.27) is fulfilled. $\square$

4.4.3 Proof of Theorem 4.4.1

By (4.11) and Fubini's theorem

$$\int_0^\infty e^{2k_- \tau}(1+\tau)^{2m_- -2}\frac{d\tau}{\Gamma^2(\tau)} < \infty.$$

Since $f \in L_2(\mathbb{R}; H_0; \Gamma)$ we obtain

$$\int_0^\infty e^{k_- \tau}(1+\tau)^{m_- -1} \| f \|_{L_2(\tau, \tau+1; H_0)} \, d\tau$$

$$\le \left(\int_0^\infty \Gamma^2(\tau) \| f \|_{L_2(\tau, \tau+1; H_0)}^2 \, d\tau \right)^{1/2}$$

$$\times \left(\int_0^\infty e^{2k_- \tau}(1+\tau)^{2m_- -2}\frac{d\tau}{\Gamma^2(\tau)} \right)^{1/2} < \infty.$$

Analogously we can verify that

$$\int_{-\infty}^0 e^{k_+ \tau}(1+|\tau|)^{m_+ -1} \| f \|_{L_2(\tau, \tau+1; H_0)} \, d\tau < \infty .$$

Therefore the function f satisfies (3.15). Due to Theorem 3.3.2, there exists a solution u of equation (2.1) corresponding to the strip $k_- < \Im \lambda < k_+$. Moreover, the estimate (3.16) holds for u.

Consider the following integral operator on $\mathbb{R}$:

$$(Sh)(t) = \int_{\mathbb{R}} \mu(\tau - t)h(\tau)d\tau ,$$

where the function μ is defined by (3.10). By (3.16) the required inequality (4.12) follows from

$$\int_{\mathbb{R}} \gamma^2(t)|Sh(t)|^2 dt \le C \int_{\mathbb{R}} \Gamma^2(t)|h(t)|^2 dt \tag{4.29}$$

by setting $h = \|f\|_{L_2(t, t+1; H_0)}$.

We decompose the operator S as the sum $S_+ + S_-$, where

$$(S_+ h)(t) = \int_{-\infty}^t e^{-k_+ (t-\tau)}(1+t-\tau)^{m_+ -1}h(\tau)d\tau$$

and
$$(S_-h)(t) = \int_t^\infty e^{-k_-(t-\tau)}(1 + \tau - t)^{m_--1} h(\tau)d\tau.$$

Using Proposition 4.4.2 we find that the inequality
$$\int_{\mathbb{R}} \gamma^2(t)|S_+h(t)|^2 dt \leq C \int_{\mathbb{R}} \Gamma^2(t)|h(t)|^2 dt \tag{4.30}$$

holds if and only if
$$\sup_{t\in\mathbb{R}} \int_{Q_+(t)} e^{2k_+(\sigma-\tau)}(1 + |\sigma - \tau|)^{2(m_++1)} \frac{\gamma^2(\tau)}{\Gamma^2(\sigma)} d\sigma d\tau < \infty.$$

Since γ, $\Gamma \in \mathbb{A}$ we can replace $1 + |\sigma - \tau|$ in the above estimate by $|\sigma - \tau|$. Therefore (4.30) follows from (4.11). Analogously, by referring to Corollary 4.4.3, we arrive at (4.30) with S_+ replaced by S_-. This implies (4.29). $\square$

4.4.4 Power-Exponential Weight Functions

We illustrate Theorem 4.4.1 with the special case of γ given by (4.9). Here, the uniqueness and existence results in Theorems 4.3.1 and 4.4.1, and the estimate (4.12) hold if the function Γ is chosen as follows:

For large positive values of t:

(i) $\Gamma(t) = \gamma(t)$ if $k_- < \alpha_0 < k_+$;
(ii) $\Gamma(t) = \gamma(t)t^{m_+}$ if $\alpha_0 = k_+$ and $\alpha_1 + m_+ < 1/2$;
(iii) $\Gamma(t) = \gamma(t)t^{m_-}$ if $\alpha_0 = k_-$ and $\alpha_1 > -1/2$.

For large negative values of t:

(iv) $\Gamma(t) = \gamma(t)$ if $k_- < \beta_0 < k_+$,
(v) $\Gamma(t) = \gamma(t)|t|^m$ if $\beta_0 = k_-$ and $\beta_1 + m_- < 1/2$,
(vi) $\Gamma(t) = \gamma(t)|t|^{m_+}$ if $\beta_0 = k_+$ and $\beta_1 > -1/2$.

This assertion follows directly from Example 4.3.2 and Theorem 4.4.1.

The special case mentioned in Sect. 4.1 can be obtained from (ii) and (vi) by setting
$$\beta = \beta_0 = \alpha_0 = k_+, \quad m = m_+, \quad \sigma^{(+)} = \alpha_1, \quad \sigma^{(-)} = \beta_1.$$

4.5 Application to the Dirichlet Problem in a Cone

Here we restrict ourselves to only one application to partial differential operators, namely, to the Dirichlet problem (3.79) in the cone K. Let us consider the weighted norm

$$\| u \|_{V_2^\ell(K;\gamma)} = \left(\int_K \gamma^2(|x|) \sum_{|\alpha|\leq\ell} |x|^{2(|\alpha|-\ell)} |D_x^\alpha u(x)|^2 dx \right)^{1/2},$$

where γ is a positive measurable function on $\mathbb{R}^1_+$ satisfying

$$c_1\gamma(r) \leq \gamma(ar) \leq c_2\gamma(r)$$

with $c_2 \geq c_1 > 0$ for all $r > 0$ and $a \in (1,2]$. The space $V^{\ell}_{2,\delta}(K)$ corresponds to $\gamma(r) = r^{\delta}$.

Let k_+ and k_- have the same meaning as in Sect. 3.9.3, i.e. these are two numbers satisfying $k_- < k_+$ such that the strip $k_- < \mathrm{Im}\lambda < k_+$ is free of eigenvalues of the pencil (1.44).

We note that the mapping (0.8) transforms the norm in $V^{\ell}_2(K;\gamma)$ to the equivalent norm

$$\left(\int_{\mathbb{R}} \gamma^2(e^{-t})e^{(2\ell-n)t} \sum_{q=0}^{\ell} \| D^{\ell-q}_t u \|^2_{W^q_2(\Omega)} \, dt \right)^{1/2}.$$

Hence, Theorem 4.3.1 gives the following uniqueness result for the problem (3.79).

Theorem 4.5.1. *Let*

$$\int_1^{\infty} r^{2k_+ +n-4m}\gamma^2(r)\frac{dr}{r} = \infty \tag{4.31}$$

$$\int_0^1 r^{2k_- +n-4m}\gamma^2(r)\frac{dr}{r} = \infty. \tag{4.32}$$

If $u \in V^{2m}_2(K;\gamma)$ satisfies (3.79) with $f = 0$ then $u = 0$.

By Theorem 4.3.3, conditions (4.31) and (4.32) are also necessary provided there exist eigenvalues of the pencil $\mathcal{A}(\lambda)$ with arbitrary large positive and negative imaginary parts.

The results of this chapter on the two-weight estimate (4.2) and the corresponding existence assertions imply similar results for the Dirichlet problem (3.79). The two-weight estimate in question takes the form

$$\| u \|_{V^{2m}_2(K;\gamma)} \leq c \| f \|_{L_2(K;\Gamma)}, \tag{4.33}$$

where

$$\| f \|_{L_2(K;\Gamma)} = \left(\int_K \Gamma^2(|x|)|f(x)|^2 dx \right)^{1/2}.$$

We restrict ourselves to the following corollary of the example in Sect. 4.4.4. Let

$$\gamma(r) = \begin{cases} r^{-\alpha_0+2m-n/2}|\log r|^{\alpha_1} & \text{for } r < \delta \\ r^{-\beta_0+2m-n/2}(\log r)^{\beta_1} & \text{for } r > \delta^{-1} \end{cases}$$

with some $\delta \in (0,1)$.

Problem (3.79) is uniquely solvable in $V^{2m}_2(K;\gamma)$ for an arbitrary $f \in L_2(K;\Gamma)$ if the weight functions γ and Γ are chosen as follows:

For $r < \delta$:

(i) $\Gamma(r) = \gamma(r)$ if $k_- < \alpha_0 < k_+$;

(ii) $\Gamma(r) = \gamma(r)|\log r|^{m_+}$ if $\alpha_0 = k_+$, $\alpha_1 + m_+ < 1/2$, where m_+ is a maximum length of Jordan chains corresponding to eigenvalues of $\mathcal{A}(\lambda)$ on the line $\Im\lambda = m_+$;

(iii) $\Gamma(r) = \gamma(r)|\log r|^{m_-}$ if $\alpha_0 = k_-$, $\alpha_1 > -1/2$, where m_- is a maximum length of Jordan chains corresponding to eigenvalues of $\mathcal{A}(\lambda)$ on the line $\Im\lambda = m_-$.

For $r > \delta^{-1}$:

(vi) $\Gamma(r) = \gamma(r)$ if $k_- < \beta < k_+$;

(v) $\Gamma(r) = \gamma(r)(\log r)^{m_-}$ if $\beta_0 = k_-$ and $\beta_1 + m_- < 1/2$;

(vi) $\Gamma(r) = \gamma(r)(\log r)^{m_+}$ if $\beta_0 = k_+$ and $\beta_1 > -1/2$.

With this choice of γ and Γ the solution satisfies (4.33). $\square$

4.6 Comments

Two weight L_2-estimates of the type (4.2) were obtained by Kozlov, Maz'ya (1985) for solutions of elliptic boundary value problems in a cone. The special case (4.3) was considered by Bagirov, Kondratiev (1991).

In this chapter we have followed Kozlov, Maz'ya (1991-1996), report 1, where the estimates from Kozlov, Maz'ya (1985) were improved by using the result of Stepanov (1989).

Part II

Differential Equations
with Variable Operator Coefficients

5. Existence, Uniqueness and "Pointwise" Estimates

5.1 Introduction

We begin to study the ordinary differential equation with variable operator coefficients

$$L(t, D_t)u = f \quad \text{on} \quad \mathbb{R},\tag{5.1}$$

where

$$L(t, D_t) = \sum_{0 \le q \le \ell} A_{\ell-q}(t)D_t^q.\tag{5.2}$$

The operator $L(t, D_t)$ is considered as a perturbation of the operator $\mathcal{A}(D_t)$ with constant coefficients studied in Part I. We characterize this perturbation by the function

$$\rho(\tau) = ||L - \mathcal{A}||_{W^\ell(\tau,\tau+1) \to L_2(\tau,\tau+1;H_0)}.\tag{5.3}$$

Throughout this chapter we denote by $k_\pm$ real numbers such that the strip $k_- < \Im\lambda < k_+$ is free of eigenvalues of the pencil $\mathcal{A}(\lambda)$. By $m_\pm$ we mean the maximum lengths of the Jordan chains corresponding to the eigenvalues of $\mathcal{A}(\lambda)$ on the lines $\Im\lambda = k_\pm$. If one of the lines $\Im\lambda = k_\pm$ does not contain eigenvalues then we set $m_\pm = 1$.

By b we denote a constant in the inequality

$$||u||_{W^\ell(t,t+1)} \le b \int_\mathbb{R} g(t - \tau)||f||_{L_2(\tau,\tau+1;H_0)}d\tau\tag{5.4}$$

(see Theorem 3.5.5). Here $u \in W_{\text{loc}}^\ell(\mathbb{R})$ is the solution of the equation $\mathcal{A}(D_t)u = f$ corresponding to the strip $k_- < \Im\lambda < k_+$ (see Sect. 3.4 and 3.5). Clearly, b depends only on the operator $\mathcal{A}(D_t)$ and the numbers k_+, k_-.

One of the main results of this chapter is a certain extension of the comparison principle for $\mathcal{A}(D_t)$ (see Theorem 3.5.5) to the operator $L(t, D_t)$. Our aim here is to show that solutions of (5.1) are majorized (modulo a constant factor) by solutions of the ordinary differential equation

$$(\mathcal{M}(\partial_t) - \omega(t))\,w(t) = ||f||_{L_2(t,t+1;H_0)}\tag{5.5}$$

on $\mathbb{R}$, where the coefficient ω is an arbitrary measurable majorant of the function $b\rho$, i.e.

$$\omega(t) \geq b\rho(t) \quad a.e. \tag{5.6}$$

The only requirement on ω that we need in order to obtain this comparison principle is the estimate

$$\sup_{t \in \mathbb{R}} \omega(t) < m_+^{m_+} m_-^{m_-} \left(\frac{k_+ - k_-}{m_+ + m_-} \right)^{m_+ + m_-}; \tag{5.7}$$

this implies the convergence of the Neumann series

$$g_\omega(t, \tau) = g(t - \tau) + \sum_{k=1}^{\infty} \int_{\mathbb{R}^k} g(t - \tau_1)\omega(\tau_1)g(\tau_1 - \tau_2)\ldots$$
$$\ldots \omega(\tau_k)g(\tau_k - \tau)d\tau_1 \ldots d\tau_k \tag{5.8}$$

for Green's function g_ω of the operator $\mathcal{M}(\partial_t) - \omega(t)$ (Kozlov, Maz'ya (1997), Ch.4).

In proving the comparison principle we assume that the right-hand side of (5.1) satisfies

$$\int_{\mathbb{R}} g_\omega(0, \tau) \|f\|_{L_2(\tau, \tau+1; H_0)} d\tau < \infty. \tag{5.9}$$

According to Kozlov, Maz'ya (1997), Sect. 5.2, condition (5.9) is necessary and sufficient for the existence of a solution to (5.5) which can be represented in the form

$$w(t) = \int_{\mathbb{R}} g_\omega(t, \tau) \|f\|_{L_2(\tau, \tau+1; H_0)} d\tau. \tag{5.10}$$

The exact statement of the comparison principle is that there exists a solution $u \in W_{\mathrm{loc}}^{\ell}(\mathbb{R})$ of (5.1) (a so-called (k_-, k_+)-solution) such that

$$\|u\|_{W^\ell(t, t+1)} \leq bw(t), \quad t \in \mathbb{R}, \tag{5.11}$$

where b is a constant as in (5.4). It is shown that this solution u satisfies

$$\|u\|_{W^\ell(t, t+1)} = o\left(\sup_{\tau \in \mathbb{R}} \frac{g_\omega(t, \tau)}{g_\omega(0, \tau)} \right) \quad \text{as} \quad t \to \pm\infty \tag{5.12}$$

(see Corollary 5.3.3).

We also obtain some uniqueness classes for solutions of (5.1). One of them is described by the relation

$$\|u\|_{W^\ell(t, t+1)} = o\left(\limsup_{\tau \to \pm\infty} \frac{g_\omega(t, \tau)}{g_\omega(0, \tau)} \right) \quad \text{as} \quad t \to \pm\infty, \tag{5.13}$$

which is similar to (5.12); it coincides with (5.12) in the case where

$$m_+ \leq 2 \quad \text{and} \quad m_- \leq 2$$

(see Theorems 5.4.2 and 5.4.5).

Furthermore, we obtain some qualitative information about the behaviour of solutions of (5.1) at infinity. In particular, the following alternative of the Phragmén-Lindelöf type is proved: Let u be a solution of $Lu = 0$ for $t > t_0$. Then either

$$\liminf_{t \to \infty} g_\omega(0, t) \|u\|_{W^\ell(t, t+1)} > 0$$

or

$$\limsup_{t \to \infty} \frac{\|u\|_{W^\ell(t, t+1)}}{g_\omega(t, 0)} < \infty.$$

The above solvability conditions and estimates of solutions are formulated in terms of g_ω. They will be used in the next chapter to obtain more explicit results under special assumptions on ω.

In Sect. 5.7, we illustrate our results with applications to partial differential equations in the cylinder and in the cone.

5.2 Auxiliary Information on the Comparison Equation

The ordinary differential equation

$$(\mathcal{M}(\partial_t) - \omega(t)) \, w(t) = h(t) \tag{5.14}$$

with ω subject to (5.7) plays an important role in the sequel. This equation was studied in detail in Kozlov, Maz'ya (1997), and here we state several results from this book which will be used in this chapter. We consider (5.14) with locally summable right-hand side h, and we call w a solution of (5.14) if $\partial_t^k w$ is locally summable for $k = 0, ..., m_+ + m_-$ and w satisfies (5.14) almost everywhere.

5.2.1 Green's Function

We here collect some auxiliary estimates for Green's function g_ω (which are obtained in Chapters 3 and 4 of Kozlov, Maz'ya (1997)).

As mentioned in the introduction, (5.7) implies the existence of Green's function $g_\omega(t, \tau)$ given by the Neumann series (5.8). From (5.8) one can easily deduce the following relations for g_ω:

$$g_\omega(t, \tau) = g(t - \tau) + \int_\mathbb{R} g(t - s)\omega(s)g_\omega(s, \tau)ds \tag{5.15}$$

and

$$g_\omega(t, \tau) = g(t - \tau) + \int_\mathbb{R} g_\omega(t, s)\omega(s)g(s - \tau)ds. \tag{5.16}$$

Let

$$\omega_0 = \sup_{t \in \mathbb{R}} \omega(t) > 0. \tag{5.17}$$

By (5.8)

$$g(t - \tau) \le g_\omega(t, \tau) \le g_{\omega_0}(t - \tau). \tag{5.18}$$

To make the upper estimate more visible, we present two-sided estimates for g_{ω_0}. First, consider roots of the polynomial $\mathcal{M}(z) - \omega_0$ with $\omega_0 > 0$. Since

$$\mathcal{M}'(z) = -(m_+ + m_-)(z + k_0)(z + k_+)^{m_+ - 1}(-z - k_-)^{m_- - 1}, \tag{5.19}$$

where

$$k_0 = \frac{m_+ k_- + m_- k_+}{m_+ + m_-}, \tag{5.20}$$

the equation $\mathcal{M}(z) = \omega_0$ has the multiple root $-k_0$ if and only if $\mathcal{M}(-k_0) = \omega_0$, or equivalently,

$$\omega_0 = m_+^{m_+} m_-^{m_-} \left(\frac{k_+ - k_-}{m_+ + m_-} \right)^{m_+ + m_-}.$$

Since

$$\omega_0 < m_+^{m_+} m_-^{m_-} \left(\frac{k_+ - k_-}{m_+ + m_-} \right)^{m_+ + m_-}, \tag{5.21}$$

all roots of $\mathcal{M}(z) = \omega_0$ are distinct and analytic with respect to ω_0.

The polynomial $\mathcal{M}(z) - \omega_0$ with $\omega_0 > 0$ has exactly two roots in the interval $(-k_+, -k_-)$. One of them lies in $(-k_0, -k_-)$ and another in $(-k_+, -k_0)$. We denote them by $-k_-(\omega_0)$ and $-k_+(\omega_0)$ respectively. They depend monotonically on ω_0 and for small ω_0 one has the following asymptotic representations:

$$k_+(\omega_0) = k_+ - (k_+ - k_-)^{-m_-/m_+} \omega_0^{1/m_+} + O(\omega_0^{2/m_+}), \tag{5.22}$$

and

$$k_-(\omega_0) = k_- + (k_+ - k_-)^{-m_+/m_-} \omega_0^{1/m_-} + O(\omega_0^{2/m_-}). \tag{5.23}$$

The remaining roots of the polynomial $\mathcal{M}(z) - \omega_0$ lie in the open circles $|z + k_+| < k_+ - k_+(\omega_0)$ and $|z + k_-| < k_-(\omega_0) - k_-$.

The Neumann series (5.8) for $\omega = \omega_0$ converges if and only if (5.21) holds. This explains the appearance of the constant in the right-hand side of the inequality (5.7). By the special case $j = 0$ and $n = 0$ in Kozlov, Maz'ya (1997), Lemma 3.3.3, we have

$$C_1 \le e^{k_+(\omega_0)t} g_{\omega_0}(t) \le C_2 \tag{5.24}$$

for $t \ge 0$, and

$$C_1 \le e^{k_-(\omega_0)t} g_{\omega_0}(t) \le C_2 \tag{5.25}$$

for $t < 0$. We note also that by (5.24) and (5.25)

$$C_1 e^{-k_\mp(\omega_0)t} \le \sup_{\tau \in \mathbb{R}} \frac{g_{\omega_0}(t - \tau)}{g_{\omega_0}(-\tau)} \le C_2 e^{-k_\mp(\omega_0)t} \tag{5.26}$$

for $t \gtrless 0$; the same lower and upper bounds are valid for

$$\limsup_{\tau \to \pm \infty} \frac{g_{\omega_0}(t - \tau)}{g_{\omega_0}(-\tau)}.$$

Estimates (5.24) and (5.25) together with (5.18) give

$$g_\omega(t, \tau) \leq \begin{cases} ce^{-k_+(\omega_0)(t-\tau)} & \text{for } t \geq \tau, \\ ce^{-k_-(\omega_0)(t-\tau)} & \text{for } t < \tau. \end{cases} \qquad (5.27)$$

This leads, in particular, to the inequality

$$g_\omega(t, \tau)g_\omega(\tau, t) \leq ce^{(k_-(\omega_0)-k_+(\omega_0))|t-\tau|},$$

this ensures the limit relation

$$g_\omega(t, \tau)g_\omega(\tau, t) \to 0 \quad \text{as } |t - \tau| \to \infty. \qquad (5.28)$$

Green's function g_ω satisfies the multiplicative estimate

$$g_\omega(t, \tau) \leq cg_\omega(t, x)g_\omega(x, \tau), \qquad (5.29)$$

for $t \leq x \leq \tau$ or $\tau \leq x \leq t$ with a positive constant c independent of t, τ, x and ω. This estimate implies, in particular, that

$$\frac{1}{g_\omega(0, t)} \leq c\frac{g_\omega(t, \tau)}{g_\omega(0, \tau)} \qquad (5.30)$$

for $0 \leq t \leq \tau$ or $\tau \leq t \leq 0$. An opposite inequality to (5.29),

$$g_\omega(t, x)g_\omega(x, \tau) \leq cg_\omega(t, \tau), \qquad (5.31)$$

is valid if x does not lie between t and τ.

We shall use the notation

$$v^{(j,k)}(t) = (\partial_t + k_+)^j(-\partial_t - k_-)^k v(t), \quad j + k \geq 0. \qquad (5.32)$$

Inequality (3.26) together with (5.15) leads to the positivity property for g_ω:

$$g_\omega^{(j,k)}(t, \tau) > 0$$

for $j = 0, ..., m_+ - 1$, $k = 0, ..., m_- - 1$.

In the sequel we will often deal with expressions involving

$$g_\omega(t_1, \tau_1)/g_\omega(t_2, \tau_2).$$

We collect some estimates for them in the following

Proposition 5.2.1. (i) *Let* $\omega \le \omega_1$ *and let* ω_1 *satisfy* (5.7). *Then*

$$\sup_{\tau \in \mathbb{R}} \frac{g_\omega^{(j,k)}(t,\tau)}{g_\omega(0,\tau)} \le \sup_{\tau \in \mathbb{R}} \frac{g_{\omega_1}^{(j,k)}(t,\tau)}{g_{\omega_1}(0,\tau)} \tag{5.33}$$

for $j \le m_+ - 1$, $k \le m_- 1$.

(ii) *For all* $t_1, t_2, \tau \in \mathbb{R}$ *and* $j \le m_+ - 1, k \le m_- - 1$, *the following inequalities hold:*

$$\frac{g_\omega(t_1,\tau)}{g_\omega(t_2,\tau)} \le \begin{cases} ce^{-k_-(t_1-t_2)} & \text{for } t_1 \ge t_2 \\ ce^{-k_+(t_1-t_2)} & \text{for } t_1 \le t_2 \end{cases} \tag{5.34}$$

and

$$\frac{g_\omega(\tau,t_1)}{g_\omega(\tau,t_2)} \le \begin{cases} ce^{-k_+(t_1-t_2)} & \text{for } t_1 \ge t_2 \\ ce^{-k_-(t_1-t_2)} & \text{for } t_1 \le t_2 \end{cases} \tag{5.35}$$

where c *does not depend on* t_1, t_2, τ *or* ω.

For estimating $g_\omega(t,\tau)$ for large $|t|$ and $|\tau|$, one only needs values of $\omega(t)$ for large $|t|$.

Lemma 5.2.2. *Let* ω *be a non-negative measurable function satisfying* (5.7). *Let* a *be a positive number.*

(i) *Let* $\omega_1(t) = \omega(t)$ *for* $t > a$ *and* $\omega_1(t) = 0$ *for* $t \le a$. *Then*

$$g_\omega(t,\tau) \le c\, g_{\omega_1}(t,\tau) \quad \text{for } t,\tau \ge a. \tag{5.36}$$

(ii) *Let* $\omega_1(t) = \omega(t)$ *for* $t < -a$ *and* $\omega_1(t) = 0$ *for* $t \ge -a$. *Then*

$$g_\omega(t,\tau) \le c\, g_{\omega_1}(t,\tau) \quad \text{for } t,\tau \le -a.$$

(iii) *Let* $\omega_2(t) = \omega(t)$ *for* $|t| > a$ *and* $v_1(t) = 0$ *for* $|t| \le a$. *Then*

$$g_\omega(t,\tau) \le c\, g_{\omega_2}(t,\tau) \quad \text{for } |t|,|\tau| \ge a.$$

In all these inequalities the constant c *depends only on* $\sup \omega(t)$, $k_\pm$ *and* $m_\pm$.

5.2.2 Existence and Uniqueness Results
for the Comparison Equation

Theorem 5.2.1 and Proposition 5.2.3 from Kozlov, Maz'ya (1997), Sect.5.2, give

Theorem 5.2.3. (*Existence*) *Let* h *be subject to*

$$\int_{\mathbb{R}} g_\omega(0,\tau)|h(\tau)|d\tau < \infty. \tag{5.37}$$

Then the equation (5.14) *has a solution* w *which can be represented as*

$$w(t) = \int_{\mathbb{R}} g_\omega(t, \tau) h(\tau) d\tau. \tag{5.38}$$

This solution satisfies

$$w^{(j,k)}(t) = o\left(\sup_{\tau \in \mathbb{R}} \frac{g_\omega^{(j,k)}(t, \tau)}{g_\omega(0, \tau)} \right) \quad \text{as } t \to \pm\infty \tag{5.39}$$

for $j = 0, ..., m_+ - 1$, $k = 0, ..., m_- 1$.

The following assertion (resulting from Theorem 5.2.3) facilitates the verification of (5.37) and (5.38) in concrete situations.

Corollary 5.2.4. *Let $\omega_1 \geq \omega$ and let (5.7) be valid for ω_1. Also let*

$$\int_{\mathbb{R}} g_{\omega_1}(0, \tau) |h(\tau)| d\tau < \infty.$$

Then the equation (5.14) has a solution w which is represented in the form (5.38) and which satisfies

$$w^{(j,k)}(t) = o\left(\sup_{\tau \in \mathbb{R}} \frac{g_{\omega_1}^{(j,k)}(t, \tau)}{g_{\omega_1}(0, \tau)} \right) \quad \text{as } t \to \pm\infty$$

for $j \leq m_+ - 1$ and $k \leq m_- - 1$.

Proof. Let ω_1 be a solution of (5.14) with ω replaced by ω_1 and h by $|h|$. By (5.8), $g_\omega \leq g_{\omega_1}$ and we have $|w| \leq w_1$. With w replaced by w_1 and ω by ω_1, the result follows by (5.39). $\square$

The following uniqueness result is a trivial consequence of Theorem 5.4.1 in Kozlov, Maz'ya (1997).

Theorem 5.2.5. *Let w be a solution of (5.14) with $h = 0$. If (5.39) is valid for $j = 0, ..., m_+ - 1$, $k = 0, ..., m_- 1$ then $w = 0$.*

Another uniqueness result to be stated here is contained in Kozlov, Maz'ya (1997), Proposition 5.4.3. It is weaker than Theorem 5.2.5 but its statement is simpler and admits a direct generalization on the infinite-dimensional case (see Theorem 5.4.2).

Theorem 5.2.6. (Uniqueness) *Let w be a solution of (5.14) with $h = 0$. If*

$$w(t) = o\left(\limsup_{\tau \to \pm\infty} \frac{g_\omega(t, \tau)}{g_\omega(0, \tau)} \right) \quad \text{as } t \to \pm\infty \tag{5.40}$$

then $w = 0$.

This assertion is monotone with respect to ω in the following sense:

Corollary 5.2.7. *Let $\omega_1 \geq \omega$ and let (5.7) be valid for ω_1. Let w be a solution of (5.14) with $h = 0$. If*

$$w(t) = o\left(\limsup_{\tau \to \pm\infty} \frac{g_{\omega_1}(t,\tau)}{g_{\omega_1}(0,\tau)} \right) \quad as \ t \to \pm\infty \qquad (5.41)$$

then $w = 0$.

Proof. Contained in Kozlov, Maz'ya (1997), Corollary 5.4.5. $\square$

Remark 5.2.8. According to Kozlov, Maz'ya (1997), Sect. 5.8-5.11, the conditions (5.40) and

$$w(t) = o\left(\sup_{\tau \in \mathbb{R}} \frac{g_\omega(t,\tau)}{g_\omega(0,\tau)} \right) \quad as \ t \to \pm\infty$$

are equivalent if $m_\pm \leq 2$. We do not know if the same is true for general $m_\pm$.

5.3 Existence

5.3.1 Assumptions on the Operator L

Here we consider the equation (5.1) with the operator L defined by (5.2). The coefficients $A_s(t)$, $0 \leq s \leq \ell$, belong to $\mathcal{L}(H_s, H_0)$ for almost all t. Furthermore, we assume that $A_s(\cdot)u(\cdot)$ is in $L_{2,\mathrm{loc}}(\mathbb{R}; H_0)$ for all $u \in W_{\mathrm{loc}}^s(\mathbb{R})$ and that the estimate

$$\|A_s(\cdot)u(\cdot)\|_{L_2(\alpha,\beta;H_0)} \leq c_{\alpha,\beta}\|u(\cdot)\|_{W^s(\alpha,\beta)}$$

holds for $\alpha, \beta \in \mathbb{R}$, $\alpha < \beta$, where

$$\|u(\cdot)\|_{W^s(\alpha,\beta)} = \left(\int_\alpha^\beta \sum_{0 \leq j \leq s} \|D^j u(t)\|_{s-j}^2 \, dt \right)^{1/2}.$$

Since the values of the function $A_s(\cdot)u(\cdot)$ on the bounded interval (α, β) are independent of its values outside (α, β), the operator $u(\cdot) \to A_s(\cdot)u(\cdot)$ is defined on the space $W^\ell(\alpha, \beta)$ and it maps $W^\ell(\alpha, \beta)$ continuously into $L_2(\alpha, \beta; H_0)$.

As we mentioned in Sect. 5.1, the difference $L - \mathcal{A}$ is characterized by the function ρ defined by (5.3). We suppose that

$$\sup_{t \in \mathbb{R}} \rho(t) < b^{-1} m_+^{m_+} m_-^{m_-} \left(\frac{k_+ - k_-}{m_+ + m_-} \right)^{m_+ + m_-}, \qquad (5.42)$$

where b is the constant in (5.4).

From our assumption on the operator functions $A_s(\cdot)$ it follows that ρ is measurable and bounded on $\mathbb{R}$. It also satisfies the estimate

$$\rho(\tau) \le c \int_{\tau-1}^{\tau+1} \rho(\sigma)d\sigma \, , \tag{5.43}$$

where c depends only on ℓ. Indeed, let u be a function in $W^\ell(\tau, \tau+1)$. We retain the notation u for the extension of u onto $(\tau-1, \tau+2)$. The extension belongs to $W^\ell(\tau-1, \tau+2)$ and satisfies the inequality

$$||u||_{W^\ell(\tau-1,\tau+2)} \le c||u||_{W^\ell(\tau,\tau+1)}$$

with c depending only on ℓ. Then

$$||(L-\mathcal{A})u||_{L_2(\tau,\tau+1;H_0)} \le \int_\tau^{\tau+1} ||(L-\mathcal{A})u||_{L_2(\sigma-1,\sigma+1;H_0)}d\sigma$$

$$\le \int_\tau^{\tau+1} (\rho(\sigma-1)+\rho(\sigma))\,||u||_{W^\ell(\sigma-1,\sigma+1)}d\sigma$$

$$\le c\int_\tau^{\tau+1} (\rho(\sigma-1)+\rho(\sigma))\,d\sigma||u||_{W^\ell(\tau,\tau+1)},$$

which is equivalent to (5.43).

5.3.2 Construction of a (k_-, k_+)-Solution

In what follows, we denote by ω a measurable majorant for $b\rho$ subject to (5.7). Due to (5.42), such a majorant exists. One can take, for example, $\omega = b\rho$. We are going to construct a certain solution of the equation (5.1) for which the right-hand side satisfying (5.9).

First, we extend the notion of a solution of (2.1) corresponding to the strip $k_- < \Im\lambda < k_+$, to the variable coefficients case. To this end, we denote by $\mathcal{A}^{-1}$ the inverse operator which associates the solution of (2.1) corresponding to the strip $k_- < \Im\lambda < k_+$ with every f satisfying

$$\int_{\mathbb{R}} g(-\tau)||f||_{L_2(\tau,\tau+1;H_0)}d\tau < \infty \tag{5.44}$$

(see Definition 3.4.1).

In equation (5.1) the role of (5.44) will be played by the stronger assumption (5.9). By (5.4), we have for the function $f_1 = (\mathcal{A}-L)\mathcal{A}^{-1}f$

$$||f_1||_{L_2(t,t+1;H_0)} \le \omega(t)\int_{\mathbb{R}} g(t-\tau)||f||_{L_2(\tau,\tau+1;H_0)}d\tau.$$

Hence, under (5.9), the function f_1 satisfies (5.44). Therefore the function $\mathcal{A}^{-1}f_1$ is well defined. We set $f_2 = [(\mathcal{A}-L)\mathcal{A}^{-1}]^2 f$.

Again using (5.4), we obtain

$$||f_2||_{L_2(t,t+1;H_0)}$$

$$\le \omega(t)\int_{\mathbb{R}^2} g(t-\tau_1)\omega(\tau_1)g(\tau_1-\tau)||f||_{L_2(\tau,\tau+1;H_0)}d\tau_1 d\tau.$$

This together with (5.9) implies (5.44) for f_2. By repeating this argument we show that $f_k = [(\mathcal{A} - L)\mathcal{A}^{-1}]^k f$ can be defined for any k, and

$$\|f_k\|_{L_2(t,t+1;H_0)} \tag{5.45}$$

$$\leq \omega(t) \int_{\mathbb{R}^k} g(t - \tau_1)\omega(\tau_1)\ldots g(\tau_{k-1} - \tau)\|f\|_{L_2(\tau,\tau+1;H_0)}d\tau_1 \ldots d\tau_{k-1}d\tau;$$

this implies (5.44) for f_k. By (5.4), the function $u_k = \mathcal{A}^{-1}f_k$ admits the estimate

$$\|u_k\|_{W^\ell(t,t+1)} \leq b \int_{\mathbb{R}^{k+1}} g(t - \tau_1)\omega(\tau_1)\ldots$$

$$\times\omega(\tau_k)g(\tau_k - \tau)\|f\|_{L_2(\tau,\tau+1;H_0)}d\tau_1 d\ldots\tau_k d\tau.$$

Thus the formal series

$$\sum_{k=0}^{\infty} \mathcal{A}^{-1}[(\mathcal{A} - L)\mathcal{A}^{-1}]^k f \tag{5.46}$$

is absolutely convergent in $W^\ell(t, t+1)$ for all $t \in \mathbb{R}$ and the limit $u \in W^\ell_{\text{loc}}(\mathbb{R})$ satisfies

$$\|u\|_{W^\ell(t,t+1)} \leq b \int_{\mathbb{R}} g_\omega(t,\tau)\|f\|_{L_2(\tau,\tau+1;H_0)}d\tau. \tag{5.47}$$

Clearly, u is a solution of (5.1). In the sequel we shall use

Definition 5.3.1. Let f satisfy (5.9). The function u defined by (5.46) is called the solution of (5.1) corresponding to the strip $k_- < \Im\lambda < k_+$, or, in short, (k_-, k_+)–solution of (5.1).

In the case when $\omega = 0$, this definition is in agreement with that given in Sect.3.4.

Thus we have arrived at the following result, which generalizes the comparison principle for solutions of (2.1) (see Theorem 3.5.5).

Theorem 5.3.2. *Let ω and f satisfy (5.7) and (5.9) respectively. Then the (k_-, k_+)–solution of (5.1) belongs to $W^\ell_{\text{loc}}(\mathbb{R})$ and satisfies (5.47) or, equivalently,*

$$\|u\|_{W^\ell(t,t+1)} \leq bw(t), \ t \in \mathbb{R}, \tag{5.48}$$

where w is the solution of (5.5) defined by (5.10).

By (5.39) with $j = k = 0$, and (5.48) we have

Corollary 5.3.3. *Let the condition (5.7) be fulfilled. Then the (k_-, k_+)–solution of (5.1) satisfies*

$$\|u\|_{W^\ell(t,t+1)} = o\left(\sup_{\tau \in \mathbb{R}} \frac{g_\omega(t,\tau)}{g_\omega(0,\tau)}\right) \quad as \quad t \to \pm\infty. \tag{5.49}$$

In the first theorem of Sect.5.5 we show that a similar relation defines a uniqueness class of solutions to the equation (5.1).

One can take ω_0 as a majorant for $b\rho$. Therefore, the estimates (5.26) combined with Theorem 5.3.2 and Corollary 5.3.3 give the following explicit result.

Corollary 5.3.4. *Let ω_0 be defined by (5.17) and satisfy (5.21). Suppose that*

$$\int_0^\infty e^{k_-(\omega_0)\tau}\|f\|_{L_2(\tau,\tau+1;H_0)}d\tau$$
$$+ \int_{-\infty}^0 e^{k_+(\omega_0)\tau}\|f\|_{L_2(\tau,\tau+1;H_0)}d\tau < \infty.$$

Then the (k_-,k_+)-solution of (5.1) satisfies

$$\|u\|_{W^\ell(t,t+1)} \le C\Big\{ \int_t^\infty e^{-k_-(\omega_0)(t-\tau)}\|f\|_{L_2(\tau,\tau+1;H_0)}d\tau$$
$$+ \int_{-\infty}^t e^{-k_+(\omega_0)(t-\tau)}\|f\|_{L_2(\tau,\tau+1;H_0)}d\tau\Big\}.$$

Moreover,
$$\|u\|_{W^\ell(t,t+1)} = o\big(e^{-k_\mp(\omega_0)t}\big) \quad \text{as } t \to \pm\infty. \tag{5.50}$$

5.4 Uniqueness Theorems

5.4.1 A Class of Uniqueness

We assume throughout that a function ω majorizing $b\rho$ is subject to (5.7). For the next uniqueness theorem we need the following auxiliary fact from Kozlov, Maz'ya (1997), Sect. 5.4.

Lemma 5.4.1. *Let ω satisfy (5.7). Let h be a non-negative, locally integrable function on $\mathbb{R}$ such that*

$$h(t) = o\left(\limsup_{\tau\to\pm\infty} \frac{g_\omega(t,\tau)}{g_\omega(0,\tau)}\right) \quad \text{as} \quad t \to \pm\infty. \tag{5.51}$$

(i) *Then the function*

$$v(t) = \int_{\mathbb{R}} g(t-\tau)\omega(\tau)h(\tau)d\tau \tag{5.52}$$

is locally bounded and

$$v^{(j,k)}(t) = o\left(\sup_{\tau\in\mathbb{R}} \frac{g_\omega^{(j,k)}(t,\tau)}{g_\omega(0,\tau)}\right) \quad \text{as} \quad t \to \pm\infty, \tag{5.53}$$

where $j = 0, \cdots, m_+ - 1$, $k = 0, 1, \cdots, m_- - 1$ and $v^{(j,k)}(t)$ is given by (5.32).
(ii) If, in addition,

$$h(t) \leq \int_{\mathbb{R}} g(t - \tau)\omega(\tau)h(\tau)d\tau, \tag{5.54}$$

then $h(t) = 0$.

The following theorem gives a condition of uniqueness which is similar to the relation (5.49) satisfied by the (k_-, k_+)-solution constructed in Theorem 5.3.2.

Theorem 5.4.2. *Let the condition (5.7) for ω be fulfilled and let $u \in W^\ell_{\text{loc}}(\mathbb{R})$ be a solution of the equation*

$$L(t, D_t)u = 0 \quad on \quad \mathbb{R}. \tag{5.55}$$

Suppose that

$$\|u\|_{W^\ell(t,t+1)} = o\left(\limsup_{\tau \to \pm\infty} \frac{g_\omega(t, \tau)}{g_\omega(0, \tau)}\right) \quad as\ t \to \pm\infty. \tag{5.56}$$

Then $u = 0$.

Proof. Put

$$f(t) = (\mathcal{A}(D_t) - L(t, D_t))\,u(t).$$

Then

$$\mathcal{A}(D_t)u(t) = f(t) \quad on \quad \mathbb{R}. \tag{5.57}$$

Using the inequality (5.34) for $t_1 = t$ and $t_2 = 0$ we get

$$\|u\|_{W^\ell(t,t+1)} = o\left(e^{-k_\mp t}\right) \quad as \quad t \to \pm\infty.$$

Since

$$\|f\|_{L_2(t,t+1;H_0)} \leq \rho(t)\|u\|_{W^\ell(t,t+1)},$$

applying the first part of Lemma 5.4.1 to $h(t) = \|u\|_{W^\ell(t,t+1)}$ we conclude that f satisfies (5.9). Applying Theorem 3.5.5 to (5.57) we arrive at (5.54). Reference to Lemma 5.4.1 completes the proof. $\square$

5.4.2 Another Class of Uniqueness

Due to (5.30), one can obtain another uniqueness result by replacing the right-hand side of (5.56) in Theorem 5.4.2 by $o(1/g_\omega(0, t))$. In the next proposition we give a stronger version of this result. It should be noted that Proposition 3.2.1 is used in the proof of the result that we obtain.

Proposition 5.4.3. *Let (5.7) be satisfied and let $u \in W^\ell_{\text{loc}}(\mathbb{R})$ be a solution of (5.55) subject to the condition*

$$\liminf_{t \to \pm\infty} g_\omega(0, t)\|u\|_{W^\ell(t,t+1)} = 0. \tag{5.58}$$

Then $u = 0$.

Proof. Due to (5.58), there exist positive numbers $T_1^{(+)}, T_2^{(+)}, \ldots$ and negative numbers $T_1^{(-)}, T_2^{(-)}, \ldots$ such that

a) $T_k^{(+)} \to +\infty$ and

$$g_\omega(0, T_k^{(+)}) \|u\|_{W^\ell(T_k^{(+)}, T_k^{(+)}+1)} \to 0 \quad \text{as} \quad k \to \infty;$$

b) $T_k^{(-)} \to -\infty$ and

$$g_\omega(0, T_k^{(-)}) \|u\|_{W^\ell(T_k^{(-)}, T_k^{(-)}+1)} \to 0 \quad \text{as} \quad k \to \infty.$$

Let k be a sufficiently large positive integer. Let η_k be a smooth function such that $\eta_k(t) = 1$ for $t \in [T_k^{(-)} + 1, T_k^{(+)}]$, $\eta_k(t) = 0$ for $t \le T_k^{(-)}$ and for $t \ge T_k^{(+)} + 1$. Suppose that

$$|D_t^j \eta_k| \le c \quad \text{on} \quad \mathbb{R} \quad \text{for} \quad j = 0, 1, \ldots, \ell, \tag{5.59}$$

where c does not depend on k and j.

We have

$$L(t, D_t)(\eta_k u) = [L(t, D_t), \eta_k]u = F \quad \text{on} \quad \mathbb{R}.$$

Since the function F has a compact support we can apply Theorems 5.3.2 to the equation $LU = F$. Hence this equation has a solution U satisfying

$$\|U\|_{W^\ell(t, t+1)} \le b \int_\mathbb{R} g_\omega(t, \tau) \|F\|_{L_2(\tau, \tau+1; H_0)} d\tau. \tag{5.60}$$

Since the functions $\eta_k u$ and U are subject to (5.56), we obtain $U = \eta_k u$ by Theorem 5.4.2 . By (5.60) and (5.35) we get

$$\|u\|_{W^\ell(t, t+1)} \le c\{g_\omega(t, T_k^{(-)}) \|F\|_{L_2(T_k^{(-)}, T_k^{(-)}+1; H_0)}$$
$$+ g_\omega(t, T_k^{(+)}) \|F\|_{L_2(T_k^{(+)}, T_k^{(+)}+1; H_0)}\}$$

for $t \in [T_k^{(-)} + 1, T_k^{(+)} - 1]$, where c does not depend on t, k and F. Using (5.34) we have

$$\|u\|_{W^\ell(t, t+1)} \le c \max\{e^{-k_- t}, \ e^{-k_+ t}\}$$
$$\times \left(g_\omega(0, T_k^{(-)}) \|u\|_{W^\ell(T_k^{(-)}, T_k^{(-)}+1)} + g_\omega(0, T_k^{(+)}) \|u\|_{W^\ell(T_k^{(+)}, T_k^{(+)}+1)} \right).$$

By taking the limit as k tends to ∞ we get $u(t) = 0$ for every t. $\square$

Remark 5.4.4. The combination of Theorem 5.3.2 and either Theorem 5.4.2 or Proposition 5.4.3 shows that under the condition (5.9), a solution of (5.1) satisfying either (5.56) or (5.58) coincides with the (k_-, k_+)-solution of (5.1). $\square$

5.4.3 The Case $m_{\pm} \leq 2$

We recall that the (k_-, k_+)-solution satisfies (5.49). By Remark 5.2.8 the relations (5.49) and (5.56) are equivalent in the case $m_+ \leq 2$ and $m_- \leq 2$. Thus the combination of Theorems 5.3.2, 5.4.2 and Corollary 5.3.3 gives

Theorem 5.4.5. *Let $m_+ \leq 2$ and $m_+ \leq 2$ and let ω satisfy (5.7). Suppose that $f \in L_2(\mathbb{R}; H_0)$ satisfies (5.9). Then the equation (5.1) has the unique solution $u \in W_{\mathrm{loc}}^l(\mathbb{R})$ subject to (5.49).*

5.4.4 An Explicit Uniqueness Condition in Terms of ω_0

Let ω_0 be defined by (5.17) and satisfy (5.21). In the case $\omega = \omega_0$, Theorem 5.4.2 leads to the uniqueness class (5.50). This follows from (5.24) and (5.25).

5.5 Behaviour of Zeros at Infinity

5.5.1 Zeros of L

The information obtained in this section will be used in Chapter 8. It is essentially a consequence of Theorems 5.4.2, 5.3.2 and Proposition 5.4.3. The first result is analogous to Proposition 3.2.4.

Proposition 5.5.1. *Let (5.7) be satisfied and let $u \in W_{\mathrm{loc}}^{\ell}(\mathbb{R})$ be a nontrivial solution of (5.55). Then the following assertions hold:*
   (i) *if*

$$\|u\|_{W^{\ell}(t,t+1)} = o\left(\limsup_{\tau \to -\infty} \frac{g_{\omega}(t,\tau)}{g_{\omega}(0,\tau)}\right) \quad as \quad t \to -\infty \tag{5.61}$$

then

$$\|u\|_{W^{\ell}(t,t+1)} \leq c g_{\omega}(t,0) \quad for \ t < 0 \tag{5.62}$$

and

$$\limsup_{t \to +\infty}\left(\liminf_{\tau \to +\infty} \frac{g_{\omega}(0,\tau)}{g_{\omega}(t,\tau)}\|u\|_{W^{\ell}(t,t+1)}\right) > 0. \tag{5.63}$$

   (ii) *Assertion* (i) *remains valid if one replaces $\pm\infty$ and $t < 0$ by $\mp\infty$ and $t > 0$.*

Proof. We prove only (i), since (ii) can be verified quite similarly. If the left-hand side in (5.63) is equal to zero then $u = 0$ by Theorem 5.4.2. Now turn to (5.62). Let η be a smooth function such that $\eta(t) = 1$ for $t < 0$ and $\eta(t) = 0$ for $t > 1$. We have

$$L(t, D_t)\eta u = [L(t, D_t), \eta]u = F \quad on \quad \mathbb{R},$$

where F has a compact support. Using Theorems 5.4.2 and 5.3.2 we obtain

$$||u||_{W^\ell(t,t+1)} \le c \int_{-1}^{1} g_\omega(t,\tau)||F||_{L_2(\tau,\tau+1;H_0)} d\tau.$$

Due to (5.35) we arrive at (5.62). $\square$

By referring to Proposition 5.4.3 instead of Theorem 5.4.2 and repeating the above proof verbatim we arrive at

Proposition 5.5.2. *Let* (5.7) *be satisfied and let* $u \in W^\ell_{\text{loc}}(\mathbb{R})$ *be a non-trivial solution of* (5.55). *Then the following assertions hold:*

(i) *If*

$$\liminf_{t \to -\infty} g_\omega(0,t)||u||_{W^\ell(t,t+1)} = 0 \tag{5.64}$$

then (5.62) *holds together with*

$$\liminf_{t \to +\infty} g_\omega(0,t)||u||_{W^\ell(t,t+1)} > 0. \tag{5.65}$$

(ii) *Assertion* (i) *remains valid if one replaces* $\pm\infty$ *and* $\mathbb{R}_-$ *by* $\mp\infty$ *and* $\mathbb{R}_+$.

5.5.2 Zeros of the Adjoint Operator

We say that $\Psi \in L_{2,\text{loc}}(\mathbb{R}; H_0)$ belongs to coker$L$ if

$$\int_{\mathbb{R}} (L(t,D_t)u(t), \Psi(t))_{H_0} dt = 0 \tag{5.66}$$

for all $u \in W^\ell(\mathbb{R})$ with compact support.

We prove an analog of Proposition 5.4.3.

Proposition 5.5.3. *Let* (5.7) *be valid and let* $\Psi \in$ coker$L$. *If*

$$\liminf_{t \to \pm\infty} g_\omega(t,0)||\Psi||_{L_2(t,t+1;H_0)} = 0 \tag{5.67}$$

then $\Psi = 0$.

Proof. By (5.67) there exist two sequences $\{T_k^{(\pm)}\}_{k \ge 1}$ such that $T_k^{(\pm)} \to \pm\infty$ as $k \to \infty$ and

$$g_\omega(T_k^{(\pm)},0)||\Psi||_{L_2(T_k^{(\pm)},T_k^{(\pm)}+1;H_0)} \to 0 \quad \text{as} \quad k \to \infty. \tag{5.68}$$

Let k be a sufficiently large integer and let η_k be a smooth function such that $\eta_k(t) = 1$ for $t \in [T_k^{(-)}+1, T_k^{(+)}]$, $\eta_k(t) = 0$ for $t \le T_k^{(-)}$ and $t > T_k^{(+)}+1$. Suppose that (5.59) holds with c independent of k and j.

Let u be the solution of (5.1) with $f(t) = \chi_T(t)\Psi(t)$ (here χ_T is the characteristic function of the interval $(T, T+1)$). Then, by Theorem 5.3.2,

$$||u||_{W^\ell(t,t+1)} \le g_\omega(t,T)||\Psi||_{L_2(T,T+1;H_0)}. \tag{5.69}$$

Let $T_k^{(-)} + 1 < T < T_k^{(+)} - 1$. By (5.66) we have

$$0 = \int_{\mathbb{R}} (L\eta_k u, \Psi)_{H_0} \, dt = \|\Psi\|_{L_2(T,T+1;H_0)}^2 + \int_{\mathbb{R}} ([L,\eta_k]u, \Psi) \, dt.$$

Therefore, using (5.69) and (5.34) we get

$$\|\Psi\|_{L_2(T,T+1;H_0)} \leq c\Big(g(T_k^{(-)},T)\|\Psi\|_{L_2(T_k^{(-)},T_k^{(-)}+1;H_0)}$$
$$+ g(T_k^{(+)},T)\|\Psi\|_{L_2(T_k^{(+)},T_k^{(+)}+1;H_0)}\Big).$$

This together with (5.68) implies $\Psi = 0$. $\square$

The following assertion is similar to Proposition 5.5.2.

Proposition 5.5.4. *Let* (5.7) *be valid and let* $\Psi \in \operatorname{coker} L$, $\Psi \neq 0$.
  (i) *If*

$$\liminf_{t \to -\infty} g_\omega(t,0)\|\Psi\|_{L_2(t,t+1;H_0)} = 0 \tag{5.70}$$

then

$$\|\Psi\|_{L_2(t,t+1;H_0)} \leq c g_\omega(0,t) \quad \text{for} \quad t \in \mathbb{R}_- \tag{5.71}$$

and

$$\liminf_{t \to +\infty} g_\omega(t,0)\|\Psi\|_{L_2(t,t+1;H_0)} > 0 . \tag{5.72}$$

  (ii) *The assertion* (i) *remains valid if one replaces* $\pm\infty$ *and* $\mathbb{R}_-$ *by* $\mp\infty$ *and* $\mathbb{R}_+$.

Proof. If the left-hand side of (5.72) is equal to 0 then $\Psi = 0$ by Proposition 5.5.3. Let us prove (5.71). We denote by χ_T the characteristic function of $(T, T+1)$, $T < 0$.
   By (5.70) there exists a sequence $\{T_k\}_{k>1}$ such that $T_k \to -\infty$ as $k \to \infty$ and

$$g_\omega(T_k,0)\|\Psi\|_{L_2(T_k,T_k+1;H_0)} \to 0 \quad \text{as} \quad k \to \infty. \tag{5.73}$$

We take a smooth function η_k such that $\eta_k(t) = 0$ for $t > 2$ and $t < T_k$, $\eta_k(t) = 1$ for $t \in (T_k + 1, 1)$ and

$$|\partial_t^j \eta_k(t)| \leq c \quad \text{for} \quad j = 0, 1, \dots, \ell,$$

where c does not depend on k.
   Denote by u the solution of (5.1) with $f(t) = \chi_T(t)\Psi(t)$. Then by (5.66) we get

$$0 = \int_{\mathbb{R}} (L\eta_k u, \Psi)_{H_0} \, dt = \|\Psi\|_{L_2(T,T+1;H_0)}^2$$
$$+ \int_{\mathbb{R}} ([L,\eta_k]u, \Psi)_{H_0} \, dt.$$

Hence

$$\|\Psi\|^2_{L_2(T;T+1;H_0)} \leq c\Big(g(0,T)\|\Psi\|_{L_2(1,2;H_0)}$$

$$+g(T_k,T)\|\Psi\|_{L_2(T_k,T_k+1;H_0)}\Big). \qquad (5.74)$$

Since $g(T_k,T) \leq cg(T_k,0)g(0,T)$ (see (5.29)), (5.70) and the estimate (5.74) imply (5.71). $\square$

5.6 Estimates for Solutions on the Semiaxis $t > t_0$

The following theorem describes the behaviour of solutions of the equation

$$L(t,D_t)u(t) = f(t) \quad \text{for} \quad t > t_0 , \qquad (5.75)$$

where t_0 is a positive constant and $f \in L_{2,\text{loc}}(t_0,\infty;H_0)$. Since we are going to apply our previous results it will be convenient to assume additionally that

$$L(t,D_t) = \mathcal{A}(D_t) \quad \text{for large negative } t.$$

Furthermore, we extend f by zero to the left of t_0. Clearly, $f \in L_{2,\text{loc}}(\mathbb{R};H_0)$ and $\rho(t) = 0$ for large negative t. As before, we denote by ω an arbitrary measurable majorant of $b\rho$.

Theorem 5.6.1. *Let (5.7) be valid and let $u \in W^\ell_{\text{loc}}(t_0,\infty)$ be a solution of (5.75) where*

$$\int_{t_0}^\infty g_\omega(0,\tau)\|f\|_{L_2(\tau,\tau+1;H_0)}d\tau < \infty. \qquad (5.76)$$

If at least one of the conditions

$$\|u\|_{W^\ell(t,t+1)} = o\left(\limsup_{\tau\to+\infty} \frac{g_\omega(t,\tau)}{g_\omega(0,\tau)}\right) \quad \text{as} \quad t \to +\infty \qquad (5.77)$$

or

$$\liminf_{t\to+\infty} g_\omega(0,t)\|u\|_{W^\ell(t,t+1)} = 0 \qquad (5.78)$$

is valid then for all $t > t_0 + 1$

$$\|u\|_{W^\ell(t,t+1)} \leq c\{\int_{t_0}^\infty g_\omega(t,\tau)\|f\|_{L_2(\tau,\tau+1;H_0)}d\tau$$

$$+g_\omega(t,t_0)\|u\|_{W^{\ell-1}(t_0,t_0+1)}\}. \qquad (5.79)$$

The constant c is independent of t_0, u and f.

Proof. Let $\zeta \in C^\infty(\mathbb{R})$, $\zeta(t) = 0$ for $t < t_0 + 1/2$ and $\zeta(t) = 1$ for $t > t_0 + 1$. Then $L(t, D_t)(\zeta u) = F$, where $F = \zeta f + [L(t, D_t), \zeta]u$. By Remark 5.4.4 the function $v = \zeta u$ is the (k_-, k_+)-solution of $Lv = F$ on $\mathbb{R}$. Using Theorem 5.3.2, we obtain

$$||\zeta u||_{W^\ell(t,t+1)} \leq b \int_{t_0-1/2}^\infty g_\omega(t,\tau)||F||_{L_2(\tau,\tau+1;H_0)}d\tau .$$

Due to (5.35) the last integral is dominated by the right-hand side in (5.79). $\square$

The following assertions of the Phragmén-Lindelöf type is a direct consequence of the previous theorem.

Corollary 5.6.2. *Let ω satisfy (5.7) and let $u \in W^\ell_{\text{loc}}(t_0, \infty)$ be a solution of (5.75) where*

$$\int_{t_0}^\infty \sup_{t>t_0} \frac{g_\omega(t,\tau)}{g_\omega(t,0)}||f||_{L_2(\tau,\tau+1;H_0)}d\tau < \infty. \tag{5.80}$$

If

$$\liminf_{t\to+\infty} g_\omega(0,t)||u||_{W^\ell(t,t+1)} = 0 \tag{5.81}$$

then

$$\limsup_{t\to+\infty} \frac{||u||_{W^\ell(t,t+1)}}{g_\omega(t,0)} < \infty. \tag{5.82}$$

Corollary 5.6.3. *Let ω satisfy (5.7) and let $u \in W^\ell_{\text{loc}}(t_0, \infty)$ be a solution of (5.75) with f subject to (5.80). If*

$$||u||_{W^\ell(t,t+1)} = o\left(\limsup_{\tau\to\infty} \frac{g_\omega(t,\tau)}{g_\omega(0,\tau)}\right)$$

as $t \to +\infty$ then the inequality (5.82) holds.

Remark 5.6.4. Theorem 5.3.2 gives the existence of a solution of the equation (5.75) (with f subject to (5.76) and $f(t) = 0$ for $t < t_0$) such that the estimate

$$||u||_{W^\ell(t,t+1)} \leq C \int_{t_0-1}^\infty g_\omega(t,\tau)||f||_{L_2(\tau,\tau+1;H_0)}d\tau \tag{5.83}$$

is valid.

5.7 Applications to Partial Differential Equations with Variable Coefficients

5.7.1 The Dirichlet Problem in a Cylinder

Here we use the same notation as in Sections 2.5 and 3.9.1. Consider the Dirichlet problem

$$
\begin{cases}
Q(x, t, D_x, D_t)u = f & \text{on } \mathcal{C} \\
\partial_\nu^j u = 0 & \text{on } \partial\mathcal{C}, \ j = 0, \dots, m-1,
\end{cases}
\tag{5.84}
$$

where

$$
Q(x, t, D_x, D_t) = \sum_{q=0}^{\ell} B_{\ell-q}(x, t, D_x) D_t^q
$$

maps $\dot{W}_{2,\mathrm{loc}}^{2m,\ell}(\mathcal{C})$ into $L_{2,\mathrm{loc}}(\mathcal{C})$. We set

$$
\rho(t) = \| Q - P \|_{\dot{W}_2^{2m,\ell}(\mathcal{C}_t) \to L_2(\mathcal{C}_t)},
\tag{5.85}
$$

where $\dot{W}_2^{2m,\ell}(\mathcal{C}_t)$ is the subspace of $W_2^{2m,\ell}(\mathcal{C}_t)$ containing the functions with zero Dirichlet conditions on $\partial\Omega \times (t, t+1)$.

In the present case, the comparison equation (5.5) can be written as

$$
(\mathcal{M}(\partial_t) - \omega(t)) w(t) = \| f \|_{L_2(\mathcal{C}_t)},
$$

where ω is a measurable function satisfying

$$
\omega(t) \geq b(P)\rho(t).
\tag{5.86}
$$

By $b(P)$ we mean a constant in the inequality (3.74). Let (5.7) be fulfilled. We thus require that the function (5.85) does not exceed a sufficiently small constant depending on P, Ω and $k_\pm$. A sufficient condition for this is the smallness in $L_\infty(\mathcal{C})$ of the coefficients of the operators $B_{\ell-q}(x, t, D_x) - A_{\ell-q}(x, D_x)$. This smallness of the coefficients can also be understood in the sense of multipliers in Sobolev spaces; this is sometimes a necessary condition (see Maz'ya, Shaposhnikova (1985)). Consider, for example, the Schrödinger operator

$$
Q(x, t, D_x, D_t) = D_t^2 - \Delta_x + p(x, t)
$$

as a perturbation of the Laplace operator

$$
P(x, D_x, D_t) = D_t^2 - \Delta_x.
$$

Clearly, $\rho(t)$ is equal to the norm of the multiplication operator by p : $\dot{W}_2^{2,2}(\mathcal{C}_t) \to L_2(\mathcal{C}_t)$. There are different possibilities for giving lower and upper estimates for this norm in analytical terms (Maz'ya, Shaposhnikova (1985)). We shall not dwell upon this, but will restrict ourselves to the simple remark that for $n = 1$

$$c_1\rho(t) \leq \int_t^{t+1} \int_\Omega |d(x)p(x,\tau)|^2 dx d\tau \leq c_2\rho(t),$$

where $d(x) = \mathrm{dist}(x, \partial\Omega)$.

With the notation just introduced, one can reformulate all the results of this chapter for the problem (5.84). For example, the result on the unique solvability of (5.84) following from Theorems 5.3.2 and 5.4.2 is

Theorem 5.7.1. *Let $f \in L_{2,\mathrm{loc}}(\mathcal{C})$ and*

$$\int_{\mathbb{R}} g_\omega(0,\tau) \, \| \, f \, \|_{L_2(\mathcal{C}_\tau)} \, d\tau < \infty,$$

where g_ω is Green's function of the operator $\mathcal{M}(\partial_t) - \omega(t)$ (see Sect. 5.1). Then there exists a solution $u \in \dot{W}_{2,\mathrm{loc}}^{2m,\ell}(\mathcal{C})$ of (5.84) such that

$$\| \, u \, \|_{W_2^{2m,\ell}(\mathcal{C}_t)} \leq b(P) \int_{\mathbb{R}} g_\omega(t,\tau) \, \| \, f \, \|_{L_2(\mathcal{C}_\tau)} \, d\tau.$$

This solution satisfies

$$\| \, u \, \|_{W_2^{2m,\ell}(\mathcal{C}_t)} = o\left(\sup_\tau \frac{g_\omega(t,\tau)}{g_\omega(0,\tau)} \right) \quad as \quad t \to \pm\infty. \tag{5.87}$$

The following similar, but more stringent condition, describes the class of uniqueness for the problem (5.84):

$$\| \, u \, \|_{W_2^{2m,\ell}(\mathcal{C}_t)} = o\left(\limsup_{\tau \to \pm\infty} \frac{g_\omega(t,\tau)}{g_\omega(0,\tau)} \right) \quad as \quad t \to \pm\infty. \tag{5.88}$$

In conclusion, we note that (5.87) and (5.88) are the same in the case $m_\pm \leq 2$ (see Theorem 5.4.5).

5.7.2 The Dirichlet Problem in a Cone

Let us consider the Dirichlet problem in the cone K

$$\begin{cases} Q(x, D_x)u = f & \text{on } K \\ \partial_\nu^j u = 0 & \text{on } \partial K \backslash \{0\}, \; j = 0, \ldots, m-1, \end{cases} \tag{5.89}$$

where

$$Q(x, D_x) = \sum_{|\alpha| \leq 2m} q_\alpha(x) D_x^\alpha.$$

We shall treat Q as a perturbation of the operator

$$P(D_x) = \sum_{|\alpha| = 2m} p_\alpha D_x^\alpha, \; p_\alpha = \text{const},$$

from Sect. 3.9.3. To this end, we introduce the function

$$\rho(\log r^{-1}) = \| Q - P \|_{\mathcal{W}_2^{2m}(K_r) \to L_2(K_r)},$$

where $\mathcal{W}_2^{2m}(K_r)$ is the subspace of $W_2^{2m}(K_r)$ (see Sect. 3.9.3) containing the functions with zero Dirichlet data on $\partial\Omega \times (e^{-1}r, r)$.

The comparison equation takes the form

$$(\mathcal{M}(\partial_t) - \omega(t)) w(t) = e^{(n/2-2m)t} \| f \|_{L_2(K_{e^{-t}})},$$

with measurable ω subject to (5.86) and $b(P)$ a constant in (3.74). We suppose that ω satisfies (5.7). This can be achieved, for example, by the requirements

$$\sum_{|\alpha|=2m} |g_\alpha(x) - p_\alpha| + \sum_{|\alpha|<2m} |x|^{2m-|\alpha|} |q_\alpha(x)| \leq c_0, \qquad (5.90)$$

where c_0 is a sufficiently small constant depending on P, K and $k_\pm$.

An analog of Theorem 5.7.1 (which follows from Theorem 5.3.2 and 5.4.2) can be formulated as

Theorem 5.7.2. *Let $f \in L_{2,\mathrm{loc}}(K)$ and*

$$\int_0^\infty g_\omega(0, \log s^{-1}) s^{2m-n/2} \| f \|_{L_2(K_s)} \frac{ds}{s} < \infty,$$

where, as before, g_ω is Green's function of the operator $\mathcal{M}(\partial_t) - \omega(t)$. Then there exists a solution $u \in V_{2,\mathrm{loc}}^{2m}(K)$ of (5.89) such that

$$\| u \|_{W_2^{2m}(K_r)}$$
$$\leq b(P) \int_0^\infty g_\omega(\log r^{-1}, \log s^{-1}) s^{2m-n/2} \| f \|_{L_2(K_s)} \frac{ds}{s}. \qquad (5.91)$$

This solution satisfies

$$\| u \|_{W_2^{2m}(K_r)} = o\left(\sup_\tau \frac{g_\omega(\log r^{-1}, \tau)}{g_\omega(0, \tau)} \right) \qquad \text{as } r \to 0 \text{ or } r \to \infty. \qquad (5.92)$$

A class of uniqueness for the problem (5.89) is given by

$$\| u \|_{W_2^{2m}(K_r)} = o\left(\limsup_{\tau \to \pm\infty} \frac{g_\omega(\log r^{-1}, \tau)}{g_\omega(0, \tau)} \right) \qquad \text{as } r \to 0 \text{ or } r \to \infty. \qquad (5.93)$$

By Theorem 5.4.5, conditions (5.92) and (5.93) are equivalent if $m_\pm \leq 2$.

As a final application, we note that Corollary 5.6.3 gives the following variant of the Phragmén-Lindelöf principle for the Dirichlet problem in K.

Corollary 5.7.3. *Let $u \in V_{2,\text{loc}}^{2m}(K)$ be a solution of (5.89) with f subject to*

$$\int_0^{r_0} \sup_{t \leq \log r_0^{-1}} \frac{g_\omega(t, \log s^{-1})}{g_\omega(t, 0)} s^{2m - n/2} \, \| f \|_{L_2(K_s)} \, \frac{ds}{s} < \infty.$$

If

$$\| u \|_{\mathcal{W}_2^{2m}(K_r)} = o \left(\limsup_{\tau \to +\infty} \frac{g_\omega(\log r^{-1}, \tau)}{g_\omega(0, \tau)} \right) \quad as \quad r \to 0$$

then the inequality

$$\limsup_{r \to 0} \frac{\| u \|_{\mathcal{W}_2^{2m}(K_r)}}{g_\omega(\log r^{-1}, 0)} < \infty$$

is valid.

6. Corollaries of Previous Results Under Special Assumptions on $L(t, D_t)$

6.1 Introduction

We devote this chapter to applications of existence and uniqueness results and to estimates on the solutions of (5.1) (as obtained in Ch.5 under rather general conditions (5.7)). We consider here several special cases where the results of Ch.5 can be made more explicit. We shall distinguish these cases by imposing different requirements on the function ρ defined by (5.3). The choice of material is dictated by the estimates for Green's function of the ordinary differential equation (5.5) given in Kozlov, Maz'ya (1997), Ch.6.

In Sect.6.2 we suppose that either the global condition

$$\sup_{\mathbb{R}} \rho(t) < b^{-1} m_+^{m_+} m_-^{m_-} \left(\frac{k_+ - k_-}{m_+ + m_-} \right)^{m_+ + m_-} \tag{6.1}$$

or its variant "at infinity"

$$\limsup_{t \to +\infty} \rho(t) < b^{-1} m_+^{m_+} m_-^{m_-} \left(\frac{k_+ - k_-}{m_+ + m_-} \right)^{m_+ + m_-} \tag{6.2}$$

is valid. We recall that b is the best constant in the inequality (5.4), for definitions of $k_\pm, m_\pm$ see Sect.5.1.

In the next section we suppose that $m_+ = m_- = 1$. In other words, the eigenvalues of the pencil $\mathcal{A}(\lambda)$ which lie on the lines $\Im \lambda = k_\pm$ have no generalized eigenvectors. (The case when one of these lines is free of the spectrum is also included).

We then turn to the case $b\rho(t) \leq \omega(t)$, where ω is a positive absolutely continuous function satisfying $\omega(t) \to 0$ and either $\left(\omega(t)^{-1/m_-} \right)' \to 0$ or $\left(\omega(t)^{-1/m_+} \right)' \to 0$ as $t \to +\infty$. This assumption on ω includes the case of power perturbations. For example, the following Phragmén-Lindelöf type theorem is contained in Theorem 6.4.3: For large positive t, let

$$\|L - \mathcal{A}\|_{W^\ell(t, t+1) \to L_2(t, t+1; H_0)} \leq \text{const } t^{-\alpha}$$

with $\alpha < m_+$. Then an arbitrary solution of (5.75) with $f = 0$ for large t, such that

$$\| u \|_{W^\ell(t, t+1)} = o\left(e^{-(k_- + \delta)t} \right) \quad \text{as } t \to \infty$$

(for some $\delta > 0$) satisfies

$$\| u \|_{W^{\ell}(t,t+1)} = O\left(\exp(-k_+ t + ct^{1-\alpha/m_+}) \right).$$

Here the exponent $1 - \alpha/m_+$ is best possible.

We conclude the chapter with several applications to partial differential equations.

6.2 General Perturbations

6.2.1 Existence and Uniqueness Theorems

Assume that (6.1) is valid. Denote

$$\rho_{\pm} = \limsup_{t \to \pm\infty} \rho(t)$$

and let ε be a positive number satisfying

$$b\rho_{\pm} + \varepsilon < m_+^{m_+} m_-^{m_-} \left(\frac{k_+ - k_-}{m_+ + m_-} \right)^{m_+ + m_-}.$$

One can choose a majorant ω of $b\rho$ such that (5.7) holds and

$$\omega = b\rho_{\pm} + \varepsilon$$

for large positive and negative t respectively. By Proposition 6.2.1 in Kozlov, Maz'ya (1997), Green's function g_{ω} admits the estimates

$$g_{\omega}(t, \tau) \leq \begin{cases} ce^{-k_+(b\rho_+ + \varepsilon)(t-\tau)} & \text{for} \quad t \geq \tau \\ ce^{-k_-(b\rho_+ + \varepsilon)(t-\tau)} & \text{for} \quad t \leq \tau \end{cases} \tag{6.3}$$

for positive t and τ, and

$$g_{\omega}(t, \tau) \leq \begin{cases} ce^{-k_+(b\rho_- + \varepsilon)(t-\tau)} & \text{for} \quad t \geq \tau \\ ce^{-k_-(b\rho_- + \varepsilon)(t-\tau)} & \text{for} \quad t \leq \tau \end{cases} \tag{6.4}$$

for negative t and τ.

We state an existence result which is similar to Corollary 5.3.4 but takes into consideration the different behaviour of $\rho(t)$ at $+\infty$ and $-\infty$.

Theorem 6.2.1. (Existence) *If*

$$\int_0^{\infty} e^{k_-(b\rho_+ + \varepsilon)\tau} \|f\|_{L_2(\tau, \tau+1; H_0)} d\tau$$

$$+ \int_{-\infty}^0 e^{k_+(b\rho_- + \varepsilon)\tau} \|f\|_{L_2(\tau, \tau+1; H_0)} d\tau < \infty \tag{6.5}$$

then there exists a solution $u \in W_{loc}^{\ell}(\mathbb{R})$ of (5.1) (the (k_-, k_+)-solution defined in Sect.5.3) satisfying

$$\|u\|_{W^{\ell}(t,t+1)} \leq c\{ \int_{0<\tau<t} e^{-k_+(b\rho_+ +\varepsilon)(t-\tau)} \|f\|_{L_2(\tau,\tau+1;H_0)} d\tau$$

$$+ \int_{t<\tau} e^{-k_-(b\rho_+ +\varepsilon)(t-\tau)} \|f\|_{L_2(\tau,\tau+1;H_0)} d\tau$$

$$+ e^{-k_+(b\rho_+ +\varepsilon)t} \int_{\mathbb{R}_-} e^{k_+(b\rho_- +\varepsilon)\tau} \|f\|_{L_2(\tau,\tau+1;H_0)} d\tau \} \qquad (6.6)$$

for $t \geq 0$. Replacing

$$0 < \tau < t, \ t < \tau, \ and \ \mathbb{R}_- \quad by \quad 0 > \tau > t, \ t > \tau \ and \ \mathbb{R}_+$$

respectively, and $k_\pm, \rho_\pm$ by $k_\mp, \rho_\mp$ we obtain a similar estimate in the case $t \leq 0$.

Proof. We restrict ourselves to the case $t \geq 0$. The required estimates for $g_\omega(t, \tau)$ with $\tau \geq 0$ are contained in (6.3). If $\tau < 0$ then by (5.29) combined with (6.3) and (6.4) we obtain

$$g_\omega(t, \tau) \leq ce^{-k_+(b\rho_+ +\varepsilon)t + k_+(b\rho_- +\varepsilon)\tau}.$$

It remains to use Theorem 5.3.2. $\square$

It is a simple exercise to verify the following result (cf. Corollary 5.3.3).

Corollary 6.2.2. *The solution u in Theorem 6.2.1 satisfies*

$$\|u\|_{W^{\ell}(t,t+1)} = o\left(e^{k_\mp(b\rho_\pm +\varepsilon)t}\right) \quad as \quad t \to \pm\infty . \qquad (6.7)$$

We also present a class of uniqueness described in terms of $k_\mp(b\rho_\pm + \varepsilon)$.

Theorem 6.2.3. (Uniqueness) *Let $u \in W_{loc}^{\ell}(\mathbb{R})$ and let*

$$\liminf_{t \to \pm\infty} e^{-k_\mp(b\rho_\pm +\varepsilon)t} \|u\|_{W^{\ell}(t,t+1)} = 0.$$

If $Lu = 0$ on $\mathbb{R}$ then $u = 0$.

Proof. The result follows from Proposition 5.4.3 together with the estimates (6.3) and (6.4). $\square$

6.2.2 Estimates for Solutions at Infinity

Consider the equation

$$L(t, D_t)u = f \quad \text{for} \quad t > t_0, \tag{6.8}$$

where $f \in L_{2,\text{loc}}(t_0, \infty; H_0)$. Let ρ satisfy (6.2). Assuming that t_0 is sufficiently large and changing, if necessary, $L(t, D_t)$ for $t < t_0$ we prove the
validity of the global condition (6.1).

By R we denote a positive constant subject to

$$\limsup_{t \to +\infty} \rho(t) < R < b^{-1} m_+^{m_+} m_-^{m_-} \left(\frac{k_+ - k_-}{m_+ + m_-} \right)^{m_+ + m_-}. \tag{6.9}$$

Theorem 6.2.4. *Let $u \in W_{\text{loc}}^\ell(t_0, \infty)$ be a solution of (6.8), where*

$$\int_{t_0}^{\infty} e^{k_- (bR)\tau} ||f||_{L_2(\tau, \tau+1; H_0)} d\tau < \infty. \tag{6.10}$$

If

$$\liminf_{t \to \infty} e^{k_- (bR)t} ||u||_{W^\ell(t, t+1)} = 0 \tag{6.11}$$

then for all $t > t_0 + 1$

$$||u||_{W^\ell(t, t+1)} \leq c \Big\{ \int_{t_0}^{t} e^{-k_+ (bR)(t-\tau)} ||f||_{L_2(\tau, \tau+1; H_0)} d\tau$$

$$+ \int_{t}^{\infty} e^{-k_- (bR)(t-\tau)} ||f||_{L_2(\tau, \tau+1; H_0)} d\tau$$

$$+ e^{-k_+ (bR)(t-t_0)} ||u||_{W^{\ell-1}(t_0, t_0+1)} \Big\}. \tag{6.12}$$

Proof. We can choose a majorant ω for $b\rho$ in such a way that estimate (5.7)
holds and

$$\omega(t) = \begin{cases} bR & \text{for large positive } t, \\ 0 & \text{for large negative } t. \end{cases}$$

Then the result follows from Theorem 5.6.1 combined with (6.3) and (6.4),
where $b\rho_+ + \varepsilon$, $b\rho_- + \varepsilon$ are replaced by bR and 0. $\square$

Corollary 6.2.5. *(The Phragmén-Lindelöf principle) Let $u \in W_{\text{loc}}^\ell(t_0, \infty)$
be a solution of (6.8), where*

$$\int_{t_0}^{\infty} e^{k_+ (bR)\tau} ||f||_{L_2(\tau, \tau+1; H_0)} d\tau < \infty.$$

Then either

$$\liminf_{t \to +\infty} e^{k_- (bR)t} ||u||_{W^\ell(t, t+1)} > 0$$

or

$$\limsup_{t \to +\infty} e^{k_+ (bR)t} ||u||_{W^\ell(t, t+1)} < \infty.$$

Proof. The result follows from Corollary 5.6.2. □

Remark 6.2.6. Under the assumption that R is small (which is frequently met in applications), we can simplify the statements of Theorem 6.2.4 and Corollary 6.2.5 by replacing $k_+(bR)$ and $k_-(bR)$ by their asymptotics (5.22) and (5.23).

Remark 6.2.7. Let $\rho \in L_1(t_0, \infty)$. Then for any $\varepsilon > 0$ there exists $t_0 = t_0(\varepsilon)$ such that

$$\int_{t_0-1}^{\infty} \rho(t)dt < \varepsilon.$$

Due to (5.43) this implies

$$\rho(t) \leq c\varepsilon \quad \text{on} \quad [t_0, \infty),$$

where c depends only on ℓ. Hence Theorem 6.2.4, Corollary 6.2.5 and Remark 6.2.6 can be applied with $R = c\,\varepsilon$.

Remark 6.2.8. Remark 5.6.4 together with the estimates (6.3) and (6.4) gives the following existence result for the equation (6.8). Let f be subject to (6.10). Then the equation (6.8) has a solution $u \in W_{loc}^{\ell}(t_0, \infty)$ satisfying

$$||u||_{W^{\ell}(t,t+1)} \leq c\{\int_{t_0-1}^{t} e^{-k_+(bR)(t-\tau)}||f||_{L_2(\tau,\tau+1;H_0)}d\tau$$

$$+ \int_{t}^{\infty} e^{-k_-(bR)(t-\tau)}||f||_{L_2(\tau,\tau+1;H_0)}d\tau \qquad (6.13)$$

for $t > t_0$. Here f is extended by zero for $t < t_0$.

6.3 The Case $m_+ = m_- = 1$

6.3.1 Equation (5.1) on $\mathbb{R}$

In this section we consider the important special case when $m_+ = m_- = 1$. The auxiliary ordinary differential equation (5.5) takes the simpler form

$$(\partial_t + k_+)(-\partial_t - k_-)w(t) - w(t)w(t) = ||f||_{L_2(t,t+1;H_0)}, \qquad (6.14)$$

where, as usual, w is a majorant of $b\rho$; we can therefore obtain a more explicit information on solutions of the operator equation (5.1). We shall use the following estimate for g_w from Kozlov, Maz'ya (1997), Sect.6.4.

Proposition 6.3.1. *Let* $m_+ = m_- = 1$ *and let* ω *be non-decreasing for* $t < 0$ *and non-increasing for* $t > 0$. *Also let*

$$\omega(\pm 0) < (k_+ - k_-)^2/4, \tag{6.15}$$

then

$$g_\omega(t, \tau) \le ce^{-\int_\tau^t k_\pm(\omega(s))ds} \quad \text{for} \quad t \gtrless \tau, \tag{6.16}$$

where the functions $k_\pm$ *introduced in Sect. 5.2.1 take the form*

$$k_\pm(x) = \frac{k_+ + k_-}{2} \pm \left(\frac{(k_+ - k_-)^2}{4} - x \right)^{1/2}. \tag{6.17}$$

Remark 6.3.2. The minimal majorant ω for $b\rho$ satisfying the monotonicity conditions in Proposition 6.3.1 is given by

$$\omega(t) = \begin{cases} b\sup_{\tau \ge t} \rho(\tau) & \text{for } t \ge 0 \\ b\sup_{\tau < t} \rho(\tau) & \text{for } t < 0 . \end{cases} \tag{6.18}$$

First we state an existence theorem which follows directly from Theorems 5.3.2 and Proposition 6.3.1.

Theorem 6.3.3. *Let* $m_+ = m_- = 1$ *and let* $\omega(t)$ *be non-decreasing for* $t < 0$, *non-increasing for* $t > 0$ *and satisfy* (6.15). *Suppose* $f \in L_{2,\text{loc}}(\mathbb{R}; H_0)$ *and*

$$\int_0^\infty e^{\int_0^\tau k_-(\omega(s))ds}||f||_{L_2(\tau,\tau+1;H_0)}d\tau < \infty, \tag{6.19}$$

$$\int_{-\infty}^0 e^{\int_0^\tau k_+(\omega(s))}||f||_{L_2(\tau,\tau+1;H_0)}d\tau < \infty. \tag{6.20}$$

Then the equation (5.1) *has a solution* $u \in W_{\text{loc}}^\ell(\mathbb{R})$ *(the* (k_-, k_+)*-solution), which satisfies*

$$||u||_{W^\ell(t,t+1)} \le c\{ \int_{-\infty}^t e^{-\int_\tau^t k_+(\omega(s))ds}||f||_{L_2(\tau,\tau+1;H_0)}d\tau$$
$$+ \int_t^\infty e^{-\int_\tau^t k_-(\omega(s))ds}||f||_{L_2(\tau,\tau+1;H_0)}d\tau \}. \tag{6.21}$$

Corollary 6.3.4. *Let all conditions of Theorem 6.3.3 be fulfilled. Then the* (k_-, k_+)*-solution satisfies*

$$||u||_{W^\ell(t,t+1)} = \begin{cases} o\left(e^{-\int_0^t k_-(\omega(s))ds} \right) & \text{for } t \to +\infty \\ o\left(e^{-\int_0^t k_+(\omega(s))ds} \right) & \text{for } t \to -\infty. \end{cases} \tag{6.22}$$

Proof. We restrict ourselves to the case $t \to +\infty$. Obviously, the integral over $(t, +\infty)$ in (6.21) satisfies (6.22) as $t \to +\infty$. Let N be an arbitrary positive number. We decompose the integral over $(-\infty, t)$ in (6.21) into the sum of two integrals over $(-\infty, N)$ and (N, t). Clearly,

$$\int_{-\infty}^{N} \dots d\tau \le c_N e^{-\int_0^t k_+(\omega(s))ds}$$

and

$$\int_{N}^{t} \dots d\tau \le e^{-\int_0^t k_-(\omega(s))ds} \int_{N}^{t} e^{\int_0^\tau k_-(\omega(s))ds} \|f\|_{L_2(\tau,\tau+1;H_0)} d\tau.$$

The result follows by the arbitrary choice of N. $\square$

Replacing (6.19) and (6.20) by the more restrictive conditions

$$\int_0^\infty e^{\int_0^\tau k_+(\omega(s))ds} \|f\|_{L_2(\tau,\tau+1;H_0)} d\tau < \infty, \tag{6.23}$$

$$\int_{-\infty}^0 e^{\int_0^\tau k_-(\omega(s))ds} \|f\|_{L_2(\tau,\tau+1;H_0)} d\tau < \infty, \tag{6.24}$$

and using (6.21) we can improve (6.22).

Corollary 6.3.5. *Let all conditions of* Theorem 6.3.3 *be fulfilled except (6.19) and (6.20) which are replaced by (6.23) and (6.24). Then the* $(k_-, k_+)-$*solution satisfies*

$$\|u\|_{W^\ell(t,t+1)} \le \begin{cases} ce^{-\int_0^t k_+(\omega(s))ds} & \text{for } t \ge 0 \\ ce^{-\int_0^t k_-(\omega(s))ds} & \text{for } t \le 0. \end{cases}$$

Proof. Similar to that of Corollary 6.3.4, and therefore omitted. $\square$

We turn to the uniqueness of solutions.

Theorem 6.3.6. *Let* $m_+ = m_- = 1$ *and let* ω *satisfy the conditions of* Theorem 6.3.3. *Also let* $u \in W^\ell_{\text{loc}}(\mathbb{R})$ *be a solution of* $L(t, D_t)u = 0$ *on* $\mathbb{R}$. *If*

$$\liminf_{t \to +\infty} e^{\int_0^t k_-(\omega(s))ds} \|u\|_{W^\ell(t,t+1)} = 0 \tag{6.25}$$

and

$$\liminf_{t \to -\infty} e^{\int_0^t k_+(\omega(s))ds} \|u\|_{W^\ell(t,t+1)} = 0 \tag{6.26}$$

then $u = 0$.

Proof. By Proposition 6.3.1 the conditions (6.25) and (6.26) imply (5.58). It remains to refer to Proposition 5.4.3. $\square$

6.3.2 Equation (6.8) on a Semiaxis

Let ρ satisfy the condition (6.2) which now takes the form

$$\limsup_{t \to +\infty} \rho(t) < \frac{(k_+ - k_-)^2}{4b}.$$

Here we are interested in the behaviour of solutions of (6.8) for large positive t. Therefore we assume that t_0 is sufficiently large. Moreover, we can change the coefficients of $L(t, D_t)$ for $t < t_0$ in such a way that the global condition (6.1), which takes the form

$$\sup_{\mathbb{R}} \rho(t) < \frac{(k_+ - k_-)^2}{4b},$$

holds.

Now we give an estimate for solutions of the equation (6.8).

Theorem 6.3.7. *Let $m_+ = m_- = 1$ and let ω be a non-increasing majorant for $b\rho$ on (t_0, ∞) such that*

$$\omega(t_0 + 0) < (k_+ - k_-)^2/4.$$

Also let $u \in W^\ell_{\mathrm{loc}}(t_0, \infty)$ be a solution of (6.8), where

$$\int_{t_0}^{\infty} e^{\int_{t_0}^{\tau} k_-(\omega(s))ds} ||f||_{L_2(\tau, \tau+1; H_0)} d\tau < \infty. \tag{6.27}$$

If

$$\liminf_{t \to +\infty} e^{\int_{t_0}^{t} k_-(\omega(s))ds} ||u||_{W^\ell(t, t+1)} = 0$$

then for all $t > t_0 + 1$

$$||u||_{W^\ell(t,t+1)} \le c \Big\{ \int_{t_0}^{t} e^{-\int_{\tau}^{t} k_+(\omega(s))ds} ||f||_{L_2(\tau, \tau+1; H_0)} d\tau$$

$$+ \int_{t}^{\infty} e^{-\int_{\tau}^{t} k_-(\omega(s))ds} ||f||_{L_2(\tau, \tau+1; H_0)} d\tau$$

$$+ e^{-\int_{t_0}^{t} k_+(\omega(s))ds} ||u||_{W^{\ell-1}(t_0, t_0+1)} \Big\}. \tag{6.28}$$

Proof. We can extend $\omega(t)$ for $t < t_0$ in order to satisfy the conditions of Proposition 6.3.1. Then the result is obtained by referring to Theorem 5.6.1 and Proposition 6.3.1. $\square$

The following variant of the Phragmén-Lindelöf principle is a consequence of Corollary 5.6.3.

Corollary 6.3.8. *Let $m_+ = m_- = 1$ and let ω be non-increasing for large positive t. Suppose that*

$$\omega(+\infty) < (k_+ - k_-)^2/4.$$

If $u \in W^\ell_{\mathrm{loc}}(\mathbb{R})$ is a solution of (6.8), where

$$\int_{t_0}^\infty e^{\int_{t_0}^\tau k_+(\omega(s))ds} \|f\|_{L_2(\tau,\tau+1;H_0)} d\tau < \infty,$$

then either

$$\liminf_{t\to+\infty} e^{\int_{t_0}^t k_-(\omega(s))ds} \|u\|_{W^\ell(t,t+1)} > 0 \tag{6.29}$$

or

$$\limsup_{t\to+\infty} e^{\int_{t_0}^t k_+(\omega(s))ds} \|u\|_{W^\ell(t,t+1)} < \infty. \tag{6.30}$$

Remark 6.3.9. From Remark 5.6.4 and the estimate (6.16) for Green's function, we have the following existence result: Let $f \in L_{2,\mathrm{loc}}(t_0,\infty;H_0)$ be subject to (6.27). Then the equation (6.8) has a solution $u \in W^\ell_{\mathrm{loc}}(t_0,\infty)$ satisfying

$$\|u\|_{W^\ell(t,t+1)} \leq c\Big\{ \int_{t_0-1}^t e^{-\int_\tau^t k_+(\omega(s))ds} \|f\|_{L_2(\tau,\tau+1;H_0)} d\tau$$
$$+ \int_t^\infty e^{-\int_\tau^t k_-(\omega(s))ds} \|f\|_{L_2(\tau,\tau+1;H_0)} d\tau \Big\} \tag{6.31}$$

for $t > t_0$. (As before, f is extended by zero to the left of t_0.)

6.4 Estimates of the Phragmén-Lindelöf Type for Solutions of (6.8) when ρ Dominates Either t^{-m_+} or t^{-m_-} for Large $t > 0$

Let, as in Sect. 5.6, ω be a majorant of $b\rho$ subject to (5.7). Let us assume that ω is absolutely continuous positive function on $[t_0, +\infty)$ subject to the condition

$$\omega(t) \to 0 \quad \text{as} \quad t \to +\infty. \tag{6.32}$$

Furthermore, we suppose that $\omega(t) = 0$ for $t < 0$. Here we give corollaries of results obtained in Sect. 5.6. We need the following theorem from Kozlov, Maz'ya (1997), Theorem 6.9.8.

Theorem 6.4.1. *Suppose that*

$$\lim_{t \to +\infty} \omega'(t)\omega(t)^{-(m_+ +1)/m_+} = 0. \tag{6.33}$$

(The function $\omega(t) = ct^{-\alpha}$, $0 < \alpha < m_+$, satisfies this condition.) Then for $t \geq \tau \geq t_0$ the following estimates hold:

$$g_\omega(t, \tau) \leq c(1 + t - \tau)^{m_+ - 1} e^{-k_+(t-\tau)} \tag{6.34}$$

when $\tau \leq t \leq \tau + \omega(\tau)^{-1/m_+}$ and

$$g_\omega(t, \tau) \tag{6.35}$$
$$\leq c\left(\omega(t)\omega(\tau)\right)^{(1-m_+)/2m_+} \exp\left\{-k_+(t-\tau) + \gamma \int_\tau^t \omega(x)^{1/m_+} dx\right\}$$

when $t \geq \tau + \omega(\tau)^{-1/m_+}$. Here γ is a constant $> (k_+ - k_-)^{-m_-/m_+}$.

From Theorem 6.4.1 applied to the adjoint operator $M(-\partial_t) - \omega$, we obtain estimates for $g_\omega(t, \tau)$ with $\tau \geq t \geq t_0$.

Theorem 6.4.2. *Suppose that*

$$\lim_{t \to +\infty} \omega'(t)\omega(t)^{-(m_- +1)/m_-} = 0 \tag{6.36}$$

(for example, $\omega(t) = ct^{-\alpha}$, $0 < \alpha < m_-$). Then for $\tau \geq t \geq t_0$ the following estimates hold:

$$g_\omega(t, \tau) \leq c(1 + \tau - t)^{m_- - 1} e^{-k_-(t-\tau)} \tag{6.37}$$

when $t \leq \tau \leq t + \omega(t)^{-1/m_-}$ and

$$g_\omega(t, \tau) \tag{6.38}$$
$$\leq c\left(\omega(t)\omega(\tau)\right)^{(1-m_-)/2m_-} \exp\left\{-k_-(t-\tau) + \gamma \int_t^\tau \omega(x)^{1/m_-} dx\right\}$$

when $\tau \geq t + \omega(t)^{-1/m_-}$. Here γ is a constant $> (k_+ - k_-)^{-m_+/m_-}$.

The following result is a variant of the Phragmén-Lindelöf principle.

Theorem 6.4.3. *Let ω satisfy (6.33) and let*

$$\int_{t_0}^\infty \frac{\exp(k_+ \tau)}{\mathcal{E}_+(\tau; \delta_1)} \|f\|_{L_2(\tau, \tau+1; H_0)} d\tau < \infty \tag{6.39}$$

for some $\delta_1 > (k_+ - k_-)^{-m_-/m_+}$, where

$$\mathcal{E}_+(t; \delta) = \exp\{\delta \int_{t_0}^t \omega(x)^{1/m_+} dx\}. \tag{6.40}$$

Also let $u \in W_{loc}^\ell(t_0, \infty)$ be a solution of (6.8) satisfying

$$\|u\|_{W^\ell(t,t+1)} = o\left(e^{-(k_- +\delta)t}\right) \quad \text{as} \quad t \to \infty \tag{6.41}$$

with some $\delta > 0$. Then

$$\|u\|_{W^\ell(t,t+1)} = O\left(\exp(-k_+ t)\mathcal{E}_+(t;\delta_2)\right) \tag{6.42}$$

with an arbitrary $\delta_2 > \delta_1$.

Proof. By (6.34) and (6.35)

$$g_\omega(t,\tau) \le c_\gamma(\omega(t)\omega(\tau))^{(1-m_+)/2m_+}e^{-k_+(t-\tau)}\frac{\mathcal{E}_+(t;\gamma)}{\mathcal{E}_+(\tau;\gamma)} \tag{6.43}$$

for $t \ge \tau \ge t_0$, where $\gamma > (k_+ - k_-)^{-m_-/m_+}$. By (6.32) it follows from (6.3) and from (5.34) and (5.35) that

$$g_\omega(t,\tau) \le c_\delta e^{-(k_- +\delta)(t-\tau)} \quad \text{for} \quad \tau \ge t \ge t_0 \tag{6.44}$$

with an arbitrary positive δ. Applying Theorem 5.6.1 and using the estimates (6.43) and (6.44) we get

$$\|u\|_{W^\ell(t,t+1)} \le c_\gamma\bigg\{ \int_{t_0}^t (\omega(t)\omega(\tau))^{(1-m_+)/2m_+} e^{-k_+(t-\tau)}$$

$$\times\frac{\mathcal{E}_+(t;\gamma)}{\mathcal{E}_+(\tau;\gamma)}\|f\|_{L_2(\tau,\tau+1;H_0)}d\tau + \int_t^\infty e^{-(k_- +\delta)(t-\tau)}\|f\|_{L_2(\tau,\tau+1;H_0)}d\tau$$

$$+\omega(t)^{(1-m_+)/2m_+}e^{-k_+ t}\mathcal{E}_+(t;\gamma)\|u\|_{W^{\ell-1}(t_0,t_0+1)}\bigg\}.$$

By observing that

$$\omega(t)^{-N} \le c_{N,\varepsilon} \exp\{\varepsilon \int_{t_0}^t \omega(x)^{1/m_+}dx\}$$

with $\varepsilon > 0$, we obtain the result by easy estimates. $\square$

The following assertion can be proved in exactly the same way.

Theorem 6.4.4. Let ω satisfy (6.36) and let

$$\int_{t_0}^\infty e^{(k_+ -\delta_1)\tau}\|f\|_{L_2(\tau,\tau+1;H_0)}d\tau < \infty$$

for some $\delta_1 > 0$. Also let $u \in W^\ell_{loc}(t_0,\infty)$ be a solution of (6.8) satisfying

$$\|u\|_{W^\ell(t,t+1)} = o\left(\frac{\exp(-k_- t)}{\mathcal{E}_-(t;\delta)}\right), \tag{6.45}$$

with some $\delta > (k_+ - k_-)^{-m_+/m_-}$, where

$$\mathcal{E}_-(t; \delta) = \exp\{\delta \int_{t_0}^t \omega(x)^{1/m_-} dx\}. \tag{6.46}$$

Then

$$\|u\|_{W^\ell(t, t+1)} = o\left(e^{-(k_+ - \delta_2)t}\right)$$

for an arbitrary $\delta_2 > \delta_1$.

The following corollary is a direct consequence of the two theorems just proved. Here we assume that both conditions (6.33) and (6.36) are fulfilled.

Corollary 6.4.5. *Let*

$$\omega'(t) = o\left(\omega(t)^{(m_0+1)/m_0}\right) \quad as \quad t \to \infty, \tag{6.47}$$

where $m_0 = \max\{m_-, m_+\}$ and let (6.39) be valid with some $\delta_1 > (k_+ - k_-)^{-m_-/m_+}$. If the solution $u \in W^\ell_{loc}(t_0, \infty)$ of (6.8) satisfies (6.45) with some $\delta > (k_+ - k_-)^{-m_+/m_-}$ then (6.42) holds.

6.5 Applications to Partial Differential Equations in a Cylinder

We illustrate the results of the present chapter with the example of the Dirichlet problem (5.84) in the cylinder $\mathcal{C}$ (this has already been treated in Sections 2.5, 3.9.1 and 5.7.1). We will use the notation introduced in these sections without further explanation.

We start with a theorem on the unique solvability valid under assumption of smallness of $\rho(t)$ and for arbitrary $m_\pm$.

Theorem 6.5.1. *Let $\rho(t) \leq \varepsilon$, where $\varepsilon \leq \varepsilon_0$ with sufficiently small ε_0 depending on P, Ω and $k_\pm$. Then there exists a positive constant $C = C(P, \Omega, k_\pm)$ such that the following assertions are valid:*

(i) If $f \in L_{2,loc}(\mathcal{C})$ and

$$\int_0^\infty e^{(k_- + C\varepsilon^{1/m_-})\tau} \| f \|_{L_2(\mathcal{C}_\tau)} d\tau$$

$$+ \int_{-\infty}^0 e^{(k_+ - C\varepsilon^{1/m_+})\tau} \| f \|_{L_2(\mathcal{C}_\tau)} d\tau < \infty$$

then there exists a solution $u \in W^{2m,\ell}_{2,loc}(\mathcal{C})$ of (5.84) satisfying

$$\| u \|_{W_2^{2m,\ell}(\mathcal{C}_t)} \leq C\Big\{ \int_{-\infty}^t e^{-(k_+ - C\varepsilon^{1/m_+})(t-\tau)} \| f \|_{L_2(\mathcal{C}_\tau)} d\tau$$

$$+ \int_t^\infty e^{-(k_- + C\varepsilon^{1/m_-})(t-\tau)} \| f \|_{L_2(\mathcal{C}_\tau)} d\tau \Big\}.$$

The following relation is valid for this solution:

$$\| u \|_{W_2^{2m,\ell}(C_t)} = o\left(e^{-(k_\mp \pm C\varepsilon^{1/m_\mp})t}\right) \qquad as \quad t \to \pm\infty. \qquad (6.48)$$

(ii) *Let* $u \in \dot{W}_{2,\mathrm{loc}}^{2m,\ell}(C)$ *be a solution of* $Q(x, D_x, D_t)u = 0$ *subject to* (6.48). *Then* $u = 0$.

This theorem is a direct corollary of Theorems 6.2.1, 6.2.3 and the estimates

$$k_+(b\varepsilon) \geq k_+ - C\varepsilon^{1/m_+}, \quad k_-(b\varepsilon) \leq k_- + C\varepsilon^{1/m_-}.$$

We shall use a measurable majorant ω for $b(P)\rho$ (where $b(P)$ is the constant in (3.74)) and assume that ω is subject to (5.7). In the case $m_+ = m_- = 1$ considered in Sect. 6.3 Theorem 6.5.1 can be improved as follows.

Theorem 6.5.2. *Let* $m_+ = m_- = 1$ *and let* $\omega(t)$ *be non-decreasing on* $\mathbb{R}_-$ *and non-increasing on* $\mathbb{R}_+$. *Then there exists a positive constant* $C = C(P, \Omega, k_\pm)$ *such that the following assertions are valid:*

(i) *If* $f \in L_{2,\mathrm{loc}}(C)$ *and*

$$\int_0^\infty e^{k_- \tau + C \int_0^\tau \omega(s)ds} \| f \|_{L_2(C_\tau)} \, d\tau$$

$$+ \int_{-\infty}^0 e^{k_+ \tau - C \int_0^\tau \omega(s)ds} \| f \|_{L_2(C_\tau)} \, d\tau < \infty$$

then there exists a solution $u \in \dot{W}_{2,\mathrm{loc}}^{2m,\ell}(C)$ *of* (5.84) *satisfying*

$$\| u \|_{W_2^{2m,\ell}(C_t)} \leq \left\{ \int_{-\infty}^t e^{-k_+(t-\tau)+C \int_\tau^t \omega(s)ds} \| f \|_{L_2(C_\tau)} \, d\tau \right.$$

$$\left. + \int_t^\infty e^{-k_-(t-\tau)-C \int_\tau^t \omega(s)ds} \| f \|_{L_2(C_\tau)} \, d\tau \right\}.$$

The following relation is valid for this solution:

$$\| u \|_{W_2^{2m,\ell}(C_t)} = o\left(e^{-k_\mp t \mp C \int_0^t \omega(s)ds}\right) \qquad as \quad t \to \pm\infty. \qquad (6.49)$$

(ii) *Let* $u \in \dot{W}_{2,\mathrm{loc}}^{2m,\ell}(C)$ *be a solution of* $Q(x, D_x, D_t)u = 0$ *subject to* (6.49). *Then* $u = 0$.

Proof. Follows directly from Theorems 6.3.3, 6.3.6 and from the estimates

$$k_+(\omega(s)) \geq k_+ - C\omega(s), \quad k_-(\omega(s)) \leq k_- + C\omega(s).$$

□

Theorem 6.5.3. *Let $\omega(t) = 0$ for $t < 0$ and let ω be absolutely continuous and positive on $[t_0, \infty)$. Suppose that $\omega(t) = o(1)$ as $t \to +\infty$ and ω satisfies (6.47) with $m_0 = \max(m_+, m_-)$. Furthermore, let*

$$\int_{t_0}^{\infty} e^{k_+ \tau - C \int_{t_0}^{\tau} \omega(x)^{1/m_+} dx} \| f \|_{L_2(C_\tau)} \, d\tau < \infty,$$

where C is a positive constant depending on P, Ω and $k_\pm$. If a solution $u \in W_{2,\mathrm{loc}}^{2m,\ell}(C)$ of (5.84) satisfies

$$\| u \|_{W_2^{2m,\ell}(C_t)} = O \left(e^{-k_- t - C \int_{t_0}^{t} \omega(\tau)^{1/m_-} d\tau} \right) \qquad as \ t \to +\infty$$

then

$$\| u \|_{W_2^{2m,\ell}(C_t)} = O \left(e^{-k_+ t + C_1 \int_{t_0}^{t} \omega(\tau)^{1/m_+} d\tau} \right) \qquad as \ t \to +\infty$$

with an arbitrary constant $C_1 > C$.

This result is a special case of Corollary 6.4.5.

6.6 Other Applications

Analogous results to those obtained in Sect. 6.5 are valid for solutions of the Dirichlet problem in the cone K which we considered in Sections 3.9.3, 5.7.2. We will not state these assertions here, but mention only that they can be obtained from corresponding statements of Sect. 6.5 by changing variables $t = \log r^{-1}$, $\tau = \log s^{-1}$ and by replacing the seminorms

$$\| u \|_{W_2^{2m,\ell}(C_t)} \qquad \text{and} \qquad \| f \|_{L_2(C_t)}$$

with

$$\| u \|_{W_2^{2m}(K_r)} \qquad \text{and} \qquad r^{2m-n/2} \| f \|_{L_2(K_r)} .$$

We give two more examples showing the usefulness of the results of this chapter.

6.6.1 Isolated Singularities

Here we discuss the behaviour of solutions to elliptic equations in a neighbourhood of an interior point of the domain. Consider an elliptic operator

$$Q(x, D_x) = \sum_{|\alpha| \leq 2m} q_\alpha(x) D_x^\alpha \tag{6.50}$$

in $B_{r_0} \backslash \{O\}$, where $B_{r_0} = \{x \in \mathbb{R}^n : 0 < |x| < r_0\}$.

We shall use the same notation as in Sections 3.9.3 and 5.7.2, except that K will denote $\mathbb{R}^n \backslash \{O\}$ and K_r will stand for the spherical layer $\{x \in \mathbb{R}^n : e^{-1}r < |x| < r\}$. Let the coefficients q_α satisfy (5.90) for $|x| \leq r_0$. A typical result arising from those in this chapter is the following Phragmén-Lindelöf principle.

Theorem 6.6.1. *Let* $n = 2m$ *and let for* $r \in (0, r_0)$

$$c_1 \geq \omega(\log r^{-1}) \geq c_2 \sup_{K_r} \left\{ \sum_{|\alpha|=2m} |q_\alpha(x) - p_\alpha| + \sum_{|\alpha|<2m} |x|^{2m-|\alpha|} |q_\alpha(x)| \right\},$$

where c_1 *and* c_2 *are constants, which depend on* P *and* Ω. *Suppose that* ω *is an absolutely continuous positive function on* $[\log r_0^{-1}, \infty)$, $\omega(t) \to 0$ *and*

$$\omega'(t) = o\left(\omega(t)^{3/2}\right) \quad as \quad t \to +\infty. \tag{6.51}$$

Let $u \in W^{2m}_{2,\text{loc}}(B_{r_0} \backslash \{0\})$ *satisfy*

$$Q(x, D_x)u = 0 \quad on \quad B_{r_0} \backslash \{0\}$$

and let

$$\| u \|_{W^{2m}_2(K_r)} = O\left(e^{-C \int_r^{r_0} \sqrt{\omega(\log \sigma^{-1})} \sigma^{-1} d\sigma}\right) \quad as \quad r \to 0,$$

where C *is a positive constant depending on* P *and* Ω. *Then*

$$\| u \|_{W^{2m}_2(K_r)} = O\left(r e^{C_1 \int_r^{r_0} \omega(\log \sigma^{-1}) \sigma^{-1} d\sigma}\right) \quad as \quad r \to 0$$

with some constant C_1, *depending on* P *and* Ω.

Proof. According to Example 1.5.6 we can take $k_- = 0, m_- = 2$ and $k_+ = 1, m_+ = 1$. The result follows from Theorem 6.5.3. $\square$

6.6.2 The Neumann Problem in a Cylinder

The next application concerns the Neumann problem

$$\begin{cases} \left(D_t^2 + E(x,t,D_x)\right) u = 0 & on \ \Omega \times (t_0, \infty) \\ \partial_\nu u = 0 & on \ \partial\Omega \times (t_0, \infty), \end{cases} \tag{6.52}$$

where Ω is a domain in $\mathbb{R}^n$ with compact closure and smooth boundary, ν is an outer normal to $\partial\Omega$ and E is an elliptic operator given by

$$E(x,t,D_x) = \sum_{i,j=1}^n e_{ij}(x,t) D_{x_i} D_{x_j} + \sum_{j=1}^n e_j(x,t) D_{x_j} + e_0(x,t).$$

We denote by $\omega(t)$ a majorant of the function

$$c \sup_{C_t} \left\{ \sum_{i,j=1}^n |e_{ij}(x,\tau) - \delta_i^j| + \sum_{j=1}^n |e_j(x,\tau)| + |e_0(x,\tau)| \right\},$$

where c depends on Ω. We assume that our operator is close to the Laplacian at $+\infty$. To be more precise, let $\omega(t) \to 0$ as $t \to +\infty$ and let (6.51) be valid.

According to the example in Sect. 1.5.1, the strip $0 < \Im\lambda < \sqrt{\gamma_1}$ (where γ_1 is the first positive Neumann eigenvalue of $-\Delta_x$ on Ω) does not contain eigenvalues of the pencil $\mathcal{A}(\lambda) = \lambda^2 - \Delta_x$. By setting $k_- = 0$ and $k_+ = \sqrt{\gamma_1}$ we note that $m_- = 2$ and $m_+ = 1$. This together with Corollary 6.4.5 implies the following Phragmén-Lindelöf principle.

Theorem 6.6.2. *If $u \in W^2_{2,\mathrm{loc}}(\mathcal{C})$ satisfies (6.52) and*

$$\| u \|_{W^2_2(\mathcal{C}_t)} = O\left(e^{-C \int_{t_0}^{t} \sqrt{\omega(\tau)} d\tau}\right)$$

then

$$\| u \|_{W^2_2(\mathcal{C}_t)} = O\left(e^{-\sqrt{\mu_1} t + C_1 \int_{t_0}^{t} \omega(\tau) d\tau}\right).$$

Here C_1 and C are positive constants depending on Ω.

6.6.3 The Dirichlet Problem for Other Nonsmooth Domains

Theorems of this chapter (as well as Ch. 5), combined with appropriate coordinate mappings enable one to obtain similar results for the Dirichlet problem in domains with certain types of singularities on the boundary.

We start with the case of a power cusp. Let $\mathcal{G}$ be a domain in $\mathbb{R}^{n+1}$ with compact closure and $\overline{\mathcal{G}} \ni 0$. Let $\partial\mathcal{G}\backslash\{O\}$ be a smooth surface and let the domain $\mathcal{G}^{(1)} = \{(x, \tau) \in \mathcal{G} : 0 < \tau < 1\}$ be given by

$$\mathcal{G}^{(1)} = \left\{(x, \tau) \in \mathbb{R}^{n+1} : \tau^{-\alpha} x \in \Omega, 0 < \tau < 1\right\},$$

where $\alpha > 1$ and Ω is a domain in $\mathbb{R}^n$.

Let $u \in W^{2m}_{2,\mathrm{loc}}(\overline{\mathcal{G}}\backslash\{O\})$ be a solution of the Dirichlet problem

$$\begin{cases} P(D_x, D_\tau)u = f & \text{on } \mathcal{G} \\ \partial^j_\nu u = 0 & \text{on } \partial\mathcal{G}, \ j = 0, \ldots, m-1, \end{cases} \tag{6.53}$$

where

$$P(D_x, D_\tau) = \sum_{|\beta|+j=2m} p_{\beta j} D_x^\beta D_\tau^j \tag{6.54}$$

is an elliptic operator with constant coefficients.

The mapping $(x, \tau) \to (\xi, t)$, where

$$t = (\alpha - 1)^{-1} \tau^{1-\alpha}, \quad \xi = \tau^{-\alpha} x, \tag{6.55}$$

transforms $\mathcal{G}^{(1)}$ onto the semicylinder

$$\mathcal{C}^{(1)} = \{(\xi, t) : \xi \in \Omega, t > 1\}.$$

By direct calculation we check that the equation $Pu = f$ becomes

$$P(D_\xi, -D_t)u + t^{-1}Q(\xi, t, D_\xi, D_t)u$$
$$= ((\alpha - 1)t)^{2m\alpha/(1-\alpha)} f \quad \text{on} \quad \mathcal{C}^{(1)}, \tag{6.56}$$

where Q is a $2m$-th order differential operator with bounded coefficients.

Hence the behaviour of solutions to the problem (6.53) near the cusp depends upon the spectrum of the pencil

$$P(D_\xi, -\lambda) : W_2^{2m}(\Omega) \cap \overset{\circ}{W}_2^m(\Omega) \to L_2(\Omega). \tag{6.57}$$

We fix a strip $k_- < \Im\lambda < k_+$ which is free of the spectrum of this pencil, and use the usual notation $m_\pm$. In order to simplify the notation, we shall restrict ourselves to the case $m_- = 1$ and $m_+ > 1$. Since $\rho(t) \leq \text{const } t^{-1}$, the next lemma follows directly from Theorem 6.4.3.

Lemma 6.6.3. *Let $m_- = 1$ and $m_+ > 1$. Let also*

$$\int_1^\infty e^{k_+ t - C t^{(m_+ - 1)/m_+}} t^{2m\alpha/(1-\alpha)} \parallel f \parallel_{L_2(\mathcal{C}_t)} dt < \infty \,,$$

where C is a positive constant depending on P, Ω and $k_\pm$. Let the solution $u \in W_{2,\text{loc}}^{2m}(\overline{\mathcal{C}}^{(1)})$ of (6.53) with zero Dirichlet data on the lateral part of $\partial\mathcal{C}^{(1)}$ satisfy

$$\parallel u \parallel_{W_2^{2m}(\mathcal{C}_t)} = O\left(e^{-(k_- + \varepsilon)t}\right) \quad as \quad \tau \to +\infty$$

with some positive ε. Then

$$\parallel u \parallel_{W_2^{2m}(\mathcal{C}_t)} = O\left(e^{-k_+ t + C_1 t^{(m_+ - 1)/m_+}}\right) \quad as \quad \tau \to +\infty$$

with $C_1 > C$.

In order to deduce a Phragmén-Lindelöf theorem from this lemma it is sufficient to return to the coordinates (x, τ). We state the result.

Theorem 6.6.4. *Let $m_- = 1$ and $m_+ > 1$. Let also*

$$\int_0^1 \exp\left\{\frac{k_+}{\alpha - 1}\tau^{1-\alpha} - C\tau^{(1-\alpha)(m_+ - 1)/m_+}\right\}$$
$$\times \tau^{\alpha(2m - (n+3)/2)} \parallel f \parallel_{L_2(\mathcal{G}_\tau)} d\tau < \infty$$

where

$$\mathcal{G}_\tau = \left\{(x, s) \in \mathcal{G}^{(1)} : \tau - \tau^\alpha < s < \tau, s^{-\alpha} x \in \Omega\right\}$$

and C is a positive constant depending on P, Ω and $k_\pm$. Let the solution $u \in W_{2,\text{loc}}^{2m}(\overline{\mathcal{G}}\backslash\{0\})$ of (6.53) with zero Dirichlet data satisfy

$$\sum_{|\gamma|+j\leq 2m} \tau^{\alpha(|\gamma|+j-(n+1)/2)} \parallel D_x^\gamma D_\tau^j u \parallel_{L_2(\mathcal{G}_\tau)}$$

$$= O\left(\exp\left\{-\frac{k_- + \varepsilon}{\alpha - 1}\tau^{1-\alpha}\right\}\right) \quad as \quad \tau \to 0$$

with some positive ε. Then

$$\sum_{|\gamma|+j\leq 2m} \tau^{\alpha(|\gamma|+j-(n+1)/2)} \parallel D_x^\gamma D_\tau^j u \parallel_{L_2(\mathcal{G}_\tau)}$$

$$= O\left(\exp\left\{-\frac{k_+}{\alpha - 1}\tau^{1-\alpha} + C_1 \tau^{(1-\alpha)(m_+ - 1)/m_+}\right\}\right) \quad as \quad \tau \to 0$$

with an arbitrary $C_1 > C$.

It is not necessary to consider only power cusps. Let, for example,

$$\mathcal{G}^{(1)} = \left\{ (x, \tau) \in \mathbb{R}^{n+1} : x\tau^{-1} |\log 2\tau|^\delta \in \Omega, \ 0 < \tau < 1 \right\}$$

with $\delta > 0$. Here the role of the coordinate change (6.55) is played by

$$t = (\delta + 1)^{-1} (\log 2\tau)^{\delta+1}, \quad \xi = \tau^{-1} |\log 2r|^\delta x.$$

A simple calculation leads to the Dirichlet problem in a semicylinder for the operator

$$P(D_\xi, -D_t) + t^{-\delta(\delta+1)^{-1}} Q(\xi, t, D_\xi, D_t),$$

where, as before, Q is a $2m$-th order differential operator with bounded operator coefficients. Our theorems on operator differential equations can now be applied directly.

Quite similar coordinate changes lead to analogous results for the Dirichlet problem in infinite paraboloidal and funnel-like domains (compare with Maz'ya, Plamenevskii (1978) or Kozlov, Maz'ya, Rossmann (1997), Ch.9). Moreover, C^1-domains with infinite curvature at the origin can be also included. We mention a few facts about the last geometry. Let the domain $\mathcal{G} \subset \mathbb{R}^n$ be described near O by the inequality

$$x_n > a \frac{|x|}{|\log|x||^\delta},$$

where a is a real constant and $\delta > 0$. We are interested in solutions of the equation

$$P(D_x)u = f \quad \text{in} \quad \mathcal{G} \quad \text{near} \quad O$$

with zero Dirichlet data on $\partial \mathcal{G}$, where P is defined by (0.7). The coordinate change

$$\xi_n = x_n - a \frac{|x|}{|\log|x||^\delta}, \quad \xi_i = x_i, \ i = 1, \ldots, n-1,$$

flattens the boundary near O and reduces the differential operator $P(D_x)$ to the form

$$P(D_\xi) + |\log|\xi||^{-\delta} S(\xi, D_\xi), \tag{6.58}$$

where

$$S(\xi, D_\xi) = \sum_{|\alpha| \leq 2m} S_\alpha(\xi) D_\xi^\alpha$$

and

$$|\xi|^{2m-\alpha} |S_\alpha(\xi)| \leq \text{const}$$

for small $|\xi|$. The new problem in $\mathbb{R}^n_+$ can be transformed by (0.8) to the Dirichlet problem in the semicylinder

$$\left\{ (\theta, t) : \theta \in S^{n-1}_+, t > t_0 \right\}, \quad t_0 = \text{const} > 0.$$

The second term in (6.58) gives rise to a perturbation $0(t^{-\delta})$. By Sect. 1.5.2, we can deal with the strips

$$m + j < \Im\lambda < m + j + 1, \quad m - n - j - 1 < \Im\lambda < m - n - j$$

with $j = 0, 1, \ldots$ and with the strip $m - n < \Im\lambda < m$. In all cases, $m_+ = m_- = 1$. In order to obtain the Phragmén-Lindelöf principle for our original problem in $\mathcal{G}$, it suffices to refer to Corollary 6.3.8.

6.7 Comments

Results of Sections 6.2-6.4 are borrowed from Kozlov, Maz'ya (1991-1996), report 5. The change of variables (6.55) was applied to the study of elliptic boundary value problems in cuspidal domains by Evgrafov (1960), Maz'ya, Plamenevskii (1977), Dauge (1996) et al.

7. Two-Weight L_2-Estimates for Equations with Variable Coefficients

7.1 Introduction

The material in this chapter is closely connected with that of Chapter 4. Here, we also consider solutions in the weighted space $W^\ell(\mathbb{R}; \gamma)$. As in Chapter 4 we use the class $\mathbb{A}$ of weight functions γ defined in Sect.4.2. We consider again the weighted spaces $W^\ell(\mathbb{R}; \gamma)$ and $L_2(\mathbb{R}; H_0; \gamma)$ with the norms

$$||u||_{W^\ell(\mathbb{R};\gamma)} = \left(\int_{\mathbb{R}} \gamma^2(\tau) \sum_{0 \leq q \leq \ell} ||D_\tau^q u(\tau)||_{\ell-q}^2 \, d\tau \right)^{1/2},$$

$$||u||_{L_2(\mathbb{R};H_0;\gamma)} = \left(\int_{\mathbb{R}} \gamma^2(\tau) \, ||u(\tau)||_0^2 \, d\tau \right)^{1/2}.$$

According to Proposition 4.2.1, the following equivalence relations are valid:

$$||u||_{W^\ell(\mathbb{R};\gamma)} \sim \left(\int_{\mathbb{R}} \gamma^2(t) \, ||u||_{W^\ell(t,t+1)}^2 \, dt \right)^{1/2}, \tag{7.1}$$

$$||u||_{L_2(\mathbb{R};H_0;\gamma)} \sim \left(\int_{\mathbb{R}} \gamma^2(t) \, ||u||_{L_2(t,t+1;H_0)}^2 \, d\tau \right)^{1/2}. \tag{7.2}$$

We start with two uniqueness theorems for solutions in the space $W^\ell(\mathbb{R}; \gamma)$ (Sect.7.2). For example, according to Theorem 7.2.2, a condition on γ which guarantees the uniqueness has the form

$$\int_{\mathbb{R}_\pm} \left(\frac{\gamma(t)}{g_\omega(0,t)} \right)^2 dt = \infty.$$

Next, in Sect.7.3, we show that the two-weight L_2-estimate

$$\int_{\mathbb{R}} |\gamma(t)w(t)|^2 dt \leq c \int_{\mathbb{R}} |\Gamma(t)h(t)|^2 dt \tag{7.3}$$

for solutions of the ordinary differential equation $(M(\partial_t) - \omega) w = h$ implies the two-weight estimate

$$||u||_{W^\ell(\mathbb{R};\gamma)} \leq c||f||_{L_2(\mathbb{R};H_0;\Gamma)} \tag{7.4}$$

for the operator equation (5.1).

As a corollary to these results, we give in Sect.7.4 conditions for the unique solvability of the equation (5.1) in the space $W^\ell_{\beta_\pm}(\mathbb{R})$ equipped with the norm

$$\left(\sum_\pm \int_{\mathbb{R}_\pm} e^{2\beta_\pm t} \sum_{k=0}^\ell \|D_t^k u\|^2_{\ell-k} dt\right)^{1/2}.$$

In Sect.7.5 we give explicit formulae for weight functions in the inequality (7.4) assuming that $m_+ = m_- = 1$.

Sect.7.6 is devoted to two-weight estimates of the type (7.4) for solutions of (6.8) in the case $\rho(t) = O(t^{-\alpha})$ for large $t > 0$, $\alpha > 0$. In particular, we show that the weight functions

$$\gamma(t) = t^\sigma \exp\{\int_{t_0}^t k_+(ax^{-\alpha})dx\} \tag{7.5}$$

and

$$\Gamma(t) = t^{1+\alpha(m_+-1)/m_+}\gamma(t),$$

$\alpha < m_+$, $2\sigma < -1 - \alpha(m_+ - 1)/m_+$, can be taken in the inequality

$$\int_{t_0+1}^\infty \gamma(\tau)^2 \sum_{j=0}^\ell \|\partial_\tau^j u(\tau)\|^2_{\ell-j} d\tau$$

$$\leq c\{\|\Gamma f\|^2_{L_2(t_0,\infty;H_0)} + \|u\|^2_{W^{\ell-1}(t_0,t_0+1)}\}. \tag{7.6}$$

Here u is a solution of

$$L(t, D_t)u(t) = f(t) \quad \text{for} \quad t > t_0 \tag{7.7}$$

satisfying a certain growth condition. (For the definition of $k_+(\cdot)$ in (7.5) see Sect.5.2.1).

In Sect.7.6 we assume that the function $b\rho(t)$ has a majorant $\omega(t)$ such that $\omega(t) = o(1)$ and $\omega'(t) = o\left(\omega(t)^{(m_++1)/m_+}\right)$ as $t \to +\infty$. We show that the weight functions

$$\Gamma(t) = \gamma(t)/\omega(t)$$

and

$$\gamma(t) = \omega(t)^\sigma \exp\{k_+ t + \nu \int_{t_0}^t \omega(x)^{1/m_+} dx\},$$

$\nu < -(k_+ - k_-)^{-m_-/m_+}$, $\sigma \in \mathbb{R}$, are admissible for (7.6).

7.2 Uniqueness Theorems in Weighted Sobolev Spaces

Here we prove two uniqueness theorems; these are based on Theorem 5.4.2 and Proposition 5.4.3 respectively. By ω we always mean a measurable majorant for the function $b\rho$.

Theorem 7.2.1. *Let (5.7) be fulfilled and let*

$$\gamma(t) \geq c \liminf_{\tau \to \pm\infty} \frac{g_\omega(0,\tau)}{g_\omega(t,\tau)} \quad for \quad t \gtrless 0. \tag{7.8}$$

If $u \in W^\ell(\mathbb{R};\gamma)$ satisfies the equation

$$L(t, D_t)u = 0 \quad on \quad \mathbb{R}, \tag{7.9}$$

then $u = 0$.

Proof. Since $u \in W^\ell(\mathbb{R};\gamma)$, we have by (4.4) that

$$\sum_{N=-\infty}^{\infty} \gamma^2(N) \int_N^{N+1} \| u \|^2_{W^\ell(\tau,\tau+1)} \, d\tau < \infty.$$

Hence

$$\gamma(\tau)\|u\|_{W^\ell(\tau,\tau+1)} \to 0 \quad as \quad \tau \to \pm\infty,$$

which, along with (7.8), implies (5.56). The proof is completed by using Theorem 5.4.2. $\square$

Theorem 7.2.2. *Let (5.7) be fulfilled and let*

$$\int_0^\infty \left(\frac{\gamma(t)}{g_\omega(0,t)} \right)^2 dt = \infty \tag{7.10}$$

and

$$\int_{-\infty}^0 \left(\frac{\gamma(t)}{g_\omega(0,t)} \right)^2 dt = \infty. \tag{7.11}$$

If $u \in W^\ell(\mathbb{R};\gamma)$ satisfies the equation

$$L(t, D_t)u = 0 \quad on \quad \mathbb{R}$$

then $u = 0$.

Proof. We have

$$\liminf_{t \to +\infty} g_\omega(0,t)\|u\|_{W^\ell(t,t+1)} = 0 \tag{7.12}$$

since in the opposite case $u \notin W^\ell(\mathbb{R};\gamma)$ by (7.10) and (5.1). Analogously

$$\liminf_{t \to -\infty} g_\omega(0,t)\|u\|_{W^\ell(t,t+1)} = 0.$$

It remains to use Proposition 5.4.3. $\square$

7.3 Existence Theorems for Solutions in Weighted Sobolev Spaces and Two-Weight Estimates

Let h be an arbitrary function from $L_2(\mathbb{R})$ with compact support and let w be the solution of the comparison equation

$$\mathcal{M}(\partial_t)w(t) - \omega(t)w(t) = h(t) \quad \text{on} \quad \mathbb{R} \tag{7.13}$$

given by (5.38).

In the next theorem we call the weight functions $\gamma, \Gamma \in \mathbb{A}$ admissible if there exists a constant c independent of h such that

$$\int_{\mathbb{R}} |\gamma(t)w(t)|^2 dt \leq c \int_{\mathbb{R}} |\Gamma(t)h(t)|^2 dt. \tag{7.14}$$

Theorem 7.3.1. *Let γ and Γ be admissible weight functions. Then for every $f \in L_2(\mathbb{R}; H_0; \Gamma)$ there exists a solution*

$$u \in W^\ell(\mathbb{R}; H_0; \Gamma)$$

of the equation (5.1) such that

$$\|u\|_{W^\ell(\mathbb{R};\gamma)} \leq c\|f\|_{L_2(\mathbb{R};H_0;\Gamma)} \tag{7.15}$$

with c independent of f.

Proof. It is sufficient to prove the existence of a solution of (5.1) satisfying (7.15) for $f \in L_2(\mathbb{R}; H_0; \Gamma)$ with compact support.

By Theorem 5.3.2 there exists the (k_-, k_+) - solution u subject to

$$\|u\|_{W^\ell(t,t+1)} \leq b\, w(t),$$

where w is the above mentioned solution of (7.13) with

$$h(t) = \|f\|_{L_2(t,t+1;H_0)}.$$

By (7.14),

$$\|u\|^2_{W^\ell(\mathbb{R};\gamma)} \leq c \int_{\mathbb{R}} (\gamma(t)w(t))^2 dt$$

$$\leq c \int_{\mathbb{R}} (\Gamma(t)\|f\|_{L_2(t,t+1;H_0)})^2 dt \leq c\|f\|^2_{L_2(\mathbb{R};H_0;\Gamma)}.$$

The proof is complete. $\square$

We show that under some complementary assumptions on γ and Γ it is sufficient to verify (7.14) for $\mathbb{R}_+$ and $\mathbb{R}_-$ separately.

Proposition 7.3.2. *Let $\gamma, \Gamma \in \mathbb{A}$. Suppose that*

$$\int_{\mathbb{R}} (g_\omega(t,0)\gamma(t))^2 \, dt + \int_{\mathbb{R}} \left(\frac{g_\omega(0,t)}{\Gamma(t)} \right)^2 dt < \infty \tag{7.16}$$

and that γ and Γ satisfy the following condition:

If $h_\pm$ are functions from $L_2(\mathbb{R})$ with compact support, $h_\pm(t) = 0$ for $t \lessgtr 0$ and if $w_\pm$ are solutions of (7.13) with $h = h_\pm$, given by (5.38), there exists a constant c independent of h such that

$$\int_{\mathbb{R}_\pm} |\gamma(t)w_\pm(t)|^2 dt \le c \int_{\mathbb{R}_\pm} |\Gamma(t)h_\pm(t)|^2 dt. \tag{7.17}$$

Then γ and Γ are admissible and therefore Theorem 7.3.1 *holds.*

Proof. Let $h \in L_2(\mathbb{R})$ have compact support. Represent h in the form $h = h_+ + h_-$, where $\mathrm{supp}\, h_\pm \subset \overline{\mathbb{R}}_\pm$. Also let $w_\pm$ be solutions of (7.13) with $h = h_\pm$. Then by (7.17)

$$\int_{\mathbb{R}} |\gamma(t)w_+(t)|^2 dt \le c \int_{\mathbb{R}_+} |\Gamma(t)h_+(t)|^2 dt$$

$$+ \int_{-\infty}^{0} |\gamma(t) \int_{0}^{\infty} g_\omega(t,\tau)h_+(\tau)d\tau|^2 dt. \tag{7.18}$$

Using (5.29) with $x = 0$ and $t < 0 < \tau$ we can estimate the last term in (7.18) by

$$c \int_{-\infty}^{0} (\gamma(t)g_\omega(t,0))^2 \, dt \int_{0}^{\infty} (g_\omega(0,\tau)/\Gamma(\tau))^2 \, d\tau \int_{0}^{\infty} |\Gamma(\tau)h_+(\tau)|^2 d\tau.$$

Analogously,

$$\int_{\mathbb{R}} |\gamma(t)w_-(t)|^2 dt \le c \int_{-\infty}^{0} |\Gamma(t)h_-(t)|^2 dt$$

$$+ \int_{0}^{\infty} |\gamma(t) \int_{-\infty}^{0} g_\omega(t,\tau)h_-(\tau)d\tau|^2 dt.$$

Due to (5.29), the last term on the right is estimated by

$$c \int_{0}^{\infty} (\gamma(t)g_\omega(t,0))^2 \, dt \int_{-\infty}^{0} (g_\omega(0,\tau)/\Gamma(\tau))^2 \, d\tau \int_{-\infty}^{0} |\Gamma(\tau)h_-(\tau)|^2 d\tau.$$

By (7.16) we get (7.14). $\square$

7.4 Unique Solvability in $W^\ell_{\beta_-,\beta_+}(\mathbb{R})$

In the case

$$\gamma(t) = \begin{cases} e^{\beta_+ t} & \text{for } t \geq 0 \\ e^{\beta_- t} & \text{for } t < 0 \end{cases}$$

$W^\ell(\mathbb{R}; \gamma)$ and $L_2(\mathbb{R}; H_0; \gamma)$ coincide with the spaces $W^\ell_{\beta_-,\beta_+}(\mathbb{R})$ and $L_{2,\beta_-,\beta_+}(\mathbb{R}; H_0)$, introduced in Sect. 2.3.2.

Theorem 7.4.1. *Let ρ satisfy* (5.42) *and let*

$$\rho_\pm = \limsup_{t \to \pm\infty} \rho(t).$$

Suppose that $\beta_\pm \in (k_-(b\rho_\pm), k_+(b\rho_\pm))$, where $k_\pm(\sigma)$ were defined in Sect. 5.2.1. Then the operator

$$L(t, D_t) : W^\ell_{\beta_\pm}(\mathbb{R}) \to L_{2,\beta_\pm}(\mathbb{R}; H_0) \tag{7.19}$$

is an isomorphism.

Proof. The continuity of $L(t, D_t)$ is trivial. Put $\omega(t) = b\rho(t)$. Using (6.3) and (6.4) we obtain for $t, \tau \gtrless 0$

$$g_\omega(t, \tau) \leq \begin{cases} c_\varepsilon e^{-(k_+(b\rho_\pm)-\varepsilon)(t-\tau)} & \text{for } t \geq \tau \\ c_\varepsilon e^{-(k_-(b\rho_\pm)+\varepsilon)(t-\tau)} & \text{for } t \leq \tau \end{cases} \tag{7.20}$$

with an arbitrary positive ε. Hence (7.10) and (7.11) hold. Therefore, the uniqueness follows from Theorem 7.2.2.

From (7.20) we get (7.16). Let us check (7.17). Let $h_\pm$ be functions in $L_2(\mathbb{R})$ with compact support on $\overline{\mathbb{R}}_\pm$ and let

$$w_\pm(t) = \int_{\mathbb{R}_\pm} g_\omega(t, \tau) h_\pm(\tau) d\tau$$

be solutions of (7.13) with $h = h_\pm$. Then

$$\int_{\mathbb{R}_\pm} e^{2\beta_\pm t} |w_\pm(t)|^2 dt$$

$$\leq \int_{\mathbb{R}_\pm} \left| \int_{\mathbb{R}_\pm} e^{\beta_\pm (t-\tau)} g_\omega(t, \tau) e^{\beta_\pm \tau} h_\pm(\tau) d\tau \right|^2 dt. \tag{7.21}$$

Due to (7.20) we have for $t, \tau \gtrless 0$

$$e^{\beta_\pm (t-\tau)} g_\omega(t, \tau) \leq c_\varepsilon e^{-\varepsilon |t-\tau|}$$

with a sufficiently small positive ε. Therefore the right-hand side of (7.21) is estimated by

$$c \int_{\mathbb{R}_\pm} e^{2\beta_\pm \tau} |h_\pm(\tau)|^2 d\tau.$$

This proves (7.17). Using Proposition 7.3.2 we obtain that the operator (7.19) is surjective. The proof is complete. $\square$

Remark 7.4.2. An important special case included in the theorem is $\rho_+ = \rho_- = 0$, i.e.

$$\|L - \mathcal{A}\|_{W^\ell(t,t+1)\to L_2(t,t+1;H_0)} \to 0 \quad \text{as } t \to \pm\infty.$$

Then the operator (7.19) is isomorphic for all $\beta_\pm \in (k_-, k_+)$, similarly to the case of constant coefficients.

Next we consider the equation (6.8) on the semi-axis $t > t_0$, where t_0 is sufficiently large. As in Sect.5.6 we assume that $L(t, D_t)$ coincides with $\mathcal{A}(D_t)$ for large negative t. Applying Theorem 7.4.1 to ζu, where ζ is the same cutoff function as in the proof of Theorem 5.6.1, we arrive at the following:

Corollary 7.4.3. *Let*

$$\rho_+ < b^{-1}m_+^{m_+}m_-^{m_-}\left(\frac{k_+ - k_-}{m_+ + m_-}\right)^{m_++m_-}.$$

Also, let $u \in W_{\mathrm{loc}}^\ell(t_0, \infty)$ *be a solution of (6.8) such that*

$$\liminf_{t\to+\infty} e^{(k_-(b\rho_+)+\varepsilon)t}\|u\|_{W^\ell(t,t+1)} = 0.$$

If $e^{\beta_+t}f \in L_2(t_0, \infty; H_0)$ *with* $\beta_+ \in (k_-(b\rho_+), k_+(b\rho_+))$ *then* $e^{\beta_+t}u \in W^\ell(t_0+1, \infty)$ *and the estimate holds:*

$$\|e^{\beta_+t}u\|_{W^\ell(t_0+1,\infty)} \le c\,\{\|e^{\beta_+t}f\|_{L_2(t_0,\infty;H_0)} + \|u\|_{W^{\ell-1}(t_0,t_0+1)}\},$$

where c *does not depend on* u *and* f. *In particular, for* $\rho_+ = 0$ *we can take an arbitrary* $\beta_+ \in (k_-, k_+)$.

Example 7.4.4. Consider the Dirichlet problem in the cone K from Sect. 5.7.2. By Corollary 7.4.3 we have the following statement. Let

$$\sum_{|\alpha|=2m} |q_\alpha(x) - p_\alpha| + \sum_{|\alpha|<2m} |x|^{2m-|\alpha|}|q_\alpha(x)| \to 0 \quad \text{as } x \to 0$$

and let $u \in W_{2,\mathrm{loc}}^\ell(\overline{K}\backslash\{0\})$ be a solution of (5.89) such that

$$\|u\|_{W_2^{2m}(K_r)} = O(r^{k_-+\varepsilon}) \quad \text{as } r \to 0$$

with a certain $\varepsilon > 0$. Then for an arbitrary $\delta \in (2m-n/2-k_+, 2m-n/2-k_-)$ we have the estimate:

$$\left(\sum_{|\alpha|\le 2m} \int_{K\cap B(e^{-2}r_0)} |x|^{2(\delta+|\alpha|-2m)}|D_x^\alpha u|^2 dx\right)^{1/2}$$

$$\le C\left(\left(\int_{K\cap B(r_0)} |x|^{2\delta}|f|^2 dx\right)^{1/2} + \|u\|_{W_2^{2m}(\mathcal{D}(r_0))}\right),$$

where $B(s) = \{x : |x| < s\}$ and $\mathcal{D}(s) = \{x \in K : e^{-3}s < |x| < s\}$ and where r_0 is a sufficiently small positive number.

7.5 The Case $m_\pm = 1$

The next result concerns the case of $\gamma(t)$ "close" to the critical exponentials $\exp(k_\pm t)$ as $t \to \pm\infty$.

Theorem 7.5.1. *Let $m_+ = m_- = 1$ and let ω be a majorant of $b\rho$, non-decreasing for $t < 0$ and non-increasing for $t > 0$. Also let*

$$\omega(\pm 0) < (k_+ - k_-)^2/4 \tag{7.22}$$

and

$$\lim_{t \to \pm\infty} \omega(t) = 0. \tag{7.23}$$

Suppose that

$$\int_0^\infty \omega(\tau)d\tau = \infty, \quad \int_{-\infty}^0 \omega(\tau)d\tau = \infty.$$

Denote

$$\gamma(t) = \sqrt{\omega(t)} \exp\left(k_\pm t + \nu_\pm \left| \int_0^t (k_\pm(\omega(\tau)) - k_\pm)d\tau \right|\right) \tag{7.24}$$

for $t \gtrless 0$ with $k_\pm(x)$ defined by (6.17) and $\nu_\pm < -1$. Then
(i) the solution $u \in W^\ell(\mathbb{R}; \gamma)$ of the equation (5.1) is unique;
(ii) for every $f \in L_2(\mathbb{R}; H_0; \Gamma)$ with

$$\Gamma(t) = \frac{\gamma(t)}{\omega(t)} \quad \text{for all} \quad t \in \mathbb{R}$$

there exists a solution $u \in W^\ell(\mathbb{R}; \gamma)$ of the equation (5.1) and the estimate (7.15) holds.

Proof. (i) By using the estimate (6.16) for g_ω we obtain

$$\int_0^\infty \left(\frac{\gamma(t)}{g_\omega(0,t)}\right)^2 dt \geq c \int_0^\infty \omega(t)e^{2\int_0^t (k_+ - k_-(\omega(\tau)))d\tau}$$

$$\times e^{2\nu_+ |\int_0^t (k_+(\omega(\tau)) - k_+)d\tau|} dt \geq c \int_0^\infty \omega(t)dt = \infty.$$

In a similar way one proves that

$$\int_{-\infty}^0 \left(\frac{\gamma(t)}{g_\omega(0,t)}\right)^2 dt = \infty$$

and the uniqueness follows from Theorem 7.2.2.

(ii) To use Proposition 7.3.2 we check first (7.16). From (6.16) and (7.24) it follows that

$$\int_0^\infty (g_\omega(t,0)\gamma(t))^2 dt \leq c \int_0^\infty e^{2(1+\nu_+) \int_0^t (k_+ - k_+(\omega(\tau)))d\tau} \omega(t)dt.$$

Since $\nu_+ < -1$, the obvious inequality

$$|k_\pm - k_\pm(\omega(\tau))| \leq \frac{2\omega(\tau)}{k_+ - k_-}, \tag{7.25}$$

implies that the last integral is finite. In the same way we verify the convergence of the integrals

$$\int_{-\infty}^0 (g_\omega(t,0)\gamma(t))^2 \, dt, \quad \int_{-\infty}^\infty \left(\frac{g_\omega(0,t)}{\Gamma(t)}\right)^2 \, dt.$$

Thus (7.16) is proved.

We turn to the verification of (7.17). Setting

$$w_+(t) = \int_0^\infty g_\omega(t,\tau)h_+(\tau)d\tau$$

and

$$\chi_\pm(t) = e^{-\int_0^t k_\pm(\omega(s))ds}$$

we have

$$|w_+(t)| \leq c \int_0^t \frac{\chi_+(t)}{\chi_+(\tau)}|h_+(\tau)|d\tau + c \int_t^\infty \frac{\chi_-(t)}{\chi_-(\tau)}|h_+(\tau)|d\tau$$

and by Proposition 4.4.2 and Corollary 4.4.3 the inequality (7.17) for w_+ holds if

$$\sup_{t>0} \int_t^\infty (\chi_+\gamma)^2 d\tau \int_0^t \frac{d\tau}{(\chi_+\Gamma)^2} < \infty \tag{7.26}$$

and

$$\sup_{t>0} \int_0^t (\chi_-\gamma)^2 d\tau \int_t^\infty \frac{d\tau}{(\chi_-\Gamma)^2} < \infty. \tag{7.27}$$

By using the function

$$\Xi_+(t) = e^{\int_0^t (k_+ - k_+(\omega(s)))ds}$$

we rewrite (7.26) in the form

$$\sup_{t>0} \int_t^\infty \omega \Xi_+^{2\nu_+ +2} d\tau \int_0^t \frac{\omega d\tau}{\Xi_+^{2\nu_+ +2}} < \infty. \tag{7.28}$$

Since

$$\Xi_+'(\tau) = (k_+ - k_+(\omega(\tau)))\Xi_+(\tau) \geq \frac{\omega(\tau)}{k_+ - k_-}\Xi_+(\tau),$$

(7.28) is equivalent to

$$\sup_{t>0} \int_t^\infty \Xi_+^{2\nu_+ +1} \Xi_+' d\tau \int_0^t \frac{\Xi_+' d\tau}{\Xi_+^{2\nu_+ +3}} < \infty,$$

which holds by the condition $\nu_+ < -1$.

To verify (7.27) we introduce the notation

$$\Lambda(s) = k_+(\omega(s)) - k_-(\omega(s)) + (\nu_+ + 1)(k_+ - k_+(\omega(s))).$$

Then the supremum in (7.27) is not greater then

$$c \sup_{t>0} \int_0^t e^{2\int_0^\tau \Lambda(s)ds} d\tau \int_t^\infty e^{-2\int_0^\tau \Lambda(s)ds} d\tau.$$

Since

$$\Lambda(s) = k_+ - k_- + o(1) \quad \text{as} \quad s \to \infty$$

it is sufficient to check the boundedness of the function

$$\int_0^t e^{2\int_0^\tau \Lambda(s)ds} \Lambda(\tau)d\tau \int_t^\infty e^{-2\int_0^\tau \Lambda(s)ds} \Lambda(\tau)d\tau,$$

which is equal to

$$\frac{1}{4}\left(e^{2\int_0^t \Lambda(s)ds} - 1\right) e^{-2\int_0^t \Lambda(s)ds}$$

and hence is less than $1/4$.

The validity of (7.17) for

$$w_-(t) = \int_{-\infty}^0 g_\omega(t,\tau)h_-(\tau)d\tau$$

is proved in the same way. We complete the proof by referring to Proposition 7.3.2. $\square$

If $\omega \in L_2(\mathbb{R})$ then by (6.16) and (6.17)

$$g_\omega(t,\tau) \le c \exp\left(-k_\pm(t-\tau) + (k_+ - k_-)^{-1}\left|\int_\tau^t \omega(s)ds\right|\right) \quad \text{for} \quad t \gtrless \tau.$$

Therefore, by duplicating the proof of Theorem 7.5.1 we arrive at its weaker but more simply formulated variant.

Theorem 7.5.2. *Let* ω *be the same as in* Theorem 7.5.1 *and let* $\omega \in L_2(\mathbb{R})$. *Then the conclusion of* Theorem 7.5.1 *holds for the weight functions*

$$\gamma(t) = \sqrt{\omega(t)} \exp\left(k_\pm t + \frac{\nu_\pm}{k_+ - k_-}\left|\int_0^t \omega(s)ds\right|\right) \quad \text{for} \quad t \gtrless 0,$$
$$\Gamma(t) = \gamma(t)/\omega(t),$$

where $\nu_\pm < -1$.

As a sample application of this result we can consider the case

$$\rho(t) \le a_\pm |t|^{-1}$$

for large $t \ge 0$, $a_\pm = \text{const} > 0$. Then we can take

$$\gamma(t) = (1 + |t|)^{\alpha_\pm} e^{k_\pm t}, \quad \Gamma(t) = (1 + |t|)^{\alpha_\pm + 1} e^{k_\pm t},$$

where

$$\alpha_\pm < -\frac{1}{2} - \frac{a_\pm b}{k_+ - k_-}.$$

Note that the loss of weight is the same here as in the constant coefficients case $\rho = 0$ (see Sect. 4.4.4 with $m_\pm = 1$).

In Theorem 7.5.1 we assume that the weight functions are close to $e^{k_\pm t}$ both at $+\infty$ and $-\infty$. One can obtain the following analogous statement in the case when γ and Γ are close to the critical exponential $\exp(k_+ t)$ only at $+\infty$.

Theorem 7.5.3. *Let $m_+ = m_- = 1$ and let ω be non-decreasing for $t < 0$ and non-increasing for $t > 0$. Also let the conditions (7.22) and (7.23) be valid. Suppose*

$$\int_0^\infty \omega(\tau)d\tau = \infty.$$

Denote

$$\gamma(t) = \begin{cases} \sqrt{\omega(t)} \exp\left(k_+ t + \nu \int_0^t (k_+ - k_+(\omega(\tau)))d\tau\right) & \text{for } t > 0 \\ \exp(\beta t) & \text{for } t \le 0 \end{cases}$$

with $\nu < -1$ and $\beta \in (k_-, k_+)$. Then

    (i) *the solution $u \in W^\ell(\mathbb{R}; \gamma)$ of the equation (5.1) is unique;*
    (ii) *for every $f \in L_2(\mathbb{R}; H_0; \Gamma)$ with*

$$\Gamma(t) = \begin{cases} \gamma(t)/\omega(t) & \text{for } t > 0 \\ \exp(\beta t) & \text{for } t \le 0 \end{cases}$$

there exists a solution $u \in W^\ell(\mathbb{R}; \gamma)$ of the equation (5.1) and the estimate (7.15) holds.

From Theorem 7.5.3 applied to ζu, where ζ is the same cutoff function as in the proof of Theorem 5.6.1, we obtain the following result on the behaviour of solutions of (6.8) at $+\infty$.

Corollary 7.5.4. *Let $\omega(t) \to 0$ as $t \to +\infty$ and let*

$$\int_{t_0}^{+\infty} \omega(\tau)d\tau = \infty .$$

Also let $u \in W_{\mathrm{loc}}^\ell(t_0, \infty)$ be a solution of (6.8) (with sufficiently large t_0) such that

$$\liminf_{t \to +\infty} e^{\int_{t_0}^t k_-(\omega(s))ds} \|u\|_{W^\ell(t,t+1)} = 0.$$

Denote

$$\gamma(t) = \sqrt{\omega(t)} \exp\left(k_+ t + \nu \int_{t_0}^t (k_+ - k_+(\omega(s)))ds\right), \qquad (7.29)$$

where $\nu < -1$, and $\Gamma = \gamma/\omega$. If $\Gamma f \in L_2(t_0, \infty; H_0)$ then

$$\left(\int_{t_0+1}^\infty \gamma(\tau)^2 \sum_{j=0}^\ell \|\partial_\tau^j u(\tau)\|_{\ell-j}^2 d\tau\right)^{1/2}$$

$$\leq c\left(\|\Gamma f\|_{L_2(t_0,\infty;H_0)} + \|u\|_{W^{\ell-1}(t_0,t_0+1)}\right), \qquad (7.30)$$

where c does not depend on u and f.

7.6 Two-Weight Estimates when ρ Dominates t^{-m_+}

Let $b\rho(t) \leq \omega(t)$ for $t \geq t_0$ where t_0 is a sufficiently large positive number and ω is an absolutely continuous, positive function on $[t_0, \infty)$ subject to (6.32), (6.33) and (5.7).

Theorem 7.6.1. *Let $u \in W_{\mathrm{loc}}^\ell(t_0, \infty)$ be a solution of (6.8) subject to (6.41). Denote $\Gamma(t) = \gamma(t)/\omega(t)$ and*

$$\gamma(t) = \omega(t)^\sigma \exp\{k_+ t + \nu \int_{t_0}^t \omega(x)^{1/m_+} dx\}$$

where $\nu < -(k_+ - k_-)^{-m_-/m_+}$, $\sigma \in \mathbb{R}$. If $\Gamma f \in L_2(t_0, \infty; H_0)$ then the estimate (7.30) holds.

Proof. Extend ω by 0 for $t < t_0$ and denote by g_ω Green's function of (7.13). By (6.44)

$$g_\omega(t, \tau) \leq c_\varepsilon e^{-(k_- +\varepsilon)(t-\tau)} \quad \text{for} \quad \tau \geq t \geq t_0 \qquad (7.31)$$

with an arbitrary $\varepsilon > 0$. From Theorem 6.4.1 the estimate

$$g_\omega(t, \tau) \leq c_\delta \left(\omega(t)\omega(\tau)\right)^{(1-m_+)/2m_+} e^{-k_+(t-\tau)} \left(\frac{\mathcal{E}(t)}{\mathcal{E}(\tau)}\right)^{1+\delta} \qquad (7.32)$$

follows where $\delta > 0$ is arbitrarily small and

$$\mathcal{E}(t) = \exp\{(k_+ - k_-)^{-m_-/m_+} \int_{t_0}^t \omega(x)^{1/m_+} dx\}.$$

The estimate (6.41) along with (7.31) imply (5.78). By using Theorem 5.6.1 we arrive at

$$||u||_{W^\ell(t,t+1)} \le c\Big(\int_{t_0}^t g_\omega(t,\tau) ||f||_{L_2(\tau,\tau+1;H_0)} d\tau$$

$$+ \int_t^\infty e^{-(k_-+\varepsilon)(t-\tau)} ||f||_{L_2(\tau,\tau+1;H_0)} d\tau + g_\omega(t,t_0) ||u||_{W^{\ell-1}(t_0,t_0+1)} \Big)$$

for $t > t_0 + 1$.

By (7.32) the function $\gamma(\cdot)g(\cdot,t_0)$ belongs to $L_2(t_0, \infty)$. It can easily be checked that

$$\int_{t_0}^\infty \gamma(t)^2 \left(\int_t^\infty e^{-(k_-+\varepsilon)(t-\tau)} h(\tau) d\tau \right)^2 dt$$

$$\le c \int_{t_0}^\infty \gamma(t)^2 h(t)^2 dt ,$$

where $h(t) = ||f||_{L_2(t,t+1;H_0)}$. Hence it is sufficient to verify the inequality

$$\int_{t_0}^\infty \left(\gamma(t)\omega(t)^{(1-m_+)/2m_+} e^{-k_+t} \mathcal{E}(t)^{1+\delta} \right)^2 \left(\int_{t_0}^t \Psi(\tau) d\tau \right)^2 dt$$

$$\le c \int_{t_0}^\infty \left(\Gamma(t)\omega(t)^{(m_+-1)/2m_+} e^{-k_+t} \mathcal{E}(t)^{1+\delta} \right)^2 \Psi(t)^2 dt , \qquad (7.33)$$

where

$$\Psi(t) = \omega(t)^{(1-m_+)/2m_+} e^{k_+t} \mathcal{E}(t)^{-1-\delta} .$$

It is well known that the estimate

$$\int_{t_0}^\infty p(t) \left(\int_{t_0}^\infty \Psi(\tau) d\tau \right)^2 dt \le c \int_{t_0}^\infty q(t)(\Psi(t))^2 dt$$

holds for all locally summable Ψ if and only if

$$\int_{t_0}^t \frac{d\tau}{q(\tau)} \int_t^\infty p(\tau) d\tau \le \text{Const}$$

for all $t \ge t_0$ (see, for example, Muckenhoupt (1972) or Maz'ya (1985), Theorem 1.3.1/4). Setting $\chi = \nu(k_+ - k_-)^{m_-/m_+} + 1 + \delta$ we see that the inequality (7.33) holds if and only if the function

$$\int_{t_0}^t \left(\frac{\omega(\tau)^{(1+m_+)/2m_+ - \sigma}}{\mathcal{E}(\tau)^\chi} \right)^2 d\tau \int_t^\infty \left(\omega(\tau)^{(1-m_+)/2m_+ + \sigma} \mathcal{E}(\tau)^\chi \right)^2 d\tau$$

is bounded. This function can be rewritten as

$$(k_+ - k_-)^{2m_-/m_+} \int_{t_0}^t \omega(\tau)^{1-2\sigma} \frac{\mathcal{E}'(\tau)d\tau}{\mathcal{E}(\tau)^{2\chi+1}}$$

$$\times \int_t^\infty \omega(\tau)^{-1+2\sigma} \mathcal{E}(\tau)^{2\chi-1} \mathcal{E}'(\tau)d\tau$$

Integrating by parts and using $\chi < 0$, we arrive at the required result. $\square$

7.7 Comments

Estimates and theorems on the unique solvability of elliptic boundary value problems in a cone in weighted spaces of the type $W_\beta^\ell(\mathbb{R})$ with $k_- < \beta < k_+$ are well known (see Kondratiev (1967), Grisvard (1985) et al). Results of the present chapter are borrowed from Kozlov, Maz'ya (1991-1996), report 5.

8. Connection of Solutions Corresponding to Different Strips

8.1 Introduction

In the present chapter we pay special attention to zeros of $L(t, D_t)$ both on $\mathbb{R}$, and the semiaxis $t > t_0$. We restrict ourselves to the zeros subject to a certain growth restriction which is formulated in terms of Green's functions of two auxiliary ordinary differential equations (see Sect. 8.3). We show, in particular, that the space $\mathfrak{X}(L)$ of the zeros on $\mathbb{R}$ is finite dimensional, find its dimension and obtain upper and lower estimates for its elements. Similar results are obtained for zeros of the adjoint operator on $\mathbb{R}$ in Sect. 8.4. The biorthogonality condition (2.13) is extended to zeros of L and L^* in Sect. 8.4.4.

The principal result of this chapter is obtained in Sect. 8.4.5, where we consider two solutions u_1 and u_2 corresponding to the strips $k_-^{(1)} < \Im\lambda < k_+^{(1)}$ and $k_-^{(2)} < \Im\lambda < k_+^{(2)}$, $k_+^{(1)} \leq k_-^{(2)}$. Under rather mild assumptions on the function ρ (as defined by (5.3)), we show that the difference $u_1 - u_2$ belongs to $\mathfrak{X}(L)$. The analog of the coefficient formula (3.59) is contained in Theorem 8.4.7. This is a generalization of a similar result related to the equation $\mathcal{A}(D_t)u = f$ (Proposition 3.8.1).

One more theorem proved in this chapter extends Proposition 3.8.2 on the asymptotics of solutions of $\mathcal{A}(D_t)u = f$, $t > t_0$, to the equation $L(t, D_t)u = f$ (Theorem 8.5.7).

8.2 Auxiliary Information

8.2.1 Notation

Let $k_-^{(1)}$, $k_+^{(1)}$, $k_-^{(2)}$, $k_+^{(2)}$ be real numbers such that

$$k_-^{(1)} < k_+^{(1)} \leq k_-^{(2)} < k_+^{(2)}$$

and let the strips

$$k_-^{(1)} < \Im\lambda < k_+^{(1)} \quad \text{and} \quad k_-^{(2)} < \Im\lambda < k_+^{(2)}$$

be free of eigenvalues of $\mathcal{A}(\lambda)$. We shall use the notation:

$m_\pm^{(j)}$ is the maximal length of Jordan chains corresponding to the eigenvalues of $\mathcal{A}(\lambda)$ on the line $\Im\lambda = k_\pm^{(j)}$, $j = 1, 2$ (in case there are no eigenvalues of $\mathcal{A}(\lambda)$ on the line $\Im\lambda = k_\pm^{(j)}$ we put, as usual, $m_\pm^{(j)} = 1$);

$g_1(t)$ and $g_2(t)$ are Green's functions of the ordinary differential operators

$$\mathcal{M}_1(\partial_t) = (\partial_t + k_+^{(1)})^{m_+^{(1)}}(-\partial_t - k_-^{(1)})^{m_-^{(1)}}$$

and

$$\mathcal{M}_2(\partial_t) = (\partial_t + k_+^{(2)})^{m_+^{(2)}}(-\partial_t - k_-^{(2)})^{m_-^{(2)}}$$

respectively (these Green's functions are given by (3.23), (3.24) where k_+, k_-, m_+, m_- have to be replaced by $k_+^{(j)}$, $k_-^{(j)}$, $m_+^{(j)}$, $m_-^{(j)}$);

b_j is a constant in the inequality

$$\|u\|_{W^\ell(t,t+1)} \le b_j \int_{\mathbb{R}} g_j(t-\tau)\|f\|_{L_2(\tau,\tau+1;H_0)}d\tau, \; j = 1, 2, \qquad (8.1)$$

where $u \in W_{\mathrm{loc}}^\ell(\mathbb{R})$ is the solution of the equation $\mathcal{A}(D_t)u = f$ corresponding to the strip $k_-^{(j)} < \Im\lambda < k_+^{(j)}$ (see Definition 3.4.1);

ω_j is a measurable majorant of the function $b_j\rho$, $j = 1, 2$, i.e.

$$b_j\rho(t) \le \omega_j(t) \;\; a.e.,$$

where ρ is defined by (5.3).

Everywhere in Sect. 8.2–8.4 the following assumption is made:

$$\sup_{t\in\mathbb{R}} \omega_j(t) < \left(m_+^{(j)}\right)^{m_+^{(j)}} \left(m_-^{(j)}\right)^{m_-^{(j)}} \left(\frac{k_+^{(j)} - k_-^{(j)}}{m_+^{(j)} + m_-^{(j)}}\right)^{m_+^{(j)}+m_-^{(j)}}. \qquad (8.2)$$

By $g_{j,\omega_j} = g_{j,\omega_j}(t,\tau)$, $j = 1, 2$, we denote Green's function of the operator $\mathcal{M}_j(\partial_t) - \omega_j(t)$. This Green's function is given by (5.8) with $g = g_j$ and $\omega = \omega_j$.

8.2.2 Estimates for Green's Functions of the Comparison Equations

Here we obtain estimates involving Green's functions g_{j,ω_j}, $j = 1, 2$; these estimates will be used later in the present chapter.

Proposition 8.2.1. (i) *The following inequalities hold:*

$$g_{2,\omega_2}(t,\tau) \le c\varepsilon_2(t,\tau)g_{1,\omega_1}(t,\tau) \quad if \quad t \ge \tau \tag{8.3}$$

and

$$g_{1,\omega_1}(t,\tau) \le c\varepsilon_1(t,\tau)g_{2,\omega_2}(t,\tau) \quad if \quad t \le \tau, \tag{8.4}$$

where

$$\varepsilon_j(t,\tau) = g_{j,\omega_j}(t,\tau)g_{j,\omega_j}(\tau,t), \; j = 1,2,$$

is a bounded function tending to 0 as $|t - \tau| \to \infty$ (see (5.28)).

(ii) *If $t \ge 0$ then*

$$\sup_{\tau \in \mathbb{R}} \frac{g_{2,\omega_2}(t,\tau)}{g_{2,\omega_2}(0,\tau)} \le cg_{1,\omega_1}(t,0), \tag{8.5}$$

If $t \le 0$ then

$$\sup_{\tau \in \mathbb{R}} \frac{g_{1,\omega_1}(t,\tau)}{g_{1,\omega_1}(0,\tau)} \le cg_{2,\omega_2}(t,0). \tag{8.6}$$

(iii) *If $t \ge 0$ then*

$$\sup_{\tau \in \mathbb{R}} \frac{g_{1,\omega_1}(\tau,t)}{g_{1,\omega_1}(\tau,0)} \le c\, g_{2,\omega_2}(0,t). \tag{8.7}$$

If $t \le 0$ then

$$\sup_{\tau \in \mathbb{R}} \frac{g_{2,\omega_2}(\tau,t)}{g_{2,\omega_2}(\tau,0)} \le c\, g_{1,\omega_1}(0,t). \tag{8.8}$$

Proof. (i). Since $g_j(t - \tau) \le g_{j,\omega_j}(t,\tau)$ we get

$$g_j(t - \tau) \le g_{j,\omega_j}(t,\tau) \le \frac{\varepsilon_j(t,\tau)}{g_j(\tau - t)}. \tag{8.9}$$

By

$$1/g_1(t) \le c\, g_2(-t) \quad for \quad t \ge 0$$
$$1/g_2(t) \le c\, g_1(-t) \quad for \quad t \le 0,$$

(which can be verified using inequality (3.25) for g_j), we deduce estimates (8.3) and (8.4) from (8.9).

(ii) We shall prove (8.5). By (5.34)

$$\sup_{\tau \in \mathbb{R}} \frac{g_{2,\omega_2}(t,\tau)}{g_{2,\omega_2}(0,\tau)} \le c\, e^{-k_-^{(2)}t}$$

for positive t. Now (8.5) follows from

$$e^{-k_-^{(2)}t} \le c\, g_1(t).$$

Estimate (8.6) is checked similarly.

(iii) By (5.35) the left-hand side of (8.7) is estimated by $c\, e^{k_+^{(1)}t}$ for $t \ge 0$. Since $e^{k_+^{(1)}t} \le c\, g_2(-t) \le c\, g_{2,\omega_2}(0,t)$ we arrive at (8.7).

Estimate (8.8) is proved analogously. $\square$

8.2.3 An Auxiliary Existence Result

The following assertion of existence for solutions to the equation (5.1) is useful in the next section.

Lemma 8.2.2. *Let $f \in L_{2,\mathrm{loc}}(\mathbb{R}; H_0)$ and*

$$\int_0^\infty g_{1,\omega_1}(0,\tau)\|f\|_{L_2(\tau,\tau+1;H_0)}d\tau$$

$$+ \int_{-\infty}^0 g_{2,\omega_2}(0,\tau)\|f\|_{L_2(\tau,\tau+1;H_0)}d\tau < \infty. \tag{8.10}$$

Then there exists a solution $u \in W_{\mathrm{loc}}^\ell(\mathbb{R})$ of (5.1) satisfying

$$\|u\|_{W^\ell(t,t+1)} \le b_2 \int_{-\infty}^0 g_{2,\omega_2}(t,\tau)\|f\|_{L_2(\tau,\tau+1;H_0)}d\tau$$

$$+ b_1 \int_{-1}^\infty g_{1,\omega_1}(t,\tau)\|f\|_{L_2(\tau,\tau+1;H_0)}d\tau. \tag{8.11}$$

Proof. We represent f as the sum $f_1 + f_2$ where $f_1 = f$ when $t \ge 0$ and $f_2 = f$ when $t < 0$. By (8.10) and Theorem 5.3.2 there exists the $(k_-^{(j)}, k_+^{(j)})$-solution $u_j \in W_{\mathrm{loc}}^\ell(\mathbb{R})$ of (5.1) with $f = f_j$, $j = 1, 2$, satisfying

$$\|u_j\|_{W^\ell(t,t+1)} \le b_j \int_{\mathbb{R}} g_{j,\omega_j}(t,\tau)\|f_j\|_{L_2(\tau,\tau+1;H_0)}d\tau.$$

Therefore, $u = u_1 + u_2$ is a solution of (5.1) and (8.11) holds. $\square$

Corollary 8.2.3. *Solution u from Theorem 8.2.2 satisfies*

$$\|u\|_{W^\ell(t,t+1)} = o\left(\sup_{\tau \in \mathbb{R}} \frac{g_{1,\omega_1}(t,\tau)}{g_{1,\omega_1}(0,\tau)}\right) \quad as \; t \to +\infty \tag{8.12}$$

and

$$\|u\|_{W^\ell(t,t+1)} = o\left(\sup_{\tau \in \mathbb{R}} \frac{g_{2,\omega_2}(t,\tau)}{g_{2,\omega_2}(0,\tau)}\right) \quad as \; t \to -\infty. \tag{8.13}$$

Proof. Since $u = u_1 + u_2$, (where u_j is the $(k_-^{(j)}, k_+^{(j)})$-solution) we have

$$\|u\|_{W^\ell(t,t+1)} = o\left(\sup_{\tau \in \mathbb{R}} \frac{g_{1,\omega_1}(t,\tau)}{g_{1,\omega_1}(0,\tau)}\right)$$

$$+ o\left(\sup_{\tau \in \mathbb{R}} \frac{g_{2,\omega_2}(t,\tau)}{g_{2,\omega_2}(0,\tau)}\right) \quad as \; t \to \pm\infty. \tag{8.14}$$

By (8.5) we arrive at (8.12). Using (8.6) we obtain (8.13). $\square$

8.3 Zeros of $L(t, D_t)$

8.3.1 The Class $\mathfrak{X}(L)$

Definition 8.3.1. By $\mathfrak{X}(L)$ we denote the set of solutions $u \in W^\ell_{\text{loc}}(\mathbb{R})$ of the equation

$$L(t, D_t)u(t) = 0 \quad \text{on} \quad \mathbb{R}, \tag{8.15}$$

subject to

$$\|u\|_{W^\ell(t,t+1)} \leq c\, g_{1,\omega_1}(t,0) \quad \text{for} \quad t \geq 0, \tag{8.16}$$

and

$$\|u\|_{W^\ell(t,t+1)} \leq c\, g_{2,\omega_2}(t,0) \quad \text{for} \quad t \leq 0. \tag{8.17}$$

We collect some simple properties of the space $\mathfrak{X}(L)$ in the following

Proposition 8.3.2. (i) *Let* $u \in W^\ell_{\text{loc}}(\mathbb{R})$ *be a solution of* (8.15) *subject to*

$$\liminf_{t \to +\infty} g_{1,\omega_1}(0,t)\|u\|_{W^\ell(t,t+1)} = 0 \tag{8.18}$$

and

$$\liminf_{t \to -\infty} g_{2,\omega_2}(0,t)\|u\|_{W^\ell(t,t+1)} = 0. \tag{8.19}$$

Then $u \in \mathfrak{X}(L)$
   (ii) *Let* u *be a nonzero element in* $\mathfrak{X}(L)$. *Then the lower estimates*

$$\liminf_{t \to +\infty} g_{2,\omega_2}(0,t)\|u\|_{W^\ell(t,t+1)} > 0, \tag{8.20}$$

$$\liminf_{t \to -\infty} g_{1,\omega_1}(0,t)\|u\|_{W^\ell(t,t+1)} > 0 \tag{8.21}$$

hold.

Proof. Follows directly from Proposition 5.5.2. $\square$

Let us mention that by (5.28) the inequalities (8.16) and (8.17) give relations (8.18) and (8.19) with $\liminf$ replaced by $\limsup$.

Example 8.3.3. Let $L = \mathcal{A}$. By Proposition 3.8.3 the space $\mathfrak{X}(\mathcal{A})$ consists of all functions $u \in W^\ell_{\text{loc}}(\mathbb{R})$ which are represented in the form (3.71). Therefore, the dimension of $\mathfrak{X}(\mathcal{A})$ is finite and equal to the total algebraic multiplicity of eigenvalues of the pencil $\mathcal{A}(\lambda)$ in the strip $k^{(1)}_+ \leq \Im\lambda \leq k^{(2)}_-$.

8.3.2 The Dimension of $\mathfrak{X}(L)$

Denote by κ the total algebraic multiplicity of the eigenvalues of $\mathcal{A}(\lambda)$ in the strip $k_+^{(1)} \leq \Im\lambda \leq k_-^{(2)}$.

Proposition 8.3.4. *The dimension of $\mathfrak{X}(L)$ equals κ.*

Proof. First consider an auxiliary operator L_* defined by

$$L_*(t, D_t) = \begin{cases} L(t, D_t) & \text{for } t \geq 0, \\ \mathcal{A}(D_t) & \text{for } t < 0. \end{cases}$$

The function

$$\rho_*(t) = \|L_* - \mathcal{A}\|_{W^\ell(t,t+1) \to L_2(t,t+1;H_0)}$$

is estimated by $\rho(t)$ when $t \geq -1$ and $\rho_*(t) = 0$ for $t < -1$. Setting $w_j^*(t) = w_j(t)$ for $t \geq -1$ and $w_j^*(t) = 0$ for $t < -1$ we have $b_j \rho_* \leq w_j^*$. It is clear that the operator L_* satisfies the assumption in Sect. 5.3.1 and w_j^* is subject to (8.2). By g_{j,w_j^*} we denote Green's function of the operator $\mathcal{M}_j(\partial_t) - w_j^*(t)$, which is given by (5.8) with $g = g_j$ and $w = w_j^*$.

We prove that $\dim \mathfrak{X}(L_*) = \kappa$. We introduce an operator S on the set of $f \in L_{2,\mathrm{loc}}(\mathbb{R}; H_0)$ subject to

$$\int_{\mathbb{R}} g_1(-\tau)\|f\|_{L_2(\tau,\tau+1;H_0)} d\tau < \infty$$

as the mapping

$$f \xrightarrow{S} u,$$

where u is the solution of $\mathcal{A}u = f$ corresponding to the strip $k_-^{(1)} < \Im\lambda < k_+^{(1)}$ (see Sect. 3.4). Then, for an arbitrary function $u \in W_{\mathrm{loc}}^\ell(\mathbb{R})$, satisfying

$$\int_{-1}^{\infty} g_1(-\tau)\rho_*(\tau)\|u\|_{W^\ell(\tau,\tau+1)} d\tau < \infty, \tag{8.22}$$

the following representation holds:

$$L_* u = \mathcal{A}\left(I - S(\mathcal{A} - L_*)\right) u. \tag{8.23}$$

If $u \in \mathfrak{X}(L_*)$ then by (8.16) the left-hand side of (8.22) does not exceed

$$c \int_{-1}^{\infty} g_1(-\tau)w_1^*(\tau)g_{1,w_1^*}(\tau, 0) d\tau.$$

Due to (5.15) it is majorized by $c\, g_{1,w_1^*}(0,0)$. Hence $u \in \mathfrak{X}(L_*)$ if and only if

$$\mathcal{A}\left(I - S(\mathcal{A} - L_*)\right) u = 0.$$

Let us verify that the equality

$$u = S(\mathcal{A} - L_*)u \tag{8.24}$$

is impossible if $u \in \mathfrak{X}(L_*)$ and $u \neq 0$. Indeed, by estimating the right-hand side of (8.24) we get

$$\|u\|_{W^\ell(t,t+1)} \leq \int_{-1}^\infty g_1(t-\tau)\omega_1^*(\tau)\|u\|_{W^\ell(\tau,\tau+1)}d\tau.$$

Using (8.16) and (5.15) we obtain

$$\|u\|_{W^\ell(t,t+1)} \leq c\, g_{1,\omega_1^*}(t,0),\; t \in \mathbb{R}\,.$$

Due to (5.28) with $\tau = 0$, this contradicts (8.21). Thus (8.24) has only trivial solutions from $\mathfrak{X}(L_*)$.

Let us check that the function $v = u - S(\mathcal{A} - L_*)u$, where $u \in \mathfrak{X}(L_*)$ belongs to $\mathfrak{X}(\mathcal{A})$. It is clear that $\mathcal{A}(D_t)v(t) = 0$. Furthermore,

$$\|v\|_{W^\ell(t,t+1)} \leq \|u\|_{W^\ell(t,t+1)} + c\, g_{1,\omega_1^*}(t,0).$$

Hence

$$g_1(0,t)\|v\|_{W^\ell(t,t+1)} \leq cg_1(0,t)g_{1,\omega_1^*}(t,0)$$

for $t \geq 0$ and

$$g_2(0,t)\|v\|_{W^\ell(t,t+1)} \leq cg_2(0,t)g_{2,\omega_2^*}(t,0)$$

for $t \leq 0$ (we have used here the inequality (8.4)). By the estimate $g_{j,\omega_j^*} \geq g_j$ and by (5.28), we arrive at (8.18) and (8.19) for the function v (with g_{j,ω_j} replaced by g_j). Hence $v \in \mathfrak{X}(\mathcal{A})$. Thus the operator $u \to v = u - S(\mathcal{A} - L_*)u$ maps $\mathfrak{X}(\mathcal{L}_*)$ into $\mathfrak{X}(\mathcal{A})$, and its kernel is trivial. The inverse is given by the Neumann series

$$u = \sum_{k=0}^\infty \left(S(\mathcal{A} - L_*)\right)^k v,$$

which is convergent in $W_{\mathrm{loc}}^\ell(\mathbb{R})$ and

$$\|u\|_{W^\ell(t,t+1)} \leq \|v\|_{W^\ell(t,t+1)}$$
$$+ \sum_{k=1}^\infty \int_{\mathbb{R}^k} g_1(t,\tau_1)\omega_1^*(\tau_1)...g_1(\tau_{k-1},\tau)\omega_1^*(\tau)\|v\|_{W^\ell(\tau,\tau+1)}d\tau_1...d\tau_{k-1}d\tau.$$

By (5.8) we can rewrite this as

$$\|u\|_{W^\ell(t,t+1)} \leq \|v\|_{W^\ell(t,t+1)} + \int_{\mathbb{R}} g_{1,\omega_1^*}(t,\tau)\omega_1^*(\tau)\|v\|_{W^\ell(\tau,\tau+1)}d\tau.$$

Since $v \in \mathfrak{X}(\mathcal{A})$ we derive from the last estimate

$$\|u\|_{W^\ell(t,t+1)} \leq c(g_1(t,0) + \int_{-1}^\infty g_{1,\omega_1^*}(t,\tau)\omega_1^*(\tau)g_1(\tau,0)d\tau$$

for $t \geq 0$. Using (5.15) we obtain (8.16). Estimate (8.17) is verified analogously (the only difference is that one should use (8.4) at the end). Thus the mapping $u \to v$ is isomorphic. Therefore, the dimensions of $\mathfrak{X}(L_*)$ and $\mathfrak{X}(\mathcal{A})$ coincide and are equal to κ by Example 8.3.3.

It remains to prove that

$$\dim \mathfrak{X}(L) = \dim \mathfrak{X}(L_*). \tag{8.25}$$

Let ζ be a smooth function such that $\zeta = 1$ for $t > 1$ and $\zeta = 0$, for $t < 0$. If $u \in \mathfrak{X}(L)$ then $f = L_*(\zeta u)$ is supported on $[0,1]$. By Theorem 5.3.2, there exists a solution of $L_* v = f$ subject to

$$||v||_{W^\ell(t,t+1)} \leq c\, g_{2,\omega_2^*}(t,0)||f||_{L_2(0,1;H_0)}. \tag{8.26}$$

It is clear that $v + \zeta u \in \mathfrak{X}(L_*)$. By (8.20)

$$||u||_{W^\ell(t,t+1)} \geq c/g_{2,\omega_2}(0,t) \quad \text{for large} \quad t > 0. \tag{8.27}$$

Furthermore, due to (8.26) and (8.27), the inequality $g_{2,\omega_2} \geq g_{2,\omega_2^*}$ and (8.4), we have

$$||v||_{W^\ell(t,t+1)} = o\left(||u||_{W^\ell(t,t+1)}\right) \quad \text{as} \quad t \to +\infty.$$

Therefore $v + \zeta u \neq 0$ if $u \neq 0$. Thus the mapping

$$\mathfrak{X}(L) \ni u \to v + \zeta u \in \mathfrak{X}(L_*)$$

is injective and hence $\dim \mathfrak{X}(L) \leq \dim \mathfrak{X}(L_*)$.

Reasoning as above (with evident changes) we get

$$\dim \mathfrak{X}(L_*) \leq \dim \mathfrak{X}(L).$$

This completes the proof. $\square$

8.3.3 The Norm in $\mathfrak{X}(L)$

Lemma 8.3.5. *The function*

$$\mathfrak{X}(L) \ni u \to ||u||_{W^\ell(0,1)}$$

is a norm in $\mathfrak{X}(L)$.

Proof. Since $\mathfrak{X}(L)$ is finite-dimensional it is enough to check that $u = 0$ on the interval $(0,1)$ implies $u = 0$ on $\mathbb{R}$. Let u_1 denote the extension of u by 0 onto $(-\infty, 0)$. Then u_1 belongs also to $\mathfrak{X}(L)$. By (8.16) and (5.28) the function u_1 satisfies the assumptions of Proposition 5.4.3. Applying this proposition we obtain $u_1 = 0$. In the same way we can show that $u - u_1 = 0$. $\square$

8.4 Solutions of (5.1) Corresponding to Different Strips

8.4.1 The Auxiliary Dual Space

We denote by $\mathfrak{F}$ the space of functions $f \in L_{2,\mathrm{loc}}(\mathbb{R}; H_0)$ subject to

$$\int_0^\infty g_{2,\omega_2}(0,\tau)\|f\|_{L_2(\tau,\tau+1;H_0)}d\tau$$
$$+ \int_{-\infty}^0 g_{1,\omega_1}(0,\tau)\|f\|_{L_2(\tau,\tau+1;H_0)}d\tau < \infty \qquad (8.28)$$

and define a norm in $\mathfrak{F}$ by the left-hand side of (8.28).

We start with a description of the dual $\mathfrak{F}^*$ of $\mathfrak{F}$.

Lemma 8.4.1. *The dual space of $\mathfrak{F}$ is a Banach space of functions $g \in L_{2,\mathrm{loc}}(\mathbb{R}; H_0)$ subject to the inequalities*

$$\|V\|_{L_2(t,t+1;H_0)} \leq c_2\, g_{2,\omega_2}(0,t) \quad \text{for} \quad t \geq 0 \qquad (8.29)$$

and

$$\|V\|_{L_2(t,t+1;H_0)} \leq c_1\, g_{1,\omega_1}(0,t) \quad \text{for} \quad t \leq 0. \qquad (8.30)$$

A norm in $\mathfrak{F}^*$ can be given by

$$\|V\|_{\mathfrak{F}^*} = \sup_{t \geq 0} \frac{\|V\|_{L_2(t,t+1;H_0)}}{g_{2,\omega_2}(0,t)} + \sup_{t \leq 0} \frac{\|V\|_{L_2(t,t+1;H_0)}}{g_{1,\omega_1}(0,t)}.$$

Proof. If $g \in \mathfrak{F}^*$ then the functional

$$\mathfrak{F} \ni f \to \int_{\mathbb{R}} (f(t), g(t))_H dt \qquad (8.31)$$

is bounded on $\mathfrak{F}$ by Minkowski's inequality.

It remains to show that every bounded linear functional

$$f \to G(f)$$

on $\mathfrak{F}$ can be represented in the form (8.31). First, let $f(t) = \eta(t)h$, where $h \in H_0$ and η is a function with compact support. Then

$$|G(\eta h)| \leq C\|h\|_0 \int_{\mathbb{R}} \Phi(t)|\eta(t)|dt,$$

where

$$\Phi(t) = \begin{cases} g_{2,\omega_2}(0,t) & \text{for } t \geq 0 \\ g_{1,\omega_1}(0,t) & \text{for } t < 0. \end{cases}$$

This implies that for every $h \in H_0$ one can represent $G(\eta h)$ as

$$G(\eta h) = \int_{\mathbb{R}} \eta(t)\bar{g}_h(\tau)d\tau, \qquad (8.32)$$

where g_h is a bounded function satisfying

$$|g_h(t)| \leq \frac{C||h||_0}{\Phi(t)}.$$

Moreover, g_h depends linearly on h. Applying Riesz's theorem we get

$$g_h(t) = (h, g(t))_H,$$

where $g(\tau) \in H_0$ for almost every τ and

$$|g(t)| \leq \frac{C}{\Phi(t)}.$$

The representation (8.32) takes the form

$$G(\eta h) = \int_{\mathbb{R}} (\eta(t)h, g(t))_{H_0} dt. \tag{8.33}$$

Using the linearity and boundedness of both functionals one extends the equality (8.33) onto all elements of $\mathfrak{F}$. $\square$

8.4.2 The Subspace $\mathfrak{X}^*(L)$

Let coker$L$ be defined as in Sect. 5.5.2.

Definition 8.4.2. By $\mathfrak{X}^*(L)$ we denote the set of all elements in coker$L$ subject to (8.29) and (8.30)

Clearly, $\mathfrak{X}^*(L) \subset \mathfrak{F}^*$. An analog of Proposition 8.3.2 runs as follows:

Proposition 8.4.3. (i) *Let V be a nonzero element in* coker$L$ *subject to*

$$\liminf_{t \to -\infty} g_{1,\omega_1}(t, 0)||V||_{L_2(t,t+1;H_0)} = 0 \tag{8.34}$$

and

$$\liminf_{t \to +\infty} g_{2,\omega_2}(t, 0)||V||_{L_2(t,t+1;H_0)} = 0. \tag{8.35}$$

Then $V \in \mathfrak{X}^*(L)$.
   (ii) *Let V be a nonzero element in $\mathfrak{X}^*(L)$. Then the lower estimates*

$$\liminf_{t \to -\infty} g_{2,\omega_2}(t, 0)||V||_{L_2(t,t+1;H_0)} > 0, \tag{8.36}$$

$$\liminf_{t \to +\infty} g_{1,\omega_1}(t, 0)||V||_{L_2(t,t+1;H_0)} > 0 \tag{8.37}$$

hold.

Proof. The result is obtained by applying Proposition 5.5.4. $\square$

8.4.3 The Difference of Two Solutions Belongs to $\mathfrak{X}(L)$

In the next lemma we assume that the right-hand side f of (5.1) is good enough that both $(k_-^{(1)}, k_+^{(1)})$- and $(k_-^{(2)}, k_+^{(2)})$- solutions exist, and describe the difference of these solutions.

Lemma 8.4.4. *Let $f \in L_{2,\mathrm{loc}}(\mathbb{R}; H_0)$ satisfy (8.28) and let u_1 and u_2 be the solutions of (5.1) corresponding to the strips $k_-^{(1)} < \Im\lambda < k_+^{(1)}$ and $k_-^{(2)} < \Im\lambda < k_+^{(2)}$ respectively. Then*

$$u_2 - u_1 \in \mathfrak{X}(L). \tag{8.38}$$

Proof. By Theorem 5.3.2

$$\|u_s\|_{W^\ell(t,t+1)} \le \int_{\mathbb{R}} g_{s,\omega_s}(t,\tau)\|f\|_{L_2(\tau,\tau+1;H_0)}d\tau. \tag{8.39}$$

Let $t \ge 0$. Then by Corollary 5.3.3

$$\|u_2\|_{W^\ell(t,t+1)} = o\left(\sup_{\tau\in\mathbb{R}} \frac{g_{2,\omega_2}(t,\tau)}{g_{2,\omega_2}(0,\tau)}\right).$$

Applying (8.5) we get

$$\|u_2\|_{W^\ell(t,t+1)} = o\big(g_{1,\omega_1}(t,0)\big). \tag{8.40}$$

Theorem 5.3.2 implies

$$\|u_1\|_{W^\ell(t,t+1)} \le c\, g_{1,\omega_1}(t,0)\bigg\{\int_{-\infty}^0 g_{1,\omega_1}(0,\tau)\|f\|_{L_2(\tau,\tau+1;H_0)}d\tau$$
$$+ \int_0^\infty \sup_{t\in\mathbb{R}_+}\frac{g_{1,\omega_1}(t,\tau)}{g_{1,\omega_1}(t,0)}\|f\|_{L_2(\tau,\tau+1;H_0)}d\tau\bigg\}. \tag{8.41}$$

Using (8.40) and (8.41) together with (8.7) we obtain

$$\|u_2 - u_1\|_{W^\ell(t,t+1)} \le c\, g_{1,\omega_1}(t,0) \quad \text{for} \quad t \ge 0.$$

Analogously we can prove that

$$\|u_2 - u_1\|_{W^\ell(t,t+1)} \le c\, g_{2,\omega_2}(t,0) \quad \text{for} \quad t \le 0.$$

Hence $u_2 - u_1 \in \mathfrak{X}(L)$. $\square$

8.4.4 A Sesquilinear Form and the Dimension of $\mathfrak{X}^*(L)$

We introduce a sesquilinear form on $\mathfrak{X}(L) \times \mathfrak{X}^*(L)$ by

$$< \Phi, \Psi >_L = - \int_{\mathbb{R}} \left(L(t, D_t)(\zeta(t)\Phi(t)), \Psi(t) \right)_H dt, \qquad (8.42)$$

where ζ is a smooth function equal to 1 for large positive t and to 0 for large negative t. It is clear that (8.42) does not depend on ζ.

Lemma 8.4.5. *The form* (8.42) *is non-degenerate, i. e.*
(i) *for every non-zero* $\Psi \in \mathfrak{X}^*(L)$ *there exists* $\Phi \in \mathfrak{X}(L)$ *such that*

$$< \Phi, \Psi >_L \neq 0; \qquad (8.43)$$

(ii) *for every non-zero* $\Phi \in \mathfrak{X}(L)$ *there exists* $\Psi \in \mathfrak{X}^*(L)$ *such that* (8.43) *holds.*

Proof. (i) Since $\Psi \neq 0$ there exists $f \in \mathfrak{F}$ such that

$$\int_{\mathbb{R}} (f(t), \Psi(t))_H dt \neq 0. \qquad (8.44)$$

Without loss of generality we suppose that f has compact support. Let u_j be the $(k_-^{(j)}, k_+^{(j)})$-solution of (5.1). Then, by Lemma 8.4.4, $\Phi := u_2 - u_1 \in \mathfrak{X}(L)$. We introduce the function $v = u_2 - (1 - \zeta)\Phi = u_1 + \zeta\Phi$. By (8.39)

$$\|v\|_{W^\ell(t,t+1)} \leq \begin{cases} g_{2,\omega_2}(t,0) & \text{for } t \geq 0 \\ g_{1,\omega_1}(t,0) & \text{for } t < 0. \end{cases} \qquad (8.45)$$

Let us show that

$$\int_{\mathbb{R}} (L(t, D_t)v(t), \Psi(t))_H dt = 0. \qquad (8.46)$$

Indeed, let ξ_N be a smooth function equal 1 for $|t| < N$ and 0 for $|t| > N+1$, and let its derivatives up to order ℓ be bounded uniformly with respect to N, $N \geq 1$. Since $\Psi \in \mathrm{coker}\,L$ we have

$$\int_{\mathbb{R}} (L(t, D_t)(\xi_N(t)v(t)), \Psi(t))_H dt = 0 \qquad (8.47)$$

for all N. Using (8.45) and compactness of $\mathrm{supp} f$ we can estimate the absolute value of the left-hand side in (8.47) by

$$C(g_{2,\omega_2}(N,0)g_{2,\omega_2}(0,N) + g_{1,\omega_1}(-N,0)g_{1,\omega_1}(0,-N));$$

this tends to zero when $N \to \infty$ by (5.28). This proves (8.46). Equality (8.46) together with (8.44) implies (8.43).

(ii) We represent $\mathfrak{X}(L)$ as the direct sum of the linear span of Φ and a $(\kappa - 1)$-dimensional subspace $\mathfrak{X}_1$ of $\mathfrak{X}(L)$. Let $f \in \mathfrak{F}$ and let u_j be the $(k_-^{(j)}, k_+^{(j)})$-solution of (5.1), $j = 1, 2$. Then by Lemma 8.4.4,

$$u_2 - u_1 = c(f)\Phi + \Phi_1,$$

where $c(f)$ is a complex number and $\Phi_1 \in \mathfrak{X}_1$. Clearly $c(f)$ is a linear functional on $\mathfrak{F}$. Since

$$\|u_2 - u_1\|_{W^\ell(0,1)} \leq c\|f\|_{\mathfrak{F}}, \tag{8.48}$$

the functional $f \to c(f)$ is continuous by Lemma 8.3.5. Hence there exists $\Psi \in \mathfrak{F}^*$ such that

$$c(f) = \int_{\mathbb{R}} (f(t), \Psi(t))_H dt.$$

If $f = L(t, D_t)u$, where $u \in W^\ell_{\text{loc}}(\mathbb{R})$ has a compact support, then $u_2 = u_1$ and therefore $c(f) = 0$. This yields $\Psi \in \text{coker} L$. Since $c(L(\zeta\Phi)) = -1$ we obtain (8.43). $\square$

The following assertion follows directly from Proposition 8.3.4 and Lemma 8.4.5.

Proposition 8.4.6. *Let κ be the same number as in* Proposition 8.3.4. *Then* $\dim \mathfrak{X}^*(L) = \kappa$.

8.4.5 Main Result

We choose a basis $\{U_j\}_{j=1}^\kappa$ in $\mathfrak{X}(L)$. By Lemma 8.4.5 there exists a unique basis $\{V_j\}_{j=1}^\kappa$ in $\mathfrak{X}^*(L)$ subject to the biorthogonality condition

$$< U_j, V_k >_L = \delta_{jk}, \quad j, k = 1, \ldots, \kappa. \tag{8.49}$$

Theorem 8.4.7. *Let f satisfy the conditions of* Theorem 8.4.4 *and let u_1 and u_2 be the solutions of (5.1) corresponding to the strips $k_-^{(1)} < \Im\lambda < k_+^{(1)}$ and $k_-^{(2)} < \Im\lambda < k_+^{(2)}$, respectively. Then*

$$u_2(t) - u_1(t) = \sum_{j=1}^\kappa \int_{\mathbb{R}} (f(\tau), V_j(\tau))_H d\tau \, U_j(t). \tag{8.50}$$

Proof. By Lemma 8.4.4

$$u_2(t) - u_1(t) = \sum_{j=1}^\kappa c_j(f)U_j(t).$$

One can show that $c_j(f)$ are bounded linear functionals on $\mathfrak{F}$ (cf. the end of the proof of Lemma 8.4.5). Hence

$$c_j(f) = \int_{\mathbb{R}} (f(t), \mathcal{V}_j(t))_H dt \tag{8.51}$$

with some $\mathcal{V}_j \in \mathfrak{F}^*$. If we set $f = Lu$ (where u has a compact support), then $c_j = 0$. By (8.51) this means that $\mathcal{V}_j \in \text{coker} L$. Thus $\mathcal{V}_j \in \mathfrak{X}^*(L)$. Taking $f = L(\zeta\mathcal{V}_j)$ we get $u_1 = \zeta\mathcal{V}_j$, $u_2 = (\zeta - 1)\mathcal{V}_j$ and $u_2 - u_1 = -\mathcal{V}_j$. This implies that the sets $\{U_j\}_{j=1}^\kappa$ and $\{\mathcal{V}_k\}_{k=1}^\kappa$ satisfy the biorthogonality condition (8.49). Hence $\mathcal{V}_j = V_j$. $\square$

8.5 Structure of Solutions of (6.8) at Infinity

8.5.1 The Spaces $\mathcal{Y}_1(L)$ and $\mathcal{Y}_2(L)$

Now we wish to generalize Proposition 3.8.2 on the asymptotics of solutions of (3.45) to equations with variable coefficients. To this end we need two auxiliary spaces $\mathcal{Y}_1(L)$ and $\mathcal{Y}_2(L)$ of zeros of $L(t, D_t)$ on the semiaxis $t > t_0$.

As in Sect. 5.6 we assume that $L(t, D_t) = \mathcal{A}(D_t)$ for large negative t. Let

$$b_j \rho(t) \leq \omega_j(t) , \quad j = 1, 2, \tag{8.52}$$

where b_j are the constants in (8.1). We suppose that ω_j satisfies (8.2) and use Green's functions g_{1,ω_1} and g_{2,ω_2} introduced in Sect. 8.2.1.

Definition 8.5.1. Let $\mathcal{Y}_j(L)$, $j = 1, 2$, be the set of solutions $v \in W^\ell_{\text{loc}}(t_0, \infty)$ of the equation

$$L(t, D_t)v = 0 \quad \text{for} \quad t > t_0 , \tag{8.53}$$

which are subject to

$$\|v\|_{W^\ell(t,t+1)} \leq C g_{j,\omega_j}(t, 0), \tag{8.54}$$

where the constant C depends on v.

We show that $\mathcal{Y}_1(L)$ is the sum of $\mathcal{Y}_2(L)$ and the space of restrictions of elements in $\mathfrak{X}(L)$ to the semiaxis $t > t_0$.

Lemma 8.5.2. Let $v \in \mathcal{Y}_1(L)$. Then there exists $u \in \mathfrak{X}(L)$ satisfying

$$\|v - u\|_{W^\ell(t,t+1)} \leq c \, g_{2,\omega_2}(t, 0) \tag{8.55}$$

for $t > t_0$. Such a function u is unique. In particular, $u = 0$ if $v \in \mathcal{Y}_2(L)$.

Proof. Let ζ be a smooth function such that $\zeta(t) = 1$ for $t > t_0 + 1$ and $\zeta(t) = 0$ for $t < t_0 + 1/2$. We seek u in the form $\zeta v + w$. Then w satisfies

$$L(t, D_t)w(t) = -\big[L(t, D_t), \zeta(t)\big]v \quad \text{on} \quad \mathbb{R} . \tag{8.56}$$

Since the right-hand side has compact support, by Theorem 5.3.2 there exists the $(k_-^{(2)}, k_+^{(2)})$-solution to (8.56) satisfying

$$\|w\|_{W^\ell(t,t+1)} \leq c \, g_{2,\omega_2}(t, 0).$$

Thus the existence of u has been proved.

We pass to the uniqueness of u. If there exists another function $u_1 \in \mathfrak{X}(L)$ satisfying (8.55) then using (8.17) for $u - u_1$ we have

$$\|u - u_1\|_{W^\ell(t,t+1)} \leq c \, g_{2,\omega_2}(t, 0)$$

for all $t \in \mathbb{R}$. Hence and by (5.28)

$$\|u - u_1\|_{W^\ell(t,t+1)} = o\left(1/g_{2,\omega_2}(0,t)\right) \quad \text{as} \quad t \to \pm\infty.$$

Applying Proposition 5.4.3 to the solution $u - u_1$ we complete the proof. $\square$

Now we present an analog of Proposition 8.3.2.

Proposition 8.5.3. (i) *If* $v \in W_{\text{loc}}^{\ell}(t_0, \infty)$ *is a solution of* (8.53) *satisfying*

$$\liminf_{t \to +\infty} g_{j,\omega_j}(0, t)\|v\|_{W_{\text{loc}}^{\ell}(t,t+1)} = 0 \tag{8.57}$$

then $v \in \mathcal{Y}_j(L)$.
   (ii) *If* $v \in \mathcal{Y}_1(L) \setminus \mathcal{Y}_2(L)$ *then*

$$\liminf_{t \to +\infty} g_{2,\omega_2}(0, t)\|v\|_{W^{\ell}(t,t+1)} > 0. \tag{8.58}$$

Proof. (i) It suffices to apply Corollary 5.6.2 with $f = 0$.
(ii) By Lemma 8.5.2 there exists $u \in \mathfrak{X}(L)$ satisfying (8.55). Since v does not belong to $\mathcal{Y}_2(L)$, $u \neq 0$ and (8.58) is valid for u, by (8.20). Now (8.58) follows from the same estimate for u and from (8.55) by using (5.28) for g_{2,ω_2}. $\square$

8.5.2 $(\mathcal{Y}_1, \mathcal{Y}_2)$-Spaces

Definition 8.5.4. A subspace $\mathbf{X}(L)$ of $\mathcal{Y}_1(L)$ is called $(\mathcal{Y}_1, \mathcal{Y}_2)$-space if the following two properties hold:
(a) $\mathbf{X}(L) \cap \mathcal{Y}_2(L) = \{O\}$.
(b) For every $v \in \mathcal{Y}_1(L)$ there exists $u \in \mathbf{X}(L)$ such that

$$v - u \in \mathcal{Y}_2(L)$$

(note that u is unique by (a)).

By Lemma 8.5.2 the space of restrictions of elements of $\mathfrak{X}(L)$ to the semi-axis $t > t_0$ serves as an example of $(\mathcal{Y}_1, \mathcal{Y}_2)$-space.

Proposition 8.5.5. *The* $(\mathcal{Y}_1, \mathcal{Y}_2)$-*space* $\mathbf{X}(L)$ *possesses the following properties:*
(i) $\dim \mathbf{X}(L) = \kappa$.
(ii) *If* $O \neq v \in \mathbf{X}(L)$ *then* (8.58) *holds.*

Proof. Since $\mathbf{X}(L) \cap \mathcal{Y}_2(L) = \{O\}$, (ii) follows from Proposition 8.5.3(ii). Let us prove (i). Consider the linear mapping

$$\mathbf{X}(L) \ni v \to u \in \mathfrak{X}(L), \tag{8.59}$$

where u is the function in $\mathfrak{X}(L)$ such that (8.55) holds. It is injective by Lemma 8.5.2. We show that it is surjective. Let $u \in \mathfrak{X}(L)$. Then the restriction of u on (t_0, ∞) belongs to $\mathcal{Y}_1(L)$. By Definition 8.5.4 there exists $v \in \mathbf{X}(L)$ such that (8.55) is valid. Thus, the operator (8.59) is bijective and $\dim \mathbf{X}(L) = \dim \mathfrak{X}(L)$. The result follows from Proposition 8.3.4. $\square$

The next assertion gives sufficient conditions for $\mathbf{X}(L)$ being a $(\mathcal{Y}_1, \mathcal{Y}_2)$-space.

Proposition 8.5.6. *Let* $\mathbf{X}(L) \subset W^\ell_{\text{loc}}(t_0, \infty)$ *be a linear space of solutions to* (8.53) *of dimension* κ. *Suppose that for every nonzero* $u \in \mathbf{X}(L)$ *the relations hold:*

$$||u||_{W^\ell(t,t+1)} = o\big(1/g_{1,\omega_1}(0,t)\big) \quad \text{as } t \to +\infty \tag{8.60}$$

and

$$\limsup_{t\to+\infty} \frac{||u(t)||_0}{g_{2,\omega_2}(t,0)} = \infty. \tag{8.61}$$

Then $\mathbf{X}(L)$ *is a* $(\mathcal{Y}_1, \mathcal{Y}_2)$-*space.*

Proof. By Proposition 8.5.3(i) $u \in \mathcal{Y}_1(L)$. Using (8.61) we obtain $\mathbf{X}(L) \cap \mathcal{Y}_2(L) = \{O\}$. Consider the mapping (8.59). It is injective by Lemma 8.5.2. Since $\dim \mathbf{X}(L) = \kappa$ the operator (8.59) is bijective. Let $v \in \mathcal{Y}_1(L)$. By Lemma 8.5.2 there exists $u \in \mathfrak{X}(L)$ such that (8.55) holds. Since the operator (8.59) is bijective there exists $w \in \mathbf{X}(L)$ such that $u - w \in \mathcal{Y}_2(L)$. Therefore $v - w \in \mathcal{Y}_2(L)$. Thus the space $\mathbf{X}(L)$ has properties (a) and (b) from Definition 8.5.4. $\square$

8.5.3 An Asymptotic Representation

Theorem 8.5.7. *Let* $\mathbf{X}(L)$ *be a* $(\mathcal{Y}_1, \mathcal{Y}_2)$-*space with a basis* $v_1, \dots, v_\kappa$. *Also let* $u \in W^\ell_{\text{loc}}(t_0, \infty)$ *be a solution of* (6.8) *where* $f \in L_{2,\text{loc}}(t_0, \infty; H_0)$ *and*

$$\int_{t_0}^\infty g_{2,\omega_2}(0,\tau)||f||_{L_2(\tau,\tau+1;H_0)}d\tau < \infty.$$

If

$$\liminf_{t\to+\infty} g_{1,\omega_1}(0,t)||u||_{W^\ell(t,t+1)} = 0 \tag{8.62}$$

then

$$u(t) = \sum_{1\le s\le\kappa} c_s v_s(t) + v(t) \tag{8.63}$$

and there exists a constant c *independent of* f *and* u *such that*

$$||v||_{W^\ell(t,t+1)} \le c\Big\{ \int_{t_0}^\infty g_{2,\omega_2}(t,\tau)||f||_{L_2(\tau,\tau+1;H_0)}d\tau$$

$$+ g_{2,\omega_2}(t,t_0)||u||_{W^{\ell-1}(t_0,t_0+1)} \Big\} \tag{8.64}$$

for all $t > t_0 + 1$. *The coefficients in* (8.63) *admit the estimate*

$$\sum_{1\le s\le\kappa} |c_s| \le c\Big(\int_{t_0}^\infty g_{2,\omega_2}(0,\tau)||f||_{L_2(\tau,\tau+1;H_0)}d\tau$$

$$+ ||u||_{W^{\ell-1}(t_0,t_0+1)} \Big). \tag{8.65}$$

Proof. Let ζ be a smooth function such that $\zeta(t) = 1$ for $t \geq t_0 + 1$ and $\zeta(t) = 0$ for $t < t_0 + 1/2$. We rewrite (6.8) as

$$L(t, D_t)(\zeta u) = \zeta f + \big[L(t, D_t), \zeta\big]u. \qquad (8.66)$$

Denote by U_s the element of $\mathfrak{X}(L)$ such that

$$\|v_s - U_s\|_{W^\ell(t,t+1)} \leq c\, g_{2,\omega_2}(t, 0)$$

for large positive t. Clearly, $U_1,\ldots, U_\kappa$ is a basis in $\mathfrak{X}(L)$.

Applying Theorem 8.4.7 we get

$$\zeta u(t) = \sum_{s=1}^{\kappa} c_s U_s(t) + w(t),$$

where w is the $(k_-^{(2)}, k_+^{(2)})$-solution of (8.66) and

$$c_s = \int_{\mathbb{R}} \big((\zeta f)(\tau) + [L(\tau, D_\tau), \zeta(\tau)]u(\tau), V_s(\tau)\big)_H d\tau. \qquad (8.67)$$

By Theorem 5.3.2

$$\|w\|_{W^\ell(t,t+1)} \leq c\bigg\{ \int_{t_0}^{\infty} g_{2,\omega_2}(t, \tau)\|f\|_{L_2(\tau,\tau+1;H_0)}$$

$$+ g_{2,\omega_2}(t, t_0)\|u\|_{W^{\ell-1}(t_0,t_0+1)}\bigg\}.$$

Using (8.67) and the estimate (8.29) for V_s we obtain (8.65). By (8.55) and

$$g_{2,\omega_2}(t, 0)g_{2,\omega_2}(0, \tau) \leq c\, g_{2,\omega_2}(t, \tau) \quad \text{for } t, \tau \geq 0,$$

(which follows from (5.31)), we arrive at (8.63) with

$$v = w + \sum_{s=1}^{\kappa} c_s(v_s - U_s)$$

subject to (8.64). $\square$

Remark 8.5.8. Let $f = 0$ for $t \geq t_0$. Formula (8.63) can be considered as an asymptotic representation of u at $+\infty$ because

$$\|u - v\|_{W^\ell(t,t+1)} \geq \frac{C}{g_{2,\omega_2}(0, t)}$$

if $u \neq v$ and

$$\|v\|_{W^\ell(t,t+1)} \leq C g_{2,\omega_2}(t, 0)$$

(see (8.64)); it remains to recall that

$$g_{2,\omega_2}(t, 0)g_{2,\omega_2}(0, t) \to 0 \quad \text{as } t \to \infty$$

by (5.28).

8.6 Comments

The inclusion (8.38) first appeared in Evgrafov (1960) for first order equations, and was extended to higher order equations by Maz'ya, Plamenevskii (1978). In this chapter we follow Kozlov, Maz'ya (1991-1996), report 6.

Formulae for coefficients which are similar to (8.50) were obtained by Maz'ya, Plamenevskii (1977) for solutions of elliptic boundary value problems in domains with conic points.

9. Applications to the Case of Perturbations Vanishing at Infinity

9.1 Introduction

In this chapter we apply general theorems of the previous chapter to two special cases. We suppose first that $\rho(t) = o(1)$ at infinity and then assume additionally that the Jordan chains corresponding to the spectrum of $\mathcal{A}(\lambda)$ in the strip $k_-^{(1)} < \Im\lambda < k_+^{(2)}$ do not contain generalized eigenvectors. In particular, one of the results relating to the first hypothesis ensures the representation

$$u(t) = \sum_k c_k v_k(t) + o\left(e^{-(k_-^{(2)}+\delta)t}\right) \tag{9.1}$$

for some positive δ. Here $\{v_k\}$ is a finite collection of zeros of $L(t, D_t)$ and u is a solution of $L(t, D_t)u = f$ for $t > t_0$, subject to the growth condition

$$\| u \|_{W^\ell(t,t+1)} = o\left(e^{-(k_-^{(1)}+\gamma)t}\right)$$

with some positive γ. The right-hand side f satisfies

$$\int_{t_0}^\infty e^{(k_-^{(2)}+\delta)\tau} \| f \|_{L_2(\tau,\tau+1;H_0)} \, d\tau < \infty.$$

Since the zeros v_k admit the lower estimate

$$\| v_j \|_{W^\ell(t,t+1)} \ge c\, e^{-(k_-^{(2)}+\varepsilon)t}$$

for large t, where ε is an arbitrary positive number, (9.1) can be considered as an asymptotic formula for u.

In Sect. 9.3.2 we prove the existence of characteristic exponents for any solution of the homogeneous equation on a semiaxis subject to lower and upper exponential estimates. We also consider the question of the so called asymptotic equivalence for two equations in the case $\rho(t) \to 0$ as $t \to \pm\infty$ (see Sect. 9.3.3).

Another application, given in Sect. 9.4, relates to the case when $k_+^{(1)} = k_-^{(2)}$ and there are no generalized eigenvectors corresponding to eigenvalues on the line $\Im\lambda = k_-^{(2)}$.

We finish the chapter with applications of our previous results to the local regularity of solutions to elliptic equations.

9.2 The Case $\rho(t) \to 0$ as $t \to \pm\infty$

We assume as before that the strips $k_-^{(j)} < \Im\lambda < k_+^{(j)}$ are free of eigenvalues of $\mathcal{A}(\lambda)$ and that $k_+^{(1)} \leq k_-^{(2)}$. By κ we denote the total algebraic multiplicity of the eigenvalues of $\mathcal{A}(\lambda)$ in the strip $k_+^{(1)} \leq \Im\lambda \leq k_-^{(2)}$.

Let us suppose that the function ρ does not exceed a sufficiently small constant on $\mathbb{R}$ and that ρ tends to zero at $\pm\infty$. In the present section we collect some simply formulated corollaries of the results in Sect. 8.4 and 8.5.

Estimates (6.3) and (6.4) imply

$$g_{j,\omega_j}(t,\tau) \leq \begin{cases} c_\varepsilon e^{-(k_+^{(j)}-\varepsilon)(t-\tau)} & \text{for } t \geq \tau \\ c_\varepsilon e^{-(k_-^{(j)}+\varepsilon)(t-\tau)} & \text{for } t \leq \tau \end{cases} \tag{9.2}$$

for an arbitrary $\varepsilon > 0$ and for t and τ of the same sign. Hence, using the estimate

$$g_{j,\omega_j}(t,\tau) \leq c\, g_{j,\omega_j}(t,0) g_{j,\omega_j}(0,\tau),$$

(which follows from (5.31) for t and τ of different signs), we extend (9.2) to t and τ of arbitrary signs.

From (5.35) we obtain

$$\sup_{\tau\in\mathbb{R}} \frac{g_{j,\omega_j}(t,\tau)}{g_{j,\omega_j}(0,\tau)} \leq \begin{cases} c\, e^{-k_-^{(j)}t} & \text{for } t \geq 0 \\ c\, e^{-k_+^{(j)}t} & \text{for } t \leq 0. \end{cases} \tag{9.3}$$

Now from Lemma 8.2.2 together with (9.2) and (9.3), we deduce:

Lemma 9.2.1. *Let $f \in L_{2,\text{loc}}(\mathbb{R}; H_0)$ and*

$$\int_0^\infty e^{(k_-^{(1)}+\delta)\tau} \| f \|_{L_2(\tau,\tau+1;H_0)} \, d\tau$$

$$+ \int_{-\infty}^0 e^{(k_+^{(2)}-\delta)\tau} \| f \|_{L_2(\tau,\tau+1;H_0)} \, d\tau < \infty$$

for some $\delta > 0$. Then there exists a solution $u \in W_{\text{loc}}^\ell(\mathbb{R})$ of (5.1) satisfying

$$\| u \|_{W^\ell(t,t+1)} = \begin{cases} o\big(e^{-k_-^{(1)}t}\big) & \text{as } t \to +\infty \\ o\big(e^{-k_+^{(2)}t}\big) & \text{as } t \to -\infty. \end{cases}$$

9.2.1 Description of the Class $\mathfrak{X}(L)$

Proposition 9.2.2. (i) *The solution $u \in W_{\text{loc}}^\ell(\mathbb{R})$ of (8.15) belongs to $\mathfrak{X}(L)$ if and only if*

$$\| u \|_{W^\ell(t,t+1)} = \begin{cases} O\big(e^{-(k_-^{(1)}+\delta)t}\big) & \text{for } t > 0 \\ O\big(e^{-(k_+^{(2)}-\delta)t}\big) & \text{for } t < 0 \end{cases} \tag{9.4}$$

for some positive δ.
   (ii) *If $u \in \mathfrak{X}(L)$ then*

$$\| u \|_{W^\ell(t,t+1)} = \begin{cases} O\big(e^{-(k_+^{(1)}-\varepsilon)t}\big) & \text{for } t > 0 \\ O\big(e^{-(k_-^{(2)}+\varepsilon)t}\big) & \text{for } t < 0 \end{cases} \tag{9.5}$$

with an arbitrary $\varepsilon > 0$.
   (iii) *If $u \in \mathfrak{X}(L)$ and $u \neq 0$ then*

$$\liminf_{t \to +\infty} e^{(k_-^{(2)}+\varepsilon)t} \| u \|_{W^\ell(t,t+1)} > 0, \tag{9.6}$$

$$\liminf_{t \to -\infty} e^{(k_+^{(1)}-\varepsilon)t} \| u \|_{W^\ell(t,t+1)} > 0 \tag{9.7}$$

with an arbitrary $\varepsilon > 0$.
   (iv) $\dim \mathfrak{X}(L) = \kappa$, *where κ is the total algebraic multiplicity of the eigenvalues of $\mathcal{A}(\lambda)$ in the strip $k_+^{(1)} \leq \Im\lambda \leq k_-^{(2)}$.*

Proof (i) By (9.2), the relation (9.4) yields (8.18) and (8.19). Hence $u \in \mathfrak{X}(L)$ by Proposition 8.3.2(i). The converse assertion results from (ii).
   (ii) and (iii) Estimates (8.16) and (8.17) together with (9.2) give (9.5). Inequalities (9.6) and (9.7) follows from Proposition 8.3.2(ii) and (9.2).
   (iv) The result follows from Proposition 8.3.4. $\square$

9.2.2 Description of the Class $\mathfrak{X}^*(L)$

Proposition 9.2.3. (i) *The solution $v \in \text{coker} L$ belongs to $\mathfrak{X}^*(L)$ if and only if*

$$\| v \|_{L_2(t,t+1;H_0)} = \begin{cases} O\big(e^{(k_+^{(2)}-\delta)t}\big) & \text{for } t > 0 \\ O\big(e^{(k_-^{(1)}+\delta)t}\big) & \text{for } t < 0 \end{cases} \tag{9.8}$$

with some positive δ.
   (ii) *If $u \in \mathfrak{X}^*(L)$ then*

$$\| v \|_{L_2(t,t+1;H_0)} = \begin{cases} O\big(e^{(k_-^{(2)}+\varepsilon)t}\big) & \text{for } t > 0 \\ O\big(e^{(k_+^{(1)}-\varepsilon)t}\big) & \text{for } t < 0 \end{cases} \tag{9.9}$$

with an arbitrary $\varepsilon > 0$.

(iii) *If $v \in \mathfrak{X}^*(L)$ then*

$$\liminf_{t \to +\infty} e^{-(k_+^{(1)} - \varepsilon)t} \| v \|_{L_2(t,t+1;H_0)} > 0, \qquad (9.10)$$

$$\liminf_{t \to -\infty} e^{-(k_-^{(2)} + \varepsilon)t} \| u \|_{L_2(t,t+1;H_0)} > 0 \qquad (9.11)$$

with an arbitrary $\varepsilon > 0$.
(iv) $\dim \mathfrak{X}^*(L) = \kappa$.

Proof As in the proof of Proposition 9.2.2 (with evident changes). $\square$

9.2.3 Characteristic Exponents

Let $u \in W_{\text{loc}}^{\ell}(\mathbb{R})$. We call the number $\gamma_{\pm}$ characteristic exponent of v at $\pm\infty$ if there exists a finite limit

$$\gamma_{\pm} = \lim_{t \to \pm\infty} t^{-1} \log \|u\|_{W^{\ell}(t,t+1)}.$$

For $v \in L_2(\mathbb{R}; H_0)$ we define the characteristic exponents $\gamma_{\pm}$ at $\pm\infty$ by

$$\gamma_{\pm} = \lim_{t \to \pm\infty} t^{-1} \log \|v\|_{L_2(t,t+1;H_0)}$$

(if the corresponding limits exist). By using this notion we can formulate Propositions 9.2.2 and 9.2.3 in the special case $k_+^{(1)} = k_-^{(2)}$ in the following (simpler) way.

Corollary 9.2.4. *Let $k_+^{(1)} = k_-^{(2)} = k_0$ and let κ be the total algebraic multiplicity of the eigenvalues of $\mathcal{A}(\lambda)$ on the line $\Im\lambda = k_0$. Then the spaces $\mathfrak{X}(L)$ and $\mathfrak{X}^*(L)$ are κ-dimensional. Every non-zero element of $\mathfrak{X}(L)$ ($\mathfrak{X}^*(L)$) has characteristic exponent $-k_0$ (k_0) at $\pm\infty$.*

Proof. For $\mathfrak{X}(L)$, the result follows from (9.5)-(9.7) which imply

$$c_{1,\varepsilon} e^{-\varepsilon|t|} \le e^{k_0 t} \| u \|_{W^{\ell}(t,t+1)} \le c_{2,\varepsilon} e^{\varepsilon|t|} \qquad (9.12)$$

for large $|t|$ with an arbitrary positive ε.

Analogously, the proof for $\mathfrak{X}^*(L)$ follows from Proposition 9.2.3. $\square$

9.2.4 The Difference of Two Solutions

Now we turn to the nonhomogeneous equation (5.1). From Theorem 8.4.7 and (9.2) we obtain the following result on solutions corresponding to different strips:

Theorem 9.2.5. *Let $f \in L_{2,\text{loc}}(\mathbb{R}; H_0)$ and*

$$\int_0^\infty e^{(k_-^{(2)}+\delta)\tau} \| f \|_{L_2(\tau,\tau+1;H_0)} \, d\tau$$

$$+ \int_{-\infty}^0 e^{(k_+^{(1)}-\delta)\tau} \| f \|_{L_2(\tau,\tau+1;H_0)} \, d\tau < \infty$$

with some $\delta > 0$. Let u_1 and u_2 be the solutions of (5.1) corresponding to the strips $k_-^{(1)} < \Im\lambda < k_+^{(1)}$ and $k_-^{(2)} < \Im\lambda < k_+^{(2)}$ respectively. Then the difference $u_2 - u_1$ belongs to $\mathfrak{X}(L)$ and admits the representation (8.50).

9.3 The Case $\rho(t) \to 0$ as $t \to +\infty$

9.3.1 Structure of Solutions at $+\infty$

By Proposition 8.5.3(i) and estimate (9.2) the space $\mathcal{Y}_j(L)$ of zeros of $L(t, D_t)$ on the semiaxis $t > t_0$ contains the zeros subject to the inequality

$$\| u \|_{W^\ell(t,t+1)} \leq c \, e^{-(k_-^{(j)}+\varepsilon)t} \quad \text{for} \quad t > t_0, \tag{9.13}$$

with some $\varepsilon > 0$. Using (9.2) and Definition 8.5.1 we note that, moreover, the elements of $\mathcal{Y}_j(L)$ satisfy (9.13) with any $\varepsilon > 0$.

By Proposition 8.5.6 and (9.2) a κ-dimensional space $\mathbf{X}(L)$ of zeros of $L(t, D_t)$ on the semiaxis $t > t_0$ is a $(\mathcal{Y}_1, \mathcal{Y}_2)$-space if its nonzero elements satisfy

$$\| u \|_{W^\ell(t,t+1)} \leq c \, e^{-(k_-^{(1)}+\varepsilon)t} \tag{9.14}$$

and

$$\limsup_{t\to+\infty} \|u(t)\|_0 e^{(k_+^{(2)}-\varepsilon)t} = \infty \tag{9.15}$$

with a positive ε.

In the following proposition we show that the relations (9.14) and (9.15) can be improved.

Proposition 9.3.1. *Let $\mathbf{X}(L)$ be a $(\mathcal{Y}_1, \mathcal{Y}_2)$-space of solutions to (8.53). If $u \in \mathbf{X}(L)$, $u \neq 0$, then the estimates*

$$c_1 \, e^{-(k_-^{(2)}+\varepsilon)t} \leq \| u \|_{W^\ell(t,t+1)} \leq c_2 \, e^{-(k_+^{(1)}-\varepsilon)t} \tag{9.16}$$

hold for large positive t, with an arbitrary $\varepsilon > 0$.

Proof. The inequalities (8.54) and (8.58) (which are valid for elements in $\mathcal{Y}_1(L) \setminus \mathcal{Y}_2(L)$) together with (9.2) imply the result. $\square$

Now we give an analog of Theorem 8.5.7.

Theorem 9.3.2. *Let* $u \in W_{\mathrm{loc}}^{\ell}(t_0, \infty)$ *be a solution of the equation* (6.8) *with* $f \in L_{2,\mathrm{loc}}(t_0, \infty; H_0)$ *subject to*

$$\int_{t_0}^{\infty} e^{(k_{-}^{(2)} + \delta)\tau} \| f \|_{L_2(\tau, \tau+1; H_0)} \, d\tau < \infty \tag{9.17}$$

for some $\delta > 0$. *Also let* $\{v_s\}_{s=1}^{\kappa}$ *be a basis in a* $(\mathcal{Y}_1, \mathcal{Y}_2)$*-space* $\mathbf{X}(L)$.
If there exists $\gamma > 0$ *such that*

$$\| u \|_{W^{\ell}(t, t+1)} \leq c \, e^{-(k_{-}^{(1)} + \gamma)t} \quad \text{for } t > t_0, \tag{9.18}$$

then

$$u(t) = \sum_{1 \leq s \leq \kappa} c_s v_s(t) + v(t) \tag{9.19}$$

holds with the remainder term v *satisfying*

$$
\begin{aligned}
\| v \|_{W^{\ell}(t, t+1)} \leq \ & c \Big\{ \int_{t_0}^{t} e^{-(k_{+}^{(2)} - \delta)(t-\tau)} \| f \|_{L_2(\tau, \tau+1; H_0)} \, d\tau \\
& + \int_{t}^{\infty} e^{-(k_{-}^{(2)} + \delta)(t-\tau)} \| f \|_{L_2(\tau, \tau+1; H_0)} \, d\tau \\
& + e^{-(k_{+}^{(2)} - \delta)(t-t_0)} \| u \|_{W^{\ell-1}(t_0, t_0+1)} \Big\},
\end{aligned} \tag{9.20}
$$

where $t > t_0 + 1$ *and* c *does not depend on* u, f. *The coefficients in* (9.19)
admit the estimate

$$
\begin{aligned}
\sum_{s=1}^{\kappa} |c_s| \leq c \ \Big(& \int_{t_0}^{\infty} e^{(k_{-}^{(2)} + \delta)\tau} \| f \|_{L_2(\tau, \tau+1; H_0)} \, d\tau \\
& + \| u \|_{W^{\ell-1}(t_0, t_0+1)} \Big).
\end{aligned}
$$

Proof. The result follows from (9.2) and Theorem 8.5.7. $\square$

Remark 9.3.3. Under the conditions of Theorem 9.3.2, the representation
(9.19) is valid with v satisfying

$$\| v \|_{W^{\ell}(t, t+1)} = o\big(e^{-(k_{-}^{(2)} + \delta)t}\big) \quad \text{as} \quad t \to +\infty. \tag{9.21}$$

This follows from (9.17) and (9.20). Comparing (9.21) with Proposition 9.3.1
we see that (9.19) is an asymptotic formula for u.

The next assertion is a direct consequence of Theorem 9.3.2.

Corollary 9.3.4. *Let* $\mathbf{X}(L)$ *be a* $(\mathcal{Y}_1, \mathcal{Y}_2)$*-space of solutions to* (8.53) *and* $u \in \mathcal{Y}_1(L)$ *Then there exists a unique* $V \in \mathbf{X}(L)$ *such that*

$$\| u - V \|_{W^{\ell}(t, t+1)} \leq C \, e^{-(k_{+}^{(2)} - \varepsilon)t} \quad \text{for} \quad t > t_0 \tag{9.22}$$

(ε *is an arbitrary positive number*). *The estimate holds:*

$$C + \|V\|_{W^{\ell}(t_0, t_0+1)} \leq c \, \|u\|_{W^{\ell}(t_0, t_0+1)} \tag{9.23}$$

with c *independent of* u.

9.3.2 Existence of Characteristic Exponents for Solutions of (8.53)

Lemma 9.3.5. *Suppose that all eigenvalues of $\mathcal{A}(\lambda)$ in the strip $k_- \leq \Im\lambda \leq k_+$ lie on the line $\Im\lambda = k_0$ with $k_0 \in (k_-, k_+)$. We denote by κ_0 the total algebraic multiplicity of eigenvalues on the line $\Im\lambda = k_0$. Then there exists a κ_0-dimensional space $\mathbf{X}(L) \subset W^\ell_{\mathrm{loc}}(t_0, \infty)$ of solutions of (8.53) with the following properties:*

(i) All nonzero elements in $\mathbf{X}(L)$ have the characteristic exponent $-k_0$ at $+\infty$.

(ii) If $U \in W^\ell_{\mathrm{loc}}(t_0, \infty)$ is a solution of (8.53) such that

$$\| U \|_{W^\ell(t, t+1)} \leq Ce^{-k_- t}$$

for large positive t, then there exists a unique $u \in \mathbf{X}(L)$ such that

$$\| U - u \|_{W^\ell(t, t+1)} \leq Ce^{-k_+ t}.$$

Proof. We take $k_+^{(1)} = k_-^{(2)} = k_0$ and choose $k_-^{(1)}$ and $k_+^{(2)}$ in such a way that $k_-^{(1)} < k_-$, $k_+^{(2)} > k_+$ and the strips $k_-^{(1)} < \Im\lambda < k_0$, $k_0 < \Im\lambda < k_+^{(2)}$ contain no eigenvalues of $\mathcal{A}(\lambda)$. Using (9.2) we obtain

$$g_{1,\omega_1}(0, t) = o(e^{-k_- t}) \quad \text{as } t \to +\infty \tag{9.24}$$

and

$$g_{2,\omega_2}(t, 0) = o(e^{-k_+ t}) \quad \text{as } t \to +\infty. \tag{9.25}$$

Let $\mathbf{X}(L)$ be a $(\mathcal{Y}_1, \mathcal{Y}_2)$-space. Then (i) and (ii) follow from Proposition 9.3.1, Corollary 9.3.4 combined with (9.24) and (9.25). $\square$

Theorem 9.3.6. *(i) Let $U \in W^\ell_{\mathrm{loc}}(t_0, \infty)$ be a solution of (8.53) such that*

$$\limsup_{t \to +\infty} \frac{1}{t} \log \|U\|_{W^\ell(t, t+1)} < +\infty \tag{9.26}$$

and

$$\liminf_{t \to +\infty} \frac{1}{t} \log \|U\|_{W^\ell(t, t+1)} > -\infty. \tag{9.27}$$

Then there exists a characteristic exponent of U.

Proof. By (9.26) and (9.27) there exists a positive constant C such that

$$c_1 e^{-Ct} \leq \|u\|_{W^\ell(t, t+1)} \leq c_2 e^{Ct} \tag{9.28}$$

for large positive t. Here c_1 and c_2 are positive constants. We can assume that the lines $\Im\lambda = \pm C$ contain no eigenvalues of the pencil $\mathcal{A}(\lambda)$. We denote by k_j, $j = 1, ..., M$, real numbers such that

$$-C < k_1 < ... < k_M < C$$

and all eigenvalues of $\mathcal{A}(\lambda)$ in the strip $-C < \Im\lambda < C$ are situated on one of the lines $\Im\lambda = k_j$. We denote by κ_j the total algebraic multiplicity of eigenvalues of $\mathcal{A}(\lambda)$ on the line $\Im\lambda = k_j$ and by $\mathbf{X}_j(L)$ the κ_j-dimensional space of solutions of (8.53) whose non-zero elements have the characteristic exponent at $+\infty$ equal to $-k_j$ (see Lemma 9.3.5(i)). Applying Lemma 9.3.5(ii) M times we obtain

$$\left\|U - \sum_{j=1}^{M} u_j\right\|_{W^\ell(t,t+1)} = o(e^{-Ct}) \quad \text{as } t \to +\infty$$

with some $u_j \in \mathbf{X}_j(L)$. By the left inequality in (9.28), we have that one of the functions $\{u_j\}_1^M$ is not zero. This gives the existence of the characteristic exponent. $\square$

9.3.3 Asymptotic Equivalence of Two Equations

Together with (8.53), let us consider the equation

$$L^{(0)}(t, D_t)v(t) = 0 \quad \text{for } t > t_1, \tag{9.29}$$

where $L^{(0)}$ is subject to the assumptions stated in Sect. 5.3.1. We introduce the function

$$\epsilon(t) = \|L^{(0)} - L\|_{W^\ell(t,t+1)\to L_2(t,t+1;H_0)}$$

on the semiaxis $t > t_0$. We show that equations (8.53) and (9.29) are asymptotically equivalent in some sense if $\epsilon(t)$ vanishes exponentially at $+\infty$.

Theorem 9.3.7. *Let u be a solution of* (8.53) *with the characteristic exponent γ at $+\infty$ and let*

$$\epsilon(t) = O(e^{-\delta t}) \quad \text{for } t > t_0 \tag{9.30}$$

with a positive constant δ. Then there exists a solution v of (9.29) *with sufficiently large t_1 such that*

$$\|u - v\|_{W^\ell(t,t+1)} = O\left(e^{(\gamma-\delta_1)t}\right) \quad \text{for } t > t_1 \tag{9.31}$$

with a certain positive δ_1.

Proof. Denote $k_+^{(1)} = k_-^{(2)} = -\gamma$ and choose the numbers $k_-^{(1)}$ and $k_+^{(2)}$ so that the strips $k_-^{(j)} < \Im\lambda < k_+^{(j)}$, $j = 1, 2$, are free of eigenvalues of $\mathcal{A}(\lambda)$. Let us seek v as the sum $u + w$, where w is a function from $W_{\text{loc}}^\ell(t_1, \infty)$ satisfying

$$L^{(0)}w = (L - L^{(0)})u \quad \text{for } t > t_1. \tag{9.32}$$

By (9.30)

$$\int_{t_0}^{\infty} e^{(k_-^{(2)}+\varepsilon_1)\tau} \|(L - L^{(0)})u\|_{W^\ell(\tau,\tau+1)} d\tau < \infty \tag{9.33}$$

if $\varepsilon_1 < \delta$; this implies (6.10). We obtain w by referring to Remark 6.2.8 with $k_\pm = k_\pm^{(2)}$ applied to the operator $L^{(0)}$. Estimate (9.31) follows directly from (6.13) with $f = (L - L^{(0)})u$. $\square$

9.4 The Case of Absence of Generalized Eigenvectors

9.4.1 Zeros of L and L^*

In order to simplify statements, we restrict ourselves here to the case $k_+^{(1)} = k_-^{(2)}$. In other words, we assume that the spectrum of $\mathcal{A}(\lambda)$ in the strip $k_- < \Im\lambda < k_+$ is concentrated on the line $\Im\lambda = k_0$, $k_0 \in (k_-, k_+)$.

We suppose that there are no generalized eigenvectors corresponding to the eigenvalues with $\Im\lambda = k_0$ and $\Im\lambda = k_\pm$.

Let

$$R(t) = \begin{cases} \sup_{\tau \geq t} \rho(\tau) & \text{for } t \geq 0 \\ \sup_{\tau \leq t} \rho(\tau) & \text{for } t < 0 \end{cases}$$

and let $R(t) \leq c_0$, where c_0 is a sufficiently small constant which depends upon the operator $\mathcal{A}(D_t)$ and the numbers $k_\pm, k_0$. Also let $\rho(t) = o(1)$ as $t \to \pm\infty$. Our aim is to specialize the general results in Sections 8.3 - 8.5.

Under the present assumptions we have

$$k_-^{(1)} = k_-, \ k_+^{(1)} = k_-^{(2)} = k_0, \ k_+^{(2)} = k_+, \ m_\pm^{(j)} = 1$$

for $j = 1, 2$.

By Proposition 6.3.1 combined with (7.25) we have

$$g_{1,\omega_1}(t, \tau) \leq \begin{cases} ce^{-k_0(t-\tau)+a\int_\tau^t R(x)dx} & \text{for } t \geq \tau \\ ce^{-k_-(t-\tau)+a\int_t^\tau R(x)dx} & \text{for } t \leq \tau \end{cases} \tag{9.34}$$

and

$$g_{2,\omega_2}(t, \tau) \leq \begin{cases} ce^{-k_+(t-\tau)+a\int_\tau^t R(x)dx} & \text{for } t \geq \tau \\ ce^{-k_0(t-\tau)+a\int_t^\tau R(x)dx} & \text{for } t \leq \tau \end{cases} \tag{9.35}$$

with some positive a and c.

Proposition 9.4.1. (i) *A solution* $u \in W_{\text{loc}}^\ell(\mathbb{R})$ *of* (8.15) *belongs to* $\mathfrak{X}(L)$ *if and only if*

$$\| u \|_{W^\ell(t,t+1)} = \begin{cases} o\Big(\exp\big(-k_-t - a\int_0^t R(x)dx\big)\Big) & \text{for } t > 0, \\ o\Big(\exp\big(-k_+t - a|\int_0^t R(x)dx|\big)\Big) & \text{for } t < 0. \end{cases} \tag{9.36}$$

(ii) *If u is a nonzero element in* $\mathfrak{X}(L)$ *then for large* $|t|$

$$c_1 e^{-a|\int_0^t R(x)dx|} \leq e^{k_0 t} \| u \|_{W^\ell(t,t+1)} \leq c_2 e^{a|\int_0^t R(x)dx|} \tag{9.37}$$

with non-negative c_1 and c_2 independent of t.

(iii) *The dimension of* $\mathfrak{X}(L)$ *is equal to the total geometric multiplicity of the eigenvalues of* $\mathcal{A}(\lambda)$ *on the line* $\Im\lambda_0 = k_0$.

Proof. Same as that of Proposition 9.2.2 with the use of (9.34) and (9.35) instead of (9.2). $\square$

An analog of Proposition 9.2.3 is stated as

Proposition 9.4.2. (i) *The solution* $v \in \mathrm{coker}L$ *belongs to* $\mathfrak{X}^*(L)$ *if and only if*

$$\| v \|_{L_2(t,t+1;H_0)} = \begin{cases} o\Big(\exp \big(k_+ t - a \int_0^t R(x)dx \big) \Big) & \text{for } t > 0, \\ o\Big(\exp \big(k_- t - a| \int_0^t R(x)dx| \big) \Big) & \text{for } t < 0. \end{cases} \quad (9.38)$$

(ii) *If* v *is a nonzero element in* $\mathfrak{X}^*(L)$ *then for large* $|t|$

$$c_1\, e^{-a| \int_0^t R(x)dx|} \leq e^{k_0 t}\, \| u \|_{L_2(t,t+1;H_0)} \leq c_2\, e^{a| \int_0^t R(x)dx|} \quad (9.39)$$

with non-negative c_1 *and* c_2 *independent of* t.
   (iii) $\dim \mathfrak{X}^*(L) = \dim \mathfrak{X}(L)$.

9.4.2 Relation of Solutions Corresponding to Different Strips

From Theorem 8.4.7 together with (9.34) and (9.35), we obtain the following result on solutions corresponding to different strips:

Theorem 9.4.3. *Let* $f \in L_{2,\mathrm{loc}}(\mathbb{R}; H_0)$ *and*

$$\int_{\mathbb{R}} e^{k_0 \tau + a| \int_0^\tau R(x)dx|} \, \| f \|_{L_2(\tau,\tau+1;H_0)} \, d\tau < \infty.$$

Let u_1 *and* u_2 *be solutions of* (5.1) *corresponding to the strips* $k_- < \Im\lambda < k_0$ *and* $k_0 < \Im\lambda < k_+$ *respectively. Then the difference* $u_2 - u_1$ *belongs to* $\mathfrak{X}(L)$ *(see Proposition 9.4.1). Moreover, the formula* (8.50) *holds, where* $\{U_j\}$ *and* $\{V_j\}$ *are bases in* $\mathfrak{X}(L)$ *and* $\mathfrak{X}^*(L)$ *subject to the biorthogonality condition* (8.49).

9.4.3 An Asymptotic Representation of Solutions at $+\infty$

By Proposition 8.5.3(i) and by (9.34) the space $\mathcal{Y}_1(L)$ of zeros of $L(t, D_t)$ on the semiaxis $t > t_0$ contains the zeros subject to

$$\| u \|_{W^\ell(t,t+1)} = o\big(\exp \big(- k_- t - a \int_{t_0}^t R(x)dx \big) \big) \quad (9.40)$$

as $t \to +\infty$. Similarly, it follows from the same proposition and (9.35) that the zeros satisfying

$$\| u \|_{W^\ell(t,t+1)} = o\big(\exp \big(- k_0 t - a \int_{t_0}^t R(x)dx \big) \big) \quad (9.41)$$

belong to $\mathcal{Y}_2(L)$.

Conversely, by Definition 8.5.1 and by (9.34), (9.35) we see that for all $u \in \mathcal{Y}_1(L)$

$$\|u\|_{W^\ell(t,t+1)} \leq c \, \exp\left(-k_0 t + a \int_{t_0}^{t} R(x) dx\right)$$

and for all $u \in \mathcal{Y}_2(L)$

$$\|u\|_{W^\ell(t,t+1)} \leq c \, \exp\left(-k_+ t + a \int_{t_0}^{t} R(x) dx\right).$$

By Proposition 8.5.6 and by (9.34), (9.35) a κ-dimensional space $\mathbf{X}(L)$ of zeros of $L(t, D_t)$ on the semiaxis $t > t_0$ is a $(\mathcal{Y}_1, \mathcal{Y}_2)$-space if for any nonzero elements

$$\|u\|_{W^\ell(t,t+1)} = o\left(\exp\left(-k_- t - a \int_{t_0}^{t} R(x) dx\right)\right)$$

as $t \to +\infty$ and

$$\limsup_{t \to +\infty} \|u(t)\|_0 \exp\left(k_+ t - a \int_{t_0}^{t} R(x) dx\right) = \infty.$$

These relations will be refined in the following

Proposition 9.4.4. *Let $\mathbf{X}(L)$ be a $(\mathcal{Y}_1, \mathcal{Y}_2)$-space of solutions to* (8.53). *If $u \in \mathbf{X}(L)$ and $u \neq 0$ then the estimates* (9.37) *hold for large positive* t.

Proof. The upper estimate follows from the definition of $\mathcal{Y}_1(L)$ and the lower one is a corollary of Proposition 8.5.5(ii). One should also take into account the estimates (9.34) and (9.35). $\square$

We give an analog of Theorem 8.5.7.

Theorem 9.4.5. *Let $u \in W_{\mathrm{loc}}^\ell(t_0, \infty)$ be a solution of* (6.8) *with f subject to*

$$\int_{t_0}^{\infty} e^{k_0 \tau + a \int_{t_0}^{t} R(x) dx} \, \| f \|_{L_2(\tau, \tau+1; H_0)} \, d\tau < \infty. \tag{9.42}$$

Also let $\{v_s\}_{s=1}^{\kappa}$ be a basis in a $(\mathcal{Y}_1, \mathcal{Y}_2)$-space $\mathbf{X}(L)$. If

$$\| u \|_{W^\ell(t,t+1)} \leq c \, e^{-k_- t - a \int_{t_0}^{t} R(x) dx} \qquad \text{for} \quad t > t_0, \tag{9.43}$$

then (9.19) *holds with v satisfying*

$$\| v \|_{W^\ell(t,t+1)} \leq c \left\{ \int_{t_0}^{t} e^{-k_+ (t-\tau) + a \int_{\tau}^{t} R(x) dx} \, \| f \|_{L_2(\tau, \tau+1; H_0)} \, d\tau \right.$$

$$+ \int_{t}^{\infty} e^{-k_0 (t-\tau) + a \int_{t}^{\tau} R(x) dx} \, \| f \|_{L_2(\tau, \tau+1; H_0)} \, d\tau$$

$$\left. + e^{-k_+ (t-t_0) + a \int_{t_0}^{t} R(x) dx} \, \| u \|_{W^{\ell-1}(t_0, t_0+1)} \right\}, \tag{9.44}$$

where $t > t_0 + 1$ and c does not depend on u, f. The coefficients in (9.19) admit the estimate

$$\sum_{s=1}^{\kappa} |c_s| \leq c \left(\int_{t_0}^{\infty} e^{k_0 \tau + a \int_{t_0}^{\tau} R(x)dx} \| f \|_{L_2(\tau,\tau+1;H_0)} \, d\tau \right.$$

$$\left. + \| u \|_{W^{\ell-1}(t_0,t_0+1)} \right). \tag{9.45}$$

Proof. The result follows from (9.34), (9.35) and Theorem 8.5.7. $\square$

Remark 9.4.6. It follows from (9.42) and (9.44) that under the conditions of Theorem 9.4.5 the remainder term v in (9.19) satisfies

$$\| v \|_{W^{\ell}(t,t+1)} = o\left(e^{-k_0 t - a \int_0^t R(x)dx}\right) \quad \text{as } t \to \infty. \tag{9.46}$$

This together with (9.37) shows that (9.19) is an asymptotic representation of the solution u in Theorem 9.4.5.

The next assertion is a direct consequence of Theorem 9.4.5.

Corollary 9.4.7. *Let $\mathbf{X}(L)$ be a $(\mathcal{Y}_1, \mathcal{Y}_2)$-space of solutions to (8.53). If $u \in W^{\ell}_{\mathrm{loc}}(t_0, \infty)$ is a solution of (8.53) subject to*

$$\| u \|_{W^{\ell}(t,t+1)} = o\left(\exp\left(- k_- t - a \int_{t_0}^{t} R(x)dx \right) \right) \tag{9.47}$$

then there exists $V \in \mathbf{X}(L)$ such that

$$\| u - V \|_{W^{\ell}(t,t+1)} \leq C \exp\left(- k_+ t + a \int_{t_0}^{t} R(x)dx \right). \tag{9.48}$$

The estimate holds:

$$C + \| V \|_{W^{\ell}(t,t+1)} \leq c \, \| u \|_{W^{\ell}(t,t+1)},$$

where c is independent of u.

The following fact will be used in the next section.

Corollary 9.4.8. *Let $u \in W^{\ell}_{\mathrm{loc}}(t_0, \infty)$ be a solution of (8.53) with the characteristic exponent $-k_0$. Then u satisfies (9.37) for large positive t.*

Proof. It is clear that the function u satisfies (9.47). By Corollary 9.4.7 there exists $V \in \mathbf{X}(L)$ such that (9.48) holds. The result follows from the estimate (9.37). $\square$

9.4.4 The Asymptotic Equivalence of Two Equations

Let the operator $L^{(0)}$ and the function $\epsilon(t)$ be as introduced in Sect. 9.3.3. We prove an analogue of Theorem 9.3.7 with the condition (9.30) replaced by

$$\epsilon(t) = O\left(R(t)e^{-2C\int_{t_0}^t R(x)dx}\right) \quad \text{for } t > t_0 \tag{9.49}$$

with a constant $C > a$, where R and a are as in Sect. 9.4.1. The result reads as follows

Theorem 9.4.9. *Suppose that*

$$\int_{t_0}^{\infty} R(x)dx = \infty$$

and that ϵ satisfies (9.49). Let u be a solution of (8.53) with the characteristic exponent $-k_0$ at $+\infty$. Then there exists a solution v of (9.29) with a sufficiently large t_1, such that

$$\|u - v\|_{W^\ell(t,t+1)} = O\left(e^{-k_0 t + (a-2C)\int_{t_0}^t R(x)dx}\right). \tag{9.50}$$

(By Corollary 9.4.8, the estimate (9.50) shows that u can be considered as an asymptotic representation for v).

Proof. As in the proof of Theorem 9.3.7 we seek v as the sum $u + w$, where w is a function from $W^\ell_{\text{loc}}(t_1, \infty)$ satisfying (9.32). By Corollary 9.4.8 the function u satisfies (9.37) for large positive t.

It follows from (9.49) that the majorant ω corresponding to the operator $L^{(0)}$ can be chosen as

$$\omega^{(0)} = R(t)\left(b + ce^{-2C\int_{t_1}^t R(x)dx}\right)$$

for $t > t_1$ and 0 for $t < t_1$. Hence Green's function $g_{\omega^{(0)}}$ satisfies the same estimates as g_ω. By (9.34), (9.35) and (9.49) together with the upper estimate in (9.37) we obtain

$$\int_{t_1}^{\infty} e^{k_0 + a\int_{t_1}^\tau R(x)dx} \|(L - L^{(0)})u\|_{L_2(\tau,\tau+1;H_0)}$$

$$\leq c\int_{t_1}^{\infty} e^{2(a-C)\int_{t_1}^\tau R(x)dx} R(\tau)d\tau < \infty. \tag{9.51}$$

Hence the existence of w follows from Remark 6.3.9. Estimate (6.31) gives the following estimate for $t > t_1$:

$$\|w\|_{W^\ell(t,t+1)} \leq c\left(\int_{t_1-1}^{\infty} e^{-k_0(t-\tau)+a\int_t^\tau R(x)dx}\epsilon(\tau)\|u\|_{W^\ell(\tau,\tau+1)}d\tau\right.$$

$$\left. + \int_{t_1-1}^{t} e^{-k_+(t-\tau)+a\int_\tau^t R(x)dx}\epsilon(\tau)\|u\|_{W^\ell(\tau,\tau+1)}d\tau\right),$$

which implies

$$\|w\|_{W^\ell(t,t+1)} \leq c\Big(e^{-k_0 t - a\int_{t_1}^t R(x)dx} \int_t^\infty R(\tau)e^{2(a-C)\int_{t_1}^\tau R(x)dx}d\tau$$
$$+ e^{-k_+ t + a\int_{t_1}^t R(x)dx} \int_{t_1-1}^t e^{(k_+ - k_0)\tau - 2C\int_{t_1}^\tau R(x)dx} R(\tau)d\tau\Big).$$

An elementary estimate leads to (9.51). □

9.5 Application to the Local Regularity of Solutions to Elliptic Equations

The examples considered in this section show the utility of Theorem 9.4.5 for the study of local properties of solutions of elliptic equations.

Example 9.5.1. Consider the elliptic operator

$$Q^\sharp(x, D_x) = \sum_{|\alpha|=2m} q_\alpha(x)D_x^\alpha \tag{9.52}$$

in $B_{r_0} \backslash \{0\}$. As in Example 5.7.2, we consider $Q^\sharp(x, D_x)$ as a perturbation of the operator $P(D_x)$ (see (0.7) and Sect. 1.5) and denote by $R^\sharp$ a nonincreasing function on $[\log r_0^{-1}, +\infty)$ such that $R^\sharp(\log r_0^{-1})$ is sufficiently small and

$$\sup_{K_r} \sum_{|\alpha|=2m} |q_\alpha(x) - p_\alpha| \leq R^\sharp(\log r^{-1}), \tag{9.53}$$

where $K_r = \{x \in \mathbb{R}^n : e^{-1}r < |x| < r\}$.

The following theorem on removable singularities of zeros of $Q^\sharp(x, D_x)$ is valid without additional assumptions on the majorant $R^\sharp$. We show that the zero growing slower at the origin than the fundamental solution of $P(D_x)$ is asymptotically equivalent to a polynomial of degree $\leq 2m - 1$.

Theorem 9.5.2. *Let* $2m < n$. *Suppose that* $u \in W^{2m}_{2,\mathrm{loc}}(B_{r_0} \backslash \{0\})$ *satisfies*

$$Q^\sharp(x, D_x)u = 0 \quad on \quad B_{r_0} \backslash \{0\} \tag{9.54}$$

and that

$$\| u \|_{W_2^{2m}(K_r)} = o\Big(r^{2m-n} e^{-C\int_r^{r_0} R^\sharp(\log\sigma^{-1})\sigma^{-1}d\sigma}\Big) \quad as \quad r \to 0,$$

where C *is a positive constant depending on the coefficients* p_α. *Then*

$$u(x) = P_{2m-1}(x) + v(x),$$

where P_{2m-1} *is a polynomial of degree* $\leq 2m - 1$ *and*

$$\| v \|_{W_2^{2m}(K_r)} = O\Big(r^{2m} e^{C\int_r^{r_0} R^\sharp(\log\sigma^{-1})\sigma^{-1}d\sigma}\Big).$$

Proof. Spectral properties of the operator pencil $\mathcal{A}(\lambda)$ corresponding to the operator P in $\mathbb{R}^n \setminus 0$ were studied in Sect. 1.5.6 (see Proposition 1.5.3). In particular, if $2m < n$ then the spectrum of $\mathcal{A}(\lambda)$ in the strip $2m - n < \Im\lambda < 2m$ consists of the eigenvalues $0, i, \ldots, (2m - 1)i$. There are no generalized eigenfunctions. The eigenfunctions of $\mathcal{A}(\lambda)$ corresponding to ik, $k = 0, 1, \ldots, 2m - 1$, are exhausted by $r^{-k} p_k(x)$, where p_k is a homogeneous polynomial.

Since there are no derivatives of order less than $2m$ in (9.52), the spaces $\mathbf{X}_k(Q^\sharp)$, $0 \leq k < 2m$, of solutions of (9.54) subject to

$$\lim_{r \to 0} r^{-k} \| u \|_{W_2^{2m}(K_r)} = k$$

consist of homogeneous polynomials of order k. We rewrite (9.54) by using the coordinates (t, ω), where $t = \log r^{-1}$ and $\omega = r^{-1} x$. It remains to apply Theorem 9.4.5 several times for the strips $2m - n < \Im\lambda < 1, \ldots, 2m - 2 < \Im\lambda < 2m$. $\square$

Example 9.5.3. We want to obtain a similar asymptotic result for solutions of the Dirichlet problem in a halfspace. We shall use the notations (9.52) and (9.53) from the previous example. Let

$$K_r^+ = \{x \in \mathbb{R}_+^n : e^{-1} r < |x| < r\} \quad \text{and} \quad B_r^+ = \{x \in \mathbb{R}_+^n : |x| < r\}.$$

Theorem 9.5.4. *Suppose that*

$$u \in W_{2,\mathrm{loc}}^{2m} \left(\overline{B_{r_0}^+} \setminus \{0\} \right)$$

satisfies

$$\begin{cases} Q^\sharp(x, D_x) u = 0 & \text{on } B_{r_0}^+ \\ \partial_\nu^j u = 0 & \text{on } \{x \in \mathbb{R}^n : x_n = 0, 0 < |x| < r_0\}, \ j = 0, \ldots, m - 1. \end{cases}$$

Let

$$\| u \|_{W_2^{2m}(K_r^+)} = o\left(r^{m-n} e^{-C \int_r^{r_0} R^\sharp (\log \sigma^{-1}) \sigma^{-1} d\sigma} \right) \quad \text{as} \quad r \to 0,$$

where C is a positive constant depending on the coefficients p_α. Then

$$u(x) = x_n^m P_{m-1}(x) + v(x),$$

where P_{m-1} is a polynomial of degree $\leq m - 1$ and

$$\| v \|_{W_2^{2m}(K_r^+)} = O\left(r^{2m} e^{C \int_r^{r_0} R^\sharp (\log \sigma^{-1}) \sigma^{-1} d\sigma} \right).$$

Proof. Similar to that of Theorem 9.5.2, but instead of Proposition 1.5.3 one should use Proposition 1.5.2.

A result of the same nature can easily be obtained for the Neumann problem in a semicylinder, as is shown by the next example.

Example 9.5.5. Consider the Neumann problem (6.52), where $e_0, e_1, \ldots, e_n$ are zero and e_{ij} are real. Let $R(t)$ be a non-increasing majorant of the function

$$\sup_{C_t} \sum_{i,j=1}^{n} |e_{ij}(x,\tau) - \delta_i^j|.$$

We assume that $R(t_0)$ is sufficiently small.

Theorem 9.5.6. *Let $u \in W_{2,\mathrm{loc}}^2(\mathcal{C})$ satisfy (6.52) and*

$$\| u \|_{W_2^2(C_t)} = o\left(e^{\sqrt{\gamma_1}t - C\int_{t_0}^t R(\tau)d\tau}\right),$$

where γ_1 is the first positive Neumann eigenvalue of $-\Delta_x$ on Ω and C is a positive constant depending on Ω. Then

$$u(x,t) = c_1 t + c_2 + v(x,t),$$

where c_1 and c_2 are constant and

$$\| v \|_{W_2^2(C_t)} = O\left(e^{-\sqrt{\gamma_1}t + C\int_{t_0}^t R(\tau)d\tau}\right).$$

9.6 Comments

Characteristic exponents discussed in Sect. 9.2.3 are often called strict characteristic exponents (see Demidovich (1967) and Dalec'kiĭ, M. Krein (1974)).

10. Variants and Extensions of the Previous Theory

10.1 Introduction

In this chapter we extend the range of the above comparison principles in different directions.

In Sect. 10.2 we consider solutions of the differential inequality

$$\| \mathcal{A}u \|_{L_2(t,t+1;H_0)} \le \rho(t) \| u \|_{W^\ell(t,t+1)} + h(t), \tag{10.1}$$

where $u \in W^\ell_{\mathrm{loc}}(\mathbb{R})$, ρ is a non-negative measurable function and $\mathcal{A}$ is the same differential operator with constant coefficients as before. Similarly to Ch. 5, we associate with $\mathcal{A}$ and ρ the ordinary differential operator $M(\partial_t) - \omega(t)$. Then the same comparison principle as developed in Ch.5 is valid for solutions of the inequality (10.1) (provided they belong to a certain class).

In the next section we consider the operator $L(t, D_t)$ as a small perturbation of a differential operator $\mathcal{A}(t, D_t)$ with variable coefficients, so that generally there is no limit operator pencil $\mathcal{A}(\infty, \lambda)$. We remark that the comparison principle for $\mathcal{A}(t, D_t)$ implies an analogous property for the perturbed operator $L(t, D_t)$. The case of a two-term second order operator $\mathcal{A}(t, D_t)$ is treated in more detail. We describe conditions ensuring the validity of comparison principles both for $\mathcal{A}(t, D_t)$ and the perturbed operator $L(t, D_t)$.

In Sect. 10.5 we extend our previous theory to the first order equation

$$\big(D_t + A(t) - N(t)\big)u(t) = f(t) \quad \text{on} \quad \mathbb{R}, \tag{10.2}$$

which imitates parabolic partial differential equations. In Sections 10.6 and 10.7 we do the same for the ℓ-th order dissipative "hyperbolic" equations. In the last case the comparison principle involves $\| u \|_{W^{\ell-1}(t,t+1)}$.

An analogous treatment of the "elliptic" equation in variational form

$$\sum_{j=0}^{n} \sum_{k=0}^{\ell-n} D_t^{n-j} \Big(A_{jk}(t) D_t^{\ell-n-k} u(t) \Big) = f(t) \tag{10.3}$$

is given in Sect. 10.8. Although the equation is written in divergence form we impose no differentiability assumptions on coefficients.

The last section (Sect. 10.9) is devoted to a generalization to operator equations in Banach spaces of the theory developed in Ch. 2–8 . The results mimic the L_p-theory for elliptic partial differential equations.

10.2 Estimates of Solutions
to Operator Differential Inequalities

Let $\mathcal{A}(D_t)$ be the same constant coefficient differential operators as before, and let h be a prescribed non-negative locally summable function. Consider the operator differential inequality (10.1). We will use the notations b, ω, g, g_ω as in Sect. 5.1 – 5.4, and we assume that the condition (5.7) is valid.

Theorem 10.2.1. *Let*

$$\int_{\mathbb{R}} g_\omega(0, \tau) h(\tau) d\tau < \infty, \tag{10.4}$$

let $u \in W_{\mathrm{loc}}^\ell(\mathbb{R})$ satisfy (10.1) and

$$\| u \|_{W^\ell(t, t+1)} = o\left(\limsup_{\tau \to \pm\infty} \frac{g_\omega(t, \tau)}{g_\omega(0, \tau)} \right) \quad as \quad t \to \pm\infty. \tag{10.5}$$

Then

$$\| u \|_{W^\ell(t, t+1)} \le bw(t), \tag{10.6}$$

where w is the solution of

$$\Big(\mathcal{M}(\partial_t) - \omega(t) \Big) w(t) = h(t) \quad on \quad \mathbb{R}, \tag{10.7}$$

which is given by

$$w(t) = \int_{\mathbb{R}} g_\omega(t, \tau) h(\tau) d\tau. \tag{10.8}$$

Proof. From (10.5) and (5.34) one obtains

$$\| u \|_{W^\ell(t, t+1)} = o\left(e^{-k_\mp t} \right) \quad as \quad t \to \pm\infty. \tag{10.9}$$

We introduce the function

$$v(t) = \int_{\mathbb{R}} g(t - \tau) \omega(\tau) \| u \|_{W^\ell(\tau, \tau+1)} \, d\tau. \tag{10.10}$$

By (10.5) and Lemma 5.4.1, the function v is locally bounded and satisfies (5.53). Hence, using (10.4) and (10.1),

$$\int_{\mathbb{R}} g(-\tau) \| \mathcal{A}u \|_{L_2(\tau, \tau+1; H_0)} \, d\tau < \infty. \tag{10.11}$$

Denote

$$f(t) = \mathcal{A}(D_t) u(t).$$

Due to (10.9) and (10.11) we can apply Theorem 3.5.5 to $\mathcal{A}u = f$. As a result we get the estimate

$$\| u \|_{W^\ell(t,t+1)} \le b \int_{\mathbb{R}} g(t - \tau) \| f \|_{L_2(\tau,\tau+1;H_0)} \, d\tau. \tag{10.12}$$

Thus (10.1) implies

$$\| u \|_{W^\ell(t,t+1)} \le \int_{\mathbb{R}} g(t - \tau) \omega(\tau) \| u \|_{W^\ell(\tau,\tau+1)} \, d\tau$$
$$+ b \int_{\mathbb{R}} g(t - \tau) h(\tau) d\tau \tag{10.13}$$

From this inequality we derive

$$\mathcal{M}(\partial_t) v(t) - \omega(t) v(t) \le H(t) \quad \text{on} \quad \mathbb{R}, \tag{10.14}$$

where

$$H(t) = b\omega(t) \int_{\mathbb{R}} g(t - \tau) h(\tau) d\tau.$$

By (5.16),

$$\int_{\mathbb{R}} g_\omega(t, \tau) H(\tau) d\tau \le b \int_{\mathbb{R}} g_\omega(t, \tau) h(\tau) d\tau. \tag{10.15}$$

Let $v = 0$. Then by (10.13)

$$\| u \|_{W^\ell(t,t+1)} \le b \int_{\mathbb{R}} g(t - \tau) h(\tau) d\tau.$$

Since $g \le g_\omega$ this gives (10.6).

If v is not identically equal to 0 then $v > 0$ by (10.10). We remark that $\mathcal{M}v \ge 0$. We replace H and ω by the functions h_1 and ω_1 satisfying

$$0 \le h_1(t) \le H(t) \quad \text{and} \quad 0 \le \omega_1(t) \le \omega(t), \tag{10.16}$$

in such a way that

$$\mathcal{M}(\partial_t) v(t) - \omega_1(t) v(t) = h_1(t) \quad \text{on} \quad \mathbb{R}. \tag{10.17}$$

Using Proposition 5.2.1(i) together with (5.53) we obtain

$$v^{(j,k)}(t) = o\left(\sup_{\tau \in \mathbb{R}} \frac{g_{\omega_1}^{(j,k)}(t, \tau)}{g_{\omega_1}(0, \tau)} \right) \quad \text{as} \quad t \to \pm\infty \tag{10.18}$$

for $j = 0, \ldots, m_+ - 1$, $k = 0, \ldots, m_- - 1$. By (10.15), (10.16) and (10.18) we can apply Theorems 5.2.3 and 5.2.5 to the equation (10.17) and we obtain

$$v(t) = \int_{\mathbb{R}} g_{\omega_1}(t, \tau) h_1(\tau) d\tau.$$

Using (10.16) and the monotonicity of g_ω with respect to ω, we arrive at

$$v(t) \le \int_{\mathbb{R}} g_\omega(t, \tau) H(\tau) d\tau.$$

Now, by (10.13)

$$\| u \|_{W^{\ell}(t,t+1)} \leq b \int_{\mathbb{R}} g(t - \tau)h(\tau)d\tau$$
$$+b \int_{\mathbb{R}} \int_{\mathbb{R}} g_{\omega}(t,\tau)\omega(\tau)g(\tau - s)h(s)dsd\tau,$$

which together with (5.16) implies (10.6). □

10.3 Perturbation of a Differential Operator with Variable Coefficients

10.3.1 General Case

Here we show that the main results of the previous chapters and their proofs can be extended to a certain class of the operator differential equations (5.1), where

$$L(t, D_t) = \mathcal{A}(t, D_t) - N(t, D_t).$$

Here $N(t, D_t)$ is a "small perturbation" of an operator $\mathcal{A}(t, D_t)$ which need not have constant coefficients.

We suppose that the operators $\mathcal{A}$ and N map $W^{\ell}_{\text{loc}}(\mathbb{R})$ into $L_{2,\text{loc}}(\mathbb{R}; H_0)$. Let $k_{\pm}$ be two real numbers, $k_- < k_+$, and let $m_{\pm}$ be positive integers. By g we denote the same Green's function as in Sect.3.5. We assume that the equation

$$\mathcal{A}(t, D_t)u(t) = f(t) \quad \text{on} \quad \mathbb{R} \tag{10.19}$$

with f subject to

$$\int_{\mathbb{R}} g(-\tau) \| f \|_{L_2(\tau,\tau+1;H_0)} \, d\tau < \infty \tag{10.20}$$

has a solution $u \in W^{\ell}_{\text{loc}}(\mathbb{R})$ satisfying (5.4) (b is a positive constant, independent of f).

Previously, the coefficients of the operator $\mathcal{A}$ did not depend on t and the assumption just made followed from the conditions in Sect.2.2. The constants $k_{\pm}, m_{\pm}$ were described by spectral properties of the pencil $\mathcal{A}(\lambda)$ (see Sect.3.1). However, in the present case the numbers $k_{\pm}, m_{\pm}$ arise (for the first time) in (10.20) and (5.4), and are not a priori connected with the spectrum of a polynomial pencil.

By ρ we shall denote the same function as before (see (5.3)). Let ω be a measurable majorant for $b\rho$. In this section we shall always assume that ω is subject to the same requirements as in analogous assertions from Chapter 5–7.

By duplicating the proof of Theorem 5.3.2 we obtain that there exists a solution of (5.1) satisfying (5.48) under condition (5.9).

Let us introduce a complementary hypothesis on $\mathcal{A}(t, D_t)$: The equation (10.19) has at most one solution subject to

$$\| u \|_{W^\ell(t,t+1)} = o(e^{-k_\mp t}) \quad \text{as} \quad t \to \pm\infty.$$

Then the uniqueness theorem 5.4.2 is valid, as is its proof for the more general operator $L(t, D_t)$ under consideration here. By making the additional assumption that the function

$$t \to \| L \|_{W^\ell(t,t+1) \to L_2(t,t+1;H_0)}$$

is bounded on $\mathbb{R}$, we can extend the uniqueness Proposition 5.4.3 to the present case. Quite analogously, all other results of Ch.5–7 can be generalized.

10.3.2 A Second Order Differential Operator

Here we consider the operator

$$\mathcal{A}(t, D_t) = D_t^2 + B(t),$$

where $B(t)$ is a measurable uniformly bounded operator function:

$$\mathbb{R} \to \mathcal{L}(H_2, H_0).$$

We suppose that for almost all t and for all $u \in H_2$

$$\Re \left(B(t)u, u \right)_{H_0} \geq k^2 \| u \|_0^2, \tag{10.21}$$

where $k > 0$. Also let the operator

$$\mathcal{A}(t, D_t) : W^2(\mathbb{R}) \to L_2(\mathbb{R}; H_0) \tag{10.22}$$

be isomorphic.

Lemma 10.3.1. *Let* $f \in L_{2,\mathrm{loc}}(\mathbb{R}; H_0)$ *and let*

$$\int_{\mathbb{R}} e^{-k|\tau|} \| f \|_{L_2(\tau,\tau+1;H_0)} \, d\tau < \infty. \tag{10.23}$$

Then the equation

$$\left(D_t^2 + B(t) \right) u(t) = f(t) \quad \text{on} \quad \mathbb{R} \tag{10.24}$$

has a solution $u \in W_{\mathrm{loc}}^2(\mathbb{R})$ *satisfying*

$$\| u \|_{W^2(t,t+1)} \leq b \int_{\mathbb{R}} e^{-k|t-\tau|} \| f \|_{L_2(\tau,\tau+1;H_0)} \, d\tau, \tag{10.25}$$

where b *is a constant independent of* f.

Proof. It is sufficient to assume that $f \in L_2(\mathbb{R}; H_0)$. The general case can then be treated by approximation. By the above assumption there exists a unique solution $u \in W^2(\mathbb{R})$ of (10.24) such that

$$\| u \|_{W^2(\mathbb{R})} \le c \| (D_t^2 + B(\cdot))u \|_{L_2(\mathbb{R}; H_0)} . \tag{10.26}$$

Since $u \in W^2(\mathbb{R})$, the function $t \to \| u(t) \|_0$ is continuous on $\mathbb{R}$ and its first derivative is absolutely continuous on the set $S = \{t : \| u(t) \|_0 > 0\}$. We have for $t \in S$

$$\partial_t \| u(t) \|_0 = \frac{\Re(u(t), u'(t))_{H_0}}{\| u(t) \|_0},$$

$$\partial_t^2 \| u(t) \|_0 = \frac{\| u'(t) \|_0^2 \| u(t) \|_0^2 - (\Re(u(t), u'(t))_{H_0})^2}{\| u(t) \|_0^3}$$

$$+ \frac{\Re(u(t), u''(t))_{H_0}}{\| u(t) \|_0}.$$

Hence

$$\partial_t^2 \| u(t) \|_0 \ge \frac{\Re(u(t), B(t)u(t) - f(t))_{H_0}}{\| u(t) \|_0}.$$

By (10.21) we arrive at

$$(-\partial_t^2 + k^2) \| u(t) \|_0 \le \| f(t) \|_0, \ t \in S. \tag{10.27}$$

Denote by I a component of S. Since $\| u(t) \|_0 = 0$ on ∂I, we can use the maximum principle

$$\| u(t) \|_0 \le w(t), \ t \in I,$$

where w satisfies

$$(-\partial_t^2 + k^2)w(t) = \| f(t) \|_0, \ t \in I,$$
$$w = 0 \quad \text{on} \quad \partial I.$$

By using the maximum principle once more, we arrive at the inequality

$$\| u(t) \|_0 \le \frac{1}{2k} \int_{\mathbb{R}} e^{-k|t-\tau|} \| f(\tau) \|_0 \, d\tau$$

for $t \in S$ which is also valid on $\mathbb{R}$. Hence

$$\| u \|_{L_2(t, t+1; H_0)} \le c \int_{\mathbb{R}} e^{-k|t-\tau|} \| f \|_{L_2(\tau, \tau+1; H_0)} \, d\tau. \tag{10.28}$$

Let ζ be a non-negative smooth function on $\mathbb{R}$, $\zeta(t) = 1$ for $t \in (\tau, \tau + 1)$ and $\zeta(t) = 0$ for t outside $(\tau - 1, \tau + 2)$. By (10.24) and by (10.21)

$$\int_{\mathbb{R}} \zeta\left(\| D_t u \|_0^2 + k^2 \| u \|_0^2 \right) dt$$

$$\leq \int_{\mathbb{R}} \left(|\partial_t^2 \zeta| \| u \|_0^2 + \zeta \| f \|_0 \| u \|_0 \right) dt.$$

Hence

$$\| D_t u \|_{L_2(\tau,\tau+1;H_0)} \leq c\Big(\| f \|_{L_2(\tau-1,\tau+2;H_0)}$$
$$+ \| u \|_{L_2(\tau-1,\tau+2;H_0)} \Big). \tag{10.29}$$

By applying (10.26) to ζu and using (10.29) we arrive at the local estimate

$$\| u \|_{W^2(\tau,\tau+1)} \leq c\Big(\| f \|_{L_2(\tau-2,\tau+3;H_0)}$$
$$+ \| u \|_{L_2(\tau-2,\tau+3;H_0)} \Big).$$

Combining this estimate with (10.28) we arrive at (10.25). □

From (10.25) and (10.23) we obtain that

$$\| u \|_{W^2(t,t+1)} = o\left(e^{-k_\mp t}\right) \quad \text{as} \quad t \to \pm\infty. \tag{10.30}$$

We now check that (10.30) describes a uniqueness class.

Lemma 10.3.2. *Let $u \in W_{\mathrm{loc}}^2(\mathbb{R})$ be a solution of (10.24) with $f = 0$ and let*

$$\liminf_{t \to \pm\infty} e^{-k|t|} \| u(t) \|_0 = 0.$$

Then $u = 0$.

Proof. By (10.27)

$$\left(-\partial_t^2 + k^2 \right) \| u(t) \|_0 \leq 0 \tag{10.31}$$

on the set $S = \{t : \| u(t) \|_0 > 0\}$. By the maximum principle, $u(t) = 0$ on every bounded component of S. Suppose that (10.31) holds on a component $(t_0, +\infty)$, $t_0 \in \mathbb{R}$. Let

$$\varepsilon_j = \| u(t_j) \|_0 \, e^{-kt_j} \to 0 \quad \text{as} \quad t_j \to +\infty.$$

By the maximum principle,

$$\| u(t) \|_0 \leq \varepsilon_j e^{kt}, \ t \in (t_0, t_j).$$

Hence $u(t) = 0$ for $t > t_0$. The components $(-\infty, t_0)$ and $\mathbb{R}$ are considered in a similar way. □

10.3.3 Perturbations of the Second Order Differential Operator

Let us turn to a perturbed operator

$$L(t, D_t) = D_t^2 + B(t) - N(t, D_t).$$

We put

$$\rho(t) = \| N \|_{W^2(t,t+1) \to L_2(t,t+1;H_0)}$$

and denote by ω a measurable majorant of $b\rho$, where b is the coefficient in (10.25).

The role of the comparison equation (5.5) will be played by

$$\left(-\partial_t^2 + k^2 - \omega(t) \right) w(t) = \| f \|_{L_2(t,t+1;H_0)}, \qquad (10.32)$$

which is a special case of (5.5) with $m_\pm = 1$, $k_+ = -k_- = k$. The condition (5.7) takes the form

$$\sup_{t \in \mathbb{R}} \omega(t) < k^2 .$$

This inequality is sufficient for the existence of a symmetric, positive Green's function g_ω of the equation (10.32). If

$$\int_{\mathbb{R}} g_\omega(0, \tau) \| f \|_{L_2(\tau,\tau+1;H_0)} \, d\tau < \infty, \qquad (10.33)$$

there exists a solution of (10.32) represented as

$$w(t) = \int_{\mathbb{R}} g_\omega(t, \tau) \| f \|_{L_2(\tau,\tau+1;H_0)} \, d\tau. \qquad (10.34)$$

By Theorem 5.2.3, w satisfies

$$w(t) = o\left(\sup_{\tau \in \mathbb{R}} \frac{g_\omega(t, \tau)}{g_\omega(0, \tau)} \right).$$

From Theorem 5.2.5 it follows that this relation describes a uniqueness class for the equation (10.32).

Based on Lemmas 10.3.1 and 10.3.2, many results of Chapters 5–7 can be carried over to the equation

$$\left(D_t^2 + B(t) - N(t, D_t) \right) u(t) = f(t), \ t \in \mathbb{R} \qquad (10.35)$$

with the same proofs. We give statements of the corresponding results.

Theorem 10.3.3. (Compare with Theorem 5.3.2). *Let $f \in L_{2,\text{loc}}(\mathbb{R}; H_0)$ and let (10.33) be valid. Then there exists a solution $u \in W_{\text{loc}}^2(\mathbb{R})$ of the equation (10.35) subject to*

$$\| u \|_{W^2(t,t+1)} \le bw(t) ,$$

where w is the solution (10.34) of (10.32).

The following assertion is analogous to Corollary 5.3.3 and Remark 5.4.4.

Corollary 10.3.4. *Under condition* (10.33), *the solution u from Theorem 10.3.3 satisfies*

$$\| u \|_{W^2(t,t+1)} = o\Big(\sup_{\tau \in \mathbb{R}} \frac{g_\omega(t,\tau)}{g_\omega(0,\tau)} \Big) \quad as \quad t \to \pm\infty \tag{10.36}$$

and (10.36) *describes a class of uniqueness for the equation* (10.35).

Theorem 10.3.5. (Compare with Theorem 6.3.3). *Let ω be non-decreasing for $t < 0$ and non-increasing for $t > 0$ and let*

$$\omega(\pm 0) < k^2.$$

Suppose that $f \in L_{2,\mathrm{loc}}(\mathbb{R}; H_0)$ and

$$\int_0^\infty e^{-\int_0^\tau \sqrt{k^2-\omega(s)}ds} \| f \|_{L_2(\tau,\tau+1;H_0)} \, d\tau < \infty,$$

$$\int_{-\infty}^0 e^{\int_0^\tau \sqrt{k^2-\omega(s)}} \| f \|_{L_2(\tau,\tau+1;H_0)} \, d\tau < \infty.$$

Then the equation (10.35) *has a solution $u \in W_{\mathrm{loc}}^2(\mathbb{R})$ which satisfies*

$$\| u \|_{W^\ell(t,t+1)}$$
$$\leq c\Big\{ \int_{-\infty}^t e^{-\int_\tau^t \sqrt{k^2-\omega(s)}ds} \| f \|_{L_2(\tau,\tau+1;H_0)} \, d\tau$$
$$+ \int_t^\infty e^{\int_\tau^t \sqrt{k^2-\omega(s)}ds} \| f \|_{L_2(\tau,\tau+1;H_0)} \, d\tau \Big\}.$$

We state an assertion corresponding to Corollary 6.3.4.

Corollary 10.3.6. *Let all conditions of Theorem 10.3.5 be fulfilled. Then the solution u satisfies*

$$\| u \|_{W^2(t,t+1)} = \begin{cases} o\Big(e^{\int_0^t \sqrt{k^2-\omega(s)}ds} \Big) & for \ t \to +\infty \\ o\Big(e^{-\int_0^t \sqrt{k^2-\omega(s)}ds} \Big) & for \ t \to -\infty. \end{cases}$$

Next, we present an analog of Corollary 6.3.5.

Corollary 10.3.7. *Let ω satisfy the conditions of Theorem 10.3.5. Suppose that $f \in L_{2,\mathrm{loc}}(\mathbb{R}; H_0)$ and*

$$\int_0^\infty e^{\int_0^\tau \sqrt{k^2-\omega(s)}ds} \| f \|_{L_2(\tau,\tau+1;H_0)} \, d\tau < \infty$$

$$\int_{-\infty}^0 e^{-\int_0^\tau \sqrt{k^2-\omega(s)}ds} \| f \|_{L_2(\tau,\tau+1;H_0)} \, d\tau < \infty.$$

Then the solution u in Theorem 10.3.5 *satisfies*

$$\| u \|_{W^2(t,t+1)} \leq \begin{cases} ce^{-\int_0^t \sqrt{k^2-\omega(s)}ds} & \text{for } t \geq 0 \\ ce^{\int_0^t \sqrt{k^2-\omega(s)}ds} & \text{for } t \leq 0. \end{cases}$$

The following uniqueness theorem corresponds to Theorem 6.3.6.

Theorem 10.3.8. *Let ω satisfy the conditions of* Theorem 10.3.5. *Also let $u \in W_{loc}^2(\mathbb{R})$ be a solution of (10.35) with $f = 0$. If*

$$\liminf_{t \to +\infty} e^{-\int_0^t \sqrt{k^2-\omega(s)}ds} \| u \|_{W^2(t,t+1)} = 0$$

and

$$\liminf_{t \to -\infty} e^{\int_0^t \sqrt{k^2-\omega(s)}ds} \| u \|_{W^2(t,t+1)} = 0$$

then $u = 0$.

We give an estimate which is analogous to (6.28) for solutions of the equation

$$\left(D_t^2 + A(t) - N(t, D_t) \right) u(t) = f(t) \quad \text{for} \quad t > t_0. \tag{10.37}$$

Theorem 10.3.9. *Let ω satisfy the conditions of* Theorem 10.3.5. *Also let $u \in W_{loc}(t_0, \infty)$ be a solution of (10.37), where*

$$\int_{t_0}^{\infty} e^{-\int_{t_0}^{\tau} \sqrt{k^2-\omega(s)}ds} \| f \|_{L_2(\tau,\tau+1;H_0)} = 0.$$

If

$$\liminf_{t \to +\infty} e^{-\int_{t_0}^{t} \sqrt{k^2-\omega(s)}ds} \| u \|_{W^2(t,t+1)} = 0,$$

then for all $t > t_0 + 1$

$$\| u \|_{W^2(t,t+1)} \leq c \Big\{ \int_{t_0}^{t} e^{-\int_{\tau}^{t} \sqrt{k^2-\omega(s)}ds} \| f \|_{L_2(\tau,\tau+1;H_0)} \, d\tau$$

$$+ \int_{t}^{\infty} e^{\int_{\tau}^{t} \sqrt{k^2-\omega(s)}ds} \| f \|_{L_2(\tau,\tau+1;H_0)} \, d\tau$$

$$+ e^{-\int_{t_0}^{t} \sqrt{k^2-\omega(s)}ds} \| u \|_{W^1(t_0,t_0+1)} \Big\}.$$

We conclude with a variant of the Phragmén-Lindelöf principle, which is an analog of Corollary 6.3.8.

Corollary 10.3.10. *Let ω be non-increasing for large positive t. Suppose that*

$$\omega(+\infty) < k^2 .$$

If $u \in W_{\text{loc}}(\mathbb{R})$ is a solution of (10.37), where

$$\int_{t_0}^{\infty} e^{\int_{t_0}^{\tau} \sqrt{k^2 - \omega(s)} ds} \parallel f \parallel_{L_2(\tau, \tau+1; H_0)} d\tau < \infty$$

then either

$$\liminf_{t \to +\infty} e^{-\int_{t_0}^{t} \sqrt{k^2 - \omega(s)} ds} \parallel u \parallel_{W^2(t,t+1)} > 0$$

or

$$\limsup_{t \to +\infty} e^{\int_{t_0}^{t} \sqrt{k^2 - \omega(s)} ds} \parallel u \parallel_{W^2(t,t+1)} < \infty.$$

10.4 Applications to Partial Differential Equations

This section is related to the previous one. We give an example from the theory of partial differential equations where the comparison principle (5.48) can be obtained under very general assumptions on the coefficients.

Example 10.4.1. Let us consider the Dirichlet problem

$$\begin{cases} u_{tt} + 2\beta(x,t)u_{xt} + \gamma(x,t)u_{xx} = f & \text{on } \mathcal{C} \\ u = 0 & \text{on } \partial\mathcal{C} , \end{cases} \qquad (10.38)$$

where f and u are real-valued and

$$\mathcal{C} = \left\{ (x,t) \in \mathbb{R}^2 : |x| \leq \ell, \ t \in \mathbb{R} \right\}, \quad \ell = \text{const} > 0.$$

We suppose that the coefficients β and γ are real, measurable and bounded. The condition of ellipticity will be written in the form

$$\mu = \inf_{\mathcal{C}} \gamma(x,t) > 0,$$

$$\nu = \sup_{\mathcal{C}} \frac{|\beta(x,t)|}{\sqrt{\gamma(x,t)}} < 1.$$

It is known that the operator of the problem (10.38) realizes an isomorphism between $W_2^2(\mathcal{C}) \cap \mathring{W}_2^1(\mathcal{C})$ and $L_2(\mathcal{C})$ (see, for example, Ladyzhenskaya, Ural'tseva (1968), Ch.III, Sect.19). In particular, we have the estimate

$$\parallel u \parallel_{W_2^2(\mathcal{C})} \leq c \parallel f \parallel_{L_2(\mathcal{C})} \qquad (10.39)$$

with positive c depending on $\sup |\beta|$, $\sup \gamma$, μ, ν and ℓ.

Although we impose no requirements of stabilization of β and γ at infinity, it is possible to obtain an analog of the comparison principle (5.48).

Theorem 10.4.2. *Let $f \in L_{2,\mathrm{loc}}(\mathcal{C})$ and*

$$\int_{\mathbb{R}} e^{-k|\tau|} \| f \|_{L_2(C_\tau)} \, d\tau < \infty , \qquad (10.40)$$

where $k = (1-\nu)\sqrt{\mu}\pi/\ell$ and $C_t = \{(x,\sigma) \in \mathcal{C} : t < \sigma < t+1 \}$. Then there exists a solution $u \in W^2_{2,\mathrm{loc}}(\overline{\mathcal{C}})$ of (10.38) such that

$$\| u \|_{W^2_2(C_t)} \le b \int_{\mathbb{R}} e^{-k|t-\tau|} \| f \|_{L_2(C_\tau)} \, d\tau. \qquad (10.41)$$

This solution satisfies

$$\| u \|_{W^2_2(C_t)} = o\left(e^{k|t|}\right) \quad as \quad t \to \pm\infty. \qquad (10.42)$$

Equality (10.42) describes a class of uniqueness for the problem (10.38).

Proof. Let $\mathrm{supp} f \subset \overline{\mathcal{C}}_T$ and let u be the solution of (10.38) in $W^2_2(\mathcal{C})$. By (10.38) we have for all $t > T+1$

$$\int_t^\infty \int_{-\ell}^\ell (u_{x\tau}^2 + 2\beta u_{x\tau} u_{xx} + \gamma u_{xx}^2 + u_{\tau\tau} u_{xx} - u_{x\tau}^2) \, dx d\tau = 0. \qquad (10.43)$$

For almost all $t > T+1$ we have

$$\int_t^\infty \int_{-\ell}^\ell (u_{\tau\tau} u_{xx} - u_{x\tau}^2) \, dx d\tau = \frac{1}{2}\frac{d}{dt} \int_{-\ell}^\ell u_x^2 dx. \qquad (10.44)$$

We note that the quadratic form

$$\xi^2 + 2\beta\xi\eta + \gamma\eta^2$$

majorizes

$$\left(1 - \frac{|\beta|}{\sqrt{\gamma}}\right) (\xi^2 + \gamma\eta^2)$$

which in its turn is not less that $(1-\nu)(\xi^2 + \mu\eta^2)$. Hence, by using (10.43) and (10.44)

$$2(1-\nu) \int_t^\infty \int_{-\ell}^\ell (u_{x\tau}^2 + \mu u_{xx}^2) \, dx d\tau + \frac{d}{dt} \int_{-\ell}^\ell u_x^2 dx \le 0.$$

Since

$$\int_{-\ell}^\ell u_{xx}^2 dx \ge \frac{\pi^2}{\ell^2} \int_{-\ell}^\ell u_x^2 dx$$

and

$$\int_0^\infty (y'(\tau)^2 + \lambda^2 y(\tau)^2) \, d\tau \ge \lambda y(0)^2$$

for $\lambda = \mathrm{const} > 0$ and for all $y \in W^1_2(0,\infty)$, it follows that

$$2(1-\nu)\sqrt{\mu}\frac{\pi}{\ell}\int_{-\ell}^{\ell}u_x^2\,dx + \frac{d}{dt}\int_{-\ell}^{\ell}u_x^2\,dx \le 0.$$

Thus for $t \ge T+1$

$$\int_{-\ell}^{\ell}u_x^2(x,t)\,dx \le e^{-2k(t-T-1)}\int_{-\ell}^{\ell}u_x^2(x,T+1)\,dx.$$

By the same argument we arrive at the estimate

$$\int_{-\ell}^{\ell}u_x^2(x,t)\,dx \le e^{-2k(T-t)}\int_{-\ell}^{\ell}u_x^2(x,T)\,dx$$

for all $t \le T$. Hence, using (10.39) we deduce that for $t \in \mathbb{R}$

$$\int_{-\ell}^{\ell}u_x^2(x,t)\,dx \le ce^{-2k|t-T|}\parallel f \parallel_{L_2(C_T)}. \tag{10.45}$$

By using the following local estimate (which can be obtained directly from (10.39))

$$\parallel u \parallel_{W_2^2(C_t)} \le c\left(\parallel u \parallel_{L_2(C_t^*)} + \parallel f \parallel_{L_2(C_t^*)}\right),$$

where

$$C_t^* = \{(x,\tau) \in C : t-1 < \tau < t+2\},$$

we obtain from (10.45)

$$\parallel u \parallel_{W_2^2(C_t)} \le ce^{-k|t-T|}\parallel f \parallel_{L_2(C_T)}. \tag{10.46}$$

Now we eliminate the condition supp $f \subset \overline{C}_T$. Denote by χ_T the characteristic function of the interval $(T, T+1)$. Let T be an integer and let u_T be the solution in $W_2^2(C)$ of the problem (10.38) with f replaced by $\chi_T f$. Estimate (10.46) implies

$$\parallel u_T \parallel_{W_2^2(C_t)} \le ce^{-k|t-T|}\parallel f \parallel_{L_2(C_T)}.$$

We introduce the series

$$u = \sum_{T\in\mathbf{Z}}u_T$$

and show that it converges in $W_{2,\mathrm{loc}}^2(\overline{C})$. In fact,

$$\sum_{T\in\mathbf{Z}}\parallel u_T \parallel_{W_2^2(C_t)} \le c\sum_{T\in\mathbf{Z}}e^{-k|t-T|}\parallel f \parallel_{L_2(C_T)}$$

$$\le b\int_{\mathbb{R}}e^{-k|t-\tau|}\parallel f \parallel_{L_2(C_\tau)}\,d\tau.$$

The finiteness of the last integral is guaranteed by (10.40) which implies both the existence of the limit u and (10.41). The estimate (10.42) is a trivial consequence of (10.41).

The proof of uniqueness duplicates the argument used in Proposition 5.4.3 where one sets

$$g_\omega(t, \tau) = \exp(-k|t - \tau|), \quad k_\pm = \pm k, \quad m_\pm = 1.$$

$\square$

Remark 10.4.3. It is perhaps worth noting that Theorem 10.4.2 is obviously best possible in the case $\beta = 0, \gamma = \text{const}$ (compare with Sect. 3.9.1).

10.5 "Parabolic" First Order Operators with a Variable Dissipative Term

10.5.1 Unperturbed Operator

We start with the first order operator

$$\mathcal{A}(t, D_t) = D_t + A(t), \tag{10.47}$$

where $A(t)$ is a measurable uniformly bounded operator function

$$A : \mathbb{R} \to \mathcal{L}(H_1, H_0).$$

Suppose that for almost all t

$$\Im\big(A(t)u, u\big)_{H_0} \leq -k_+ \| u \|_0^2 \quad \text{for all} \quad u \in H_1, \tag{10.48}$$

where $k_+ = \text{const}, k_+ > 0$.
   We assume that the operator

$$\mathcal{A}(t, D_t) : W^1(\mathbb{R}) \to L_2(\mathbb{R}; H_0) \tag{10.49}$$

is an isomorphism.

Lemma 10.5.1. *Let $f \in L_{2,\text{loc}}(\mathbb{R}; H_0)$ and let*

$$\int_{-\infty}^0 e^{k_+\tau} \| f \|_{L_2(\tau,\tau+1;H_0)} \, d\tau < \infty. \tag{10.50}$$

Then the equation

$$\big(D_t + A(t)\big)u(t) = f(t) \quad \text{on} \quad \mathbb{R} \tag{10.51}$$

has a solution $u \in W_{\text{loc}}^1(\mathbb{R})$ satisfying

$$\| u \|_{W^1(t,t+1)} \leq b\Big(\| f \|_{L_2(t,t+1;H_0)}$$
$$+ \int_{-\infty}^t e^{-k_+(t-\tau)} \| f \|_{L_2(\tau,\tau+1;H_0)} \, d\tau\Big) \tag{10.52}$$

Proof. It is sufficient to assume that $f \in L_2(\mathbb{R}; H_0)$. Since the operator (10.49) is an isomorphism there exists a unique solution $u \in W^1(\mathbb{R})$ of (10.51) subject to

$$\| u \|_{W^1(\mathbb{R})} \leq c \, \| \, (D_t + A(\cdot))u \, \|_{L_2(\mathbb{R};H_0)} \, . \tag{10.53}$$

By (10.51),

$$\frac{1}{2}\partial_t \| u(t) \|_0^2 - \Im\big(A(t)u(t), u(t)\big)_{H_0}$$
$$= -\Im\big(f(t), u(t)\big)_{H_0} \quad \text{for} \quad a.e. \ t.$$

Hence by (10.48),

$$\partial_t \| u(t) \|_0 + k_+ \| u(t) \|_0 \leq \| f(t) \|_0 \, . \tag{10.54}$$

Since $u \in W^1(\mathbb{R})$ we can write

$$\| u(t) \|_0 \leq \int_{-\infty}^{t} e^{-k_+(t-\tau)} \| f(\tau) \|_0 \, d\tau, \tag{10.55}$$

which implies that $u(t)$ depends only on the values of $f(\tau)$ with $\tau < t$.

The right-hand side of (10.55) is majorized by

$$c \int_{-\infty}^{t} e^{-k_+(t-\tau)} \int_{\tau}^{\tau+1} \| f(x) \|_0 \, dx d\tau,$$

which increases when the inner integral is replaced by $\|f\|_{L_2(\tau,\tau+1;H_0)}$. Therefore

$$\| u(t) \|_0 \leq c \int_{-\infty}^{t} e^{-k_+(t-\tau)} \| f \|_{L_2(\tau,\tau+1;H_0)} \, d\tau. \tag{10.56}$$

We fix $\tau \in \mathbb{R}$. Consider the equation (10.51), with $f(t) = 0$ for $t > \tau$. As noted already, this assumption does not influence the values of u for $t < \tau$. Let $\eta \in C^\infty(\mathbb{R})$, $\eta(\tau) = 1$ for $\tau > t - 1$ and $\eta(\tau) = 0$ for $\tau < t - 2$. Since

$$(D_t + A)(\eta u) = \eta f + (D_t \eta)u,$$

it follows by (10.53) that we have the local estimate:

$$\| u \|_{W^1(\tau-1,\tau)} \leq c\Big(\| f \|_{L_2(\tau-2,\tau,H_0)} + \| u \|_{L_2(\tau-2,\tau-1;H_0)} \Big).$$

Along with (10.56), this implies that

$$\| u \|_{W^1(t,t+1)} \leq c\Big(\| f \|_{L_2(t-1,t+1,H_0)}$$
$$+ \int_{-\infty}^{t} e^{-k_+(t-\tau)} \| f \|_{L_2(\tau,\tau+1;H_0)} \, d\tau\Big).$$

It remains to note that

$$\| f \|_{L_2(t-1,t;H_0)} \leq \int_{t-1}^{t} \| f \|_{L_2(\tau-1,\tau+1;H_0)} \, d\tau$$

$$\leq \sqrt{2} \int_{t-2}^{t} \| f \|_{L_2(\tau,\tau+1;H_0)} \, d\tau.$$

$\square$

It follows from (10.50) and (10.52) that

$$\| u \|_{W^1(t,t+1)} = o\left(e^{-k_+ t}\right) \quad \text{as} \quad t \to -\infty.$$

Now we show that this relation defines a uniqueness class for the equation (10.51).

Lemma 10.5.2. *Let $u \in W_{\mathrm{loc}}^1(\mathbb{R})$ be a solution of (10.51) with $f = 0$ and let*

$$\liminf_{t \to -\infty} e^{k_+ t} \| u(t) \|_0 = 0. \tag{10.57}$$

Then $u = 0$.

Proof. By (10.54),

$$\partial_t \| u(t) \|_0 + k_+ \| u(t) \|_0 \leq 0.$$

Hence

$$\| u(t) \|_0 \leq e^{-k_+(t-t_1)} \| u(t_1) \|_0,$$

which together with (10.57) leads to $u = 0$. $\square$

10.5.2 Perturbed Operator

Here we consider equation (10.2), and characterize the perturbation operator N by the function

$$\rho(t) = \| N \|_{W^1(t,t+1) \to L_2(t,t+1;H_0)} . \tag{10.58}$$

We suppose that

$$\sup_{t \in \mathbb{R}} \rho(t) = \rho_0 \leq (3b)^{-1}$$

where b is the constant in (10.52).

Theorem 10.5.3. *Let $f \in L_{2,\mathrm{loc}}(\mathbb{R})$. If*

$$\int_{-\infty}^{0} e^{k_+ \tau + a \int_{\tau}^{0} \rho(x) dx} \| f \|_{L_2(\tau,\tau+1;H_0)} \, d\tau < \infty \tag{10.59}$$

with

$$a = b(1 - b\rho_0)^{-1}, \tag{10.60}$$

then there exists a solution $u \in W_{\mathrm{loc}}^1(\mathbb{R})$ of (10.2) satisfying the estimate

$$\| u \|_{W^1(t,t+1)} \le c\Big(\| f \|_{L_2(t,t+1;H_0)}$$

$$+ \int_{-\infty}^t e^{-k_+(t-\tau)+a \int_\tau^t \rho(x)dx} \| f \|_{L_2(\tau,\tau+1;H_0)} \, d\tau \Big), \qquad (10.61)$$

where

$$c = \frac{b}{1 - 2b\rho_0}.$$

Proof. In order to construct a solution of (10.2) we use the iteration procedure

$$\big(D_t + A(t)\big)u_{k+1}(t) = N(t, D_t)u_k(t) + f(t), \quad t \in \mathbb{R}, \qquad (10.62)$$

$$k = 0, 1, \dots, \quad \text{and} \quad u_0 = 0.$$

First we find a majorant for u_k. Let

$$h(t) = \| f \|_{L_2(t,t+1;H_0)} \cdot$$

Consider the inequality

$$w(t) \ge b\Big(h(t) + \rho_0 w(t)$$

$$+ \int_{-\infty}^t e^{-k_+(t-\tau)}\big(h(\tau) + \rho(\tau)w(\tau)\big)d\tau\Big). \qquad (10.63)$$

We seek a solution of (10.63) in the form

$$w(t) = \frac{b}{1 - 2b\rho_0}\Big(h(t) + \int_{-\infty}^t e^{-k_+(t-\tau)+a \int_\tau^t \rho(x)dx} h(\tau)d\tau\Big),$$

where a is a positive constant to be determined. Integrating by parts we obtain

$$\int_{-\infty}^t e^{-k_+(t-\tau)}\rho(\tau) \int_{-\infty}^\tau e^{-k_+(\tau-z)+a \int_z^\tau \rho(x)dx} h(z)dzd\tau$$

$$= \frac{1}{a} \int_{-\infty}^t e^{-k_+(t-\tau)+a \int_\tau^t \rho(x)dx} h(\tau)d\tau$$

$$- \frac{1}{a} \int_{-\infty}^t e^{-k_+(t-\tau)} h(\tau)d\tau . \qquad (10.64)$$

Hence the inequality (10.63) holds if a is chosen as in (10.60). Moreover, the relation (10.63) together with (10.59) implies

$$\int_{-\infty}^0 e^{k_+\tau}\big(h(\tau) + \rho(\tau)w(\tau)\big)d\tau < \infty. \qquad (10.65)$$

Let us introduce the auxiliary iterative procedure

$$w_{k+1}(t) = b\Big(h(t) + \rho_0 w_k(t)$$
$$+ \int_{-\infty}^t e^{-k_+(t-\tau)}\big(h(\tau) + \rho(\tau)w_k(\tau)\big)d\tau\Big), \qquad (10.66)$$

when $k = 0, 1, \ldots$, and $w_0 = 0$. One easily sees that

$$w_k(t) \le w_{k+1}(t) \le w(t), \quad k = 0, 1, \ldots \qquad (10.67)$$

Hence the integral in (10.66) converges.

Let us prove that the equation (10.62) is solvable for every k and

$$\| u_k \|_{W^1(t,t+1)} \le w_k(t), \quad k = 0, 1, \ldots \qquad (10.68)$$

For $k = 0$ this estimate is trivial. Suppose (10.68) is proved for k. Since

$$\| N u_k + f \|_{L_2(t,t+1;H_0)} \le \rho(t)w_k(t) + h(t),$$

due to (10.65) and (10.67), the right-hand side of (10.62) satisfies (10.50). By Lemma 10.5.1 there exists a solution of (10.62) such that

$$\| u_{k+1} \|_{W^1(t,t+1)} \le b\Big(h(t) + \rho_0 w_k(t)$$
$$+ \int_{-\infty}^t e^{-k_+(t-\tau)}\big(h(\tau) + \rho(\tau)w_k(\tau)\big)d\tau\Big) = w_{k+1}(t).$$

Thus the estimate (10.68) is proved.

To check the convergence of $\{u_k\}$ in $W^1_{\mathrm{loc}}(\mathbb{R})$ we introduce

$$v_k(t) = \| u_k - u_{k-1} \|_{W^1(t,t+1)}.$$

From (10.62) and Lemma 10.5.1 we get

$$v_{k+1}(t) \le b\Big(\rho_0 v_k(t) + \int_{-\infty}^t e^{-k_+(t-\tau)} v_k(\tau)d\tau\Big).$$

One can verify by induction that

$$v_k(t) \le w_k(t) - w_{k-1}(t).$$

Hence

$$\sum_{k=1}^{\infty} v_k(t) \le w(t).$$

This proves the convergence of $\{u_k\}$ and the estimate (10.61). $\square$

The following assertion is a direct consequence of (10.61).

Corollary 10.5.4. *The solution u from* Theorem 10.47 *satisfies*

$$\| u \|_{W^1(t,t+1)} = o\Big(e^{-k_+ t - a\int_t^0 \rho(x)dx}\Big) \quad as \quad t \to -\infty. \qquad (10.69)$$

We show that the last relation describes a uniqueness class.

Theorem 10.5.5. (Uniqueness). *Let* $u \in W^1_{\text{loc}}(\mathbb{R})$ *be a solution of* (10.2) *when* $f = 0$ *and* (10.69) *is valid. Then* $u = 0$.

Proof. By (10.69)

$$\int_{-\infty}^0 e^{k+\tau} \rho(\tau) \, \| \, u \, \|_{W^1(\tau,\tau+1)} \, d\tau < \infty.$$

Applying Lemmas 10.5.1 and 10.5.2 to

$$\big(D_t + A(t)\big)u(t) = N(t, D_t)u(t) \quad \text{on} \quad \mathbb{R},$$

we arrive at

$$\| \, u \, \|_{W^1(t,t+1)} \leq b\Big(\rho_0 \, \| \, u \, \|_{W^1(t,t+1)}$$
$$+ \int_{-\infty}^t e^{-k_+(t-\tau)} \rho(\tau) \, \| \, u \, \|_{W^1(\tau,\tau+1)} \, d\tau\Big). \tag{10.70}$$

Put

$$v(t) = \int_{-\infty}^t e^{-k_+(t-\tau)} \rho(\tau) \, \| \, u \, \|_{W^1(\tau,\tau+1)} \, d\tau.$$

Then (10.70) implies

$$\big(\partial_t + k_+\big)v(t) \leq \frac{b}{1 - b\rho_0}\rho(t)v(t), \quad t \in \mathbb{R} \, .$$

Furthermore, we deduce from (10.69) that

$$v(t) = o\Big(e^{-k_+t - a\int_t^0 \rho(x)dx}\Big). \tag{10.71}$$

Since the function

$$w(t) = e^{-k_+t - a\int_t^0 \rho(x)dx}$$

satisfies

$$\big(\partial_t + k_+\big)w(t) = \frac{b}{1 - b\rho_0}w(t), \tag{10.72}$$

it follows by (10.71) and the maximum principle for the equation (10.72) that $v = 0$ and hence $u = 0$. $\square$

Combining Theorems 10.5.3 and 10.5.5 we arrive at the following.

Corollary 10.5.6. *Let* $f \in L_{2,\text{loc}}(\mathbb{R}; H_0)$ *be subject to* (10.59). *Then there exists one and only one solution* $u \in W^1_{\text{loc}}(\mathbb{R})$ *of* (10.2) *satisfying* (10.69).

We give another corollary of the estimate (10.61).

Corollary 10.5.7. *Let* $f \in L_{2,\text{loc}}(\mathbb{R}; H_0)$ *be subject to*

$$\int_{\mathbb{R}} e^{k_+ \tau + a \int_\tau^0 \rho(x)dx} \| f \|_{L_2(\tau, \tau+1; H_0)} \, d\tau < \infty.$$

Then the solution of (10.2) *from Corollary 10.5.6 satisfies*

$$\| u \|_{W^1(t,t+1)} \le c e^{-k_+ t + a \int_0^t \rho(x)dx} \quad \text{for} \quad t > 0.$$

10.6 "Hyperbolic" Operator Equations with Constant Coefficients

10.6.1 Assumptions on the Pencil $\mathcal{A}(\lambda)$

Condition II from Sect. 2.2 on the pencil $\mathcal{A}(\lambda)$ is rather restrictive in a certain sense: while it includes partial differential equations of elliptic and parabolic type, it excludes hyperbolic equations. In this section we start with a modification of this condition which, in particular, enables one to extend the previous theory to hyperbolic equations with dissipation. As a simple example, we shall be able to consider the boundary value problem

$$\left(-\frac{\partial^2}{\partial t^2} + 2c_0 \frac{\partial}{\partial t} + \Delta_x \right) u = f \tag{10.73}$$

on $\Omega \times \mathbb{R}$, $\Omega \subset \mathbb{R}^n$, and

$$u = 0 \quad \text{on } \partial\Omega \times \mathbb{R} \tag{10.74}$$

where c_0 is a real constant.

Let $\mathcal{A}(\lambda)$ be the operator pencil (1.10) with the coefficients A_q continuously mapping H_q into H_0. We suppose that

$$\mathcal{A}(\lambda) : H_\ell \to H_0$$

is Fredholm for all λ in the strip $s_- < \Im\lambda < s_+$ and that it is invertible for at least one value of λ in the strip. We replace Condition II (Sect. 2.2) by the following

Condition $\mathfrak{H}$. *There exists* $r > 0$ *such that*

$$\sum_{q=1}^{\ell} |\lambda|^{q-1} \| \varphi \|_{\ell-q} \le c \| \mathcal{A}(\lambda)\varphi \|_0 \tag{10.75}$$

for all $\varphi \in H_\ell$ *and for all* λ *in the strip* $s_- < \Im\lambda < s_+$, $|\lambda| \ge r$.

By this condition and Proposition A.8.4, every strip $s_- + \varepsilon < \Im\lambda < s_+ - \varepsilon$, $\varepsilon > 0$, contains only a finite number of eigenvalues of $\mathcal{A}(\lambda)$, each of which has finite algebraic multiplicity.

10.6.2 Example of a Second Order Differential Operator

Example 10.6.1. Consider the differential operator

$$\mathcal{A}(D_t) = D_t^2 + 2iAD_t - B,$$

where A and B are selfadjoint operators in H_0 which continuously map H_1 and H_2 into H_0 respectively. (This example includes the Dirichlet problem (10.73), (10.74).)

We suppose that for all λ in the half-plane $\Im\lambda > c_0$, the operator pencil

$$\mathcal{A}(\lambda) : H_2 \to H_0$$

is Fredholm and that it is invertible at least for one value of λ. We assume that

$$(Bu, u)_{H_0} \geq c_1 \parallel u \parallel_1^2 - c_2 \parallel u \parallel_0^2, \tag{10.76}$$

where c_1, c_2 are positive constants, and that

$$(Au, u)_{H_0} \geq -c_0 \parallel u \parallel_0^2$$

for all $u \in H_2$.

Since

$$\Im\big(\mathcal{A}(\lambda)u, u\big)_{H_0} = 2\Re\lambda\big(\Im\lambda \parallel u \parallel_0^2 + (Au, u)_{H_0}\big),$$

it follows that

$$|\Re\lambda| \parallel u \parallel_0 \leq \frac{1}{2}\big(\Im\lambda - c_0\big)^{-1} \parallel \mathcal{A}(\lambda)u \parallel_0 \tag{10.77}$$

for $\Im\lambda > c_0$. Using the identity

$$(Bu, u) = \big((\Re\lambda)^2 - (\Im\lambda)^2\big) \parallel u \parallel_0^2 - \Re\big(\mathcal{A}(\lambda)u, u\big)_{H_0},$$

(10.76) and (10.77), we get for sufficiently large $|\Re\lambda|$ that

$$\parallel u \parallel_1 \leq c\big(1 + \frac{1}{\Im\lambda - c_0}\big) \parallel \mathcal{A}(\lambda) \parallel_0$$

(c does not depend on λ). We thus arrive at the estimate (10.75) in an arbitrary strip $s_- < \Im\lambda < s_+$, $s_- > c_0$ when $\ell = 2$. In other words, the pencil $\mathcal{A}(\lambda)$ satisfies Condition $\mathfrak{H}$.

In particular, let $A = -c_0$ and let the spectrum of B be discrete and have a limit point at $+\infty$. Denote by $\{\gamma_j\}_{j\geq 1}$ the sequence of eigenvalues of $B : \gamma_1 \leq \gamma_2 \leq \ldots$ Then the eigenvalues of the pencil $\mathcal{A}(\lambda)$ are given by

$$\lambda_j^{(\pm)} = c_0 i \pm \sqrt{\gamma_j - c_0^2}. \tag{10.78}$$

Generalized eigenvectors can appear if and only if $\gamma_j = c_0^2$ for some j. In this case one can verify directly that a canonical system is given by $(\varphi_k, 0)$, $k = 1, \ldots, J$, where $J = \dim\ker(B - c_0^2)$ and $\varphi_1, \ldots, \varphi_J$ is a basis in $\ker(B - c_0^2)$. In other words, power-exponential solutions of $\mathcal{A}(D_t)u = 0$ are

$$\exp((-c_0 \pm \sqrt{c_0^2 - \gamma_j})t)\varphi_j \,,$$

where φ_j is an eigenvector of B corresponding to the eigenvalue γ_j. In the case $\gamma_j = c_0^2$, these solutions are completed by $te^{-c_0 t}\varphi_j$. By using the spectral decomposition of B we show that the estimate (10.75) is valid for all λ in the strip $s_- < \Im\lambda < s_+$ if and only if either $c_0 < s_-$ or $c_0 > s_+$. Such strips can contain only the eigenvalues (10.78) with $\gamma_j < c_0^2$.

10.6.3 Solutions in Weighted Sobolev Spaces

We introduce the distribution space $\mathcal{D}'(\mathbb{R}; H_k)$; i.e. the space of bounded linear functionals on $\mathcal{D}(\mathbb{R})$ which take values in H_k (see Lions, Magenes (1972), Ch.I). This space is isomorphic to the dual of $\mathcal{D}(\mathbb{R}; H_k^*)$, where $\mathcal{D}(\mathbb{R}; H_k^*)$ is the space of infinitely differentiable functions in $\mathbb{R}$ with values in H_k^* and compact support. Obviously, $\mathcal{A}^*(D_t)$ maps $\mathcal{D}(\mathbb{R}; H_0)$ into $\mathcal{D}(\mathbb{R}; H_\ell^*)$.

Consider the equation (2.1) with $f \in L_{2,\mathrm{loc}}(\mathbb{R}; H_0)$. By its solution we mean a distribution $u \in \mathcal{D}'(\mathbb{R}; H_\ell)$ satisfying (2.1), i.e. the equality

$$\int_{\mathbb{R}} \left(u(t), \mathcal{A}^*(D_t)v(t)\right)_{H_0} dt = \int_{\mathbb{R}} \left(f(t), v(t)\right)_{H_0} dt \qquad (10.79)$$

holds for all $v \in \mathcal{D}(\mathbb{R}; H_0)$.

We need the spaces

$$W_\beta^{\ell-1}(\mathbb{R}), \quad W^{\ell-1}(a,b), \quad W_{\mathrm{loc}}^{\ell-1}(\mathbb{R}), \quad W_{\beta_-,\beta_+}^{\ell-1}(\mathbb{R}),$$

whose definitions can be obtained from Definitions 2.3.6, 2.3.2, 2.3.4, 2.3.7 by changing ℓ for $\ell - 1$. Clearly, $W_{\mathrm{loc}}^{\ell-1}(\mathbb{R}) \in \mathcal{D}'(\mathbb{R}; H_{\ell-1})$.

Lemma 10.6.2. *Let $f_j \in L_{2,\mathrm{loc}}(\mathbb{R}; H_0)$ and let $u_j \in W_{\mathrm{loc}}^{\ell-1}(\mathbb{R}) \cap \mathcal{D}'(\mathbb{R}; H_\ell)$ satisfy*

$$\mathcal{A}(D_t)u_j = f_j \quad on \quad \mathbb{R} \,. \qquad (10.80)$$

Suppose that

$$u_j \to u \quad as \quad j \to \infty \quad in \quad W_{\mathrm{loc}}^{\ell-1}(\mathbb{R}) \qquad (10.81)$$

and

$$f_j \to f \quad as \quad j \to \infty \quad in \quad L_{2,\mathrm{loc}}(\mathbb{R}; H_0).$$

Then u is a solution of (2.1).

Proof. Let λ_0 be a regular point of $\mathcal{A}(\lambda)$. We rewrite (10.79) for u_j, f_j in the form

$$\int_{\mathbb{R}} \left(u_j, (\mathcal{A}^*(\overline{\lambda}_0) + (D_t - \overline{\lambda}_0)\mathcal{B}^*(\overline{\lambda}_0, D_t))v\right)_{H_0} dt = \int_{\mathbb{R}} (f_j, v)_{H_0} dt,$$

where

$$\mathcal{B}(\lambda, z) = \frac{\mathcal{A}(\lambda) - \mathcal{A}(z)}{\lambda - z}.$$

We set $v(t) = \left(\mathcal{A}^*(\bar{\lambda}_0)\right)^{-1} w(t)$, where $w \in \mathcal{D}(\mathbb{R}; H_\ell^*)$, and obtain

$$
\int_{\mathbb{R}} (u_j, w)_{H_0} dt = \int_{\mathbb{R}} \left(\mathcal{A}^{-1}(\lambda_0) f_j, w\right)_{H_0} dt
$$
$$
- \int_{\mathbb{R}} \left(\mathcal{A}^{-1}(\lambda_0) \mathcal{B}(\lambda_0, D_t) u_j, (D_t - \bar{\lambda}_0) w\right)_{H_0} dt. \qquad (10.82)
$$

Since $\mathcal{B}(\lambda_0, D_t)$ continuously maps $W_{\text{loc}}^{\ell-1}(\mathbb{R})$ into $L_{2,\text{loc}}(\mathbb{R}; H_0)$, the right-hand side in (10.82) converges, which means that u_j has a limit in $\mathcal{D}'(\mathbb{R}; H_\ell)$. By (10.81), $u_j \to u$ in $\mathcal{D}'(\mathbb{R}; H_{\ell-1})$. Hence $u \in \mathcal{D}'(\mathbb{R}; H_\ell)$. Passing to the limit in (10.82) we conclude that u satisfies (10.79). $\square$

We prove that the equation (2.1) is uniquely solvable in a certain weighted Sobolev space.

Theorem 10.6.3. *Let $\beta \in (s_-, s_+)$. Suppose that there are no eigenvalues of $\mathcal{A}(\lambda)$ on the line $\Im \lambda = \beta$.*
  (i) (Existence) If $f \in L_{2,\beta}(\mathbb{R}; H_0)$ then there exists a solution $u \in W_\beta^{\ell-1}(\mathbb{R})$ of (2.1) given by (2.19) such that the following estimate holds:

$$
\| u \|_{W_\beta^{\ell-1}(\mathbb{R})} \leq c \| f \|_{L_{2,\beta}(\mathbb{R}; H_0)}. \qquad (10.83)
$$

  (ii) (Uniqueness) If u is a solution of (2.1) with $f = 0$ and $u \in W_\beta^{\ell-1}(\mathbb{R})$ then $u = 0$.

Proof. (i) We show that the function u defined by (2.19) is the required solution. Condition $\mathfrak{H}$ implies that $u \in W_\beta^{\ell-1}(\mathbb{R})$ and (10.83) is valid. It remains to show that $u \in \mathcal{D}'(\mathbb{R}; H_\ell)$ and that (2.1) holds.

Let λ_0 be a regular point of the operator pencil $\mathcal{A}(\lambda)$. We write (2.19) as

$$
u(t) = u_0(t) + (D_t - \lambda_0) u_1(t),
$$

where

$$
u_0(t) = \mathcal{A}^{-1}(\lambda_0) f(t)
$$

and

$$
u_1(t) = -\frac{1}{2\pi} \mathcal{A}^{-1}(\lambda_0) \int_{\Im \lambda = \beta} e^{it\lambda} \mathcal{B}(\lambda, \lambda_0) \mathcal{A}^{-1}(\lambda) \tilde{f}(\lambda) d\lambda.
$$

Since $\mathcal{A}^{-1}(\lambda_0) : H_0 \to H_\ell$ is a bounded operator, and since by Condition $\mathfrak{H}$

$$
\| \mathcal{B}(\lambda, \lambda_0) \mathcal{A}^{-1}(\lambda) \|_{H_0 \to H_0} \leq c, \quad \Im \lambda = \beta,
$$

it follows that u_0 and u_1 belong to $L_{2,\beta}(\mathbb{R}; H_\ell)$. Hence, $u \in \mathcal{D}'(\mathbb{R}; H_\ell)$.

The relation (10.79) is verified by using the Fourier transform.

  (ii) Let η be a real-valued function in $\mathcal{D}(\mathbb{R})$. By (10.79), we have for all $v \in \mathcal{D}(\mathbb{R}; H_0)$

$$0 = \int_{\mathbb{R}} \left(u, \mathcal{A}^*(D_t)(\bar{\eta}v) \right)_{H_0} dt$$
$$= \int_{\mathbb{R}} \left\{ (\eta u, \mathcal{A}^*(D_t)v)_{H_0} - (u, [\mathcal{A}^*(D_t), \bar{\eta}]v)_{H_0} \right\} dt.$$

The inclusion $u \in W_\beta^{\ell-1}(\mathbb{R})$ implies

$$\int_{\mathbb{R}} (\eta u, \mathcal{A}^*(D_t)v)_{H_0} dt = \int_{\mathbb{R}} ([\mathcal{A}(D_t), \eta]u, v) dt.$$

Since ηu is a distribution with compact support we can use the Parseval formula to obtain

$$\mathcal{A}(\lambda)(\widetilde{\eta u})(\lambda) = \tilde{f}(\lambda),$$

where $f = [\mathcal{A}(D_t), \eta]u$. Hence

$$(\eta u)(t) = \frac{1}{2\pi} \int_{\Im\lambda=\beta} e^{it\lambda} \mathcal{A}^{-1}(\lambda)\tilde{f}(\lambda) d\lambda$$

and by Condition $\mathfrak{H}$:

$$\| \eta u \|_{W_\beta^{\ell-1}(\mathbb{R})} \le c \, \| [\mathcal{A}(D_t), \eta]u \|_{L_{2,\beta}(\mathbb{R};H_0)} . \tag{10.84}$$

We set $\eta(t) = \zeta(t/T)$, where $\zeta \in \mathcal{D}(\mathbb{R})$, $\zeta(t) = 1$ in a neighbourhood of $t = 0$. Since $u \in W_\beta^{\ell-1}(\mathbb{R})$, the right-hand side in (10.84) tends to 0 as $T \to \infty$. Therefore $u = 0$. $\square$

As before, by $U_\sigma^{(\nu)}(t), V_\sigma^{(\nu)}(t)$ we designate special solutions of $\mathcal{A}(D_t)u = 0$ and $\mathcal{A}^*(D_t)v = 0$ defined in Sect.2.2. These solutions satisfy the biorthogonality condition (2.13). Proposition 2.8.1 has the following counterpart.

Proposition 10.6.4. *Let $\beta, \gamma \in (s_-, s_+)$ and $\beta < \gamma$. Suppose that there are no eigenvalues on the lines $\Im\lambda = \beta$, $\Im\lambda = \gamma$. Also let $f \in L_{2,\beta}(\mathbb{R}; H_0) \cap L_{2,\gamma}(\mathbb{R}; H_0)$ and u_1, u_2 be the solutions of (2.1) from the spaces $W_\beta^{\ell-1}(\mathbb{R})$, $W_\gamma^{\ell-1}(\mathbb{R})$ respectively. Then the relations (2.50)–(2.52) hold.*

Proof. By Theorem 10.6.3,

$$u_2(t) - u_1(t) = \frac{1}{2\pi} \int_\Gamma e^{it\lambda} \mathcal{A}^{-1}(\lambda)\tilde{f}(\lambda) d\lambda ,$$

where Γ is a closed contour located in the strip $\beta \le \Im\lambda \le \gamma$ which surrounds all the eigenvalues of $\mathcal{A}(\lambda)$ in this strip. Let P_ν be defined by (2.39). By deforming Γ we arrive at (2.50). Now (2.51) and (2.52) follow from (2.47), (2.11) and (2.12). $\square$

The next proposition (used in the sequel) can be proved in the same way as Proposition 2.8.3.

Proposition 10.6.5. *Let $\beta, \gamma \in (s_-, s_+)$, $\beta < \gamma$, and let $u \in W_{\gamma,\beta}^{\ell-1}(\mathbb{R})$ satisfy (2.53). Then (2.54) holds. In particular, if the strip $\beta < \Im\lambda < \gamma$ is free of the spectrum of $\mathcal{A}(\lambda)$ then $u = 0$.*

Let $k_\pm$ be real numbers satisfying $s_- < k_-$, $s_+ > k_+$ such that the strip $k_- < \Im\lambda < k_+$ is free of eigenvalues of $\mathcal{A}(\lambda)$. By $m_\pm$ we denote the maximum lengths of the Jordan chains corresponding to the eigenvalues of $\mathcal{A}(\lambda)$ on the lines $\Im\lambda = k_\pm$. If the line $\Im\lambda = k_\pm$ does not contain eigenvalues of $\mathcal{A}(\lambda)$ then we put $m_\pm = 1$.

The following improvement of the uniqueness part of Theorem 10.6.3 is proved in the same way as Proposition 3.2.1 (one should only replace ℓ by $\ell - 1$).

Proposition 10.6.6. *Let $u \in W_{\text{loc}}^{\ell-1}(\mathbb{R})$ be a solution of (2.53) satisfying*

$$\liminf_{t \to \pm\infty} e^{k_\mp t} \| u \|_{W^{\ell-1}(t,t+\delta)} = 0$$

with some $\delta > 0$. Then $u = 0$.

10.6.4 Comparison Principle and Solvability in $W_{\text{loc}}^{\ell-1}(\mathbb{R})$

Now we are in a position to obtain a result similar to Theorem 3.3.2. In particular, this result contains a comparison principle for the function

$$t \to \| u \|_{W^{\ell-1}(t,t+1)}$$

and a solution of the ordinary differential equation (3.21). In the following statement, g denotes Green's function (3.23), (3.24) of the operator $\mathcal{M}(\partial_t)$.

Theorem 10.6.7. *(i)(Existence) Let $f \in L_{2,\text{loc}}(\mathbb{R}; H_0)$ and*

$$\int_{\mathbb{R}} g(-\tau) \| f \|_{L_2(\tau,\tau+1;H_0)} \, d\tau < \infty. \tag{10.85}$$

There exists a solution $u \in W_{\text{loc}}^{\ell-1}(\mathbb{R})$ of (2.1) such that

$$\| u \|_{W^{\ell-1}(t,t+1)} \leq b \int_{\mathbb{R}} g(t - \tau) \| f \|_{L_2(\tau,\tau+1;H_0)} \, d\tau. \tag{10.86}$$

This solution also satisfies

$$\| u \|_{W^{\ell-1}(t,t+1)} = o\left(e^{-k_\mp t}\right) \quad as \quad t \to \pm\infty. \tag{10.87}$$

(ii) *(Uniqueness) Let $f \in L_{2,\text{loc}}(\mathbb{R}; H_0)$. If u_1, u_2 are solutions of (2.1) in $W_{\text{loc}}^{\ell-1}(\mathbb{R})$ subject to (10.87) then $u_1 = u_2$.*

Proof. (i) We fix β in the interval (k_-, k_+). First, we assume that $\mathrm{supp}f \subset [0,1]$ and denote by u the solution of (2.1) from $W_\beta^{\ell-1}(\mathbb{R})$ (see Theorem 10.6.3).

We fix numbers $\beta_\pm$ satisfying $\beta_+ \in (k_+, s_+)$, $\beta_- \in (s_-, k_-)$, and such that the strips $\beta_- \leq \Im\lambda < k_-$, $k_+ < \Im\lambda \leq \beta_+$ are free of the eigenvalues of $\mathcal{A}(\lambda)$. Denote by $u_\pm$ the solutions of (2.1) in the spaces $W_{\beta_\pm}^{\ell-1}(\mathbb{R})$. By (10.83)

$$\| u_\pm \|_{W_{\beta_\pm}^{\ell-1}(\mathbb{R})} \leq c \, \| f \|_{L_2(0,1;H_0)} \, . \tag{10.88}$$

Furthermore, using Proposition 10.6.4 we obtain

$$u(t) = \sum_{\Im\lambda_\nu = k_-} \int_0^1 e^{i\lambda_\nu(t-\tau)} P_\nu(t-\tau) f(\tau) d\tau + u_-(t) \tag{10.89}$$

and

$$u(t) = - \sum_{\Im\lambda_\nu = k_+} \int_0^1 e^{i\lambda_\nu(t-\tau)} P_\nu(t-\tau) f(\tau) d\tau + u_+(t). \tag{10.90}$$

The estimate (10.88) implies

$$\| u_\pm \|_{W^{\ell-1}(t,t+1)} \leq c \, e^{-\beta_\pm t} \, \| f \|_{L_2(0,1;H_0)},$$

which together with (10.89) and (10.90) gives

$$\| u \|_{W^{\ell-1}(t,t+1)} \leq c \, g(t) \, \| f \|_{L_2(0,1;H_0)} \, .$$

Now, let f be an arbitrary function in $L_{2,\mathrm{loc}}(\mathbb{R}; H_0)$ which is subject to (10.85). We represent f as the series

$$f(t) = \sum_{j=-\infty}^{+\infty} f_j(t) \, ,$$

where $f_j(t) = f(t)$ if $t \in (j, j+1]$ and 0 otherwise. Denote by u_j the solution of (2.1) with $f = f_j$ in $W_\beta^{\ell-1}(\mathbb{R})$. Using the translation invariance of the operator $\mathcal{A}(D_t)$, we obtain the estimate

$$\| u_j \|_{W^{\ell-1}(t,t+1)} \leq c \, g(t-j) \, \| f \|_{L_2(j,j+1;H_0)} \, .$$

Hence

$$\sum_{j=-\infty}^{\infty} \| u_j \|_{W^{\ell-1}(t,t+1)} \leq c \int_{\mathbb{R}} g(t-\tau) \, \| f \|_{L_2(\tau,\tau+1;H_0)} \, d\tau.$$

This together with (10.85) implies the convergence of

$$\sum_{j=-\infty}^{\infty} u_j$$

in $W_{\mathrm{loc}}^{\ell-1}(\mathbb{R})$, as well as estimate (10.86) for the limit u. It remains to apply Lemma 10.6.2.

(ii) The uniqueness follows from Proposition 10.6.6. $\square$

Remark 10.6.8. In a similar way, the Phragmen-Lindelöf estimates in Sect. 3.6 and results on the asymptotics from Sect. 3.8, can be carried over to the hyperbolic case. As an example, we formulate here an analog of Corollary 3.8.2 preserving the notation $k_\pm^{(j)}, m_\pm^{(j)}, \ j = 1, 2$, and $\mu^{(2)}$.

Theorem 10.6.9. *Let $u \in W_{loc}^{\ell-1}(\mathbb{R})$ be a solution of (3.45) satisfying*

$$\liminf_{t \to +\infty} e^{k_-^{(1)} t} \| u \|_{W^{\ell-1}(t,t+1)} = 0,$$

and let (3.63) hold. Then for $t > t_0 + 1$, the asymptotic representation (3.65) is valid where

$$\| v \|_{W^{\ell-1}(t,t+1)} \leq C \Big\{ \int_{t_0}^{\infty} \mu^{(2)}(\tau - t) \| f \|_{L_2(\tau,\tau+1;H_0)} \, d\tau$$

$$+ \mu^{(2)}(t_0 - t) \| u \|_{W^{\ell-1}(t_0,t_0+1)} \Big\}.$$

Example 10.6.10. Consider the equation

$$\left(D_t^2 - 2ic_0 D_t - B \right) u = f \quad \text{on} \quad (t_0, \infty), \tag{10.91}$$

where $c_0 \in \mathbb{R}$ and B is the same as in Example 10.6.1. Theorem 10.6.9 leads to the following asymptotic statement.

Let

$$\int_{t_0}^{\infty} e^{c_0 \tau} \| f \|_{L_2(\tau,\tau+1;H_0)} \, d\tau < \infty$$

and let $u \in W_{loc}^1(t_0, \infty)$ be a solution of (10.91) satisfying

$$\| u \|_{W^1(t,t+1)} = O \left(e^{Ct} \right) \quad \text{as} \quad t \to +\infty$$

with some $C > 0$. Then

$$u(t) = \sum_{\gamma_j < c_0^2} e^{(-c_0 + \sqrt{c_0^2 - \gamma_j})t} \varphi_j + v(t),$$

where $\varphi_j \in \ker(B - \gamma_j)$ and

$$\| v \|_{W^1(t,t+1)} = O \left(e^{(-c_0 + \varepsilon)t} \right) \quad \text{as} \quad t \to +\infty,$$

with an arbitrary $\varepsilon > 0$.

This is a direct consequence of Example 10.6.1 and Theorem 10.6.9, where $k_-^{(2)}$ is sufficiently large and $c_0 < k_-^{(1)} < k_+^{(1)} < c_0 + \varepsilon$.

10.7 "Hyperbolic" Operator Equation with Variable Coefficients

Now we consider the operator (5.2) as a perturbation of the operator $\mathcal{A}(D_t)$ with constant coefficients.

We introduce the equation

$$\Big(\mathcal{A}(D_t) - N(t, D_t)\Big)u(t) = f(t) \quad \text{on} \quad \mathbb{R}, \tag{10.92}$$

where

$$N(t, D_t) = \sum_{q=0}^{\ell-1} N_{\ell-1-q}(t)D_t^q.$$

Let $N_j(t)$ be bounded operators mapping H_j into H_0 for almost all t. Furthermore, we assume that $N(\cdot, D)u(\cdot)$ is in $L_{2,\text{loc}}(\mathbb{R}; H_0)$ irrespective of the function $u \in W_{\text{loc}}^{\ell-1}(\mathbb{R})$, and that

$$\| N(\cdot, D)u(\cdot) \|_{L_2(a,b;H_0)} \leq c_{a,b} \| u \|_{W^{\ell-1}(a,b)} .$$

Similar to Sect.5.1 we introduce the function

$$\rho(t) = \| N \|_{W^{\ell-1}(t,t+1) \to L_2(t,t+1;H_0)},$$

which is measurable, locally bounded and satisfies (5.43).

As before, let $\omega(t)$ denote a measurable majorant of $b\rho(t)$, where b is the constant in (10.86). We assume that ω satisfies (5.7). Then the Neumann series (5.8) is convergent and Green's function g_ω of (5.5) is well defined.

Since N maps $W_{\text{loc}}^{\ell-1}(\mathbb{R})$ into $L_{2,\text{loc}}(\mathbb{R}; H_0)$, we can easily define the notion of a solution to (10.92) with $f \in L_{2,\text{loc}}(\mathbb{R}; H_0)$. We say that $u \in \mathcal{D}'(\mathbb{R}; H_\ell) \cap W_{\text{loc}}^{\ell-1}(\mathbb{R})$ is a solution of (10.92) if

$$\int_\mathbb{R} \Big(\big(u(t), \mathcal{A}^*(D_t)v(t)\big)_{H_0} - \big(N(t, D_t)u(t), v(t)\big)_{H_0} \Big) dt$$

$$= \int_\mathbb{R} \big(f(t), v(t)\big)_{H_0} dt \tag{10.93}$$

for an arbitrary $v \in \mathcal{D}(\mathbb{R}; H_0)$.

The following result is an analog of the comparison principle in Sect.5.3.

Theorem 10.7.1. *Let* $f \in L_{2,\text{loc}}(\mathbb{R}; H_0)$ *be subject to* (5.9). *Then there exists a solution* $u \in \mathcal{D}'(\mathbb{R}; H_\ell) \cap W_{\text{loc}}^{\ell-1}(\mathbb{R})$ *of* (10.92) *such that*

$$\| u \|_{W^{\ell-1}(t,t+1)} \leq bw(t), \ t \in \mathbb{R}, \tag{10.94}$$

where w *is the solution of* (5.5) *defined by* (5.10).

Proof. By using the inverse operator $\mathcal{A}^{-1}$ constructed in Theorem 10.6.7 we literally repeat the argument in the beginning of Sect.5.3. We thus obtain the convergence of the series (5.46) in $W_{\mathrm{loc}}^{\ell-1}(\mathbb{R})$, as well as the estimate

$$\| u \|_{W^{\ell-1}(t,t+1)} \le b \int_{\mathbb{R}} g_\omega(t,\tau) \| f \|_{L_2(\tau,\tau+1;H_0)} \, d\tau$$

(which is similar to (5.47) and equivalent to (10.94)). The jth sum u_j of the series (5.46) satisfies the equation (10.92) with $f_j = Nu_{j-1} + f$. Since $\{f_j\}$ converges to $Nu + f$ in $L_{2,\mathrm{loc}}(\mathbb{R}; H_0)$, u is a solution of (10.92) by Lemma 10.6.2. $\square$

Replacing ℓ by $\ell-1$ in the proofs of Theorem 5.4.2 and Proposition 5.4.3 we arrive at the following uniqueness result.

Theorem 10.7.2. *Let $u \in \mathcal{D}'(\mathbb{R}; H_\ell) \cap W_{\mathrm{loc}}^{\ell-1}(\mathbb{R})$ be a solution of (10.92) with $f = 0$. Suppose that either*

$$\| u \|_{W^{\ell-1}(t,t+1)} = o\left(\limsup_{\tau \to \pm\infty} \frac{g_\omega(t,\tau)}{g_\omega(0,\tau)} \right) \quad \text{as} \quad t \to \pm\infty$$

or

$$\liminf_{t \to \pm\infty} g_\omega(0,t) \| u \|_{W^{\ell-1}(t,t+1)} = 0.$$

Then $u = 0$.

Similarly, all other results in Sect.5.4–9 and Ch.6–7 have direct analogs, which are valid under the assumptions of the present section.

10.8 The Operator L in Variational Form

10.8.1 Function Spaces

Let $q = 0, 1, \ldots, \ell$ and let $W_{\mathrm{comp}}^q(\mathbb{R})$ be the set of all vector valued functions $u \in W^q(\mathbb{R})$ with compact supports. A sequence $\{u_k\}_{k \ge 1}$ of elements in $W_{\mathrm{comp}}^q(\mathbb{R})$ converges to $u \in W_{\mathrm{comp}}^q(\mathbb{R})$ if the supports of u_k are uniformly bounded and $u_k \to u$ in $W^q(\mathbb{R})$. It is a simple exercise to show that $W_{\mathrm{comp}}^q(\mathbb{R})$ is a complete locally convex space. We denote by $W_{\mathrm{loc}}^{-q}(\mathbb{R})$ the spaces of antilinear bounded functionals on $W_{\mathrm{comp}}^q(\mathbb{R})$ endowed with the seminorms

$$\| f \|_{W^{-q}(a,b)} = \sup_v |f\{v\}|,$$

where a, b are real numbers, $a < b$, and supremum is taken over all $v \in W_{\mathrm{comp}}^q(\mathbb{R})$ such that

$$\mathrm{supp}\, v \subset [a,b], \quad \| v \|_{W^q(\mathbb{R})} = 1. \tag{10.95}$$

Lemma 10.8.1. *The following inequality holds:*

$$\| f \|_{W^{-q}(a,b)} \leq c \int_{a-1}^{b} \| f \|_{W^{-q}(t,t+1)} \, dt,$$

where c is independent of a, b, f.

Proof. First, we remark that the function

$$t \rightarrow \| f \|_{W^{-q}(t,t+1)}$$

is measurable. This follows from the equality

$$\| f \|_{W^{-q}(t,t+1)} = \sup_{v} |f\{v(\cdot - t)\}|,$$

where the supremum is taken over all $v \in W^{q}_{\mathrm{comp}}(\mathbb{R})$ which are normalized by $\| v \|_{W^{q}(\mathbb{R})} = 1$ and have supports in $[0, 1]$.

Let $\eta \in C_0^\infty(0, 1)$ and

$$\int_{\mathbb{R}} \eta(t) dt = 1.$$

Subject to (10.95), we have for any v ,

$$|f\{v\}| = \left| \int_{\mathbb{R}} f\{\eta(\cdot - \tau)v\} d\tau \right|$$

$$\leq c \int_{a-1}^{b} \| f \|_{W^{-q}(\tau,\tau+1)} \| v \|_{W^{q}(\tau,\tau+1)} \, d\tau.$$

The result follows. □

We say that the vector valued distribution $f \in \mathcal{D}'(\mathbb{R}; H_q^*)$ is equal to zero on an open set $\mathcal{O} \in \mathbb{R}$ if $f\{v\} = 0$ for every $v \in \mathcal{D}(\mathcal{O}; H_q)$. Let $\cup \mathcal{O}$ be the union of all open $\mathcal{O}$ in which f is equal to zero. The complement of $\cup \mathcal{O}$ is called the support of f.

If $f_j \in L_{2,\mathrm{loc}}(\mathbb{R}; H_j^*)$, $j = 0, \ldots, q$, then the sum

$$f(t) = \sum_{j=0}^{q} D_t^j f_{q-j}(t), \tag{10.96}$$

where $D_t^j \varphi$ is a distributional derivative of $\varphi \in L_{2,\mathrm{loc}}(\mathbb{R}; H_{q-j}^*)$, is an element of $W^{-q}_{\mathrm{loc}}(\mathbb{R})$ and

$$\| f \|_{W^{-q}(t,t+1)} \leq \sum_{j=0}^{q} \| f_j \|_{L_2(t,t+1;H_j^*)} .$$

In the next lemma we show that every element of $W^{-q}_{\mathrm{loc}}(\mathbb{R})$ can be represented in the form (10.96). The statement will be preceded by definitions of some spaces of vector valued functions.

Let $\mathring{W}^q(a,b)$ denote the subspace of $W^q(\mathbb{R})$ consisting of functions with supports on $[a,b]$. Let also

$$\mathbf{L}_2(a,b) = \prod_{k=0}^{q} L_2(a,b;H_{q-k}).$$

Its dual with respect to the scalar product

$$\sum_{k=0}^{q} \int_a^b \big(F_k(t), w_k(t)\big)_{H_0} dt$$

(where $w_k \in L_2(a,b;H_{q-k})$) is given by

$$\big(\mathbf{L}_2(a,b)\big)^* = \prod_{k=0}^{q} L_2(a,b;H_{q-k}^*).$$

Let $\mathcal{D}_q(a,b)$ denote the set of vectors $\mathrm{col}(D_t^k v)_{k=0}^q$ with $v \in \mathring{W}^q(a,b)$. One can directly verify that $\mathcal{D}_q(a,b)$ is a closed subspace of

$$\mathbf{L}_{2,\mathrm{comp}}(\mathbb{R}) = \{V \in \mathbf{L}_2(\mathbb{R}) : \mathrm{supp}\, V \text{ is compact}\}.$$

Lemma 10.8.2. *If $f \in W_{\mathrm{loc}}^{-q}(\mathbb{R})$ there exist functions $f_k \in L_{2,\mathrm{loc}}(\mathbb{R};H_k^*)$, $k = 0,\dots,q$, such that (10.96) holds and*

$$\sum_{k=0}^{q} \| f_k \|_{L_2(t,t+1;H_k^*)} \le c \| f \|_{W^{-q}(t-1,t+2)}, \tag{10.97}$$

where c does not depend on f and t.

Proof. For every $j = 0,\pm 1,\dots$ we can consider f as an antilinear bounded functional on $\mathcal{D}_q(j/2,1+j/2)$. By the Hahn-Banach theorem (Rudin (1973), Ch. 3) there exists a vector function

$$\mathrm{col}\big(f_k^{(j)}(t)\big)_{k=0}^{q} \in \big(\mathbf{L}_2(j/2,1+j/2)\big)^*$$

such that for all $v \in \mathring{W}^q(j/2,1+j/2)$

$$f\{v\} = \sum_{k=0}^{q} \int_{j/2}^{1+j/2} \big(f_{q-k}^{(j)}(t), D_t^k v(t)\big)_{H_0} dt \tag{10.98}$$

and

$$\Big(\sum_{k=0}^{q} \| f_k^{(j)} \|_{L_2(j/2,1+j/2;H_{q-k})}^2 \Big)^{1/2} = \| f \|_{W^{-q}(j/2,1+j/2)} . \tag{10.99}$$

Let $\{\eta_j\}_{j=-\infty}^{+\infty}$ be a partition of unity in $\mathbb{R}$ such that

$$\text{supp } \eta_j \in [j/2, 1 + j/2]$$

and

$$|\partial_t^k \eta_j(t)| \le c \text{ on } \mathbb{R}, \quad k = 0, 1, \ldots, q.$$

For $v \in W_{\text{comp}}^q(\mathbb{R})$ we put

$$F\{v\} = \sum_{j=-\infty}^{+\infty} \sum_{k=0}^{q} \int_{\mathbb{R}} \left(f_k^{(j)}(t), D_t^k(\eta_j v)(t)\right)_{H_0} dt. \tag{10.100}$$

The sum on the right is finite and by (10.98) we get

$$F\{v\} = f\{v\}$$

for all $v \in W_{\text{comp}}^q(\mathbb{R})$. If we set

$$f_s(t) = \sum_{j=-\infty}^{+\infty} \sum_{k=s}^{q} \binom{k}{s} (-D_t)^{k-s} \left(\overline{\eta}_j(t)\right) f_k^{(j)}(t), \tag{10.101}$$

then the formula (10.100) can be rewritten as

$$F\{v\} = \sum_{s=0}^{q} \int_{\mathbb{R}} \left(f_s(t), D_t^s v(t)\right)_{H_0} dt.$$

By (10.101) and (10.99) we arrive at (10.97). $\square$

We introduce several spaces of vector valued distributions to be used in the sequel.

By $W_\beta^{-q}(\mathbb{R})$, $\beta \in \mathbb{R}$, we denote the dual of $W_{-\beta}^q(\mathbb{R})$. It is clear that

$$W_\beta^{-q}(\mathbb{R}) \subset W_{\text{loc}}^{-q}(\mathbb{R}).$$

Let $\mathcal{D}'(\mathbb{R}; H_q^*)$ denote the space of bounded linear functionals on $\mathcal{D}(\mathbb{R})$ taking values in H_q^*.

The Schwartz space of scalar functions f such that

$$\sup_{\mathbb{R}} |t^k D_t^n f(t)| < \infty,$$

for all $k, n = 0, 1, \ldots$, will be denoted by $\mathcal{S}(\mathbb{R})$. We shall also use the distribution space $\mathcal{S}'(\mathbb{R}; H_q^*)$, consisting of bounded linear functionals on $\mathcal{S}(\mathbb{R})$ which take values in H_q^*.

It is well known that the Fourier transform maps $\mathcal{S}'(\mathbb{R}; H_q^*)$ onto itself (see Lions, Magenes (1972), Ch.I).

We also need the space

$$\mathcal{S}_\beta'(\mathbb{R}; H_q^*) = \left\{u \in \mathcal{D}'(\mathbb{R}; H_q^*) : e^{\beta t} u \in \mathcal{S}'(\mathbb{R}; H_q^*)\right\}.$$

The range of the Fourier transform defined on $\mathcal{S}'_\beta(\mathbb{R}; H^*_q)$ is the space $\mathcal{S}'$ of distributions on the line $\Im\lambda = \beta$ with values in H^*_q.

We show that $W^{-q}_\beta(\mathbb{R}) \subset \mathcal{S}'_\beta(\mathbb{R}; H^*_q)$. In fact, let $f \in W^{-q}_\beta(\mathbb{R})$. Then for all $h \in \mathcal{S}(\mathbb{R})$, $v \in H_q$ we have

$$\left| \int_\mathbb{R} e^{\beta t}(f, hv)_{H_0} dt \right| \le c \parallel v \parallel_q \left(\int_\mathbb{R} \sum_{k=0}^q |D^k_t h(t)|^2 dt \right)^{1/2}$$

and hence

$$\int_\mathbb{R} e^{\beta t}(f, hv)_{H_0} dt = (F, v)_{H_0},$$

where $F \in H^*_q$. Therefore $f \in \mathcal{S}'_\beta(\mathbb{R}; H^*_q)$.

We introduce a norm in H^*_q which depends on the parameter $\lambda \in \mathbb{C}$. Let $f \in H^*_q$. We put

$$\parallel f \parallel_{\lambda, H^*_q} = \sup_v |(f, v)_{H_0}|, \tag{10.102}$$

the supremum being taken over all $v \in H_q$ such that

$$\sum_{j=0}^q |\lambda|^{2j} \parallel v \parallel^2_{q-j} = 1. \tag{10.103}$$

By $\mathcal{F}W^q(\mathbb{R})$ we denote the space of functions $w \in L_{2,\mathrm{loc}}(\mathbb{R}; H_q)$ subject to

$$\int_\mathbb{R} \sum_{j=0}^q |\lambda|^{2j} \parallel w(\lambda) \parallel^2_{q-j} d\lambda < \infty.$$

The Fourier transform maps $W^q(\mathbb{R})$ isomorphically onto $\mathcal{F}W^q(\mathbb{R})$.

Lemma 10.8.3. *The dual $(\mathcal{F}W^q(\mathbb{R}))^*$ of $\mathcal{F}W^q(\mathbb{R})$ with respect to the sesquilinear form*

$$\int_\mathbb{R} (h(\lambda), w(\lambda))_{H_0} d\lambda, \quad w \in \mathcal{F}W^q(\mathbb{R}), \tag{10.104}$$

consists of the functions $h \in L_{2,\mathrm{loc}}(\mathbb{R}; H^*_q)$ which satisfy

$$\int_\mathbb{R} \parallel h(\lambda) \parallel^2_{\lambda, H^*_q} d\lambda < \infty. \tag{10.105}$$

Proof. Clearly, any functional (10.104) with h satisfying (10.105) belongs to $(\mathcal{F}W^q(\mathbb{R}))^*$.

We prove the converse statement. Since $\mathcal{F}W^q(\mathbb{R})$ is a Hilbert space, any antilinear bounded functional on this space can be given by

$$w \to \int_\mathbb{R} \sum_{j=0}^q |\lambda|^{2j} (u(\lambda), w(\lambda))_{q-j} d\lambda,$$

where $u \in W^q(\mathbb{R})$. There exists $h(\lambda) \in H^*_q$ such that for all $v \in H_q$

$$\left(h(\lambda), v\right)_{H_0} = \sum_{j=0}^{q} |\lambda|^{2j} \left(u(\lambda), v\right)_{q-j}.$$

By (10.102),

$$\| h(\lambda) \|_{\lambda, H_q^*} = \left(\sum_{j=0}^{q} |\lambda|^{2j} \| u(\lambda) \|_{q-j}^2 \right)^{1/2}.$$

This implies (10.105). $\square$

The space $W_\beta^{-q}(\mathbb{R})$ admits the following description.

Lemma 10.8.4. *In order that* $f \in W_\beta^{-q}(\mathbb{R})$, *it is necessary and sufficient that the Fourier transform* $\tilde{f}(\lambda)$ *belongs to* $L_{2,\mathrm{loc}}$ *on the line* $\Im\lambda = \beta$ *with values in* H_q^* *and satisfies*

$$\int_{\Im\lambda=\beta} \| \tilde{f}(\lambda) \|_{\lambda, H_q^*}^2 \, d\lambda < \infty. \tag{10.106}$$

The square root of the left-hand side represents a norm in $W_\beta^{-q}(\mathbb{R})$.

Proof. It is sufficient to put $\beta = 0$. Then the result follows from Parseval's theorem and Lemma 10.8.3. $\square$

10.8.2 An Equivalent Norm in $W_\beta^{-q}(\mathbb{R})$

Lemma 10.8.5. *The norm in* $W_\beta^{-q}(\mathbb{R})$ *is equivalent to the norm*

$$|||f||| = \left(\int_\mathbb{R} e^{2\beta t} \| f \|_{W^{-q}(t,t+1)}^2 \, dt \right)^{1/2}.$$

Proof. Let $\{\eta_j\}$ be the same partition of unity on $\mathbb{R}$ as in Lemma 10.8.1. We represent an arbitrary function $v \in W_{-\beta}^q(\mathbb{R})$ as

$$v = \sum_{j=-\infty}^{+\infty} v_j, \quad v_j(t) = \eta_j(\tau + t)v(t),$$

where $\tau \in [0, 1)$. By Proposition 4.2.1 the norm of v in $W_{-\beta}^q(\mathbb{R})$ is equivalent to the norm

$$\left(\int_\mathbb{R} e^{-2\beta t} \| v \|_{W^q(t,t+1)}^2 \, dt \right)^{1/2}.$$

Clearly,

$$\left| \int_\mathbb{R} \left(f(t), v(t) \right)_{H_0} dt \right|$$

$$\leq c \sum_{j=-\infty}^{+\infty} \| f \|_{W^{-q}(\tau+j/2,\tau+1+j/2)} \| v \|_{W^q(\tau+j/2,\tau+1+j/2)}.$$

Integrating this inequality with respect to τ from 0 to 1 and applying the Cauchy inequality, we get

$$\left| \int_{\mathbb{R}} (f(t), v(t))_{H_0} dt \right|$$

$$\leq c \left(\int_{\mathbb{R}} e^{2\beta t} \| f \|^2_{W^{-q}(t,t+1)} dt \right)^{1/2} \| v \|_{W^q_{-\beta}(\mathbb{R})},$$

which implies

$$\| f \|_{W^{-q}_\beta(\mathbb{R})} \leq c \, |||f|||.$$

Let $w_j \in W^q(\mathbb{R})$, supp $w_j \in [\tau + j, \tau + j + 1]$, $j = 0, \pm 1, \dots$ We put

$$\| w_j \|_{W^q(\tau+j, \tau+j+1)} = \alpha_j$$

and assume that

$$\sum_{j=-\infty}^{+\infty} e^{-2\beta(\tau+j)} \alpha_j^2(\tau) = 1. \tag{10.107}$$

We choose w_j satisfying

$$\int_{\mathbb{R}} (f, w_j)_{H_0} dt \geq \frac{\alpha_j}{2} \| f \|_{W^{-q}(\tau+j, \tau+j+1)}$$

and set

$$w = \sum_{j=-\infty}^{+\infty} w_j.$$

Due to (10.107)

$$c_1 \geq \| w \|_{W^q_{-\beta}(\mathbb{R})} \geq c_2 > 0,$$

where c_1 and c_2 do not depend on τ and $\lambda_j(\tau)$. We have

$$\int_{\mathbb{R}} (f, w)_{H_0} dt \leq \| f \|_{W^{-q}_\beta(\mathbb{R})} \| w \|_{W^q_{-\beta}(\mathbb{R})}$$

and

$$\int_{\mathbb{R}} (f, w)_{H_0} dt \geq \frac{1}{2} \sum_{j=-\infty}^{+\infty} \alpha_j \| f \|_{W^{-q}(\tau+j, \tau+j+1)} \cdot$$

By (10.107) these inequalities imply

$$\| f \|_{W^{-q}_\beta(\mathbb{R})} \geq c \sum_{j=-\infty}^{+\infty} \alpha_j \| f \|_{W^{-q}(\tau+j, \tau+j+1)} \cdot$$

Taking the supremum on the right with respect to those $\{\alpha_j\}$ satisfying (10.107), we arrive at

$$\| f \|^2_{W_\beta^{-q}(\mathbb{R})} \geq c \sum_{j=-\infty}^{\infty} e^{2\beta(\tau+j)} \| f \|^2_{W^{-q}(\tau+j,\tau+j+1)} \cdot$$

Integrating the last estimate with respect to τ from 0 to 1, we obtain

$$\| f \|_{W_\beta^{-q}(\mathbb{R})} \geq c \|\|f\|\|.$$

$\square$

10.8.3 Assumptions on the Pencil $\mathcal{A}(\lambda)$

Let

$$A_{jk} : H_k \to H_j^*, \ j = 0, \ldots, q, \ k = 0, \ldots, \ell - q$$

be bounded linear operators. Put

$$\mathcal{A}(\lambda) = \sum_{j=0}^{q} \sum_{k=0}^{\ell-q} \lambda^{\ell-j-k} A_{jk}.$$

We suppose that the mapping

$$\mathcal{A}(\lambda) : H_{\ell-q} \to H_q^*$$

is Fredholm for all $\lambda \in \mathbb{C}$ and that it is invertible for at least one value of λ in this strip. The analog of Condition II in Sect. 2.2 runs as follows.

Condition $\mathfrak{V}$ *There exists $r > 0$ such that*

$$\sum_{k=0}^{\ell-q} |\lambda|^{2k} \| u \|^2_{\ell-q-k} \leq c \| \mathcal{A}(\lambda)u \|^2_{\lambda,H_q^*} \qquad (10.108)$$

for all $u \in H_{\ell-q}$ and for all real λ with $|\lambda| \geq r$.

Proposition 10.8.6. *There exists $\theta \in (0, \pi/2)$ such that for all*

$$\lambda \in S_{r,\theta} = \left\{ z \in \mathbb{C} : |\arg(\pm z)| \leq \theta, \ |\Re z| \geq r \right\}$$

the inequality (10.108) holds.

Proof. Along the same lines as that of Proposition 2.2.1. $\square$

We state a sufficient condition for (10.108) in the case $\ell = 2q$; this condition is often met in applications.

Proposition 10.8.7. *Let ℓ be even and let $\ell = 2q$. Suppose that the following estimate holds for all $u \in H_q$ and $\tau \in \mathbb{R}$ with sufficiently large $|\tau|$:*

$$\sum_{j,k=0}^{q} \tau^{2q-j-k} \Re(A_{jk}u, u)_{H_0} \geq c \sum_{j=0}^{q} \tau^{2(q-j)} \| u \|_j^2 . \tag{10.109}$$

Then the above Condition $\mathfrak{V}$ *is valid.*

Proof. The inequality (10.109) implies

$$\Re\big(\mathcal{A}(\lambda)u, u\big)_{H_0} \geq c \sum_{j=0}^{q} |\lambda|^{2j} \| u \|_{q-j}^2 \tag{10.110}$$

for real λ with $|\lambda| \geq r$, (r is a sufficiently large positive number).

Let $f \in H_q^*$. By the Riesz representation theorem there exists a unique element $u_0 \in H_q$ such that for all $v \in H$

$$(f, v)_{H_0} = \sum_{j=0}^{q} |\lambda|^{2j} (u_0, v)_{H_{q-j}} . \tag{10.111}$$

By (10.102)

$$\| f \|_{\lambda, H_q^*}^2 = \sum_{j=0}^{q} |\lambda|^{2j} \| u_0 \|_{q-j}^2 . \tag{10.112}$$

Let $f = \mathcal{A}(\lambda)u$ with $u \in H_q$. It follows from (10.111) and (10.110) that

$$\Re \sum_{j=0}^{q} |\lambda|^{2j} (u_0, u)_{H_{q-j}} = \Re\big(\mathcal{A}(\lambda)u, u\big)_{H_0}$$

$$\geq c \sum_{j=0}^{q} |\lambda|^{2j} \| u \|_{q-j}^2 .$$

Hence

$$\sum_{j=0}^{q} |\lambda|^{2j} \| u_0 \|_{q-j}^2 \geq c \sum_{j=0}^{q} |\lambda|^{2j} \| u \|_{q-j}^2 .$$

This together with (10.112) gives (10.108). $\square$

10.8.4 Equation with Constant Coefficients

We shall study a variational form of the equation $\mathcal{A}(D_t)u = f$ on $\mathbb{R}$, where $f \in W_{\text{loc}}^{-q}(\mathbb{R})$. By a solution of this equation we mean a function $u \in W_{\text{loc}}^{\ell-q}(\mathbb{R})$ satisfying

$$\sum_{j=0}^{q} \sum_{k=0}^{\ell-q} \int_{\mathbb{R}} \left(A_{jk} D_t^{\ell-q-k} u(t), D_t^{q-j} v(t) \right)_{H_0} dt$$

$$= \int_{\mathbb{R}} \left(f(t), v(t) \right)_{H_0} dt \tag{10.113}$$

for all $v \in W_{\text{comp}}^{q}(\mathbb{R})$.

The left-hand side of (10.113) generates a bounded linear operator

$$\mathcal{A}(D_t) : W_{\text{loc}}^{\ell-q}(\mathbb{R}) \to W_{\text{loc}}^{-q}(\mathbb{R}).$$

Theorem 10.8.8. *Let $\beta \in \mathbb{R}$. Suppose that there are no eigenvalues of $\mathcal{A}(\lambda)$ on the line $\Im\lambda = \beta$. If $f \in W_\beta^{-q}(\mathbb{R}; H_0)$ then there exists a solution $u \in W_\beta^{\ell-q}(\mathbb{R})$ of (10.113). This solution is unique in $W_\beta^{\ell-q}(\mathbb{R})$ and is given by (2.19).*

Proof. Applying the Fourier transform to both sides of (10.113) we get

$$\int_{\mathbb{R}} \left(\mathcal{A}(\tau + i\beta)\tilde{u}(\tau + i\beta), \tilde{v}(\tau - i\beta) \right)_{H_0} d\tau$$

$$= \int_{\mathbb{R}} \left(\tilde{f}(\tau + i\beta), \tilde{v}(\tau - i\beta) \right)_{H_0} d\tau$$

Since $v \in W_{\text{comp}}^{q}(\mathbb{R})$ is arbitrary, we get

$$\mathcal{A}(\tau + i\beta)\tilde{u}(\tau + i\beta) = \tilde{f}(\tau + i\beta). \tag{10.114}$$

As there are no eigenvalues of $\mathcal{A}(\lambda)$ on the line $\Im\lambda = \beta$, equation (10.114) is uniquely solvable, and by Condition $\mathfrak{V}$:

$$\sum_{j=0}^{\ell-q} |\lambda|^{2j} \| \tilde{u}(\lambda) \|_{\ell-q}^2 \le c \| \tilde{f}(\lambda) \|_{\lambda, H_q^*}^2, \quad \Im\lambda = \beta.$$

By the Parseval theorem, Lemma 10.8.4 and (2.16), we get $u \in W_\beta^{\ell-q}(\mathbb{R})$ and

$$\| u \|_{W_\beta^{\ell-q}(\mathbb{R})} \le c \| f \|_{W_\beta^{-q}(\mathbb{R})}.$$

$\square$

Analogs of Propositions 2.8.1 and 3.2.1 run as follows (they are proved in the same way).

Proposition 10.8.9. *Let $\beta, \gamma \in \mathbb{R}$ and $\beta < \gamma$. Suppose that there are no eigenvalues on the lines $\Im\lambda = \beta$ and $\Im\lambda = \gamma$. Also let $f \in W_\beta^{-q}(\mathbb{R}) \cap W_\gamma^{-q}(\mathbb{R})$ and u_1, u_2 be the solutions of (10.113) from the spaces $W_\beta^{\ell-q}(\mathbb{R})$, $W_\gamma^{\ell-q}(\mathbb{R})$ respectively. Then the relations (2.50)–(2.52) hold.*

Proposition 10.8.10. *Let $k_- < k_+$ and let the strip $k_- < \Im\lambda < k_+$ be free of eigenvalues of $\mathcal{A}(\lambda)$. Let $u \in W_{\mathrm{loc}}^{\ell-q}(\mathbb{R})$ be a solution of (10.113) satisfying*

$$\liminf_{t \to \pm\infty} e^{k_\mp t} \parallel u \parallel_{W^{\ell-q}(t,t+\delta)} = 0$$

with some $\delta > 0$. Then $u = 0$.

The next result is similar to Theorem 3.3.2. By g we denote the same Green's function of the operator

$$\mathcal{M}(\partial_t) = (\partial_t + k_+)^{m_+}(-\partial_t - k_-)^{m_-}$$

as in Sect.3.5.

Theorem 10.8.11. (i) *Let $f \in W_{\mathrm{loc}}^{-q}(\mathbb{R})$ and*

$$\int_{\mathbb{R}} g(-\tau) \parallel f \parallel_{W^{-q}(\tau,\tau+1)} d\tau < \infty. \tag{10.115}$$

There exists a solution $u \in W_{\mathrm{loc}}^{\ell-q}(\mathbb{R})$ of (10.113) such that

$$\parallel u \parallel_{W^{\ell-q}(t,t+1)} \le b \int_{\mathbb{R}} g(t-\tau) \parallel f \parallel_{W^{-q}(\tau,\tau+1)} d\tau. \tag{10.116}$$

This solution also satisfies

$$\parallel u \parallel_{W^{\ell-q}(t,t+1)} = o(e^{-k_\mp t}) \quad as \quad t \to \pm\infty. \tag{10.117}$$

(ii) *The solution $u \in W_{\mathrm{loc}}^{\ell-q}(\mathbb{R})$ of (10.113) satisfying (10.117) is unique.*

Before proving this theorem we remark that as in Sect.3.5, the estimate (10.116) can be interpreted as the comparison principle

$$\parallel u \parallel_{W^{\ell-q}(t,t+1)} \le bw(t),$$

where w is the solution of the ordinary differential equation

$$\mathcal{M}(\partial_t)w(t) = \parallel f \parallel_{W^{-q}(t,t+1)} \quad on \quad \mathbb{R} \tag{10.118}$$

is subject to

$$|w(t)| = o(e^{-k_\mp t}) \quad as \quad t \to \pm\infty. \tag{10.119}$$

(According to Proposition 3.5.3 there exists one and only one solution of (10.118) satisfying (10.119)).

Proof. (i) We fix $\beta \in (k_-, k_+)$ and consider $f \in W_{\mathrm{loc}}^{-q}(\mathbb{R})$ with $\operatorname{supp} f \subset$ [0, 1]. Then $f \in W_\beta^{-q}(\mathbb{R})$ and we denote by u the solution of (10.113) in the space $W_\beta^{\ell-q}(\mathbb{R})$ (see Theorem 10.8.8). Let $\beta_\pm$ be the same as in the proof of Theorem 10.6.7 (i), where $s_\pm = \pm\infty$. Denote by $u_\pm$ the solutions of (10.113) in $W_{\beta_\pm}^{\ell-q}(\mathbb{R})$. By Proposition 10.8.7, the formulae (10.108) and (10.109) hold where

$$\| u_\pm \|_{W_{\beta_\pm}^{\ell-q}(\mathbb{R})} \le c \, \| f \|_{W^{-q}(0,1)} .$$

This together with (2.47) gives

$$\| u \|_{W^{\ell-q}(t,t+1)} \le c \, g(t) \, \| f \|_{W^{-q}(0,1)} . \tag{10.120}$$

Now let f be an arbitrary function in $W_{\mathrm{loc}}^{-q}(\mathbb{R})$ which is subject to (10.115). We fix the same partition of unity $\{\eta_j\}_{j=-\infty}^{+\infty}$ as in Lemma 10.8.2 and represent f as

$$f = \sum_{j=-\infty}^{+\infty} f_j, \quad f_j = \eta_j f.$$

It is clear that $\operatorname{supp} f_j \in [j/2, j/2 + 1]$. Denote by $u_j \in W_\beta^{\ell-q}(\mathbb{R})$ the solution of (10.113) with $f = f_j$. Using the translation invariance of $\mathcal{A}(D_t)$ we derive from (10.120) that

$$\| u_j \|_{W^{\ell-q}(t,t+1)} \le c \, g(t - j/2) \, \| f \|_{W^{-q}(j/2, j/2+1)} .$$

Hence

$$\sum_{j=-\infty}^{\infty} \| u_j \|_{W^{\ell-q}(t,t+1)} \le c \int_{\mathbb{R}} g(t - \tau) \, \| f \|_{W^{-q}(\tau,\tau+1)} \, d\tau.$$

This, together with (10.115), gives the convergence of

$$\sum_{j=-\infty}^{+\infty} u_j$$

in $W_{\mathrm{loc}}^{\ell-q}(\mathbb{R})$. The limit u satisfies (10.116) and (10.113).

(ii) The uniqueness follows from Proposition 10.8.10. $\square$

10.8.5 Equation with Variable Coefficients

Now we turn to the equation (10.3) in the case when $A_{jk}(t)$ is a bounded linear operator: $H_k \to H_j^*$ for almost all t and satisfying

$$A_{jk}(\cdot)v(\cdot) \in L_{2,\mathrm{loc}}(\mathbb{R}; H_j^*)$$

for every $v \in L_{2,\mathrm{loc}}(\mathbb{R}; H_k)$. There are no differentiability requirements on the coefficients A_{jk}. We will interpret (10.3) as a variational equation

$$\sum_{j=0}^{q} \sum_{k=0}^{\ell-q} \int_{\mathbb{R}} \left(A_{jk}(t)D_t^{\ell-q-k}u(t), D_t^{q-j}v(t)\right)_{H_0} dt$$

$$= \int_{\mathbb{R}} \left(f(t), v(t)\right)_{H_0} dt, \tag{10.121}$$

where v is an arbitrary function in $W_{\mathrm{comp}}^q(\mathbb{R})$.

Let

$$N_{jk}(t) = A_{jk} - A_{jk}(t).$$

The role of the function (5.3) will be played by a function $\rho(t)$ satisfying

$$\left| \sum_{j=0}^{q} \sum_{k=0}^{\ell-q} \int_t^{t+1} \left(N_{jk}(\tau)D_\tau^{\ell-q-k}u(\tau), D_\tau^{q-j}v(\tau)\right)_{H_0} d\tau \right|$$

$$\leq \rho(t) \| u \|_{W^{\ell-q}(t,t+1)} \| v \|_{W^q(t,t+1)} \tag{10.122}$$

for arbitrary functions $u \in W^{\ell-q}(t, t+1), v \in \mathring{W}^q(t, t+1)$. In other words, the sesquilinear form on the left-hand side of (10.122) generates the operator

$$N : W_{\mathrm{loc}}^{\ell-q}(\mathbb{R}) \to W_{\mathrm{loc}}^{-q}(\mathbb{R}),$$

and

$$\| Nu \|_{W^{-q}(t,t+1)} \leq \rho(t) \| u \|_{W^{\ell-q}(t,t+1)} . \tag{10.123}$$

We suppose that there exists a majorant ω of the function $b\rho$ satisfying (5.7). We are going to prove a comparison principle between solutions of (10.121) and the ordinary differential equation

$$\Big(\mathcal{M}(\partial_t) - \omega(t)\Big)w(t) = \| f \|_{W^{-q}(t,t+1)} \tag{10.124}$$

(cf. Sect.5.3). By g_ω we denote Green's function of this last equation (see Sect. 5.2).

Theorem 10.8.12. (Existence) *Let $f \in W_{\mathrm{loc}}^{-q}(\mathbb{R})$ be subject to*

$$\int_{\mathbb{R}} g_\omega(0, \tau) \| f \|_{W^{-q}(\tau,\tau+1)} d\tau < \infty.$$

Then there exists a solution $u \in W_{\mathrm{loc}}^{\ell-q}(\mathbb{R})$ of (10.121) such that

$$\| u \|_{W^{\ell-q}(t,t+1)} \leq bw(t), \ t \in \mathbb{R}, \tag{10.125}$$

where w is the solution of (10.124) given by

$$w(t) = \int_{\mathbb{R}} g_\omega(t, \tau) \| f \|_{W^{-q}(\tau,\tau+1)} d\tau.$$

Proof. Let $\mathcal{A}^{-1}$ be the inverse operator constructed in Theorem 10.8.11. The convergence of the series (5.46) in $W_{\text{loc}}^{\ell-q}(\mathbb{R})$ as well as the estimate (10.125), follows as at the beginning of Sect.5.3. $\square$

Theorem 10.8.13. (Uniqueness) *Let* $u \in W_{\text{loc}}^{\ell-q}(\mathbb{R})$ *be a solution of* (10.121) *with* $f = 0$. *Suppose that either* (5.56) *or* (5.58) *holds, where* ℓ *is replaced by* $\ell - q$. *Then* $u = 0$.

Proof. (i) The uniqueness of u satisfying (5.56) is proved by the argument used in Theorem 5.4.2.

(ii) Suppose that (5.58) holds. Let η be the same cut-off function as in Theorem 5.4.3. The function ηu is a solution of (10.121), where f is given by

$$
v \to \sum_{j=0}^{q} \sum_{k=0}^{\ell-q} \int_{\mathbb{R}} \Big\{ \big((A_{jk}(t)[D_t^{\ell-q-k}, \eta]u(t), D_t^{q-j}v(t)) \big)_{H_0}
$$
$$
- \big(A_{jk}(t)D_t^{\ell-q-k}u(t), [D_t^{q-j}, \overline{\eta}]v(t) \big)_{H_0} \Big\} dt.
$$

Clearly, $\mathrm{supp} f \subset (T_j, T_{j+1}) \cup (S_j, S_{j+1})$ and

$$
\| f \|_{W^q(t,t+1)} \leq \begin{cases} c\,\| u \|_{W^{\ell-q}(T_j,T_j+1)} & \text{if } t \in (T_j - 1, T_j + 1) \\ c\,\| u \|_{W^{\ell-q}(S_j,S_j+1)} & \text{if } t \in (S_j - 1, S_j + 1) \\ 0 & \text{otherwise.} \end{cases}
$$

In accordance with (i), and by Theorem 10.8.12,

$$
\| \eta v \|_{W^\ell(t,t+1)} \leq c \Big\{ g_\omega(t, T_j)\,\| u \|_{W^{\ell-q}(T_j,T_j+1)}
$$
$$
+ g_\omega(t, S_j)\,\| u \|_{W^{\ell-q}(S_j,S_j+1)} \Big\}.
$$

Due to (5.58) and the choice of T_j, S_j (see Theorem 5.4.2), this implies $u = 0$.
$\square$

Remark 10.8.14. Since analogs of the key results of Sect.5.3 and 5.4 (Theorems 5.3.2, 5.4.2, 5.4.3) have been carried over to the variational form of equation (10.3), one can easily obtain analogs of all results from Sect.5.4–5.6 and Ch.6–9.

10.8.6 The Variational Form of the Dirichlet Problem in a Cone

As an application of the theory developed in the present section, we consider the Dirichlet problem (3.79) in the cone K defined by (0.10), where Ω is an arbitrary proper subdomain of S^{n-1} with smooth boundary.

As in Example 1.5.2, we assume that the polynomial $P(\xi)$ is homogeneous of degree $2m$ and that $\Re P(\xi) > 0$ for $\xi \neq 0$.

Let $\overset{\circ}{V}_2^m(K)$ denote the completion of $C_0^\infty(K)$ in the norm

$$\|u\|_{V_2^m(K)} = \left(\sum_{|\delta|=m} \int_K |D_x^\delta u(x)|^2 dx \right)^{1/2}.$$

By (1.46) this norm is equivalent to the norm

$$\left(\sum_{|\delta|\leq m} \int_K |x|^{2(|\delta|-m)} |D_x^\delta u(x)|^2 dx \right)^{1/2}.$$

We say that the function u belongs to the space $\overset{\circ}{V}_{2,\mathrm{loc}}^m(K)$ if $\eta u \in \overset{\circ}{V}_2^m(K)$ for an arbitrary $\eta \in C_0^\infty(\overline{K}\backslash\{0\})$. We supply the space $\overset{\circ}{V}_{2,\mathrm{loc}}^m(K)$ with the family of seminorms $\|u\|_{W_2^m(K_r)}$, $r \in (0,\infty)$, where

$$K_r = \{x \in K : e^{-1}r < |x| < r\}$$

and

$$\|u\|_{W_2^m(K_r)} = \left(\sum_{|\alpha|\leq m} r^{2|\alpha|-n} \|D_x^\alpha u\|_{L_2(K_r)}^2 \right)^{1/2}.$$

By definition, the space $\overset{\circ}{V}_{2,\mathrm{comp}}^m(K)$ contains those functions from $\overset{\circ}{V}_2^m(K)$ with compact supports in $\overline{K}\backslash\{0\}$. The dual space of $\overset{\circ}{V}_{2,\mathrm{comp}}^m(K)$ with respect to the scalar product in $L_2(K)$ will be denoted by $V_{2,\mathrm{loc}}^{-m}(K)$; this dual will be supplied with the family of seminorms

$$\|f\|_{W_2^{-m}(K_r)} = r^{-n} \sup \left| \int_K f\overline{z}dx \right|, \quad r > 0,$$

where the supremum is taken over all $z \in \overset{\circ}{V}_2^m(K)$ which are supported by $\overline{K}_r$ and satisfy the normalization condition:

$$\|z\|_{W_2^m(K_r)} = 1.$$

We write the operator $P(D_x)$ as

$$P(D_x) = \sum_{|\gamma|=|\delta|=m} p_{\gamma\delta} D_x^{\gamma+\delta}.$$

We are now in a position to give a variational formulation of the Dirichlet problem (3.79). The function $u \in \overset{\circ}{V}_{2,\mathrm{loc}}^m(K)$ is called a solution of the problem (3.79) with $f \in V_{2,\mathrm{loc}}^{-m}(K)$ if

$$\int_K \sum_{|\gamma|=|\delta|=m} p_{\gamma\delta} D_x^\gamma u \overline{D_x^\delta z} dx = \int_K f\overline{z}dx \tag{10.126}$$

for all $z \in \overset{\circ}{V}_{2,\mathrm{comp}}^m(K)$. By setting $t = \log r^{-1}$ and

$$v(t,\theta) = e^{(2m-n)t} z(e^{-t}\theta),$$

we arrive at the equality

$$\int_{\mathcal{C}} \sum_{j,k \leq m} \Phi_k(\theta, D_\theta) D_t^{m-k} u \cdot \overline{\Psi_j(\theta, D_\theta) D_t^{m-j} v} \, dt \, d\theta$$

$$= \int_{\mathcal{C}} e^{-2mt} f \overline{v} \, dt \, d\theta, \tag{10.127}$$

where Φ_k and Ψ_j are differential operators on S^{n-1} with smooth coefficients, ord $\Phi_k \leq j$ and ord $\Psi_j \leq j$.

To include problem (10.127) in the abstract scheme of this section, we put $H_k = \mathring{W}_2^k(\Omega)$, $k = 1, \ldots, m$; $H_0 = L_2(\Omega)$.

Furthermore, let $\mathring{W}_{2,\mathrm{loc}}^m(\mathcal{C})$, $\mathring{W}_{2,\mathrm{comp}}^m(\mathcal{C})$ and $W_{2,\mathrm{loc}}^{-m}(\mathcal{C})$ be the images of $\mathring{W}_{2,\mathrm{loc}}^m(K)$, $\mathring{W}_{2,\mathrm{comp}}^m(\mathcal{C})$ and $W_{2,\mathrm{loc}}^{-m}(\mathcal{C})$ under the mapping

$$K \ni x \to (t, \theta) \in \mathcal{C}, \quad t = \log|x|^{-1}, \quad \theta = x/|x|.$$

These spaces can easily be introduced directly, but we do not dwell upon this here. They are readily interpreted as the spaces of vector valued functions

$$W_{\mathrm{loc}}^m(\mathbb{R}), \quad W_{\mathrm{comp}}^m(\mathbb{R}) \quad \text{and} \quad W_{\mathrm{loc}}^{-m}(\mathbb{R}).$$

Since the operators

$$\Phi_k : H_k \to H_0 \;, \quad \Psi_j : H_j \to H_0$$

are continuous, the same is true for the operator

$$A_{jk} = \Psi_j^* \Phi_k \;:\; H_k \to H_j^*,$$

where $H_j^* = W_2^{-j}(\Omega)$ (i.e. the dual of $\mathring{W}_2^j(\Omega)$ with respect to the scalar product in $L_2(\Omega)$). Now, the equation (10.127) is written as (10.113) (with $e^{-2mt} f$ playing the role of f). The corresponding operator pencil

$$\mathcal{A}(\lambda) = \sum_{j=0}^{m} \sum_{k=0}^{m} \lambda^{2m-j-k} A_{jk}$$

maps H_m continuously into H_m^*. The validity of Condition $\mathfrak{V}$ in 10.8.3 is a consequence of the following assertion (which was checked in the proof of Lemma 1.5.1).

Proposition 10.8.15. *For all real λ the operator*

$$\mathcal{A}(\lambda + i(m - n/2)) \;:\; \mathring{W}_2^m(\Omega) \to W_2^{-m}(\Omega)$$

is isomorphic, and the following estimate holds:

$$\Re \int_\Omega \mathcal{A}(\lambda + i(m - n/2))\varphi \cdot \overline{\varphi} d\omega$$

$$\geq c \sum_{j=0}^m |\lambda|^{2j} \|\varphi\|^2_{W_2^{m-j}(\Omega)}$$

for all $\varphi \in \mathring{W}_2^m(\Omega)$.

Let us turn to the case of variable coefficients. Consider the Dirichlet problem

$$\sum_{|\gamma|,|\delta|\leq m} D_x^\delta \left(q_{\gamma\delta}(x)D_x^\gamma u\right) = f \quad \text{in } K , \tag{10.128}$$

where $u \in \mathring{V}^m_{2,\mathrm{loc}}(K)$ and $f \in V^{-m}_{2,\mathrm{loc}}(K)$. This problem will be understood as the equality

$$\int_K \sum_{|\gamma|,|\delta|\leq m} q_{\gamma\delta}(x)D_x^\gamma u \cdot \overline{D_x^\delta z}dx = \int_K f\overline{z}dx, \tag{10.129}$$

which should hold for all $z \in \mathring{V}^m_{2,\mathrm{comp}}(K)$. We consider the sesquilinear form on the left-hand side as a perturbation of the form on the left in (10.126). This perturbation will be characterized by the function $S(r)$ subject to the inequality

$$\left| \int_K \sum_{|\gamma|,|\delta|\leq m} (q_{\gamma\delta}(x) - p_{\gamma\delta}) D_x^\gamma u \overline{D_x^\delta z}dx \right|$$

$$\leq S(r)\|u\|_{W_2^m(K_r)}\|z\|_{W_2^m(K_r)},$$

valid for all

$$u \in \mathring{V}^m_{2,\mathrm{loc}}(K), \quad z \in \mathring{V}^m_{2,\mathrm{comp}}(K), \ \mathrm{supp}\, z \in \overline{K_r}.$$

We assume that $p_{\gamma\delta} = 0$ if $|\gamma| + |\delta| < 2m$. Let $S(r)$ be majorized by a sufficiently small constant.

One can use Hardy's inequality to check that the role of $S(r)$ can be played by the function

$$r \to c \sum_{|\gamma|,|\delta|\leq m} \sup_{K_r} d(x)^{2m-|\gamma|-|\delta|}|q_{\gamma\delta}(x) - p_{\gamma\delta}|$$

where $d(x)$ is the distance from x to ∂K.

It is an easy matter to show that one can take $cS(r)$ as a substitute for $\rho(\log r^{-1})$, where ρ is defined by (10.122) with $\ell = 2m$ and $q = m$.

Let $k_- < \Im\lambda < k_+$ be a strip free of eigenvalues of $\mathcal{A}(\lambda)$ which contains the line $\Im\lambda = m-n/2$. By $m_\pm$ we denote the maximum lengths of the Jordan chains of $\mathcal{A}(\lambda)$ corresponding to the eigenvalues on the lines $\Im\lambda = k_\pm$. If the line $\Im\lambda = k_\pm$ is free of eigenvalues we put $m_\pm = 1$. The comparison equation (10.124) takes the form

$$(\mathcal{M}(\partial_t) - \omega(t))\, w = e^{-2mt} \|f\|_{W_2^{-m}(K_{e^{-t}})} \tag{10.130}$$

where $\omega(t) = \text{const } S(e^{-t})$.

Now, Theorems 10.8.12 and 10.8.13 imply the following existence and uniqueness statements.

Theorem 10.8.16. *Let $f \in V_{2,\text{loc}}^{-m}(K)$ satisfy*

$$\int_0^\infty g_\omega(0, \log s^{-1}) \|f\|_{W_2^{-m}(K_s)} s^{2m-1} ds < \infty.$$

Then there exists a solution u of the problem (10.129) such that

$$\|u\|_{W_2^m(K_r)} \le \text{const } w(\log r^{-1}),$$

where w is the solution of (10.130) given by

$$w(t) = \int_{\mathbb{R}} g_\omega(t, \tau) e^{-2m\tau} \|f\|_{W_2^{-m}(K_{e^{-\tau}})}\, d\tau.$$

Theorem 10.8.17. *Let $u \in \mathring{V}_{2,\text{loc}}^m(K)$ be a solution of (10.129) with $f = 0$ which satisfies either (5.56) or (5.58) with the norm $\|u\|_{W^\ell(t,t+1)}$ replaced by $\|u\|_{W_2^m(K_{e^{-t}})}$. Then $u = 0$.*

These two general results, together with the estimates for g_ω in Sect. 6.2, 6.3 and 6.4, lead to more transparent existence and uniqueness theorems, as well as to estimates for solutions under additional requirements on $S(r)$ (compare with Sect. 6.6).

10.9 Ordinary Differential Equations in Banach Spaces

10.9.1 Assumptions on the Operator with Constant Coefficients

Let $\{B_j\}_{j=0}^\ell$ be a collection of Banach spaces with norms $\|\cdot\|_j$ such that

$$B_\ell \subset B_{\ell-1} \subset \cdots \subset B_0$$

and

$$\| u \|_j \le \| u \|_{j+1} \quad \text{for} \quad j = 0, 1, \ldots, \ell - 1.$$

We suppose that B_ℓ is dense in B_0.

Also let A_j be a linear continuous operator from B_j into B_0. Consider the operator pencil

$$\mathcal{A}(\lambda) = \sum_{j=0}^\ell A_{\ell-j} \lambda^j : B_\ell \to B_0.$$

We assume that $\mathcal{A}(\lambda)$ is a Fredholm operator for all $\lambda \in \mathbb{C}$, invertible for at least one value of λ. By Proposition A.8.4, the spectrum of $\mathcal{A}(\lambda)$ consists of isolated eigenvalues of finite algebraic multiplicity. We suppose that the following analog of Condition II, in Sect. 2.2.1, is fulfilled:

Condition II' *There exists $r > 0$ such that*

$$\sum_{j=0}^{\ell} |\lambda|^j \|\varphi\|_{\ell - j} \leq c \|\mathcal{A}(\lambda)\varphi\|_0 \tag{10.131}$$

is valid for all $\varphi \in B_\ell$ and for all real λ, $|\lambda| \geq r$.

Similarly to Proposition 2.2.1, it can be proved that Condition II' implies that the estimate (10.131) is valid for all $\lambda \in S_{r,\theta}$ with some positive θ.
By

$$\langle \varphi, \psi \rangle, \quad \text{where } \varphi \in B_0, \ \psi \in B_0^*,$$

we denote a duality satisfying properties (i)-(iii) of Sect. A.2. This duality can be extended by continuity onto $B_j \times B_j^*, j = 1, \ldots, \ell$.

We introduce the spaces $L_p(a, b; B_j)$ and $W_p^\ell(a, b)$, $1 < p < \infty$, endowed with the norms

$$\| f \|_{L_p(a,b;B_j)} = \left(\int_a^b \| f(t) \|_j^p \, dt \right)^{1/p},$$

$$\| u \|_{W_p^\ell(a,b)} = \left(\sum_{j=0}^{\ell} \int_a^b \| D_t^j u(t) \|_{\ell-j}^p \, dt \right)^{1/p}.$$

Let $L_{p,\text{loc}}(\mathbb{R}; B_j)$ and $W_{p,\text{loc}}^\ell(\mathbb{R})$ be the spaces of functions belonging to $L_p(a, b; B_j)$ and $W_p^\ell(a, b)$ for all real a and $b, a < b$. We equip these spaces with the seminorms

$$\| f \|_{L_p(t,t+1;B_j)} \quad \text{and} \quad \| u \|_{W_p^\ell(t,t+1)}, \ t \in \mathbb{R},$$

respectively.

Lemma 10.9.1. *Let $a \in \mathbb{R}$ and $n = 1, 2, \ldots$ Also let*

$$f \in L_p(a - 1, a + n; B_j).$$

Then

$$\| f \|_{L_p(a,a+n;B_j)} \leq \int_{a-1}^{a+n} \| f \|_{L_p(\tau,\tau+1;B_j)} \, d\tau. \tag{10.132}$$

Proof. We have the inequalities

$$\| f \|_{L_p(a,a+n;B_j)} \leq \int_a^{a+1} \| f \|_{L_p(\tau-1,\tau+n;B_j)} \, d\tau$$

$$\leq \int_a^{a+1} \sum_{i=0}^{n} \| f \|_{L_p(\tau-1+i,\tau+i;B_j)} \, d\tau \leq \int_{a-1}^{a+n} \| f \|_{L_p(\tau,\tau+1;B_j)} \, d\tau,$$

and (10.132) follows. $\square$

By $L_{p,\beta}(\mathbb{R}; B_j)$, $W_{p,\beta}^\ell(\mathbb{R})$ we denote the spaces of vector valued functions with finite norms

$$\| f \|_{L_{p,\beta}(\mathbb{R};B_j)}= \left(\int_\mathbb{R} e^{p\beta t} \| f(t) \|_j^p \, dt \right)^{1/p}, \tag{10.133}$$

$$\| u \|_{W_{p,\beta}^\ell(\mathbb{R})}= \left(\sum_{j=0}^\ell \int_\mathbb{R} e^{p\beta t} \| D_t^j u(t) \|_{\ell-j}^p \, dt \right)^{1/p}. \tag{10.134}$$

In the same way as Proposition 4.2.1 one can prove

Proposition 10.9.2. *The norms*

$$\left(\int_\mathbb{R} e^{p\beta t} \| f \|_{L_p(t,t+1;B_j)}^p \, dt \right)^{1/p}$$

and

$$\left(\int_\mathbb{R} e^{p\beta t} \| u \|_{W_p^\ell(t,t+1)}^p \, dt \right)^{1/2}$$

are equivalent to (10.133) *and* (10.134) *respectively.*

We assume additionally that the following condition is valid:

Condition $\mathfrak{C}$ *For every $u \in W_p^\ell(-1,2)$ the local estimate*

$$\| u \|_{W_p^\ell(0,1)}\le c\left(\| \mathcal{A}(D_t)u \|_{L_p(-1,2;B_0)} + \| u \|_{W_p^{\ell-1}(-1,2)} \right) \tag{10.135}$$

holds.

Using the translation invariance of the operator $\mathcal{A}$ and (10.135) one easily derives the following assertion:

Lemma 10.9.3. *Let $t \in \mathbb{R}$ and $u \in W_p^\ell(t-1,t+2)$. Then*

$$\| u \|_{W_p^\ell(t,t+1)} \le c\big(\| \mathcal{A}(D_t)u \|_{L_p(t-1,t+2;B_0)}$$
$$+ \| u \|_{W_p^{\ell-1}(t-1,t+2)} \big). \tag{10.136}$$

10.9.2 Solvability of the Equation with Constant Coefficients

We look for a solution of (2.1) in the form (2.24), where

$$G(t) = \frac{1}{2\pi} \lim_{N \to +\infty} \int_{\substack{\Im\lambda=\beta \\ |\Re\lambda|<N}} e^{it\lambda} \mathcal{A}^{-1}(\lambda) d\lambda \qquad (10.137)$$

and the limit is taken in the space $\mathcal{L}(B_0, B_{\ell-1})$.

Duplicating the proof of Proposition 2.6.2 we arrive at the following result.

Proposition 10.9.4. *All assertions of* Proposition 2.6.2 *are valid if the spaces H_j are replaced by B_j, $j = 0, \dots, \ell$, in its formulation.*

Remark 10.9.5. We add to Proposition 10.9.4 that all assertions of Sect. 2.7 about the asymptotics of G remain true after the replacement of H_j by B_j and $(\cdot, \cdot)_{H_0}$ by $\langle \cdot, \cdot \rangle$. The proofs remain unchanged.

From Proposition 10.9.4 and Remark 10.9.5 we obtain:

Lemma 10.9.6. *Let the strip $k_- < \Im\lambda < k_+$ be free of eigenvalues of $\mathcal{A}(\lambda)$ and let $m_\pm$ be the maximum length of the Jordan chains corresponding to the eigenvalues of $\mathcal{A}(\lambda)$ on the line $\Im\lambda = k_\pm$. Also let G be defined by (10.137) with $\beta \in (k_-, k_+)$ and g be Green's function of (3.21) given by (3.23) and (3.24). Then*

$$\| D_t^s G(t) \|_{B_0 \to B_\ell} \leq c_s g(t) \quad for \quad |t| \geq 1, \qquad (10.138)$$

where $s = 0, 1, \dots$, and

$$\| D_t^j G(t) \|_{B_0 \to B_{\ell-j-1}} \leq c(|\log|t|| + 1) \quad for \quad |t| \leq 1, \qquad (10.139)$$

where $j = 0, \dots, \ell - 1$.

We are now in a position to prove the existence result which generalizes Theorem 3.5.5.

Theorem 10.9.7. *Let $k_\pm, m_\pm$ be the same numbers as in Lemma 10.9.6. Let $f \in L_{p,\text{loc}}(\mathbb{R}; B_0)$ and*

$$\int_{\mathbb{R}} g(-\tau) \| f \|_{L_p(\tau,\tau+1;B_0)} d\tau < \infty. \qquad (10.140)$$

Then the following assertions are valid:
(i) The integral

$$u(t) = \int_{\mathbb{R}} G(t - \tau) f(\tau) d\tau \qquad (10.141)$$

is absolutely convergent in the norm of the space $H_{\ell-1}$ for all $t \in \mathbb{R}$.
(ii) The function u, defined by (10.141), belongs to $W_{p,\text{loc}}^\ell(\mathbb{R})$, satisfies (2.1) and

$$\| u \|_{W_p^\ell(t,t+1)} \leq b \int_{\mathbb{R}} g(t-\tau) \| f \|_{L_p(\tau,\tau+1;B_0)} \, d\tau, \qquad (10.142)$$

where b is a constant independent of f.
   (iii) *The following relation holds:*

$$\| u \|_{W_p^\ell(t,t+1)} = o(e^{-k_\mp t}) \quad as \quad t \to \pm\infty. \qquad (10.143)$$

Proof. (i) We have

$$\int_{\mathbb{R}} \| G(t-\tau)f(\tau) \|_{\ell-1} \, d\tau = \int_{\mathbb{R}} \int_{\tau}^{\tau+1} \| G(t-s)f(s) \|_{\ell-1} \, ds d\tau$$

$$\leq \int_{\mathbb{R}} K(t,\tau) \| f \|_{L_p(\tau,\tau+1;B_0)} \, d\tau, \qquad (10.144)$$

where

$$K(t,\tau) = \left(\int_{\tau}^{\tau+1} \| G(t-s) \|_{B_0 \to B_{\ell-1}}^{p'} \, ds \right)^{1/p'}, \quad \frac{1}{p'} = 1 - \frac{1}{p}.$$

It follows from (10.138) and (10.139) that

$$K(t,\tau) \leq c \, g(t-\tau). \qquad (10.145)$$

Using Lemma 3.5.2 one can check that

$$g(t-\tau) \leq c \, \max\{e^{-k_- t}, e^{-k_+ t}\} g(-\tau). \qquad (10.146)$$

Hence, by (10.140), we get the required convergence. From (10.144) and (10.145) we arrive at

$$\| u(t) \|_{\ell-1} \leq c \int_{\mathbb{R}} g(t-\tau) \| f \|_{L_p(\tau,\tau+1;B_0)} \, d\tau.$$

Analogously, using (10.138) and (10.139), we obtain

$$\| D_t^j u(t) \|_{\ell-1-j} \leq c \int_{\mathbb{R}} g(t-\tau) \| f \|_{L_p(\tau,\tau+1;B_0)} \, d\tau$$

for $j = 0, 1, \ldots, \ell - 1$. Hence

$$\| u \|_{W_p^{\ell-1}(t,t+1)} \leq c \int_{\mathbb{R}} g(t-\tau) \| f \|_{L_p(\tau,\tau+1;B_0)} \, d\tau. \qquad (10.147)$$

   (ii) First, we assume that $f \in C_0^\infty(\mathbb{R}; B_0)$. Consider the function

$$v(t) = \frac{1}{2\pi} \int_{\Im\lambda=\beta} e^{i\lambda t} \mathcal{A}^{-1}(\lambda) \tilde{f}(\lambda) d\lambda, \qquad (10.148)$$

where

$$\tilde{f}(\lambda) = \int_{\mathbb{R}} e^{-i\lambda\tau} f(\tau) d\tau. \qquad (10.149)$$

The integral in (10.149) is absolutely convergent in the norm of the space B_0 and

$$\| \tilde{f}(\lambda) \|_0 \leq |\lambda|^{-s} \int_{\mathbb{R}} e^{\beta\tau} \| D_\tau^s f(\tau) \|_0 \, d\tau \tag{10.150}$$

for $s = 0, 1, \ldots$ and for $\Im\lambda = \beta$. Hence, by Condition II' in 10.9.1, the integral in (10.148) is absolutely convergent in the norm of B_ℓ and

$$\| D_t^k v(t) \|_\ell \leq c_k e^{-\beta t} \int_{\Im\lambda=\beta} \| \tilde{f}(\lambda) \|_0 \, |d\lambda|$$

for $k = 0, 1, \ldots$ Therefore $u \in W_{\text{loc}}^\ell(\mathbb{R})$ and

$$\mathcal{A}(D_t)v(t) = \frac{1}{2\pi} \int_{\Im\lambda=\beta} e^{i\lambda t} \tilde{f}(\lambda) d\lambda = f(t).$$

Reasoning as in the proof of Theorem 2.6.3 we obtain

$$v(t) = \int_{\mathbb{R}} G(t-\tau) f(\tau) d\tau.$$

(The finiteness of the integral (2.36) used in the proof is a consequence of (10.150)). Hence $v = u$. Applying the local estimate (10.136) to v we get

$$\| v \|_{W_p^\ell(t,t+1)} \leq c \Big(\| f \|_{L_p(t-1,t+2;B_0)} + \| v \|_{W_p^{\ell-1}(t-1,t+2)} \Big).$$

Using (10.132) and (10.147) we arrive at (10.142). Hence (ii) has been proved for $f \in C_0^\infty(\mathbb{R}; B_0)$.

Now let $f \in L_{p,\text{loc}}(\mathbb{R}; B_0)$ satisfy (10.140). We introduce the Banach space Λ consisting of all $f \in L_{p,\text{loc}}(\mathbb{R}; B_0)$ subject to (10.140) with finite norm

$$\int_{\mathbb{R}} g(-\tau) \| f \|_{L_p(\tau,\tau+1;B_0)} \, d\tau.$$

We approximate f by $f_j \in C_0^\infty(\mathbb{R}; B_0)$, $j = 1, \ldots,$ in Λ and denote by $u_j \in W_{p,\text{loc}}^\ell(\mathbb{R})$ the solution of (2.1) given by

$$u_j(t) = \int_{\mathbb{R}} G(t-\tau) f_j(\tau) d\tau. \tag{10.151}$$

Making use of (10.142) we get

$$\| u_j - u_k \|_{W_p^\ell(t,t+1)} \leq c \int_{\mathbb{R}} g(t-\tau) \| f_j - f_k \|_{L_p(\tau,\tau+1;B_0)} \, d\tau.$$

By (10.146) this implies

$$\| u_j - u_k \|_{W^\ell(t,t+1)}$$
$$\leq c \max\{e^{-k-t}, e^{-k+t}\} \int_{\mathbb{R}} g(-\tau) \| f_j - f_k \|_{L_p(\tau,\tau+1;B_0)} \, d\tau.$$

Hence the sequence $\{u_j\}_{j\geq 1}$ converges in $W_{p,\mathrm{loc}}^{\ell}(\mathbb{R})$. We denote its limit by $u \in W_{p,\mathrm{loc}}^{\ell}(\mathbb{R})$. Then u satisfies (2.1) and (10.142). In order to prove (10.141) it suffices to note that by (10.138) and (10.139) one can take the limit in (10.151) as $j \to \infty$.

(iii) The relation (10.143) follows from (10.142) and (10.140). $\square$

10.9.3 Uniqueness for the Equation with Constant Coefficients

We generalize Proposition 3.2.1.

Theorem 10.9.8. *Let $u \in W_{p,\mathrm{loc}}^{\ell}(\mathbb{R})$ be a solution of (2.1) with $f = 0$ subject to*

$$\liminf_{t\to\pm\infty} e^{k\mp t}\,\|\,u\,\|_{W_p^{\ell-1}(t,t+1)} = 0. \tag{10.152}$$

Then $u = 0$.

Proof. By (10.152) there exist sequences $\{T_j^{(\pm)}\}_{j\geq 1}$ such that $T_j^{(\pm)} \to \pm\infty$ as $j \to \infty$ and

$$e^{k\mp T_j^{(\pm)}}\,\|\,u\,\|_{W_p^{\ell-1}\left(T_j^{(\pm)},\,T_j^{(\pm)}+1\right)} \to 0 \quad \text{as} \quad j \to \infty. \tag{10.153}$$

It is convenient to assume that $T_1^{(\pm)} = 0$.

We introduce a cut-off function $\eta \in C_0^{\infty}(\mathbb{R})$, $\eta(t) = 1$ for $t > 1$ and $\eta(t) = 0$ for $t < 0$. We put

$$\eta_j^{(-)}(t) = \eta(t - T_j^{(-)}),\quad \eta_j^{(+)}(t) = 1 - \eta(t - T_j^{(+)}),$$

and

$$u_{jk}(t) = \eta_j^{(-)}(t)\eta_k^{(+)}(t)u(t).$$

Then

$$\mathcal{A}(D_t)u_{jk}(t) = f_{jk}(t) \quad \text{on}\quad \mathbb{R}, \tag{10.154}$$

where

$$f_{jk} = \left[\mathcal{A}(D_t), \eta_j^{(-)}\eta_k^{(+)}\right]u. \tag{10.155}$$

Since both functions in (10.154) have compact supports, we can apply the Fourier transform to (10.154)

$$\mathcal{A}(\lambda)\tilde{u}_{jk}(\lambda) = \tilde{f}_{jk}(\lambda) \quad \lambda \in \mathbb{C}.$$

Hence

$$u_{jk}(t) = \frac{1}{2\pi}\int_{\Im\lambda=\beta} e^{it\lambda}\mathcal{A}^{-1}(\lambda)\tilde{f}_{jk}(\lambda)d\lambda, \tag{10.156}$$

where β is a real number such that the line $\Im\lambda = \beta$ is free of eigenvalues of $\mathcal{A}^{-1}(\lambda)$. By (10.155) we have $D_t f_{jk} \in L_p(\mathbb{R}; B_0)$. Therefore

$$\|\,\tilde{f}_{jk}(\lambda)\,\|_0 \leq c_\beta(1 + |\lambda|)^{-1} \quad \text{for}\quad \Im\lambda = \beta. \tag{10.157}$$

Reasoning an in the proof of Theorem 10.9.7 (ii) and using the estimate (10.157) to treat the integral (10.156), we arrive at

$$u_{jk}(t) = \int_{\mathbb{R}} G_\beta(t - \tau) f_{jk}(\tau) d\tau, \tag{10.158}$$

where G_β is given by (10.137).

First let $\beta = \beta_+ = k_+ + \varepsilon$, where ε is a small positive number. By (10.155) and (10.153),

$$\| f_{jk} \|_{L_p(\tau,\tau+1)} \leq \begin{cases} \varepsilon_j^{(-)} e^{-k_+ T_j^{(-)}} & \text{for } \tau \in (T_j^{(-)} - 1, T_j^{(-)} + 1) \\ \varepsilon_k^{(+)} e^{-k_- T_j^{(+)}} & \text{for } \tau \in (T_k^{(+)} - 1, T_k^{(+)} + 1) \\ 0 & \text{for other values of } \tau, \end{cases}$$

where $\varepsilon_j^{(-)}, \varepsilon_k^{(+)} \to 0$ when $j, k \to \infty$. By using

$$\| G_{\beta_+}(t) \|_{H_0 \to H_{\ell-1}} \leq c\, e^{-(k_+ + \varepsilon)t} \quad \text{for} \quad t < -1,$$

we take the limit in (10.158) as $j \to +\infty$ and arrive at

$$\eta_1^{(+)}(t)u(t) = \int_{\mathbb{R}} G_{\beta_+}(t - \tau)[\mathcal{A}(D_\tau), \eta_1^{(+)}(\tau)] u(\tau) d\tau.$$

Analogously,

$$\eta_1^{(-)}(t)u(t) = \int_{\mathbb{R}} G_{\beta_-}(t - \tau)[\mathcal{A}(D_\tau), \eta_1^{(-)}(\tau)] u(\tau) d\tau$$

where $\beta_- = k_- - \varepsilon$.

Now let $\beta \in (k_-, k_+)$. By Remark 10.9.5 Theorem 2.7.3 we obtain

$$\eta_1^{(\pm)}(t)u(t) = \int_{\mathbb{R}} G_\beta(t - \tau)[\mathcal{A}(D_\tau), \eta_1^{(\pm)}(\tau)] u(\tau) d\tau$$

$$+ \sum_{\Im\lambda = k_\pm} \sum_{\sigma=1}^{\kappa_\nu} c_{\nu,\sigma}^{(\pm)} U_\sigma^{(\nu)}(t), \tag{10.159}$$

where $U_\sigma^{(\nu)}$ are special solutions of $\mathcal{A}(D_t)u = 0$ given by (2.9) and (2.11). Summing up two equalities, with $+$ and $-$, in (10.159) and using the identity $\eta_1^{(+)} + \eta_1^{(-)} = 1$ we arrive at

$$u(t) = \sum_{\Im\lambda = k_\pm} \sum_{\sigma=1}^{\kappa_\nu} c_{\nu\sigma}^{(\pm)} U_\sigma^{(\nu)}(t).$$

Along with (10.152), this implies $c_{\nu\sigma}^{(\pm)} = 0$. □

10.9.4 Comparison Principle for the Equation with Constant Coefficients

As in Sect.3.5 (and elsewhere), (10.142) can be written as the comparison principle

$$\| u \|_{W_p^\ell(t,t+1)} \leq b \, w(t),$$

where w is a solution of

$$\left(\partial_t + k_+\right)^{m_+}\left(-\partial_t - k_-\right)^{m_-} w(t) = \| f \|_{L_p(t,t+1;B_0)} \, .$$

This fact is a foundation of the L_p-theory which is parallel to that developed in Ch.5–8. In the next subsection we state some basic facts of this L_p-theory which relate to operator differential equations with variable coefficients.

10.9.5 Equation with Variable Coefficients

Let

$$L(t, D_t) = \sum_{0 \leq q \leq \ell} A_{\ell-q}(t) D_t^q.$$

Here, $A_q(t)$ is a linear bounded mapping: $B_q \to B_0$ for almost all t such that

$$A_q(\cdot)v(\cdot) \in L_{p,\mathrm{loc}}(\mathbb{R}; B_0)$$

for every $v \in L_{p,\mathrm{loc}}(\mathbb{R}; B_q)$. We consider L as a perturbation of $\mathcal{A}$ and characterize their difference by the function

$$\rho(t) = \| L - \mathcal{A} \|_{W_p^\ell(t,t+1) \to L_p(t,t+1;B_0)} \, .$$

Let ω be a measurable majorant of $b\rho$, where b is a constant in (10.142). We suppose that ω is subject to (5.7).

In the next theorem we formulate a comparison principle between a solution of $Lu = f$ and a solution of the ordinary differential equation

$$\left(\left(\partial_t + k_+\right)^{m_+}\left(-\partial_t - k_-\right)^{m_-} - \omega(t)\right)w(t) = \| f \|_{L_p(t,t+1;B_0)} \, . \tag{10.160}$$

This is a generalization of Theorem 5.3.2.

Theorem 10.9.9. *Let $f \in L_{p,\mathrm{loc}}(\mathbb{R}; B_0)$ be subject to*

$$\int_{\mathbb{R}} g_\omega(0, \tau) \| f \|_{L_p(\tau,\tau+1;B_0)} \, d\tau < \infty.$$

Then there exists a solution $u \in W_{p,\mathrm{loc}}^\ell(\mathbb{R})$ of $Lu = f$ on $\mathbb{R}$ such that

$$\| u \|_{W_p^\ell(t,t+1)} \leq b \, w(t), \quad t \in \mathbb{R} \, ,$$

where w is the solution of (10.160) *given by*

$$w(t) = \int_{\mathbb{R}} g_\omega(t, \tau) \| f \|_{L_p(\tau,\tau+1;B_0)} \, d\tau.$$

Proof. Same as Theorem 5.3.2. $\square$

By repeating the proof of Corollary 5.3.3 one obtains

$$\| u \|_{W_p^\ell(t,t+1)} = o\Big(\sup_{\tau \in \mathbb{R}} \frac{g_\omega(t,\tau)}{g_\omega(0,\tau)} \Big) \quad as \quad t \to \pm\infty.$$

The next assertion extends Theorem 5.4.2 and Proposition 5.4.3 (the proofs are similar).

Theorem 10.9.10. (Uniqueness) *Let* $u \in W_{p,\mathrm{loc}}^\ell(\mathbb{R})$ *be a solution of* $Lu = 0$ *on* $\mathbb{R}$. *Suppose that either*

$$\| u \|_{W_p^\ell(t,t+1)} = o\Big(\limsup_{\tau \to \pm\infty} \frac{g_\omega(t,\tau)}{g_\omega(0,\tau)} \Big) \quad as \quad t \to \pm\infty$$

or

$$\liminf_{t \to \pm\infty} g_\omega(0,t) \, \| u \|_{W_p^\ell(t,t+1)} = 0.$$

Then $u = 0$.

All other results of Ch.5 - 9 have counterparts in the present context. The formulations and the proofs are the same except for the replacement of $p = 2$ by an arbitrary $p \in (1, \infty)$ and H_j by B_j.

10.9.6 Application to the Dirichlet Problem in a Cylinder

We shortly describe an application of the above L_p-theory to partial differential equations.

Example 10.9.11. Consider the Dirichlet problem (2.22) for an elliptic operator of order $2m$ in a cylinder $\mathcal{C} = \{(x,t) : x \in \Omega, \, t \in \mathbb{R}\}$, where Ω is a bounded domain in $\mathbb{R}^n$ with smooth boundary. We shall use the notation from Sect. 2.5 with $\ell = 2m$. The operator pencil $\mathcal{A}(\lambda) = P(x, D_x, \lambda)$ will act from

$$B_\ell = \{\varphi \in W_p^{2m}(\Omega) ; \, \partial_\nu^j \varphi = 0 \text{ on } \partial\Omega, \, j = 0, \ldots, m - 1\}$$

to $B_0 = L_p(\Omega)$, $p \in (1, \infty)$. Here $W_p^s(\Omega)$ is the Sobolev space

$$\{\varphi \in L_p(\Omega) : D_x^\alpha \varphi \in L_p(\Omega) \text{ for all multi-indices } \alpha \text{ with } 0 < |\alpha| \le s\}$$

supplied with the norm

$$\|\varphi\|_{W_p^s(\Omega)} = \Big(\sum_{|\alpha| \le s} \int_\Omega |D_x^\alpha \varphi|^p \, dx \Big)^{1/p}.$$

The coefficients $A_k(x, D_x)$ will be considered as operators from

$$B_k = W_p^k(\Omega) \cap B_\ell \quad \text{to} \quad B_0.$$

The following result, which implies Condition II$'$, Sect. 10.9.1, is proved in Agmon, Nirenberg (1963) (Theorems 5.2 and 5.3).

Theorem 10.9.12. *The mapping*

$$\mathcal{A}(\lambda) \; : \; B_\ell \to B_0$$

is Fredholm for all $\lambda \in \mathbb{C}$ *and one to one for all real* λ *which have sufficiently large absolute values. For these* λ *the estimate* (10.131) *holds for all* $\varphi \in B_\ell$.

It is a well known consequence of the theory of elliptic boundary value problems that the spectrum of $\mathcal{A}(\lambda)$ is independent of p and that eigenfunctions and generalized eigenfunctions are smooth (Agmon, Douglis and Nirenberg (1959, 1964)).

Condition $\mathfrak{C}$ in Sect. 10.9.1 follows from the local estimate

$$\|u\|_{W_p^{2m}(\Omega \times (0,1))} \leq C\Big(\|Pu\|_{L_p(\Omega \times (-1,2))} + \|u\|_{L_p(\Omega \times (-1,2))}\Big),$$

which is proved in Agmon, Douglis and Nirenberg (1959, 1964).

The role of the function $t \to \|u\|_{W_p^\ell(t,t+1)}$ in Sect. 3.9.1 will be played by

$$t \to \|u\|_{W_p^{2m}(\Omega \times (t,t+1))}$$

$$= \Big(\int_t^{t+1} \int_\Omega \sum_{|\alpha|+k \leq 2m} |D_x^\alpha D_\tau^k u(x,\tau)|^p \, dx d\tau \Big)^{1/p}.$$

The right-hand side f of (2.22) will be characterized by the function

$$\mathbb{R} \ni t \to \|f\|_{L_p(\Omega \times (t,t+1))}.$$

In the same way, replacing 2 by p in Sect. 3.9.1 we obtain definitions of the spaces $L_{p,\mathrm{loc}}(\mathcal{C})$ and $\mathring{W}_{p,\mathrm{loc}}^{2m}(\mathcal{C})$. These two spaces will be interpreted as the abstract spaces $L_{p,\mathrm{loc}}(\mathbb{R})$ and $W_{p,\mathrm{loc}}^\ell(\mathbb{R})$.

Our principal object here is the Dirichlet problem (5.84) with t-dependent coefficients. Now we have

$$\rho(t) = \|Q - P\|_{\mathring{W}_p^{2m}(\mathcal{C}_t) \to L_p(\mathcal{C}_t)},$$

where $\mathring{W}_p^{2m}(\mathcal{C}_t)$ is the subspace of $W_p^{2m}(\mathcal{C}_t)$ containing the functions with zero Dirichlet conditions on $\partial\Omega \times (t,t+1)$.

With the above notation, all abstract results in this section can be specified for the problem (5.84).

Since the study of the Dirichlet problem in the cone K is reduced to the case just considered, the L_p-version of Example 5.7.2 is also contained in the present example.

10.10 Applications to Elliptic Boundary Value Problems in a Cone

During last thirty years, a detailed theory of elliptic boundary value problems for domains with angle and conic points has been developed (see Kondratiev (1967), Maz'ya, Plamenevskii (1978) et al.). Fundamental results of this theory are the Fredholm property in different functional spaces, estimates and asymptotic representations of solutions in a neighbourhood of the boundary singularity.

For the Dirichlet problem in a cone, the question can easily be reduced to the study of an ordinary operator differential equation which satisfies all of the conditions formulated in Sections 2.2.1 and 5.3.1. However, for an equation with general boundary conditions, such a reduction seems to be inconvenient. Therefore it makes sense to develop a theory parallel to that in parts I and II directly for general elliptic boundary value problems in a cone. We sketch such a theory in the present section (for more details one can consult Kozlov, Maz'ya (1985 and 1988).

10.10.1 The Comparison Principle for the Model Problem

Let K be an open cone in $\mathbb{R}^n$ with the vertex at the origin and with the boundary ∂K cutting out a region Ω with smooth boundary in the unit sphere S^{n-1}. Let (r, ω) be spherical coordinates of the point $x \in \mathbb{R}^n$.

Consider the elliptic boundary value problem with normal boundary conditions

$$\begin{cases} P_0(x, D_x)u = f & \text{on } K \\ \mathcal{L}_{j0}(x, D_x)u = f_j & \text{on } \partial K \backslash 0, \; j = 1, \ldots, m, \end{cases} \tag{10.161}$$

(see Lions, Magenes (1972)), where $D_x = i^{-1}\mathrm{grad}$ and $P_0, \mathcal{L}_{j0}$ are differential operators of orders $2m, m_j, \; m_j < m$, respectively, which can be represented in the form

$$P_0(x, D_x) = r^{-2m}\mathcal{A}(rD_r)$$
$$\mathcal{L}_{j0}(x, D_x) = r^{-m_j}\mathcal{B}_j(rD_r).$$

By $\mathcal{A}(\lambda)$ and $\mathcal{B}_j(\lambda)$ we denote polynomials in λ, whose operator coefficients are differential operators on $\overline{\Omega}$ with smooth coefficients.

We shall characterize the behaviour of the solution u by the function

$$M_K^{\ell,p}(u; \rho) = \left(\sum_{|\alpha| \leq \ell} \rho^{p|\alpha|} \int_{e^{-1}\rho}^{\rho} \int_{\Omega} |D_x^\alpha u|^p d\omega \frac{dr}{r} \right)^{1/p},$$

where $p \in (1, \infty)$ and $\ell = 0, 1, \dots$ We say that $u \in W^{\ell,p}_{\text{loc}}(\overline{K} \backslash O)$ if $M^{\ell,p}_K(u; \rho) < \infty$ for every $\rho > 0$.

Let

$$W^{\ell-1/p,p}_{\text{loc}}(\partial K \backslash O), \quad \ell = 1, 2, \dots ,$$

be the trace space on $\partial K \backslash 0$ for the space $W^{\ell,p}_{\text{loc}}(\partial K \backslash 0)$. We introduce the family of the seminorms

$$M^{\ell-1/p,p}_{\partial K}(v; r) = \inf M^{\ell,p}_K(u; r), \quad r > 0,$$

where inf is taken over all $u \in W^{\ell,p}_{\text{loc}}(\overline{K} \backslash O)$ satisfying $u = v$ on $(e^{-1}r, r) \times \partial \Omega$.

The functions $f, f_1, \dots, f_m$ in the right-hand side of (10.161) will be characterized by

$$F^{\ell,p}(r) = r^{2m} M^{\ell,p}_K(f; r) + \sum_{j=1}^{m} r^{m_j} M^{\ell+2m-m_j-1/p,p}_{\partial K}(f_j; r).$$

Consider the operator pencil

$$\mathcal{P}(\lambda) = \{\mathcal{A}(\lambda); \partial \mathcal{B}_j(\lambda), 1 \leq j \leq m\}, \quad \lambda \in \mathbb{C},$$

where ∂ is the trace operator on $\partial \Omega$. It is well known (see Kondratiev (1967) and Maz'ya, Plamenevskii (1977)) that the operator

$$\mathcal{P}(\lambda) : W^{2m+\ell,p}(\Omega) \to W^{\ell,p}(\Omega) \times \prod_{j=1}^{m} W^{\ell+2m-m_j-1/p,p}(\partial \Omega)$$

is Fredholm for every $\lambda \in \mathbb{C}$. Moreover, for all $\lambda \in \mathbb{C}$, except perhaps for a countable set, $\mathcal{P}(\lambda)$ is isomorphic. The countable set just mentioned, consists of eigenvalues of finite algebraic multiplicity and it is situated, except perhaps for a finite set, in a double angle less than π containing the imaginary axis.

We fix real numbers $k_-, k_+, k_- < k_+$, such that the strip

$$k_- < \Im \lambda < k_+$$

does not contain eigenvalues of $\mathcal{P}(\lambda)$. By $m_\pm$ we denote an arbitrary integer upper bound for the maximal lengths of the Jordan chains corresponding to the eigenvalues of $\mathcal{P}(\lambda)$ on the lines $\Im \lambda = k_\pm$ (if there are no eigenvalues on the line $\Im \lambda = k_\pm$ we put $m_\pm = 1$).

The following existence result was obtained in Kozlov, Maz'ya (1985).

Theorem 10.10.1. *Let $\ell = 0, 1, \dots$ and $1 < p < \infty$. Suppose that*

$$f \in W^{\ell,p}_{\text{loc}}(\overline{K} \backslash 0), \ f_j \in W^{\ell+2m-m_j-1/p,p}_{\text{loc}}(\partial K \backslash 0), \ j = 1, \dots, m,$$

and

$$\int_0^1 \rho^{k_+}(1+|\log\rho|)^{m_+-1}F^{\ell,p}(\rho)\frac{d\rho}{\rho}$$

$$+\int_1^\infty \rho^{k_-}(1+|\log\rho|)^{m_--1}F^{\ell,p}(\rho)\frac{d\rho}{\rho} < \infty. \qquad (10.162)$$

Then the problem (10.161) has a solution $u \in W_{\mathrm{loc}}^{2m+\ell,p}(\overline{K}\backslash 0)$ *satisfying*

$$M_K^{2m+\ell,p}(u;r) \le c\Big\{ \int_0^r (\tfrac{\rho}{r})^{k_+}(1+\log\tfrac{r}{\rho})^{m_+-1}F^{\ell,p}(\rho)\frac{d\rho}{\rho}$$

$$+ \int_r^\infty (\tfrac{\rho}{r})^{k_-}(1+\log\tfrac{\rho}{r})^{m_--1}F^{\ell,p}(\rho)\frac{d\rho}{\rho}\Big\} \qquad (10.163)$$

with c independent of $f, f_1, \ldots, f_m$.

It follows directly from (10.163) that

$$M_K^{2m+\ell,p}(u;r) = \begin{cases} o(r^{-k_+}) & \text{as } r \to 0 \\ o(r^{-k_-}) & \text{as } r \to \infty. \end{cases} \qquad (10.164)$$

According to Kozlov, Maz'ya (1985) a solution $u \in W^{2m+\ell,p}(\overline{K}\backslash 0)$ of the problem (10.161) subject to (10.164) is unique.

Let us introduce new variables t and τ by

$$r = e^t \quad \text{and} \quad \rho = e^\tau.$$

Then (10.163) takes the form

$$M_K^{2m+\ell,p}(u;e^t) \le c\Big\{ \int_t^\infty e^{k_-(\tau-t)}(1+\tau-t)^{m_+-1}F^{\ell,p}(e^\tau)d\tau$$

$$+ \int_{-\infty}^t e^{k_+(\tau-t)}(1+t-\tau)^{m_--1}F^{\ell,p}(e^\tau)d\tau\Big\}.$$

By Lemma 3.5.2(i) the last estimate is equivalent to

$$M_K^{2m+\ell,p}(u;e^t) \le b\int_{-\infty}^{+\infty} g(t-\tau)F^{\ell,p}(e^\tau)d\tau ,$$

where b is a positive constant independent of u and $f, f_1, \ldots, f_m$, and, as before, g is Green's function of the operator $\mathcal{M}(\partial_t)$. In other words, the function

$$\mathbb{R}_+ \ni r \to M_K^{2m+\ell,p}(u;r)$$

satisfies the comparison principle

$$M_K^{2m+\ell,p}(u;r) \le b\, w(\log r) \qquad (10.165)$$

where w is a solution of the equation

$$\mathcal{M}(\partial_t)w(t) = F^{\ell,p}(e^t) \quad \text{on } \mathbb{R}$$

satisfying $w(t) = o(e^{-k_\mp t})$ as $t \to \pm\infty$.

Condition (10.162) (which ensures this comparison principle) cannot be relaxed because it is both necessary and sufficient for the existence of w (see Proposition 3.5.3)

10.10.2 The Comparison Principle for the Boundary Value Problem with Variable Coefficients

We introduce the boundary value problem

$$\begin{cases} P(x, D_x)u = f & \text{on } K \\ \mathcal{L}_j(x, D_x)u = f_j & \text{on } \partial K \backslash 0, \ j = 1, \ldots, m, \end{cases} \tag{10.166}$$

where P and $\mathcal{L}_j$ are differential operators of orders $2m$ and m_j respectively.

We consider this boundary value problem as a perturbation of the model problem (10.161), and define the function

$$w(\log r) = b\Big(r^{2m} \parallel P - P_0 \parallel_{W_p^{\ell+2m}(K_r) \to W_p^{\ell}(K_r)}$$

$$+ \sum_{j=1}^{m} r^{m_j} \parallel \mathcal{L}_j - \mathcal{L}_{j0} \parallel_{W_p^{\ell+2m}(K_r) \to W_p^{\ell+2m-m_j-1/p}(\Gamma_r)} \Big),$$

where b is the constant in (10.165),

$$K_r = \{x \in K : e^{-1}r < |x| < r\},$$
$$\Gamma_r = \{\in \partial K : e^{-1}r < |x| < r\},$$

and $W_p^s(K_r)$ and $W_p^{s-1/p}(\Gamma_r)$ are Sobolev spaces with the norms $M_K^{s,p}(u; r)$ and $M_{\partial K}^{s-1/p,p}(u; r)$. One can easily give upper estimates for w in terms of coefficients of the operators $P, \mathcal{L}_j$ (see, for example, Kozlov, Maz'ya (1985)).

We shall assume that the function w satisfies the inequality (5.7), which guarantees the existence of Green's function g_w of the equation (5.14).

The following existence theorem is valid.

Theorem 10.10.2. *Let*

$$\int_{\mathbb{R}} g_w(0, \tau) F^{\ell,p}(e^\tau) d\tau < \infty.$$

Then there exists a solution $u \in W_{\text{loc}}^{2m+\ell,p}(\overline{K} \backslash 0)$ *of the boundary value problem* (10.166) *such that*

$$M_K^{\ell+2m,p}(u; r) \leq b\, w(\log r), \tag{10.167}$$

where w *is the solution of* (5.14) *with*

$$h(t) = F^{\ell,p}(e^t). \tag{10.168}$$

By Theorem 5.2.3 the solution u in the above theorem satisfies

$$M_K^{\ell+2m,p}(u; e^t) = o\left(\sup_\tau \frac{g_w(t, \tau)}{g_w(0, \tau)} \right) \quad \text{as } t \to \pm\infty \tag{10.169}$$

The following assertion shows that a similar relation describes a uniqueness class.

Theorem 10.10.3. *If a solution* $u \in W_{\mathrm{loc}}^{\ell+2m,p}(\overline{K}\backslash 0)$ *of* (10.166) *with* $f = 0$
and $f_j = 0$, $j = 1,\ldots,m$, *satisfies*

$$M_K^{\ell+2m,p}(u; e^t) = o\left(\limsup_{\tau\to\pm\infty} \frac{g_\omega(t,\tau)}{g_\omega(0,\tau)}\right) \quad as\ t \to \pm\infty \tag{10.170}$$

then $u = 0$.

By Remark 5.2.8 relations (10.169) and (10.170) are equivalent if m_+ and m_- do not exceed 2.

These results are direct analogs of Theorems 5.3.2 and 5.4.2 and their proofs are the same as the proofs of these theorems.

Inequality (10.167) is a comparison principle similar to (5.48). It enables one to obtain various information on solutions of the boundary value problem (10.166) by using estimates for solution of the ordinary differential equation (5.14) with h given by (10.168).

We will not formulate special cases of Theorems 10.10.2 and 10.10.3 analogous to the results of Sect. 5–9. One can easily obtain these statements replacing $\| u \|_{W^\ell(t,t+1)}$ and $\| f \|_{L_2(t,t+1;H_0)}$ by $M_K^{\ell+2m,p}(u; e^t)$ and $F^{\ell,p}(e^t)$.

Remark 10.10.4. Estimates (10.165) and (10.167) are somewhat unusual for the existing theory of elliptic boundary value problems in domains with conic points at the boundary (see Kondratiev (1967), Maz'ya, Plamenevskii (1978) etc). In this theory, the solutions are sought in various weighted Sobolev spaces. Since the comparison principles (10.165) and (10.167) give more precise "pointwise" information on solutions, they lead, in particular, to "integral" estimates obtained in the just mentioned papers.

Part III

Asymptotic Theory
of Operator Differential Equations

11. Complete Asymptotic Expansions Under Exponential and Power Perturbations of $\mathcal{A}(D_t)$

11.1 Introduction

We pass to asymptotic representations of solutions to the equation

$$L(t, D_t)u(t) = f(t) \quad \text{for} \quad t > t_0, \tag{11.1}$$

where (as before)

$$L(t, D_t) = \sum_{j=0}^{\ell} A_j(t) D_t^{\ell-j}.$$

The operator L is considered as a small perturbation of the operator $\mathcal{A}(D_t)$ with constant coefficients A_j.

In Sect.11.2 and 11.3 we assume that the differences $A_j(t) - A_j$, $j = 0, \ldots, \ell$, decay exponentially at $+\infty$, and we obtain an asymptotic representation for solutions of (11.1) in the form of power-exponential series. Since the perturbation is very weak in this case, the general term has the same form as the operator $\mathcal{A}(D_t)$ with constant coefficients.

This is not the case for the situation considered in Sect. 11.4 and 11.5; for these we suppose that the coefficients can be decomposed as

$$A_j(t) \sim A_j + \sum_{k=1}^{\infty} t^{-k} A_{jk}, \quad t \to +\infty. \tag{11.2}$$

We find asymptotic formulae for solutions of (11.1) in the form

$$t^\alpha e^{\beta t} \sum_{k=0}^{\infty} t^{-k} \varphi_k.$$

First we construct the formal asymptotic series and then we estimate the remainder term by using the comparison principle established in Theorem 5.3.2.

11.2 Perturbation with Exponential Decay (Homogeneous Equation)

Here we suppose that the coefficients of $L(t, D_t)$ satisfy

$$\| A_j(t) - A_j - \sum_{k=1}^{M} e^{-\alpha_k t} A_{jk}(t) \|_{H_j \to H_0} = O\left(e^{-(\Re \alpha_M + \varepsilon)t}\right), \qquad (11.3)$$

where $M < \infty$, $\varepsilon > 0$,

$$0 < \Re \alpha_1 \leq \Re \alpha_2 \leq \ldots \leq \Re \alpha_M$$

and $A_{jk}(t)$ are operator polynomials with coefficients in $\mathcal{L}(H_j, H_0)$.

Lemma 11.2.1. *Let $f \in L_{2,\mathrm{loc}}(t_0, \infty; H_0)$ and*

$$\| f \|_{L_2(t,t+1;H_0)} \leq c \, e^{-\varkappa t}, \qquad (11.4)$$

where κ is a real number such that the line $\Im \lambda = \kappa$ contains no eigenvalues of $\mathcal{A}(\lambda)$. Then for sufficiently large t_0, equation (11.1) has a solution $u \in W_{\mathrm{loc}}^{\ell}(t_0, \infty)$ subject to

$$\| u \|_{W^{\ell}(t,t+1)} \leq c \, e^{-\varkappa t}. \qquad (11.5)$$

Proof. Let k_- and k_+ be real numbers such that the strip $k_- < \Im \lambda < k_+$ is free of the eigenvalues of $\mathcal{A}(\lambda)$ and $\varkappa \in (k_-, k_+)$. By (11.3) the function ρ given by (5.3) admits the estimate

$$\rho(t) \leq c e^{-qt}$$

for large t, where q is a positive constant. By Remark 6.2.8 there exists a solution $u \in W^{\ell}(t_0, \infty)$ such that (6.13) holds. Since $\rho(t) = o(1)$ as $t \to \infty$, the number R subject to (6.9) can be made arbitrarily small. Therefore one can suppose that

$$k_-(bR) < \varkappa < k_+(bR).$$

Now (11.5) follows from (6.13) and (11.4). $\square$

We shall denote by $\mathfrak{S}$ the set of all sums

$$\gamma = \sum_s \alpha_{k_s}$$

satisfying

$$\Re \gamma \leq \Re \alpha_M.$$

Theorem 11.2.2. *Let*

$$U_\sigma^{(\nu)}(t) = e^{i\lambda_\nu t} P_\sigma^{(\nu)}(t)$$

be a solution of (1.21) *defined by* (2.11). *There exists a solution* $u_\sigma^{(\nu)}$ *of*

$$L(t, D_t)u(t) = 0 \quad \text{for } t > t_0 \tag{11.6}$$

represented by

$$u_\sigma^{(\nu)}(t) = e^{i\lambda_\nu t}\Big(P_\sigma^{(\nu)}(t) + \sum_{\gamma \in \mathfrak{S}} e^{-\gamma t} P_{\sigma,\gamma}^{(\nu)}(t)\Big) + v_\sigma^{(\nu)}(t). \tag{11.7}$$

Here $P_{\sigma,\gamma}$ *are polynomials with coefficients in* H_ℓ *and*

$$\| v_\sigma^{(\nu)} \|_{W^\ell(t,t+1)} \le c\, e^{-(\Im\lambda_\nu + \Re\alpha_M + \delta)t} \tag{11.8}$$

for some positive δ.

Proof. Let $\mathfrak{S}_m = \{\gamma \in \mathfrak{S} : \gamma \le m\Re\alpha_1\}$. Suppose that we have constructed the function

$$u_{\sigma,m}^{(\nu)}(t) = e^{i\lambda_\nu t}\Big(P_\sigma^{(\nu)}(t) + \sum_{\gamma \in \mathfrak{S}_m} e^{-\gamma t} P_{\sigma,\gamma}^{(\nu)}(t)\Big),$$

such that

$$L(t, D_t)u_{\sigma,m}^{(\nu)}(t) = R_m(t) \quad \text{for} \quad t > t_0 \tag{11.9}$$

with

$$\| R_m(t) - e^{i\lambda_\nu t} \sum_{\gamma \in \mathfrak{S}\backslash\mathfrak{S}_m} e^{-\gamma t} r_{m,\gamma}(t) \|_0 \le c\, e^{-(\Im\lambda_\nu + \Re\alpha_M + \delta)t}, \tag{11.10}$$

where $r_{m,\gamma}$ are polynomials and δ is positive.

By Theorem 1.4.1, the equation

$$\mathcal{A}(D_t)v(t) = e^{i\lambda_\nu t} \sum_{\gamma \in \mathfrak{S}_{m+1}\backslash\mathfrak{S}_m} e^{-\gamma t} r_{m,\gamma}(t)$$

has a solution of the form

$$v(t) = e^{i\lambda_\nu t} \sum_{\gamma \in \mathfrak{S}_{m+1}\backslash\mathfrak{S}_m} e^{-\gamma t} v_\gamma(t),$$

where v_γ are polynomials with coefficients in H_ℓ. By setting

$$u_{\sigma,m+1}^{(\nu)}(t) = u_{\sigma,m}^{(\nu)}(t) - v(t)$$

we arrive at

$$L(t, D_t)u_{\sigma,m+1}^{(\nu)}(t) = R_{m+1}(t) \quad \text{for} \quad t > t_0$$

with

$$R_{m+1}(t) = R_m(t) - e^{i\lambda_\nu t} \sum_{\gamma \in \mathfrak{S}_{m+1} \setminus \mathfrak{S}_m} e^{-\gamma t} r_{m,\gamma}(t)$$

$$+ \Big(A(D_t) - L(t, D_t)\Big) v(t).$$

By (11.10) and (11.3) the vector valued function R_{m+1} satisfies (11.10) with m replaced by $m+1$.

Continuing this procedure we get a function $u_{\sigma,N}^{(\nu)}$ such that

$$L(t, D_t) u_{\sigma,N}^{(\nu)}(t) = R_N(t) \quad \text{for} \quad t > t_0$$

and

$$\| R_N(t) \|_0 \le c\, e^{-(\Im\lambda_\nu + \Re\alpha_M + \delta)t}$$

By Lemma 11.2.1, there exists a solution of

$$L(t, D_t) v(t) = R_N(t) \quad \text{for} \quad t > t_0$$

satisfying

$$\| v \|_{W^\ell(t,t+1)} \le c\, e^{-(\Im\lambda_\nu + \Re\alpha_M + \delta)t} \quad \text{for} \quad t > t_0$$

with some positive δ. Putting $u_\sigma^{(\nu)} = u_{\sigma,N}^{(\nu)} - v$ and $v_\sigma^{(\nu)} = v$ we complete the proof. $\square$

Corollary 11.2.3. *Let $k_\pm^{(1)}$ and $k_\pm^{(2)}$ be real numbers such that*

$$k_-^{(1)} < k_+^{(1)} \le k_-^{(2)} < k_+^{(2)}$$

and let the strips

$$k_-^{(1)} < \Im\lambda < k_+^{(1)} \quad \text{and} \quad k_-^{(2)} < \Im\lambda < k_+^{(2)}$$

be free of eigenvalues of $A(\lambda)$.

The linear hull of the vector functions $u_\sigma^{(\nu)}$ with $k_+^{(1)} \le \Re\lambda_\nu \le k_-^{(2)}$ is a $(\mathcal{Y}_1, \mathcal{Y}_2)$-space.

Proof. Follows from Theorem 11.2.2 and from the description of $(\mathcal{Y}_1, \mathcal{Y}_2)$-spaces given in the beginning of Sect. 9.3.1. $\square$

11.3 Perturbation with Exponential Decay (Nonhomogeneous Equation)

Theorem 11.3.1. *Let*

$$f(t) = e^{-\beta t} Q(t) \tag{11.11}$$

where Q is a polynomial with coefficients in H_0. Then there exists a solution $u \in W^\ell_{\text{loc}}(t_0, \infty)$ such that

$$u(t) = e^{-\beta t}\left(u_0(t) + \sum_{\gamma \in \mathfrak{S}} e^{-\gamma t} u_\gamma(t)\right) + v(t), \tag{11.12}$$

where u_0, u_γ are polynomials in t with coefficients in H_ℓ and the remainder v satisfies

$$\| v \|_{W^\ell(t, t+1)} \le c\, e^{-[\Re(\beta + \alpha_M) + \delta]t} \tag{11.13}$$

for some positive δ.

Proof. Let $\mathfrak{S}_m$ be the same as in the proof of Theorem 11.2.2. Reasoning by induction as in the proof of Theorem 11.2.2, we construct the functions

$$u_m(t) = e^{-\beta t}\left(u_0(t) + \sum_{\gamma \in \mathfrak{S}_m} e^{-\gamma t} u_\gamma(t)\right)$$

such that

$$L(t, D_t) u_m(t) = R_m(t) \quad \text{for} \quad t > t_0$$

with

$$\| R_m(t) - e^{-\beta t} \sum_{\gamma \in \mathfrak{S} \backslash \mathfrak{S}_m} e^{-\gamma t} r_{m,\gamma}(t) \|_0$$

$$\le c\, e^{-(\Re(\beta + \alpha_M) + \delta)t},$$

where $r_{m,\gamma}$ are polynomials. Hence for sufficiently large N we get

$$L(t, D_t) u_N(t) = R_N(t) \quad \text{for} \quad t > t_0,$$

where

$$\| R_N(t) \|_0 \le c\, e^{-(\Re(\beta + \alpha_M) + \delta)t}.$$

By Lemma 11.2.1 there exists a solution of

$$L(t, D_t) v(t) = r_N(t) \quad \text{for} \quad t > t_0$$

satisfying (11.13). $\square$

Remark 11.3.2. Suppose that the right-hand side f in (11.1) is of the form

$$f(t) = \sum_{k=0}^{N} e^{-\beta_k t} Q_k(t) + f_N(t), \tag{11.14}$$

where Q_k are polynomials in t with coefficients in H_0,

$$\Re \beta_0 \le \Re \beta_1 \le \ldots \le \Re \beta_N$$

and

$$\| f_N(t) \|_0 \le c\, e^{-(\Re \beta_N + \varepsilon)t}.$$

Then by Theorem 11.3.1 and Lemma 11.2.1 there exists a solution $u \in W^\ell_{\text{loc}}(t_0, \infty)$ such that

$$u(t) = \sum_{k=0}^{N} e^{-\beta_k t} \left(u_k(t) + \sum_{\gamma \in \mathfrak{S}} e^{-\gamma t} u_{k,\gamma}(t) \right) + v(t), \tag{11.15}$$

where u_k, $u_{k,\gamma}$ are polynomials in t with coefficients in H_ℓ. The remainder v satisfies (11.13) with β replaced by β_0.

The next theorem follows directly from Theorem 11.2.2 and Remark 11.3.2 along with Theorem 8.5.7 and Corollary 11.2.3. This result contains a description of the asymptotics of an arbitrary solution to the equation (11.1) subject to

$$\| u \|_{W^\ell(t,t+1)} = O(e^{-\varkappa t}) \quad \text{as} \quad t \to +\infty. \tag{11.16}$$

Here $\varkappa$ is a real number such that there are no eigenvalues of $\mathcal{A}(\lambda)$ on the line $\Im \lambda = \varkappa$.

Theorem 11.3.3. *Let $u \in W^\ell_{\text{loc}}(t_0, \infty)$ be a solution of (11.1) with f represented by (11.14) with $\Re \beta_0 > \varkappa$. Suppose that (11.3) holds. If u is subject to (11.16) then*

$$u(t) = \sum_{k=0}^{N} e^{-\beta_k t} \left(u_k(t) + \sum_{\gamma \in \mathfrak{S}} e^{-\gamma t} u_{k\gamma}(t) \right)$$

$$+ \sum_{\varkappa < \Im \lambda_\nu \le \varkappa + \Re \alpha_M} \sum_{\sigma=1}^{\kappa_\nu} c_{\nu\sigma} \left(U_\sigma^{(\nu)}(t) + \sum_{\gamma \in \mathfrak{S}} e^{(i\,\lambda_\nu - \gamma)t} P_{\sigma\gamma}^{(\nu)}(t) \right) + v(t),$$

where $u_k, u_{k\gamma}, P_{\sigma\gamma}^{(\nu)}$ are polynomials, $c_{\nu\sigma}$ are constants and

$$\| v \|_{W^\ell(t,t+1)} \le c\, e^{-(\varkappa + \Re \alpha_M)t}$$

for large t.

11.4 Perturbation in the Form of Laurent Series (Homogeneous Equation)

Assume that the coefficients of $L(t, D_t)$ admit the asymptotic expansions

$$A_j(t) \sim \sum_{k=0}^{\infty} t^{-k} A_{jk} \tag{11.17}$$

for large positive t, where $A_{j,0}$ is equal to the coefficient A_j in (1.10) and $A_{jk} \in \mathfrak{L}(H_j, H_0)$. We understand (11.17) in the sense that

$$\| A_j(t) - \sum_{k=0}^{M-1} t^{-k} A_{jk} \|_{H_j \to H_0} \leq c_M t^{-M} \tag{11.18}$$

for $t \geq t_0$ and $M = 1, 2, \ldots$.

Let $k_0 \in (k_-, k_+)$. We suppose that the spectrum in the strip $k_- \leq \Im \lambda \leq k_+$ lies on the line $\Im \lambda = k_0$ and consists of eigenvalues $\lambda_1, \ldots, \lambda_n$ of $\mathcal{A}(\lambda)$ with geometric multiplicities $J_1, \ldots, J_n$. We assume that corresponding to these eigenvalues λ_ν are the linearly independent eigenvectors

$$\varphi_1^{(\nu)}, \ldots, \varphi_{J_\nu}^{(\nu)}$$

and that there are no generalized eigenvectors. We denote by

$$\psi_1^{(\nu)}, \ldots, \psi_{J_\nu}^{(\nu)}$$

the eigenvectors of $\mathcal{A}^*(\lambda)$ corresponding to the eigenvalue $\overline{\lambda}_\nu$, which are subject to the biorthogonality condition

$$\left(\mathcal{A}'(\lambda_\nu) \varphi_j^{(\nu)}, \psi_k^{(\nu)} \right)_{H_0} = \delta_j^k \tag{11.19}$$

(see (1.20)).

We introduce the operators $\mathcal{A}_0(D_t) = \mathcal{A}(D_t)$ and

$$\mathcal{A}_k(D_t) = \sum_{j=1}^{\ell} A_{jk} D_t^{\ell-j}, \quad k = 1, \ldots,$$

so that

$$L(t, D_t) \sim \sum_{k=0}^{\infty} t^{-k} \mathcal{A}_k(D_t).$$

In the next theorem, $\alpha_1^{(\nu)}, \ldots, \alpha_{J_\nu}^{(\nu)}$ are eigenvalues of the matrix

$$- i \left\{ \left(\mathcal{A}_1(\lambda_\nu) \varphi_h^{(\nu)}, \psi_k^{(\nu)} \right)_{H_0} \right\}_{h,k=1}^{J_\nu}, \tag{11.20}$$

where h and k index columns and rows respectively.

Theorem 11.4.1. *Suppose that*

$$\alpha_j^{(\nu)} - \alpha_k^{(\nu)} \neq \text{integer for } j \neq k. \tag{11.21}$$

There exist solutions $u_j^{(\nu)} \in W_{\text{loc}}^\ell(t, \infty)$, $\nu = 1, \ldots, n$, $j = 1, \ldots, J_\nu$, *of the equation* (11.6) *such that*

$$u_j^{(\nu)}(t) \sim t^{\alpha_j^{(\nu)}} e^{i\lambda_\nu t} \sum_{k=0}^{\infty} \varphi_{jk}^{(\nu)} t^{-k}, \quad t > t_0, \tag{11.22}$$

where $\varphi_{jk}^{(\nu)}$ *are elements of* H_ℓ *and* $\varphi_{j0}^{(\nu)}$ *is a non-zero linear combination of* $\varphi_1^{(\nu)}, \ldots \varphi_{J_\nu}^{(\nu)}$. *The remainder term*

$$v_{jm}^{(\nu)}(t) = u_j^{(\nu)}(t) - t^{\alpha_j^{(\nu)}} e^{i\lambda_\nu t} \sum_{k=0}^{m} \varphi_{jk}^{(\nu)} t^{-k}$$

admits the estimate

$$\| v_{jm}^{(\nu)} \|_{W^\ell(t,t+1)} \leq c_m t^{\Re \alpha_j^{(\nu)} - m - 1} e^{-k_0 t}, \quad t > t_0. \tag{11.23}$$

Proof. Let

$$u_m(t) = t^\alpha e^{i\lambda_\nu t} \sum_{k=0}^{m} \varphi_k t^{-k},$$

where φ_k are elements of H_ℓ. By the identity

$$\mathcal{A}_j(\lambda_\nu + D_t)(t^\sigma \varphi) = \sum_{s \geq 0} (-i)^s \binom{\sigma}{s} t^{\sigma - s} \mathcal{A}_j^{(s)}(\lambda_\nu)\varphi$$

(where $\varphi \in H_\ell$ and $\mathcal{A}_j^{(s)}(\lambda) = \partial_\lambda^s \mathcal{A}_j(\lambda)$), we have

$$L(t, D_t)u_m(t) = t^\alpha e^{i\lambda_\nu t} \tag{11.24}$$

$$\times \left(\sum_{n=0}^{m} t^{-n} \sum_{j=0}^{n} \sum_{s=0}^{n-j} (-i)^s \binom{\alpha + j + s - n}{s} \mathcal{A}_j^{(s)}(\lambda_\nu)\varphi_{n-j-s} + R_m(t) \right).$$

The remainder R_m admits the estimate

$$\| R_m(t) \|_0 \leq c_m t^{-m-1} \quad \text{for} \quad t \geq t_0. \tag{11.25}$$

The coefficient in t^{-0} in (11.24) vanishes if

$$\varphi_0 = \sum_{h=1}^{J_\nu} c_h^{(0)} \varphi_h^{(\nu)}. \tag{11.26}$$

The coefficient in t^{-1} in (11.24) is zero provided

$$\mathcal{A}_0(\lambda_\nu)\varphi_1 = i\alpha\mathcal{A}_0^{(1)}(\lambda_\nu)\varphi_0 - \mathcal{A}_1(\lambda_\nu)\varphi_0. \tag{11.27}$$

By (11.19), the solvability condition for this equation is

$$\alpha\, c_k^{(0)} = -i\sum_{h=1}^{J_\nu}\left(\mathcal{A}_1(\lambda_\nu)\varphi_h^{(\nu)}, \psi_k^{(\nu)}\right)_{H_0} c_h^{(0)},$$

where $k = 1, \ldots, J_\nu$. In other words, α is an eigenvalue of the matrix (11.20) and $\mathrm{col}(c_h^{(0)})_{h=1}^{J_\nu}$ is the corresponding eigenvector. With this choice, the solution φ_1 of (11.27) exists and is determined up to a linear combination of $\varphi_1^{(\nu)}, \ldots, \varphi_{J_\nu}^{(\nu)}$.

Let $n > 1$. Then the coefficient in t^{-n} in (11.24) vanishes if

$$\mathcal{A}_0(\lambda_\nu)\varphi_n = -\sum_{0<s+j\leq n}(-i)^s\binom{\alpha+j+s-n}{s}\mathcal{A}_j^{(s)}(\lambda_\nu)\varphi_{n-j-s}.$$

It is convenient to rewrite this in the form

$$\mathcal{A}_0(\lambda_\nu)\varphi_n = \left(i(\alpha+1-n)\mathcal{A}_0^{(1)}(\lambda_\nu) - \mathcal{A}_1(\lambda_\nu)\right)\varphi_{n-1} + G_{n-1}, \tag{11.28}$$

where G_{n-1} depends on $\varphi_0, \ldots \varphi_{n-2}$. Suppose that the function φ_{n-1} has been found up to a linear combination of $\varphi_1^{(\nu)}, \ldots, \varphi_{J_\nu}^{(\nu)}$:

$$\varphi_{n-1} = \varphi_{n-1,0} + \sum_{h=1}^{J_\nu}c_h^{(n-1)}\varphi_h^{(\nu)}, \tag{11.29}$$

where $c_i^{(n-1)}$ are unknown constants. In order to solve (11.28) with respect to φ_n, the right-hand side must be orthogonal to $\psi_k^{(\nu)}$:

$$\left(\left(i(\alpha+1-n)\mathcal{A}_0^{(1)}(\lambda_\nu) - \mathcal{A}_1(\lambda_\nu)\right)\varphi_{n-1}, \psi_k^{(\nu)}\right)_{H_0} = -\left(G_{n-1}, \psi_k^{(\nu)}\right)_{H_0}.$$

By (11.29) and (11.19) this can be rewritten as

$$i(\alpha+1-n)c_k^{(n-1)} - \sum_{h=1}^{J_\nu}\left(\mathcal{A}_1(\lambda_\nu)\varphi_h^{(\nu)}, \psi_k^{(\nu)}\right)c_h^{(n-1)}$$

$$= \left(\left(\mathcal{A}_1(\lambda_\nu) - i(\alpha+1-n)\mathcal{A}_0^{(1)}(\lambda_\nu)\right)\varphi_{n-1,0}, \psi_k^{(\nu)}\right)_{H_0} - \left(G_{n-1}, \psi_k^{(\nu)}\right)_{H_0}.$$

The unique solvability of this equation with respect to $\mathrm{col}\left(c_k^{(n-1)}\right)_{k=1}^{J_\nu}$ follows from (11.21).

Thus, all the terms of the asymptotic series (11.22) have been uniquely determined and

$$L(t, D_t)u_m(t) = t^\alpha e^{i\lambda_\nu t}R_m(t), \quad t > t_0,$$

where R_m satisfies (11.25).

We look for a solution of (11.1) in the form

$$u(t) = u_m(t) + v_m(t).$$

Then

$$L(t, D_t)v_m = -t^\alpha e^{i\lambda_\nu t} R_m \quad \text{for } t > t_0. \tag{11.30}$$

Denote the right-hand side by f. We shall apply Remark 6.3.9 to obtain the existence of v_m. By (11.17) we can choose

$$\omega(t) = C t^{-1}$$

with a positive constant C. The role of the strip $k_- < \Im\lambda < k_+$ in Ch. 6 will be played by the strip $k_0 < \Im\lambda < k_+$. It is free of the spectrum of $\mathcal{A}(\lambda)$ and $m_0 = 1$. Since there are no eigenvalues of the pencil $\mathcal{A}(\lambda)$ on the line $\Im\lambda = k_+$ we can take $m_+ = 1$. Thus we can apply Remark 6.3.9 with

$$k_+(\omega(s)) = k_+ - \frac{C}{(k_+ - k_0)s} + O\left(\frac{1}{s^2}\right) \tag{11.31}$$

and

$$k_-(\omega(s)) = k_0 + \frac{C}{(k_+ - k_0)s} + O\left(\frac{1}{s^2}\right) \tag{11.32}$$

(see (6.17)). Then the inequality (6.27) takes the form

$$\int_{t_0}^\infty e^{k_0\tau} \tau^N \|f\|_{L_2(\tau, \tau+1; H_0)} d\tau < \infty , \tag{11.33}$$

where N is a positive constant. By (11.25) the right-hand side in (11.30) satisfies (11.33) if m is sufficiently large. Thus, by Remark 6.3.9 the problem (11.30) has a solution $v_m \in W_{\mathrm{loc}}^\ell(t_0, \infty)$ subject to

$$\|v_m\|_{W^\ell(t,t+1)} \le c\left(\int_{t_0-1}^t e^{-k_+(t-\tau)} \left(\frac{t}{\tau}\right)^N \|f\|_{L_2(\tau, \tau+1; H_0)} d\tau \right.$$
$$\left. + \int_t^\infty e^{-k_0(t-\tau)} \left(\frac{\tau}{t}\right)^N \|f\|_{L_2(\tau, \tau+1; H_0)} d\tau \right).$$

(As usual we assume that $f(t) = 0$ for $t < t_0$.) Hence using (11.25) we get

$$\| v_m \|_{W^\ell(t,t+1)} \le \left\{ t^N e^{-k_+ t} \int_{t_0-1}^t e^{(k_+ - k_0)\tau} \tau^{-N-m-1+\alpha} d\tau \right.$$
$$\left. + t^{-N} e^{-k_0 t} \int_t^\infty \tau^{N-m-1+\alpha} d\tau \right\} \le c\, t^{\alpha-m} e^{-k_0 t}. \tag{11.34}$$

Since

$$v_m(t) = v_{m+1}(t) - t^{\alpha-m-1} e^{i\lambda_\nu t} \varphi_{m+1} ,$$

the estimate (11.23) follows by applying (11.34) to v_{m+1}. $\square$

Corollary 11.4.2. *Let $k_\pm^{(1)}$ and $k_\pm^{(2)}$ be real numbers such that*

$$k_-^{(1)} < k_-, \quad k_+^{(1)} = k_-^{(2)} = k_0, \quad k_+^{(2)} > k_+$$

and let the strips

$$k_-^{(1)} < \Im\lambda < k_0 \quad and \quad k_0 < \Im\lambda < k_+^{(2)}$$

be free of eigenvalues of $\mathcal{A}(\lambda)$.

The linear hull of the vector functions $u_j^{(\nu)}$ with $\nu = 1, ..., n$, $j = 1, ..., J_\nu$, constructed in Theorem 11.4.1, is a $(\mathcal{Y}_1, \mathcal{Y}_2)$-space. If $u \in W_{\text{loc}}^\ell(t_0, \infty)$ is a solution of (11.6) subject to

$$\|u\|_{W^\ell(t,t+1)} \leq ce^{-k_-t} \quad for \ t > t_0$$

then

$$u = \sum_{\nu=1}^n \sum_{j=1}^{J_\nu} c_j^{(\nu)} u_j^{(\nu)} + v,$$

where $c_j^{(\nu)} = $ const. The remainder term v satisfies

$$\|v\|_{W^\ell(t,t+1)} \leq ce^{-k_+t} \quad for \ t > t_0.$$

Proof. We note that the characteristic exponent of any nonzero element in the linear hull of $\{u_j^{(\nu)}\}$ equals $-k_0$. Furthermore, the number of $u_j^{(\nu)}$ equals the total algebraic multiplicity of the eigenvalues λ_ν with $\Im\lambda_\nu = k_0$. Hence the linear hull of $\{u_j^{(\nu)}\}$ is a $(\mathcal{Y}_1, \mathcal{Y}_2)$-space (see the beginning of Sect. 9.3.1). The remaining assertion follows from Corollary 9.3.4. □

11.5 Perturbation in the Form of Laurent Series (Nonhomogeneous Equation)

We use the same notation as in Sect.11.4.

Theorem 11.5.1. *Let $L(t, D_t)$ be the same operator as in Theorem 11.4.1. Also let f be represented in the form of the asymptotic series*

$$f(t) \sim t^\beta e^{i\lambda t} \sum_{k=0}^\infty f_k t^{-k}, \tag{11.35}$$

where $f_k \in H_0$, β and λ are complex numbers, $\Im\lambda \in (k_-, k_+)$ and relation (11.35) means that the remainder

$$F_m(t) = f(t) - t^\beta e^{i\lambda t} \sum_{k=0}^m f_k t^{-k}$$

admits the estimate

$$\| F_m(t) \|_0 \le c \, t^{\Re\beta - m - 1} e^{-\Im\lambda t} \tag{11.36}$$

for large positive t.

(i) *If λ is not an eigenvalue of the pencil $\mathcal{A}$ then there exists a solution $u \in W_{\text{loc}}^\ell(t_0, \infty)$ of (11.1) such that*

$$u(t) \sim t^\beta e^{i\lambda t} \sum_{k=0}^\infty \varphi_k t^{-k},$$

where $\varphi_k \in H_\ell$ and the remainder

$$v_m(t) = u(t) - t^\beta e^{i\lambda t} \sum_{k=0}^m \varphi_k t^{-k} \tag{11.37}$$

is estimated as follows

$$\| v_m \|_{W^\ell(t, t+1)} \le c \, t^{\Re\beta - m - 1} e^{-\Im\lambda t}. \tag{11.38}$$

(ii) *If $\lambda = \lambda_\nu$ and*

$$\beta + 1 - k \ne \alpha_j^{(\nu)} \tag{11.39}$$

for all $k = 0, 1, \ldots$ and $j = 1, \ldots, J_\nu$ then there exists a solution $u \in W_{\text{loc}}^\ell(t_0, \infty)$ of (11.1) such that

$$u(t) \sim t^{\beta+1} e^{i\lambda t} \sum_{k=0}^\infty \varphi_k t^{-k}, \tag{11.40}$$

where $\varphi_k \in H_\ell$ and the remainder

$$v_m(t) = u(t) - t^{\beta+1} e^{i\lambda t} \sum_{k=0}^m \varphi_k t^{-k}$$

satisfies

$$\| v_m \|_{W^\ell(t, t+1)} \le c \, t^{\Re\beta - m} e^{-\Im\lambda t}.$$

Proof. (i) Let

$$u_m(t) = t^\beta e^{i\lambda t} \sum_{k=0}^m \varphi_k t^{-m}. \tag{11.41}$$

We have

$$L(t, D_t) u_m = t^\beta e^{i\lambda t} \tag{11.42}$$

$$\times \left(\sum_{n=0}^m t^{-n} \sum_{j=0}^n \sum_{s=0}^{n-j} (-i)^s \binom{\beta + j + s - n}{s} \mathcal{A}_j^{(s)}(\lambda) \varphi_{n-j-s} + R_m(t) \right),$$

where R_m satisfies (11.25). Comparing the right-hand side with (11.35) we obtain

$$A_0(\lambda)\varphi_0 = f_0$$

and

$$A_0(\lambda)\varphi_n = f_n - \sum_{0 < s+j \le n} (-i)^s \binom{\beta + j + s - n}{s} A_j^{(s)}(\lambda)\varphi_{n-j-s}$$

for $n \ge 1$. Since λ is a regular point of the pencil A, $\varphi_0, \ldots, \varphi_m$ are uniquely determined. The remainder (11.37) solves the equation

$$L(t, D_t)v_m(t) = F_m(t) - t^\beta e^{i\lambda t} R_m(t) \tag{11.43}$$

for $t > t_0$. By (11.25) and (11.36) we have

$$\| F_m(t) - t^\beta e^{i\lambda t} R_m(t) \|_0 \le c\, t^{\Re\beta - m - 1} e^{-\Im\lambda t}. \tag{11.44}$$

Let $\Im\lambda \in (k_-, k_0)$. Applying Remark 6.3.9 with k_+ replaced by k_0 and using that

$$k_+(\omega(s)) = k_0 - \frac{C}{(k_0 - k_-)s} + O\left(\frac{1}{s^2}\right)$$

and

$$k_-(\omega(s)) = k_- + \frac{C}{(k_0 - k_-)s} + O\left(\frac{1}{s^2}\right),$$

where C is the same constant as in (11.31) and (11.32), we obtain the existence of solution v_m to the equation (11.43) and the estimate

$$\| v_m \|_{W^\ell(t,t+1)} \le c_1 \left(\int_t^\infty \left(\frac{\tau}{t}\right)^N e^{-k_-(t-\tau)} e^{-\Im\lambda\tau} \tau^{\Re\beta - m - 1} d\tau \right.$$
$$\left. + \int_{t_0-1}^t \left(\frac{t}{\tau}\right)^N e^{-k_0(t-\tau)} e^{-\Im\lambda\tau} \tau^{\Re\beta - m - 1} d\tau \right)$$
$$\le c_2 t^{\Re\beta - m - 1} e^{-\Im\lambda t}.$$

If $\Im\lambda \in [k_0, k_+)$ and m is sufficiently large then by Remark 6.3.9 the (k_0, k_+)-solution of (11.43) exists and satisfies

$$\| v_m \|_{W^\ell(t,t+1)} \le c_1 \left(\int_t^\infty \left(\frac{\tau}{t}\right)^N e^{-k_0(t-\tau)} e^{-\Im\lambda\tau} \tau^{\Re\beta - m - 1} d\tau \right.$$
$$\left. + \int_{t_0-1}^t \left(\frac{t}{\tau}\right) e^{-k_+(t-\tau)} e^{-\Im\lambda\tau} \tau^{\Re\beta - m - 1} d\tau \right)$$
$$\le c_2 t^{\Re\beta - m} e^{-\Im\lambda t}.$$

Since

$$\| v_{m+1} - v_m \|_{W^\ell(t,t+1)} \le c\, t^{\Re\beta - m - 1} e^{-\Im\lambda t},$$

we arrive at (11.38) in both cases $\Im\lambda \in (k_-, k_0)$ and $\Im\lambda \in [k_0, k_+)$.

(ii). We replace β by $\beta + 1$ in (11.41) and (11.42). Then $\mathcal{A}_0(\lambda_\nu)\varphi_0 = 0$, i.e. φ_0 is given by (11.26). The vector φ_1 can be found from

$$\mathcal{A}_0(\lambda_\nu)\varphi_1 = i(\beta + 1)\mathcal{A}_0^{(1)}(\lambda_\nu)\varphi_0 - \mathcal{A}_1(\lambda_\nu)\varphi_0 + f_0.$$

The right-hand side should be orthogonal to $\psi_1^{(\nu)}, \ldots, \psi_{J_\nu}^{(\nu)}$; this gives

$$(\beta + 1)c_k^{(0)} + i\sum_{h=1}^{J_\nu} \left(\mathcal{A}_1(\lambda_\nu)\varphi_{\nu h}, \psi_k^\nu\right)_{H_0} c_h^{(0)} = i(f_0, \psi_k^\nu).$$

This algebraic system is uniquely solvable by (11.39).

For $n > 1$ we have

$$\mathcal{A}_0(\lambda_\nu)\varphi_n = - \sum_{0 < s+j \leq n} (-i)^s \binom{\beta + 1 + j + s - n}{s} \mathcal{A}_j^{(s)}(\lambda_\nu)\varphi_{n-j-s} + f_{n-1},$$

which can be rewritten as

$$\mathcal{A}_0(\lambda_\nu)\varphi_n = \left(i(\beta + 2 - n)\mathcal{A}_0^{(1)}(\lambda_\nu) - \mathcal{A}_1(\lambda_\nu)\right)\varphi_{n-1} + G_{n-1}, \qquad (11.45)$$

where G_{n-1} depends on $\varphi_0, \ldots, \varphi_{n-2}$ and f_{n-1}. By assuming that φ_{n-1} is given by (11.29) with arbitrary constants $c_h^{(n-1)}$, $h = 1, \ldots, J_\nu$, we write the solvability condition for (11.45) in the form of the algebraic system with respect to $\{c_h^{(n-1)}\}_{h=1}^{J_\nu}$:

$$(\beta + 2 - n)c_k^{(n-1)} + i\sum_{h=1}^{J_\nu} \left(\mathcal{A}_1(\lambda_\nu)\varphi_h^{(\nu)}, \psi_k^{(\nu)}\right)_{H_0} c_h^{(n-1)} = H_k,$$

where H_k depends on $\varphi_0, \ldots, \varphi_{n-2}$ and f_{n-1}, $\varphi_{n-1,0}$. By (11.39) this system is uniquely solvable.

Therefore, all terms of the asymptotic series (11.40) can be found, and the remainder v_m satisfies (11.43) with β is replaced by $\beta + 1$ and $\lambda = \lambda_\nu$. The right-hand side in (11.43) is subject to (11.44) with this replacement. If m is large enough then, by Remark 6.3.9, equation (11.43) has the (k_0, k_+)-solution satisfying

$$\| v_m \|_{W^\ell(t,t+1)} \leq c_1 \left(\int_t^\infty \left(\frac{\tau}{t}\right)^N e^{-k_0(t-\tau)} e^{-k_0\tau} \tau^{\Re\beta - m} d\tau \right.$$

$$\left. + \int_{t_0-1}^t \left(\frac{t}{\tau}\right)^N e^{-k_+(t-\tau)} e^{-k_0\tau} \tau^{\Re\beta - m} d\tau \right) \leq c_2 t^{\Re\beta + 1 - m} e^{-k_0 t}.$$

Since

$$\| v_{m+1} - v_m \|_{W^\ell(t,t+1)} \leq c\, t^{\Re\beta - m} e^{-k_0 t},$$

the result follows. $\square$

The combination of Theorems 11.5.1 and 11.4.1 with Corollary 11.4.2 leads to the following asymptotic representation for an arbitrary solution of (10.1) subject to the growth condition

$$\| u \|_{W^\ell(t,t+1)} = O(e^{-k-t}) \quad \text{as} \quad t \to +\infty. \tag{11.46}$$

Theorem 11.5.2. *Let $u \in W^\ell_{\text{loc}}(t_0, \infty)$ be a solution of (11.1) with f represented in the form of asymptotic series (11.35), where $\Im \lambda = k_0$ and λ is a regular point of the pencil $\mathcal{A}$. Suppose that the operator $L(t, D_t)$ is the same as in Theorem 11.4.1. If u is subject to (11.46) then the asymptotic decomposition*

$$u(t) \sim \sum_{\nu=1}^{n} e^{i\lambda_\nu t} \sum_{j=1}^{J_\nu} c_j^{(\nu)} t^{\alpha_j^{(\nu)}} \sum_{k=0}^{\infty} \varphi_{jk}^{(\nu)} t^{-k}$$
$$+ t^\beta e^{i\lambda t} \sum_{k=0}^{\infty} \varphi_k t^{-k}$$

is valid. Here $c_j^{(\nu)} = \text{const}$ and λ_ν, $\alpha_j^{(\nu)}$, $\varphi_{jk}^{(\nu)}$, φ_k are as in Theorems 11.4.1, 11.5.1 (i).

The same assertion holds with β replaced by $\beta + 1$ holds if λ and β satisfy the conditions of Theorem 11.5.1 (ii).

The following observation concerns all results of this and the previous section.

Remark 11.5.3. If at least one of the "non-resonance" conditions (11.21), (11.39) is violated then, in general, the form of the asymptotics of solutions in Theorems 11.4.1, 11.5.1 (ii) and 11.5.2 changes. In such a case the coefficients $\varphi_{jk}^{(\nu)}$, φ_k may depend polynomially on $\log t$. We have excluded this situation to avoid cumbersome calculations. After the formal asymptotic decompositions are constructed they can be justified in the same way as in Theorems 11.4.1 and 11.5.1.

11.6 The Dirichlet Problem for a Cuspidal Domain

Let $\mathcal{G}$ be the same domain with the power cusp as in Sect. 6.6.3. We shall consider the Dirichlet problem (6.53) under the assumption $f = 0$ on $\mathcal{G}^{(1)}$. Using the mapping (6.55), we transform $\mathcal{G}^{(1)}$ onto the semicylinder $\mathcal{C}^{(1)} = \{(\xi, t) : \xi \in \Omega, t > 1\}$. Since

$$\partial_x = (t(\alpha - 1))^{\frac{\alpha}{\alpha-1}} \partial_\xi,$$
$$\partial_\tau = (t(\alpha - 1))^{\frac{\alpha}{\alpha-1}} \left(-\partial_t - \frac{\alpha}{(\alpha - 1)t} \xi \cdot \partial_\xi \right),$$

it follows that

$$\partial_x^\beta = (t(\alpha - 1))^{\frac{|\beta|\alpha}{\alpha - 1}} \partial_\xi^\beta,$$

$$\partial_\tau^j = (-1)^j((\alpha - 1)t)^{\frac{\alpha}{\alpha - 1}j}\Big\{\partial_t^j + \frac{j(j-1)}{2t}\frac{\alpha}{\alpha - 1}\partial_t^{j-1}$$

$$+ j\frac{\alpha}{(\alpha - 1)t}\partial_t^{j-1}\xi \cdot \partial_\xi + \dots\Big\},$$

with the dots standing for the terms $O(t^{-2})$. Hence

$$P(D_x, D_\tau) = (t(\alpha - 1))^{2m\frac{\alpha}{\alpha - 1}}\Big\{P(D_\xi, -D_t)$$

$$+ t^{-1}P_1(\xi, D_\xi, D_t) + \sum_{k=2}^{2m-1} t^{-k}P_k(\xi, D_\xi, D_t)\Big\},$$

where P_k are differential operators with polynomial coefficients in ξ and,

$$P_1(\xi, D_\xi, D_t) = \frac{\alpha}{\alpha - 1}(-1)^m \sum_{|\beta|+j=2m} p_{\beta j}(-1)^j j\partial_\xi^\beta \left(\frac{j-1}{2} + \xi \cdot \partial_\xi\right)\partial_t^{j-1}.$$

By using the identity

$$\partial_\xi^\beta \xi \cdot \partial_\xi = \frac{|\beta| - n}{2}\partial_\xi^\beta + \frac{1}{2}\left(\partial_\xi^\beta \partial_\xi \cdot \xi + \xi \cdot \partial_\xi \partial_\xi^\beta\right),$$

we arrive at the representation

$$P_1(\xi, D_\xi, D_t) = \frac{\alpha i}{\alpha - 1}\left(m - \frac{n+1}{2}\right)P'(D_\xi, -D_t)$$

$$- \frac{\alpha}{2(\alpha - 1)}(P'(D_\xi, -D_t)D_\xi \cdot \xi + \xi \cdot D_\xi P'(D_\xi, -D_t)),$$

where $P'(z, \lambda) = \frac{\partial}{\partial \lambda}P(z, \lambda)$.

As in Sect. 11.4, let all the eigenvalues $\lambda_1, \dots, \lambda_N$ of the pencil (6.57) in the strip $k_- \leq \Im\lambda \leq k_+$ be on the line $\Im\lambda = k_0$, $k_0 \in (k_-, k_+)$. We assume that they are simple and denote the corresponding eigenfunctions by $\varphi^{(1)}, \dots, \varphi^{(N)}$. Let $\psi^{(1)}, \dots, \psi^{(N)}$ be eigenfunctions of the adjoint operator pencil

$$\sum_{|\beta|+j=2m} \overline{p}_{\beta j} D_x^\beta(-\lambda)^j : W_2^{2m}(\Omega) \cap \mathring{W}^m(\Omega) \to L_2(\Omega) \tag{11.47}$$

corresponding to the eigenvalues $\overline{\lambda}_1, \dots, \overline{\lambda}_N$ and normalized by

$$\int_\Omega P'(D_\xi, -\lambda_\sigma)\varphi^{(\sigma)}(\xi) \cdot \overline{\psi^{(\sigma)}(\xi)}d\xi = -1. \tag{11.48}$$

Now Theorem 11.4.1 implies the following asymptotic result.

Theorem 11.6.1. *There exist solutions $u^{(1)}, \ldots, u^{(N)}$ in $W_{2,\text{loc}}^{2m}(\overline{\mathcal{G}^{(1)}}\setminus\{0\})$ of the problem*

$$P(D_x, D_\tau)u = 0 \quad \text{in } \mathcal{G}^{(1)}, \tag{11.49}$$

$$\partial_\nu^j u = 0 \quad \text{on } \{(x, \tau) \in \partial\mathcal{G},\ 0 < \tau < 1\},\ j = 0, \ldots, m-1, \tag{11.50}$$

such that

$$u^{(\sigma)}(x, \tau) \sim \tau^{\alpha s_\sigma} e^{i\lambda_\sigma \tau^{1-\alpha}/(\alpha-1)} \sum_{k=0}^{\infty} \varphi_k^{(\sigma)}(\tau^{-\alpha}x)\tau^{k(\alpha-1)}, \tag{11.51}$$

where $\varphi_0^{(\sigma)}(\xi) = \varphi^{(\sigma)}(\xi)$ and

$$s_\sigma = m - \frac{n+1}{2}$$

$$-\frac{i}{2}\int_\Omega (P'(D_\xi, -\lambda_\sigma)D_\xi \cdot \xi + \xi \cdot D_\xi P'(D_\xi, -\lambda_\nu))\varphi^{(\sigma)} \cdot \overline{\psi^{(\sigma)}}d\xi.$$

The estimate for the remainder term can easily be extracted from (11.23) and (6.55).

Furthermore, Theorem 11.5.2 shows that an arbitrary solution of (11.49), (11.50) satisfying

$$\sum_{|\gamma|+j\leq 2m} \tau^{\alpha(|\gamma|+j-(n+1)/2)}\|D_x^\gamma D_\tau^j u\|_{L_2(\mathcal{G}_\tau)}$$

$$= o\left(e^{-k-\tau^{1-\alpha}/(\alpha-1)}\right) \quad \text{as} \quad \tau \to 0$$

(the domain $\mathcal{G}_\tau$ was introduced in Sect. 6.6.3), is asymptotically equal to a linear combination of the right-hand sides of (11.51).

Remark 11.6.2. We mention two cases, where the constant s_σ is equal to $m - (n+1)/2$.

1) The polynomial $P(z, \lambda)$ contains only even powers of λ. Then

$$\int_\Gamma P'(D_\xi, -\lambda_\sigma)D_\xi(\xi\varphi^{(\sigma)}) \cdot \overline{\psi^{(\sigma)}}d\xi$$

$$= \int_\Omega \varphi^{(\sigma)} \cdot \overline{\xi D_\xi(P'(D_\xi, -\lambda_\sigma))^*\psi^{(\sigma)}}d\xi$$

$$= -\int_\Omega \varphi^{(\sigma)}\xi D_\xi P'(D_\xi, -\lambda_\sigma)\overline{\psi^{(\sigma)}}d\xi. \tag{11.52}$$

We note that $\overline{\varphi^{(\sigma)}}$ is an eigenfunction of the pencil (11.47) corresponding to the eigenvalue $\overline{\lambda_\sigma}$. Indeed,

$$P^*(D_\xi, -\overline{\lambda}_\sigma)\overline{\varphi^{(\sigma)}(\xi)} = \sum_{|\beta|+j=2m} \overline{p}_{\beta j}(-\overline{\lambda}_\sigma)^j D_\xi^\beta \overline{\varphi^{(\sigma)}(\xi)}$$

$$= \sum_{|\beta|+j=2m} \overline{p_{\beta j}(-\lambda_\sigma)^j(-1)^{|\beta|} D_\xi^\beta \varphi^{(\sigma)}(\xi)} = \overline{P(D_\xi, -\lambda_\sigma)\varphi^{(\sigma)}(\xi)} = 0.$$

Hence $\overline{\varphi^{(\sigma)}} = \text{const } \psi^{(\sigma)}$ and by (11.48) and (11.52), we obtain

$$\int_\Omega P'(D_\xi, -\lambda_\sigma)D_\xi(\xi\varphi^{(\sigma)}) \cdot \overline{\psi^{(\sigma)}}d\xi$$

$$= -\int_\Omega \xi D_\xi P'(D_\xi, -\lambda_\sigma)\varphi^{(\sigma)} \cdot \overline{\psi^{(\sigma)}}d\xi.$$

This results in $s_\sigma = m - (n+1)/2$.

2) The domain Ω is centrally symmetric. In this case $\varphi^{(\sigma)}(-\xi)$ is an eigenfunction of the pencil (11.47) corresponding to the eigenvalue $\overline{\lambda}_\sigma$. Hence

$$\psi^{(\sigma)}(\xi) = \text{const } \overline{\varphi^{(\sigma)}(-\xi)}.$$

Since

$$\int_\Omega P'(D_\xi, -\lambda_\sigma)D_\xi(\xi\varphi^{(\delta)}(\xi)) \cdot \varphi^{(\sigma)}(-\xi)d\xi$$

$$= \int_\Omega \varphi^{(\sigma)}(\xi)\xi D_\xi P'(\xi, -\lambda_\sigma)\varphi^{(\sigma)}(-\xi)d\xi$$

$$= -\int \xi D_\xi P'(D_\xi, -\lambda_\sigma)\varphi^{(\sigma)}(\xi) \cdot \varphi^{(\sigma)}(-\xi)d\xi,$$

it follows that $s_\sigma = m - (n+1)/2$. Thus, in these two cases, (11.51) takes the form

$$u^{(\sigma)}(x,\tau) \sim \tau^{\alpha(m-(n+1)/2)}e^{i\lambda_\sigma\tau^{1-\alpha}/(\alpha-1)} \sum_{k=0}^\infty \varphi_k^{(\sigma)}(\tau^{-\alpha}x)\tau^{k(\alpha-1)}.$$

11.7 Comments

Theorem 11.3.1 is due to Pazy (1967) who extended the asymptotic formula (2.62) to equations whose coefficients differ from constants by exponentially decreasing terms.

Theorem 11.4.1 contains an extension to the infinite dimensional case of Poincaré's asymptotic formula for scalar ordinary differential equations (see Poincaré (1886)). This theorem was obtained in Maz'ya, Plamenevskii (1972) under the assumption that there is only one simple eigenvalue of the pencil (6.57) on the line $\Im\lambda = k_0$. It was generalized by Plamenevskii (1972) to the case of one eigenvalue λ_0 on the line $\Im\lambda = k_0$ of algebraic multiplicity κ and geometric multiplicity 1. He found an asymptotic expansion of solution $u(t)$ in the form

$$Ct^\alpha \exp\left(i\sum_{\nu=0}^{\kappa-1}\mu_\nu t^{(\kappa-\nu)/\kappa}\right)\sum_{k=0}^{\infty}t^{-k/\kappa}\Phi_k, \tag{11.53}$$

where C is a constant which depends on the solution, $\mu_0 = \lambda_0$, Φ_0 is the corresponding eigenvector and the constants α, μ_1, ..., $\mu_{\kappa-1}$ as well as the vectors Φ_1, Φ_2, ... do not depend on u.

The application of Theorem 11.4.1 to the Dirichlet problem in cuspidal domains in Sect. 11.6 is borrowed from Maz'ya, Plamenevskii (1977).

12. Reduction to a First Order System

12.1 Introduction

In the previous chapter we constructed complete asymptotic expansions of solutions to equation (11.1) assuming that the coefficients also admit certain asymptotic expansions. The asymptotic procedures used were direct and rather simple. If one is interested in asymptotic formulae which are valid under weak restrictions to the coefficients, the same techniques do not apply. A remedy used in the classical theory of ordinary differential equations is a reduction of the higher order equation to a first order system. The present chapter is devoted to a generalization of this approach to equations with unbounded operator coefficients.

We begin with corollaries of the results obtained in Chapters 5–9 which will be used in the foundation of the asymptotic theory to be developed in the remaining part of the book.

In the subsequent analysis of asymptotic behaviour of solutions of higher order ordinary differential equations with operator coefficients we shall use a transformation of the equation (2.1) to the first-order system

$$(I_\ell D_t + \mathfrak{A})U = F \quad \text{on} \quad \mathbb{R}, \tag{12.1}$$

where I_ℓ is the identity $\ell \times \ell$-matrix operator and $\mathfrak{A}$ is an $\ell \times \ell$-matrix operator generated by the coefficients $A_0, \ldots, A_\ell$. The reduction of the polynomial pencil $\mathcal{A}(\lambda)$ to the linear pencil $I_\ell \lambda + \mathfrak{A}$ is called *linearization*.

In the present chapter we collect information on the pencil $I_\ell \lambda + \mathfrak{A}$ and on solutions of equation (12.1) which will be used in the sequel. In the last section, we reduce the equation (5.1) with variable coefficients to the system

$$(I_\ell D_t + \mathfrak{A} - \mathcal{N}(t))U = F \quad \text{on} \quad \mathbb{R}, \tag{12.2}$$

where $\mathcal{N}$ is a matrix operator which is small in some sense.

12.2 Prerequisites for the Subsequent Asymptotic Theory

In the next chapters such results about the equation (2.1) with constant coefficients as Theorem 3.3.2 and Proposition 3.8.3 will be repeatedly referred to. As for the equation (5.1) with variable coefficients Proposition 9.2.2 will be of special use. Now we collect auxiliary information on the equation (11.1) on the semiaxis $t > t_0$ which follows from general theorems proved in Ch. 5-9 and is used in the sequel.

Let k_-, k_+ be real numbers, $k_- < k_+$. We suppose that the spectrum of the pencil $\mathcal{A}(\lambda)$ in the strip $k_- \leq \Im\lambda \leq k_+$ is situated on the line $\Im\lambda = k_0$, $k_0 \in (k_-, k_+)$. Denote by m_0 the maximal partial multiplicity of the eigenvalues of $\mathcal{A}(\lambda)$ lying on the line $\Im\lambda = k_0$ and by κ the total algebraic multiplicity of these eigenvalues. We assume that

$$\rho(t) \to 0 \quad \text{as} \quad t \to +\infty, \tag{12.3}$$

where ρ is introduced in (5.3).

We consider the equation (11.1), where t_0 is a sufficiently large positive number. We start with an existence result and an estimate for a solution subject to

$$\| u \|_{W^\ell(t,t+1)} = O(e^{-k_- t}) \quad \text{as} \quad t \to +\infty. \tag{12.4}$$

Theorem 12.2.1. *Let $f \in L_{2,\mathrm{loc}}(t_0, \infty; H_0)$ and let*

$$\int_{t_0}^\infty e^{k-\tau} \| f \|_{L_2(\tau,\tau+1;H_0)} \, d\tau < \infty. \tag{12.5}$$

Then

(i) *the equation (11.1) has a solution $u \in W_{\mathrm{loc}}^\ell(t_0, \infty)$ satisfying*

$$\| u \|_{W^\ell(t,t+1)} \leq c_\varepsilon \Big(\int_{t_0-1}^t e^{-(k_0-\varepsilon)(t-\tau)} \| f \|_{L_2(\tau,\tau+1;H_0)} \, d\tau$$
$$+ \int_t^\infty e^{-k_-(t-\tau)} \| f \|_{L_2(\tau,\tau+1;H_0)} \, d\tau \Big) \quad \text{for} \quad t > t_0, \tag{12.6}$$

with an arbitrary positive ε and the constant c_ε independent of f. (The function f is supposed to be extended by zero for $t < t_0$.) The solution u satisfies (12.4).

(ii) *Let $u \in W_{\mathrm{loc}}^\ell(t_0, \infty)$ be a solution of (11.1) subject to (12.4). Then*

$$\| u \|_{W^\ell(t,t+1)} \leq c_\varepsilon \Big(\int_{t_0}^t e^{-(k_0-\varepsilon)(t-\tau)} \| f \|_{L_2(\tau,\tau+1;H_0)} \, d\tau \tag{12.7}$$
$$+ \int_t^\infty e^{-k_-(t-\tau)} \| f \|_{L_2(\tau,\tau+1;H_0)} \, d\tau + \| u \|_{W^{\ell-1}(t_0,t_0+1)} \, e^{-(k_0-\varepsilon)t} \Big)$$

for $t > t_0 + 1$, where c_ε does not depend on u and f.

Proof. (i) Let δ be a small positive number such that the strip $k_- - \delta \leq \Im\lambda < k_0$ is free of eigenvalues of $\mathcal{A}(\lambda)$. Applying Remark 6.2.8 with k_- and k_+ replaced by $k_- - \delta$ and k_0 we obtain (i). Theorem 6.2.4 implies (ii). $\square$

A similar result for solutions satisfying

$$\| u \|_{W^\ell(t,t+1)} = O\big(e^{-(k_0+\delta)t}\big) \quad \text{as} \quad t \to +\infty \tag{12.8}$$

with some positive δ is given in the following theorem.

Theorem 12.2.2. *Let $f \in L_{2,\text{loc}}(t_0, \infty; H_0)$ be subject to*

$$\int_{t_0}^\infty e^{(k_0+\varepsilon)} \| f \|_{L_2(\tau,\tau+1;H_0)} \, d\tau < \infty$$

with some positive ε. Then
  (i) *the equation* (11.1) *has a solution $u \in W^\ell_{\text{loc}}(t_0, \infty)$ satisfying*

$$\| u \|_{W^\ell(t,t+1)} \le c_\varepsilon \Big(\int_{t_0-1}^t e^{-k_+(t-\tau)} \| f \|_{L_2(\tau,\tau+1;H_0)} \, d\tau$$
$$+ \int_t^\infty e^{-(k_0+\varepsilon)(t-\tau)} \| f \|_{L_2(\tau,\tau+1;H_0)} \, d\tau \Big) \quad \text{for } t > t_0$$

with c_ε independent of f. Here $f(t) = 0$ for $t < t_0$.
  (ii) *Let $u \in W^\ell_{\text{loc}}(t_0, \infty)$ be a solution of* (11.1) *subject to* (12.8) *with some $\delta > 0$. Then*

$$\| u \|_{W^\ell(t,t+1)} \le c_\varepsilon \Big(\int_{t_0}^t e^{-k_+(t-\tau)} \| f \|_{L_2(\tau,\tau+1;H_0)} \, d\tau$$
$$+ \int_t^\infty e^{-(k_0+\varepsilon)(t-\tau)} \| f \|_{L_2(\tau,\tau+1;H_0)} \, d\tau + \| u \|_{W^{\ell-1}(t_0,t_0+1)} \, e^{-k_+t} \Big)$$

for $t > t_0 + 1$, where c_ε independent of u and f.

Proof. One should apply Theorem 6.2.4 and Remark 6.2.8 to the strip $k_0 < \Im\lambda < k_+ + \delta$ with a positive δ (compare with the proof of Theorem 12.2.1. $\square$

We pass to the homogeneous equation

$$L(t, D_t)u(t) = 0 \quad \text{for} \quad t > t_0 \tag{12.9}$$

and collect in the following theorem the information about its solutions which will be used in the sequel.

Theorem 12.2.3. *Let $\mathbf{X}(L)$ be a linear space of solutions to* (12.9) *of dimension κ. Suppose that for every non-zero $u \in \mathbf{X}(L)$ the relations hold:*

$$\|u\|_{W^\ell(t,t+1)} \le c_2 e^{-k_-t} \tag{12.10}$$

for large positive t and

$$\limsup_{t \to +\infty} e^{k_+t}\|u(t)\|_0 = \infty. \tag{12.11}$$

Then the following assertions are valid:
  (i) *For every non-zero $u \in \mathbf{X}(L)$ the inequalities*

$$C_1 e^{-\varepsilon t} \le e^{k_0 t} \parallel u \parallel_{W^\ell(t,t+1)} \le C_2 e^{\varepsilon t} \tag{12.12}$$

hold for large t with an arbitrary $\varepsilon > 0$ and with positive constants C_1 and C_2 depending on u and ε. (In other words, the characteristic exponent of u is equal to $-k_0$.)

(ii) *If $u \in W^\ell_{loc}(t_0, \infty)$ is a solution of (12.9) subject to*

$$\|u\|_{W^\ell(t,t+1)} \le c e^{-k_- t} \quad \text{for } t > t_0. \tag{12.13}$$

then there exists a unique $V \in \mathbf{X}(L)$ such that

$$\| u - V \|_{W^\ell(t,t+1)} \le C e^{-k_+ t} \quad \text{for } t \ge t_0.$$

The function V and the constant C admit the estimate

$$C + \| V \|_{W^\ell(t_0,t_0+1)} \le c_* \| u \|_{W^\ell(t_0,t_0+1)}, \tag{12.14}$$

where c_* does not depend on u.

Proof. Let $k_\pm^{(j)}$, $j = 1, 2$, be the same real numbers as in the proof of Corollary 9.3.5. Then (12.10) and (12.11) imply (9.14) and (9.15) with some positive ε. Hence $\mathbf{X}(L)$ is a $(\mathcal{Y}_1, \mathcal{Y}_2)$-space (see Sect. 9.3.1).

Assertion (i) follows from Proposition 9.3.1, whereas (ii) is a direct consequence of Corollary 9.3.4. $\square$

12.3 Linearization of the Pencil $\mathcal{A}(\lambda)$

Consider the differential equation (2.1) with constant coefficients. Due to the assumptions on $\mathcal{A}(\lambda)$ (see Sect. 2.2.1) the operator A_0 is invertible in H_0. Therefore, without loss of generality, we suppose that $A_0 = I$, where I is the identity operator.

Equation (2.1) can be transformed in different ways to the first order system. We shall use the following standard construction (see, for example, Gohberg, Krein (1969)). Put $U = \text{col}(U_j)_{j=1}^\ell$, where

$$U_1 = u, \ U_j = D^{j-1}u, \ j = 2, \ldots, \ell. \tag{12.15}$$

The right-hand side in (12.1) is defined by $F = \text{col}(F_j)_{j=1}^\ell$, where

$$F_1 = \ldots = F_{\ell-1} = 0, \quad F_\ell = f. \tag{12.16}$$

The matrix-operator $\mathfrak{A}$ takes the form

$$\mathfrak{A} = \begin{pmatrix} 0 & -I & 0 & \cdots & 0 \\ 0 & 0 & -I & \cdots & 0 \\ \vdots & \vdots & \vdots & & \vdots \\ A_\ell & A_{\ell-1} & A_{\ell-2} & \cdots & A_1 \end{pmatrix}. \tag{12.17}$$

It is clear that the operator

$$\mathfrak{A} : \mathcal{H}_1 \to \mathcal{H}_0,$$

where

$$\mathcal{H}_0 = H_{\ell-1} \times \ldots \times H_0, \quad \mathcal{H}_1 = H_\ell \times \ldots \times H_1, \tag{12.18}$$

is bounded.

To establish connections between properties of solutions of the operator equation (2.1) and the operator system (12.1) we will use the identity

$$\mathcal{E}(\lambda)(\lambda I_\ell + \mathfrak{A}) = \begin{pmatrix} A(\lambda) & 0 & \ldots & 0 \\ 0 & I & \ldots & 0 \\ \vdots & \vdots & \ddots & \vdots \\ 0 & 0 & \ldots & I \end{pmatrix} J(\lambda), \tag{12.19}$$

where

$$J(\lambda) = \begin{pmatrix} I & 0 & \ldots & 0 & 0 \\ -\lambda & I & \ldots & 0 & 0 \\ \vdots & \vdots & & \vdots & \vdots \\ 0 & 0 & \ldots & I & 0 \\ 0 & 0 & \ldots & -\lambda & I \end{pmatrix} \tag{12.20}$$

and

$$\mathcal{E}(\lambda) = \begin{pmatrix} E_1(\lambda) & E_2(\lambda) & \ldots & E_{\ell-1}(\lambda) & I \\ -I & 0 & \ldots & 0 & 0 \\ 0 & -I & \ldots & 0 & 0 \\ \vdots & \vdots & & \vdots & \vdots \\ 0 & 0 & \ldots & -I & 0 \end{pmatrix} \tag{12.21}$$

We use the notation

$$E_\ell(\lambda) = I,$$

and

$$E_{\ell-j}(\lambda) = A_j + \lambda E_{\ell-j+1}, \quad j = 1, \ldots, \ell - 1,$$

or equivalently,

$$E_j(\lambda) = \sum_{k=0}^{\ell-j} A_{\ell-j-k} \lambda^k, \quad j = 1, \ldots, \ell. \tag{12.22}$$

The crucial property of the representation (12.19) is that the inverse matrices to $\mathcal{E}(\lambda)$ and $J(\lambda)$ are also polynomial:

$$\mathcal{E}(\lambda)^{-1} = \begin{pmatrix} 0 & -I & 0 & \ldots & 0 \\ 0 & 0 & -I & \ldots & 0 \\ \vdots & \vdots & \vdots & & -I \\ I & E_1(\lambda) & E_2(\lambda) & \ldots & E_{\ell-1}(\lambda) \end{pmatrix} \tag{12.23}$$

and

$$J(\lambda)^{-1} = \begin{pmatrix} I & 0 & 0 & \cdots & 0 & 0 \\ \lambda & I & 0 & \cdots & \vdots & \vdots \\ \lambda^2 & \lambda & I & \cdots & \vdots & \vdots \\ \vdots & \vdots & \vdots & & I & 0 \\ \lambda^{\ell-1} & \lambda^{\ell-2} & \lambda^{\ell-3} & \cdots & \lambda & I \end{pmatrix}. \tag{12.24}$$

Let $\mathcal{H}_j^*$, $j = 0, 1$, be the dual of $\mathcal{H}_j$ with respect to the scalar product in $(H_0)^\ell$, i.e.

$$\mathcal{H}_1^* = H_\ell^* \times \ldots \times H_1^*, \quad \mathcal{H}_0^* = H_{\ell-1}^* \times \ldots \times H_0.$$

In what follows we also need the adjoint operator pencil

$$I_\ell \lambda + \mathfrak{A}^* : \mathcal{H}_0^* \to \mathcal{H}_1^*. \tag{12.25}$$

Formula (12.19) implies that

$$(I_\ell \lambda + \mathfrak{A}^*)\mathcal{E}^*(\lambda) = J^*(\lambda) \begin{pmatrix} \mathcal{A}^*(\lambda) & 0 & \cdots & 0 \\ 0 & I & \cdots & 0 \\ \vdots & \vdots & \ddots & \vdots \\ 0 & 0 & \cdots & I \end{pmatrix}. \tag{12.26}$$

We show that $I_\ell \lambda + \mathfrak{A}$ has the same properties as $\mathcal{A}(\lambda)$.

Proposition 12.3.1. (i) *The operator pencil $I_\ell \lambda + \mathfrak{A}$ is Fredholm.*

(ii) *The spectra of the operator pencils $I_\ell \lambda + \mathfrak{A}$ and $\mathcal{A}(\lambda)$ coincide and consist of eigenvalues of the same geometric, partial and algebraic multiplicities.*

Proof. (i) Since the operator $\mathcal{A}(\lambda) : H_\ell \to H_0$ is Fredholm for every $\lambda \in \mathbb{C}$, formula (12.19) and the invertibility of $\mathcal{E}(\lambda)$ and $J(\lambda)$ imply that $I_\ell \lambda + \mathfrak{A}$ is a Fredholm operator for every $\lambda \in \mathbb{C}$.

(ii) The coincidence of the spectra follows from (12.19). Furthermore, the relation (12.19) implies the equivalence of $\lambda I_\ell + \mathfrak{A}$ and

$$\begin{pmatrix} \mathcal{A}(\lambda) & & 0 \\ & \ddots & \\ 0 & & I \end{pmatrix} \tag{12.27}$$

in a neighbourhood of every eigenvalue (see Sect. A.5). By Proposition A.5.1, we obtain the coincidence of the geometric, partial and algebraic multiplicities of $I_\ell \lambda + \mathfrak{A}$ and (12.27). Since all multiplicities of (12.27) are equal to the corresponding multiplicities of $\mathcal{A}(\lambda)$, the proof is complete. $\square$

12.4 Canonical Sets of Jordan Chains of $I_\ell\lambda + \mathfrak{A}$

Here, and in what follows, we fix canonical sets of Jordan chains

$$\{\varphi_{kj}^{(\nu)}\} \quad \text{and} \quad \{\psi_{kj}^{(\nu)}\}, \tag{12.28}$$

$k = 1, \ldots, J_\nu$, $j = 0, \ldots, m_{\nu k} - 1$, corresponding to the eigenvalues λ_ν and $\overline{\lambda}_\nu$ of the operator pencils $\mathcal{A}(\lambda)$ and $\mathcal{A}^*(\lambda)$, and subject to the biorthogonality relation (1.20).

We return to the triple numeration (ν, k, h) for $U_\sigma^{(\nu)}(t)$ and $V_\sigma^{(\nu)}(t)$ (see (2.11) and (2.12)). Then (2.13) can be rewritten as

$$\int_{\mathbb{R}} \left(\mathcal{A}(D_t)\eta(t)U_{kj}^{(\nu)}(t), \, V_{pq}^{(\mu)}(t)\right)_{H_0} dt = \delta_\mu^\nu \, \delta_p^k \, \delta_q^j, \tag{12.29}$$

where η is a smooth function on $\mathbb{R}$ equal to 1 in a neighbourhood of $-\infty$ and 0 in a neighbourhood of $+\infty$.

We introduce the vector-functions

$$\mathcal{U}_{kj}^{(\nu)}(t) = \operatorname{col}\left(U_{kj,s}^{(\nu)}(t)\right)_{s=1}^\ell, \tag{12.30}$$

where

$$U_{kj,s}^{(\nu)}(t) = D_t^{s-1} U_{kj}^{(\nu)}(t). \tag{12.31}$$

This is equivalent to

$$\mathcal{U}_{kj}^{(\nu)}(t) = J(D_t)^{-1}\operatorname{col}\left(U_{kj}^{(\nu)}(t), 0, \ldots, 0\right). \tag{12.32}$$

We denote

$$\mathcal{V}_{kj}^{(\nu)}(t) = \operatorname{col}\left(V_{kj,s}^{(\nu)}(t)\right)_{s=1}^\ell, \tag{12.33}$$

where

$$V_{kj,s}^{(\nu)}(t) = E_s^*(D_t)V_{kj}^{(\nu)}(t) \quad \text{for} \quad s = 1, \ldots, \ell - 1 \tag{12.34}$$

and

$$V_{kj,\ell}^{(\nu)}(t) = V_{kj}^{(\nu)}(t). \tag{12.35}$$

This can be rewritten as one vector relation

$$\mathcal{V}_{kj}^{(\nu)}(t) = \mathcal{E}^*(D_t)\operatorname{col}\left(V_{kj}^{(\nu)}(t), 0, \ldots, 0\right). \tag{12.36}$$

By (12.31), (12.34) and (12.35)

$$\mathcal{U}_{kj}^{(\nu)}(t) \in \mathcal{H}_1, \quad \mathcal{V}_{kj}^{(\nu)}(t) \in \mathcal{H}_0^* \tag{12.37}$$

for any $t \in \mathbb{R}$ (the spaces $\mathcal{H}_0$ and $\mathcal{H}_1$ were introduced by (12.18)). Using (12.19) and (12.26) one easily verifies that

$$\left(I_\ell D_t + \mathfrak{A}\right)\mathcal{U}_{kj}^{(\nu)}(t) = 0, \tag{12.38}$$

$$\left(I_\ell D_t + \mathfrak{A}^*\right)\mathcal{V}_{kj}^{(\nu)}(t) = 0. \tag{12.39}$$

We show that the vector functions $\mathcal{U}_{kj}^{(\nu)}$, $\mathcal{V}_{kj}^{(\nu)}$ satisfy the biorthogonality condition (12.29) with respect to $I_\ell D_t + \mathfrak{A}$.

Lemma 12.4.1. *Let η be the same function as in (12.29). Then*

$$\int_{\mathbb{R}} \left((I_\ell D_t + \mathfrak{A})(\eta(t) \mathcal{U}_{kj}^{(\nu)}(t)), \; V_{pq}^{(\mu)}(t) \right)_{(H_0)^\ell} dt = \delta_\mu^\nu \, \delta_p^k \, \delta_q^j. \qquad (12.40)$$

Proof. By (12.32) and (12.36) the left-hand side of (12.40) is equal to

$$\int_{\mathbb{R}} \left((I_\ell D_t + \mathfrak{A}) \left(\eta(t) J(D_t)^{-1} \mathrm{col}(U_{kj}^{(\nu)}(t), 0, \ldots, 0) \right) \right.$$
$$\left. \mathcal{E}^*(D_t) \mathrm{col}(V_{pq}^{(\mu)}(t), 0, \ldots, 0) \right)_{(H_0)^\ell} dt.$$

Since the commutator $[J(D_t)^{-1}, \eta]$ has a compact support and since

$$(I_\ell D_t + \mathfrak{A}^*) \mathcal{E}^*(D_t) \mathrm{col}\left(V_{pq}^{(\mu)}(t), 0, \ldots, 0 \right) = 0$$

(due to (12.26)), the last integral can be transformed as

$$\int_{\mathbb{R}} \left((I_\ell D_t + \mathfrak{A}) J(D_t)^{-1} (\eta(t) \, \mathrm{col} \, (U_{kj}^{(\nu)}(t), 0, \ldots, 0)), \right.$$
$$\left. \mathcal{E}^*(D_t) \, \mathrm{col} \, (V_{pq}^{(\mu)}(t), 0, \ldots, 0) \right)_{(H_0)^\ell} dt.$$

By (12.19) this integral is equal to

$$\int_{\mathbb{R}} \left(\mathcal{E}(D_t)^{-1} \, \mathrm{col} \, (\mathcal{A}(D_t)(\eta(t) U_{kj}^{(\nu)}(t)), 0, \ldots, 0), \right.$$
$$\left. \mathcal{E}^*(D_t) \, \mathrm{col} \, (V_{pq}^{(\mu)}(t), 0, \ldots, 0) \right)_{(H_0)^\ell} dt.$$

Since the function $\mathcal{A}(D_t) \eta(t) U_{kj}^{(\nu)}(t)$ has a compact support, by partial integration we can rewrite the last integral as

$$\int_{\mathbb{R}} \left(\mathcal{A}(D_t)(\eta(t) U_{kj}^{(\nu)}(t)), \; V_{pq}^{(\mu)}(t) \right)_{H_0} dt.$$

It remains to use (12.29). $\square$

Proposition 12.4.2. *For every $\tau \in \mathbb{R}$ the systems*

$$\{ \mathcal{U}_{kj}^{(\nu)}(\tau) \}, \; \{ \mathcal{V}_{k, m_{\mu k} - 1 - j}^{(\nu)}(\tau) \},$$

$k = 1, \ldots, J_\nu$, $j = 0, \ldots, m_{\nu k} - 1$, *form canonical sets of Jordan chains corresponding to the eigenvalues λ_ν and $\overline{\lambda}_\nu$ of the operator pencils $\lambda I_\ell + \mathfrak{A}$ and $\lambda I_\ell + \mathfrak{A}^*$. The biorthogonality relation holds:*

$$\left(\mathcal{U}_{kj}^{(\mu)}(\tau), \; \mathcal{V}_{sq}^{(\nu)}(\tau) \right)_{(H_0)^\ell} = -i \delta_\nu^\mu \, \delta_s^k \, \delta_q^j. \qquad (12.41)$$

Proof. By the definition of $U_{k,h}^{(\nu)}$ and $V_{k,h}^{(\nu)}$ (see the end of Sect. 2.2),

$$U_{kh}^{(\nu)}(t) = (D_t - \lambda_\nu)^{m_{\nu k}-1-h} U_{k,m_{\nu k}-1}^{(\nu)}(t) \tag{12.42}$$

and

$$V_{kh}^{(\nu)}(t) = (D_t - \overline{\lambda}_\nu)^h V_{k0}^{(\nu)}(t). \tag{12.43}$$

Due to (12.32) and (12.36), the formulae (12.42) and (12.43) are valid with
U and V replaced by $\mathcal{U}$ and $\mathcal{V}$. By (12.31) and (12.35) the degrees of the
polynomials

$$\left\{ e^{-i\lambda_\nu t} \mathcal{U}_{k,m_{\nu k}-1}^{(\nu)}(t) \right\}, \quad \left\{ e^{-i\lambda_\nu t} \mathcal{V}_{k0}(t) \right\}$$

are equal to $m_{\nu k} - 1$ and the leading coefficients of these polynomials are
linearly independent. Reference to Lemma 1.3.7 and Proposition 12.3.1(ii)
completes the proof of the first statement.

Formula (12.41) follows from (12.40) by the arbitrary choice of η. $\square$

The following assertion is contained in Proposition 12.4.2 (and it can
easily be verified directly).

Corollary 12.4.3. (i) *The equalities*

$$(I_\ell \lambda_\nu + \mathfrak{A}) \mathcal{U}_{k0}^{(\nu)}(\tau) = 0, \tag{12.44}$$

$$(I_\ell \lambda_\nu + \mathfrak{A}) \mathcal{U}_{kh}^{(\nu)}(\tau) + \mathcal{U}_{k,h-1}^{(\nu)}(\tau) = 0 \tag{12.45}$$

hold for $h = 1, \ldots, m_{\nu k} - 1$.
(ii) *The vectors* $\mathcal{V}_{kh}^{(\nu)}$ *satisfy*

$$(I_\ell \overline{\lambda}_\nu + \mathfrak{A}^*) \mathcal{V}_{k,m_{\nu k}-1}^{(\nu)}(\tau) = 0, \tag{12.46}$$

$$(I_\ell \overline{\lambda}_\nu + \mathfrak{A}^*) \mathcal{V}_{kh}^{(\nu)}(\tau) + \mathcal{V}_{k,h+1}^{(\nu)}(\tau) = 0 \tag{12.47}$$

for $h = 0, 1, \ldots, m_{\nu k} - 2$.

12.5 The Riesz Projector and Its Properties

Let C be a smooth simple contour consisting of regular points of the operator
pencil $I_\ell \lambda + \mathfrak{A}$ which encloses the eigenvalues λ_ν with $\Im \lambda_\nu = k_0$. The operator

$$P = \frac{1}{2\pi i} \int_C (I_\ell \lambda + \mathfrak{A})^{-1} d\lambda \tag{12.48}$$

is called the Riesz projector corresponding to the spectrum of $-\mathfrak{A}$ on the
line $\Im \lambda = k_0$.

Proposition 12.5.1. *For every* $\tau \in \mathbb{R}$ *the formula*

$$P = i \sum_{\Im\lambda_\nu = k_0} \sum_{k=1}^{J_\nu} \sum_{j=0}^{m_{\nu k}-1} (\cdot, \mathcal{V}_{kj}^{(\nu)}(\tau))_{(H_0)^\ell} \, \mathcal{U}_{kj}^{(\nu)}(\tau) \tag{12.49}$$

holds, where $\mathcal{U}_{kj}^{(\nu)}$ *and* $\mathcal{V}_{kj}^{(\nu)}$ *are given by* (12.32) *and* (12.36). *(In particular, the right-hand side in* (12.49) *does not depend on* τ*.)*

Proof. Let λ_ν be an eigenvalue of the pencil $I_\ell \lambda + \mathfrak{A}$ with $\Im\lambda_\nu = k_0$. Using Theorem 1.2.1 and Proposition 12.4.2 we obtain

$$(I_\ell \lambda + \mathfrak{A})^{-1} = \sum_{k=1}^{J_\nu} \sum_{h=0}^{m_{\nu k}-1} \frac{P_{kh}^{(\nu)}}{(\lambda - \lambda_\nu)^{m_{\nu k}-h}} + \Gamma_\nu(\lambda), \tag{12.50}$$

where Γ_ν is holomorphic in a neighbourhood of λ_ν and

$$P_{kh}^{(\nu)} = i \sum_{j=0}^{h} (\cdot, \mathcal{V}_{k,m_{\nu k}-1-j}^{(\nu)}(\tau))_{(H_0)^\ell} \, \mathcal{U}_{k,h-j}^{(\nu)}(\tau).$$

Let C_ν be a small circle centered at λ_ν. By (12.50) and the Cauchy integral formula, we get

$$\frac{1}{2\pi i} \int_{C_\nu} (I_\ell \lambda + \mathfrak{A})^{-1} d\lambda = \sum_{k=0}^{J_\nu} P_{k,m_{\nu k}-1}^{(\nu)}$$

$$= i \sum_{k=1}^{J_\nu} \sum_{j=0}^{m_{\nu k}} (\cdot, \mathcal{V}_{kj}^{(\nu)}(\tau))_{(H_0)^\ell} \, \mathcal{U}_{kj}^{(\nu)}(\tau). \tag{12.51}$$

Using

$$\int_C (I_\ell \lambda + \mathfrak{A})^{-1} d\lambda = \sum_{\Im\lambda_\nu = k_0} \int_{C_\nu} (I_\ell \lambda + \mathfrak{A})^{-1} d\lambda$$

and representation (12.51), we arrive at (12.49). $\square$

The following properties of the operator P are well known and can easily be verified.

Proposition 12.5.2. (i) *The operator P maps $\mathcal{H}_0$ into $\mathcal{H}_1$ and $P^2 = P$.*
(ii) *For an arbitrary* $\mathcal{U} \in \mathcal{H}_1$,

$$\mathfrak{A}P\mathcal{U} = P\mathfrak{A}\mathcal{U}. \tag{12.52}$$

(iii) *The following equalities hold:*

$$P\mathcal{U}_{kj}^{(\nu)}(\tau) = \mathcal{U}_{kj}^{(\nu)}(\tau), \tag{12.53}$$

$$P^*\mathcal{V}_{kj}^{(\nu)}(\tau) = \mathcal{V}_{kj}^{(\nu)}(\tau), \tag{12.54}$$

where P^* is the adjoint operator of P.

We conclude this section with the following auxiliary result.

Proposition 12.5.3. *The operator function*

$$(t,\tau) \rightarrow \sum_{k=1}^{J_\nu} \sum_{j=0}^{m_{\nu k}-1} \left(\cdot\,,\mathcal{V}_{kj}^{(\nu)}(\tau)\right)_{(H_0)_\ell}\mathcal{U}_{kj}^{(\nu)}(t) \tag{12.55}$$

depends only on $t-\tau$.

Proof. By (2.44), the operator function

$$(t,\tau) \rightarrow \sum_{k=1}^{J_\nu} \sum_{s=0}^{m_{\nu k}-1} \left(\cdot\,,D_\tau^s\Psi_k^{(\nu)}(i\tau)\right)_{H_0}D_t^{m_{\nu s}-1-s}\Phi_k^{(\nu)}(it)$$

depends only on $t-\tau$. Hence using (2.9) and (2.10), the same is true for the function

$$(t,\tau) \rightarrow \sum_{k=1}^{J_\nu} \sum_{j=0}^{m_{\nu k}-1} \left(\cdot\,,V_{kj}^{(\nu)}(\tau)\right)_{H_0}U_{kj}^{(\nu)}(t).$$

Therefore

$$\sum_{k=1}^{J_\nu} \sum_{j=0}^{m_{\nu k}-1} \left(\cdot\,,E_s^*(D_\tau)V_{kj}^{(\nu)}(\tau)\right)_{H_0}D_t^{m-1}U_{kj}^{(\nu)}(t)$$

also depends only on $t-\tau$. It remains to use the definitions of $\mathcal{U}_{kj}^{(\nu)}$ and $\mathcal{V}_{kj}^{(\nu)}$.
□

12.6 The Vector Function Spaces $\mathbb{S}_{\mathrm{loc}}(\mathbb{R})$, $\mathbb{X}_{\mathrm{loc}}(\mathbb{R})$ and $\mathbb{Y}_{\mathrm{loc}}(\mathbb{R})$

Here we collect definitions of some spaces which are used in the sequel. Let $\mathcal{H}_1$ and $\mathcal{H}_0$ be the spaces defined by (12.18). We introduce the space $\mathbb{S}(a,b)$ of vector-functions $U = \mathrm{col}(U_j)_{j=1}^\ell$ with values in $\mathcal{H}_1$, and endowed with the norm

$$\| U \|_{\mathbb{S}(a,b)} = \left(\int_a^b \sum_{j=1}^\ell \left(\| U_j(\tau) \|_{\ell-j+1}^2 + \| D_\tau U_j(\tau) \|_{\ell-j}^2 \right)d\tau\right)^{1/2}.$$

This definition is equivalent to

$$\mathbb{S}(a,b) = \{U : U \in L_2(a,b;\mathcal{H}_1), D_tU \in L_2(a,b;\mathcal{H}_0)\}. \tag{12.56}$$

By $\mathbb{S}_{\mathrm{loc}}(\mathbb{R})$ we denote the space of vector functions whose restrictions to an arbitrary finite interval (a,b) belong to $\mathbb{S}(a,b)$. Analogously, we say that $U \in \mathbb{S}_{\mathrm{loc}}(a,\infty)$ provided $U \in \mathbb{S}(a,b)$ for any finite $b > a$.

Obviously, if $U = \text{col}\{D_t^{j-1}u\}_{j=1}^{\ell}$, where $u \in W^{\ell}(a, b)$, then $U \in \mathbb{S}(a, b)$ and the estimates

$$\|U\|_{\mathbb{S}(a,b)} \leq \|u\|_{W^{\ell}(a,b)} \leq \sqrt{2}\|U\|_{\mathbb{S}(a,b)}$$

are valid.

Lemma 12.6.1. *Let P be the Riesz projector (12.49). If $U \in \mathbb{S}(a, b)$ then $PU \in \mathbb{S}(a, b)$ and*

$$\| PU \|_{\mathbb{S}(a,b)} \leq c \| U \|_{\mathbb{S}(a,b)},$$

where c does not depend on a, b.

Proof. By (12.49) with $\tau = 0$,

$$PU(t) = i \sum_{\Im\lambda_{\nu}=k_0} \sum_{k=1}^{J_{\nu}} \sum_{j=0}^{m_{\nu k}-1} \left(U(t), V_{kj}^{(\nu)}(0)\right)_{(H_0)_{\ell}} \mathcal{U}_{kj}^{(\nu)}(0).$$

It suffices to estimate the norms

$$\| PU \|_{L_2(a,b;\mathcal{H}_1)}, \| D_t PU \|_{L_2(a,b;\mathcal{H}_0)}.$$

This is achieved by using (12.37) with $t = 0$. $\square$

By $\mathbb{X}(a, b)$ we denote the space of all vector functions $\mathcal{U}(t)$ represented in the form

$$\mathcal{U}(t) = (I_{\ell} - P) \text{ col } \left(D_t^{j-1}u(t)\right)_{j=1}^{\ell} \tag{12.57}$$

with some $u \in W^{\ell}(a, b)$. We supply $\mathbb{X}(a, b)$ with the norm

$$\|\mathcal{U}\|_{\mathbb{X}(a,b)} = \inf \|u\|_{W^{\ell}(a,b)}, \tag{12.58}$$

where the infimum is taken over all u in (12.57). The space $\mathbb{X}_{\text{loc}}(a, \infty)$ consists of all vector valued functions $\mathcal{U}$ which can be represented in the form (12.57) with $u \in W_{\text{loc}}^{\ell}(a, \infty)$.

Lemma 12.6.2. *The space $\mathbb{X}(a, b)$ is continuously embedded into $\mathbb{S}(a, b)$ and the estimate*

$$\| \mathcal{U} \|_{\mathbb{S}(a,b)} \leq c \| \mathcal{U} \|_{\mathbb{X}(a,b)} \tag{12.59}$$

holds, where c does not depend on a, b and $\mathcal{U} \in \mathbb{X}(a, b)$.

Proof. It is sufficient to check (12.59). Let $u \in W^{\ell}(a, b)$ be a function in (12.57). Using Lemma 12.6.1 we get

$$\| \mathcal{U} \|_{\mathbb{S}(a,b)} \leq c \sum_{j=0}^{\ell-1} \left(\| D_t^j u \|_{L_2(a,b;H_{\ell-j})} \right.$$

$$\left. + \| D_t^{j+1}u \|_{L_2(a,b;H_{\ell-1-j})} \right) \leq c \| u \|_{W^{\ell}(a,b)};$$

this implies (12.59). □

We define the space $\mathbb{X}_{\mathrm{loc}}(\mathbb{R})$ of all vector valued functions on $\mathbb{R}$ which can be represented in the form (12.57) with a certain $u \in W^{\ell}_{\mathrm{loc}}(\mathbb{R})$.

We shall use the space $\mathbb{Y}_{\mathrm{loc}}(\mathbb{R})$ of vector functions $\mathcal{F}(t) = \mathrm{col}\big(\mathcal{F}_j(t)\big)_{j=1}$ represented in the form

$$\mathcal{F}(t) = (I_{\ell} - P)\mathrm{col}\big(0, \ldots, 0, f(t)\big) \tag{12.60}$$

with some $f \in L_{2,\mathrm{loc}}(\mathbb{R}; H_0)$. Replacing $\mathbb{R}$ by (t_0, ∞) in this definition we obtain the definition of $\mathbb{Y}_{\mathrm{loc}}(t_0, \infty)$.

12.7 Existence of Solutions in $\mathbb{X}_{\mathrm{loc}}(\mathbb{R})$

Let g be Green's function of the ordinary differential operator $(\partial_t + k_+)(-\partial_t - k_-)$. General formulae ((3.23) and (3.24)) give

$$g(t) = \begin{cases} (k_+ - k_-)^{-1}e^{-k_+t} & \text{for } t \geq 0 \\ (k_+ - k_-)^{-1}e^{-k_-t} & \text{for } t < 0. \end{cases} \tag{12.61}$$

Theorem 12.7.1. *Let $\mathcal{F} \in \mathbb{Y}_{\mathrm{loc}}(\mathbb{R})$ satisfy*

$$\int_{\mathbb{R}} g(-\tau) \, \| f \|_{L_2(\tau, \tau+1; H_0)} \, d\tau < \infty , \tag{12.62}$$

where f is the function in (12.60). Then there exists a solution $U \in \mathbb{X}_{\mathrm{loc}}(\mathbb{R})$ of (12.1) such that

$$\| U \|_{\mathbb{X}(t,t+1)} \leq c \int_{\mathbb{R}} g(t - \tau) \, \| f \|_{L_2(\tau, \tau+1; H_0)} \, d\tau, \tag{12.63}$$

where c depends only on the coefficients $A_1, \ldots, A_{\ell}$ of the pencil $\mathcal{A}(\lambda)$ and $k_{\pm}$.

Proof. We fix real numbers q_+ and q_- such that $k_- < q_+ < k_0 < q_- < k_+$. Then the strips $k_- < \Im\lambda < q_+$, $q_- < \Im\lambda < k_+$ are free of eigenvalues of $\mathcal{A}(\lambda)$. Let g_- and g_+ be Green's functions of the operators

$$(\partial_t + q_+)(-\partial_t - k_-)$$

and

$$(\partial_t + k_+)(-\partial_t - q_-)$$

respectively. They are given by (12.61) with $k_+ = q_+$, $k_- = k_-$ and with $k_+ = k_+$, $k_- = q_-$. Using the explicit representations for g and $g_{\pm}$ we obtain the estimates

$$c_1 g_-(t) \leq g(t) \leq c_2 g_-(t) \quad \text{for} \quad t \leq 0 \tag{12.64}$$

and

$$c_1 g_+(t) \le g(t) \le c_2 g_+(t) \quad \text{for} \quad t \ge 0. \tag{12.65}$$

Let ζ be a smooth function on $\mathbb{R}$ which is equal to 1 for $t \ge 1$ and 0 for $t \le 0$. For a fixed $a \in \mathbb{R}$ we represent f as $f = f_a^{(-)} + f_a^{(+)}$, where

$$f_a^{(-)}(t) = \zeta(t-a)f(t), \quad f_a^{(+)}(t) = (1 - \zeta(t-a))f(t).$$

Then by (12.62), (12.64) and (12.65)

$$\int_{\mathbb{R}} g_{\mp}(-t) \, \| f_a^{(\mp)} \|_{L_2(t,t+1;H_0)} \, dt < \infty.$$

Hence by Theorem 3.3.2 there exist the (k_-, q_+)-solution $u_a^{(-)}$ and the (q_-, k_+)-solution $u_a^{(+)}$ of the problem (2.1) with $f = f_a^{(-)}$ and $f = f_a^{(+)}$ respectively. By the same theorem, these solutions satisfy

$$\| u_a^{(+)} \|_{W^\ell(t,t+1)} \le c \int_{\mathbb{R}} g_+(t-\tau) \, \| f_a^{(+)} \|_{L_2(\tau,\tau+1;H_0)} \, d\tau, \tag{12.66}$$

$$\| u_a^{(-)} \|_{W^\ell(t,t+1)} \le c \int_{\mathbb{R}} g_-(t-\tau) \, \| f^{(-)} \|_{L_2(\tau,\tau+1;H_0)} \, d\tau \tag{12.67}$$

with a positive constant c depending only on $A_1, \ldots, A_\ell$, $k_\pm$ and $q_\pm$. Moreover,

$$\| u_a^{(\mp)} \|_{W^\ell(t,t+1)} = o\!\left(e^{-q_\pm t}\right) \quad \text{as} \quad t \to \mp\infty, \tag{12.68}$$

$$\| u_a^{(\mp)} \|_{W^\ell(t,t+1)} = o\!\left(e^{-k_\mp t}\right) \quad \text{as} \quad t \to \pm\infty. \tag{12.69}$$

We put $u_a = u_a^{(+)} + u_a^{(-)}$. It is clear that $u_a \in W_{\text{loc}}^\ell(\mathbb{R})$ and $\mathcal{A}(D_t)u_a = f$ on $\mathbb{R}$.

By (12.66), (12.67), (12.64) and (12.65) we get the estimate

$$\| u_a \|_{W^\ell(a,a+1)} \le c \int_{\mathbb{R}} g(a-\tau) \, \| f \|_{L_2(\tau,\tau+1;H_0)} \, d\tau. \tag{12.70}$$

We introduce the function

$$U_a(t) = (I_\ell - P)\text{col}\!\left(D_t^{j-1} u_a(t)\right)_{j=1}^\ell. \tag{12.71}$$

First, we show that U_a does not depend upon a. In fact, let

$$\mathcal{A}(D_t)(u_a - u_b) = 0 \quad \text{on} \quad \mathbb{R}.$$

From (12.68) and (12.69) it follows that

$$\| u_a - u_b \|_{W^\ell(t,t+1)} = o\!\left(e^{-k_\mp t}\right) \quad \text{as} \quad t \to \mp\infty.$$

Along with Proposition 3.8.3, this implies

$$(u_a - u_b)(t) = \sum_{\Im \lambda_\nu = k_0} \sum_{k=1}^{J_\nu} \sum_{j=0}^{m_{\nu k}-1} c_{kj}^{(\nu)} U_{kj}^{(\nu)}(t).$$

Hence, by using (12.71), (12.31) and (12.53) we obtain $U_a = U_b$. We now denote U_a by U.

Since $\mathcal{A}(D_t)u_a = f$,

$$(D_t + \mathfrak{A})\text{col}(D_t^{j-1}u_a)_{j=1}^\ell = \text{col}(0,\ldots,0,f) \quad \text{on} \quad \mathbb{R}. \tag{12.72}$$

By applying $I_\ell - P$ to both sides of (12.72) and using (12.52) we conclude that U satisfies (12.1).

Since, by (12.58),

$$\| U \|_{\mathbb{X}(a,a+1)} \leq \| u_a \|_{W^\ell(a,a+1)},$$

reference to (12.70) completes the proof. $\square$

Corollary 12.7.2. *Estimate* (12.63) *implies*

$$\| U \|_{\mathbb{X}(t,t+1)} = o(e^{-k_\mp t}) \quad \text{as} \quad t \to \pm\infty. \tag{12.73}$$

Proof. Duplicates that of (3.17). $\square$

12.8 Uniqueness of Solutions in $\mathbb{X}_{\text{loc}}(\mathbb{R})$

Lemma 12.8.1. *Let $U \in \mathbb{S}_{\text{loc}}(\mathbb{R})$ be a solution of*

$$(I_\ell D_t + \mathfrak{A})U = 0 \quad \text{on} \quad \mathbb{R}. \tag{12.74}$$

Also let

$$\| U \|_{\mathbb{S}(t,t+1)} = O(e^{-k_\mp t}) \quad \text{for } t \gtrless 0. \tag{12.75}$$

Then

$$U(t) = \sum_{\Im \lambda_\nu = k_0} \sum_{k=1}^{J_\nu} \sum_{j=0}^{m_{\nu k}-1} c_{kj}^{(\nu)} \mathcal{U}_{kj}^{(\nu)}(t), \tag{12.76}$$

where $c_{kj}^{(\nu)}$ are constants.

Proof. Equality (12.74) yields

$$D_t U_j = U_{j+1}, \quad j = 1,\ldots,\ell-1. \tag{12.77}$$

Using $U \in \mathbb{S}_{\text{loc}}(\mathbb{R})$, we get

$$D_t^k U_j \in L_{2,\text{loc}}(\mathbb{R}; H_{\ell-j-k+1}), \quad j = 1,\ldots,\ell-k+1, \tag{12.78}$$

and

$$\| D_t^k U_j \|_{L_2(t,t+1;H_{\ell-j-k+1})} \leq c \| U \|_{\mathbb{S}(t,t+1)}.$$

This implies, in particular, that

$$U_1 \in W^\ell_{\text{loc}}(\mathbb{R}) \quad \text{and} \quad \| U_1 \|_{W^\ell(t,t+1)} \leq c \, \| U \|_{S(t,t+1)} . \tag{12.79}$$

By (12.77) and (12.78) we obtain

$$U_j = D_t^{j-1} U_1 \quad \text{for} \quad j = 2, \ldots, \ell. \tag{12.80}$$

Therefore, the last equation in (12.74) takes the form

$$\mathcal{A}(D_t)U_1 = 0.$$

Using (12.79) and (12.75) we arrive at

$$\| U_1 \|_{W^\ell(t,t+1)} = O(e^{-k_\mp t}) \quad \text{for} \quad t \geq 0.$$

Taking into account that there are no eigenvalues with $\Im \lambda = k_\pm$ we can apply Proposition 3.8.3 (with k_+ and k_- replaced by $k_+ + \varepsilon$ and $k_- - \varepsilon$ for sufficiently small ε) to U_1. Then

$$U_1(t) = \sum_{\Im \lambda_\nu = k_0} \sum_{k=1}^{J_\nu} \sum_{j=0}^{m_{\nu k}-1} c_{kj}^{(\nu)} U_{kj}^{(\nu)}(t).$$

Applying (12.80) and (12.31) we obtain (12.76). $\square$

Theorem 12.8.2. *Let $U \in \mathbb{X}_{\text{loc}}(\mathbb{R})$ satisfy (12.74) and (12.75). Then $U = 0$.*

Proof. By Lemma 12.8.1 we have (12.76). Applying $I_\ell - P$ to both sides of (12.76) and using (12.53) together with $(I_\ell - P)U = U$, we complete the proof. $\square$

By $\mathfrak{L}$ we denote the operator $\mathcal{F} \to U$, where

$$\mathcal{F} = (I_\ell - P)\text{col}(0, \ldots, 0, f) \in \mathbb{Y}_{\text{loc}}(\mathbb{R}) \tag{12.81}$$

with f subject to (12.62) and $U \in \mathbb{X}_{\text{loc}}(\mathbb{R})$ being a solution of (12.1) satisfying (12.75). Such a solution exists and is unique by Theorems 12.7.1 and 12.8.2. Thus, (12.63) can be rewritten as

$$\| \mathfrak{L}\mathcal{F} \|_{\mathbb{X}(t,t+1)} \leq c \int_{\mathbb{R}} g(t - \tau) \, \| f \|_{L_2(\tau,\tau+1;H_0)} \, d\tau. \tag{12.82}$$

12.9 From the Equation with Variable Coefficients to a First Order System

We transform the equation (5.1) to a system of the first order in a similar way as in the case of the operator $\mathcal{A}(D_t)$ with constant coefficients (see Sect. 12.3). We assume that $A_0(t) = I$ and put

$$N_j(t) = A_j - A_j(t), \quad j = 1, \ldots, \ell. \tag{12.83}$$

We then rewrite (5.1) in the form

$$\Big(\mathcal{A}(D_t) - N(t, D_t)\Big)u(t) = f(t) \quad \text{on} \quad \mathbb{R}, \tag{12.84}$$

where

$$N(t, D_t) = \sum_{j=1}^{\ell} N_j(t) D_t^{\ell-j}. \tag{12.85}$$

As in Chapter 5, we shall measure the value of the perturbation operator $N(t, D_t)$ by the function

$$\rho(t) = \| N(\cdot, D) \|_{W^\ell(t,t+1) \to L_2(t,t+1;H_0)} . \tag{12.86}$$

We put

$$U_1 = u, \quad U_j = D_t^{j-1} u, \quad j = 2, \ldots, \ell. \tag{12.87}$$

Then the vector $U = \mathrm{col}(U_j)_{j=1}^{\ell}$ satisfies the system

$$\big(I_\ell D_t + \mathfrak{A} - \mathcal{N}(t)\big) U(t) = F(t) \quad \text{on} \quad \mathbb{R}, \tag{12.88}$$

where

$$F(t) = \mathrm{col}(0, \ldots, 0, f(t)). \tag{12.89}$$

The matrix operator $\mathfrak{A}$ is defined by (12.17) and

$$\mathcal{N}(t) = \begin{pmatrix} 0 & 0 & \cdots & 0 \\ \vdots & \vdots & & \vdots \\ 0 & 0 & \cdots & 0 \\ N_\ell(t) & N_{\ell-1}(t) & \cdots & N_1(t) \end{pmatrix}. \tag{12.90}$$

The following estimate will be used to solve system (12.88).

Proposition 12.9.1. *Let* $\mathcal{U} \in \mathbb{X}_{\mathrm{loc}}(\mathbb{R})$. *Then* $(\mathcal{N}\mathcal{U})_k = 0$ *for* $k = 1, \ldots, \ell - 1$ *and*

$$\| (\mathcal{N}\mathcal{U})_\ell \|_{L_2(t,t+1;H_0)} \leq c\rho(t) \| \mathcal{U} \|_{\mathbb{X}(t,t+1)}, \tag{12.91}$$

where c *does not depend on* t *and* $\mathcal{U}$.

Proof. Let

$$\mathcal{U}(\tau) = (I_\ell - P)\mathrm{col}(D_\tau^{j-1}u(\tau))_{j=1}^\ell \quad \text{for} \quad \tau \in (t, t+1) \tag{12.92}$$

with some $u \in W^\ell(t, t+1)$. We denote

$$U(\tau) = \mathrm{col}(D_\tau^{j-1}u(\tau))_{j=1}^\ell.$$

Then

$$(\mathcal{N}\mathcal{U})_\ell(\tau) = N(\tau, D_\tau)u(\tau) - (\mathcal{N}(\tau)PU)_\ell. \tag{12.93}$$

Using (12.49) we write

$$PU(\tau) = \sum_{k_- < \Im\lambda_\nu < k_+} \sum_{k=1}^{J_\nu} \sum_{j=0}^{m_{\nu k}-1} (U(\tau), \mathcal{V}_{kj}^{(\nu)}(\tau - t))_{(H_0)^\ell}\, \mathcal{U}_{kj}^{(\nu)}(\tau - t).$$

Hence, by (12.31), we obtain

$$\big(\mathcal{N}(\tau)PU(\tau)\big)_\ell$$

$$= \sum_{k_- < \Im\lambda_\nu < k_+} \sum_{k=1}^{J_\nu} \sum_{j=0}^{m_{\nu k}-1} (U(\tau), \mathcal{V}_{kj}^{(\nu)}(\tau - t))_{(H_0)^\ell} N(\tau, D_t)U_{kj}^{(\nu)}(\tau - t).$$

Since by the definitions of $U_{kj}^{(\nu)}$ and $V_{kj}^{(\nu)}$

$$\| U_{kj}^{(\nu)} \|_{W^\ell(0,1)} \le c \quad \text{and} \quad \| V_{kj}^{(\nu)} \|_{L_\infty(0,1;\mathcal{H}_0^*)} \le c,$$

it follows that the inequalities

$$\| N(\cdot, D_\tau)U_{kj}^{(\nu)}(\cdot, -t) \|_{L_2(t,t+1;H_0)} \le c\rho(t) \tag{12.94}$$

and

$$\left| (U(\tau), V_{kj}^{(\nu)}(\tau - t))_{(H_0)^\ell} \right| \le c \sum_{j=1}^\ell \| D_\tau^{j-1}u(\tau) \|_{\ell-j}$$

hold. Furthermore, by Proposition 2.3.5

$$\sum_{j=1}^\ell \| D_\tau^{j-1}u(\tau) \|_{\ell-j} \le c \| u \|_{W^\ell(t,t+1)}. \tag{12.95}$$

Using (12.94) and (12.95) we estimate the second term in (12.93) and obtain

$$\| (\mathcal{N}\mathcal{U})_\ell \|_{L_2(t,t+1;H_0)} \le c\rho(t) \| u \|_{W^\ell(t,t+1)}.$$

Finally, we take the infimum of the right-hand side over all $u \in W^\ell(t, t+1)$ in representation (12.92) and arrive at (12.91). $\square$

Corollary 12.9.2. *If $\mathcal{F}$ is given by (12.81) where f satisfies (12.62), then*

$$\| \mathcal{N}\mathfrak{L}\mathcal{F} \|_{L_2(t,t+1;H_0)}$$

$$\le c_0\rho(t) \int_{\mathbb{R}} g(t - \tau) \| f \|_{L_2(\tau,\tau+1;H_0)}\, d\tau \tag{12.96}$$

with c_0 depending only on $A_1, \cdots, A_\ell$ and $k_\pm$.

Proof. Follows from (12.82) and Proposition 12.9.1. $\square$

12.10 Comments

For higher order ordinary differential equations, reductions to first order systems (which are simplified by certain diagonalizing transformations) date back to Birkhoff (1908). This approach, which has become standard in the asymptotic theory of ordinary differential equations, is presented in the books Gantmaher (1959), Wasov (1976) and Eastham (1989). The relation (12.19) is borrowed from Gohberg, Goldberg and Kaashoek (1990).

The functional spaces in Sect. 12.6 seem to be new; as is the material in Sect. 12.7.

13. General Asymptotic Representation

13.1 Introduction

Here we consider the equation (12.84). As in the previous chapter we suppose that the eigenvalues of $\mathcal{A}(\lambda)$ placed in the strip $k_- \leq \Im\lambda \leq k_+$ are concentrated on the line $\Im\lambda = k_0$ with $k_0 \in (k_-, k_+)$.

We assume that the function ρ given by (12.86) vanishes at $\pm\infty$ and

$$\rho(t) \leq \rho_0 \quad \text{on} \quad \mathbb{R}, \tag{13.1}$$

where ρ_0 is a constant depending on $A_1, \cdots, A_\ell$ and $k_\pm$, k_0.

In Sect. 13.6 we show that solutions of (12.84) and their derivatives admit the representation

$$\mathrm{col}\left(u(t), \cdots, D_t^{\ell-1}u(t)\right) = \mathbf{u}(t) + \mathbf{v}(t), \tag{13.2}$$

where $\mathbf{u}$ is a vector function taking values in some κ-dimensional space, and $\mathbf{v}$ is dominated by $\mathbf{u}$ at $\pm\infty$. The dimension κ is equal to the total algebraic multiplicity of the eigenvalues of $\mathcal{A}(\lambda)$ on the line $\Im\lambda = k_0$. The vector function $\mathbf{u}$ satisfies a κ-dimensional system of ordinary differential equations perturbed by a "weak" non-local operator.

The representation (13.2) plays a crucial role in the following chapters, since it is a source of explicit asymptotic formulae for $\mathrm{col}\left(D_t^{j-1}u(t)\right)_{j=1}^\ell$ under various complementary assumptions on the perturbation $N(t, D_t)$ and the spectral properties of $\mathcal{A}(\lambda)$.

In the present chapter we derive (13.2) from the system (12.88). We show in Sect. 13.2 that (12.88) can be equivalently split into two "weakly coupled" equations (κ-dimensional and infinite dimensional) for the pair $\mathbf{u}, \mathbf{v}$. The analysis of the second equation performed in Sect. 13.3 enables one to resolve it with respect to $\mathbf{v}$, and then to estimate $\mathbf{v}$ by $\mathbf{u}$ (Sect. 13.6).

Previously we characterized the perturbation N by the function $\rho(t)$. Starting in Sect. 13.6, an important role is played by a function $\alpha(t)$ which is a more subtle characteristic of the perturbation operator. Whereas $\rho(t)$ is a norm of N, the function $\alpha(t)$ is a sum of the norms of the values of N on power-exponential zeros of $\mathcal{A}(D_t)$. Estimates for α to be used in the sequel are obtained in Sect. 13.5.

In Sect. 13.6 we arrive at the above mentioned $\kappa \times \kappa$ first order system of ordinary differential equations for u with a non-local perturbation term. This term is estimated in Sect. 13.7, where another characterization $\beta(t)$ of the perturbation operator appears. This function is similar to $\alpha(t)$ but its definition is more complicated. Estimates for β are collected in Sect. 13.8.

Starting in Sect. 13.9, we turn to the κ-dimensional system for u obtained in Sect. 13.6. In order to find concrete asymptotic representations for u in subsequent chapters, we use techniques from the asymptotic theory of ordinary differential equations. To this end it is convenient to rewrite the equation for u in a certain basis; this is done in Sect. 13.9. In the next section we consider (12.63) in a neighbourhood of $+\infty$ and reformulate previous results for this case.

In Sect. 13.11 a collection of solutions of the above mentioned finite dimensional system with zero right-hand side is constructed. Finally in Sect. 13.12 we obtain an estimate for the sum of a certain Neumann series.

13.2 Spectral Decomposition of the First Order System

In what follows, the derivation of asymptotic formulae for solutions of equation (12.84) will be based on a certain spectral splitting of the system (12.88). We describe this splitting in the present section.

Here (and elsewhere in this chapter) we assume that the requirements on the operators $\mathcal{A}(D_t)$ and $N(t, D_t)$ stated at the beginning of Introduction are satisfied.

Denote by P the Riesz projector of $\mathfrak{A}$, as given by (12.49). We transform (12.84) to the system (12.88) with

$$F(t) = \operatorname{col}\big(0, \ldots, 0, f(t)\big) \tag{13.3}$$

(as in Sect. 12.9). Applying the operators P and $I_\ell - P$ to (12.88) and using (12.52) we obtain the equations

$$\Big(D_t + \mathfrak{A} - P\mathcal{N}(t)\Big)\mathbf{u}(t) - P\mathcal{N}(t)\mathbf{v}(t) = PF(t), \tag{13.4}$$

$$\Big(D_t + \mathfrak{A} - (I_\ell - P)\mathcal{N}(t)\Big)\mathbf{v}(t) - (I_\ell - P)\mathcal{N}\mathbf{u}(t) = (I_\ell - P)F(t), \tag{13.5}$$

where

$$\mathbf{u}(t) = P\operatorname{col}\big(D_t^{j-1}u(t)\big)_{j=1}^\ell, \quad \mathbf{v}(t) = (I_\ell - P)\operatorname{col}\big(D_t^{j-1}u(t)\big)_{j=1}^\ell. \tag{13.6}$$

Theorem 13.2.1. *Let $f \in L_{2,\mathrm{loc}}(\mathbb{R})$. Then the following assertions hold:*
(i) If $u \in W_{\mathrm{loc}}^\ell(\mathbb{R})$ is a solution of (12.84) then the vector functions (13.6) belong to $\mathbb{S}_{\mathrm{loc}}(\mathbb{R})$ and satisfy (13.4) and (13.5) with F given by (13.3).
(ii) Let $\mathbf{u}, \mathbf{v} \in \mathbb{S}_{\mathrm{loc}}(\mathbb{R})$ and let

$$Pu(t) = u(t), \; Pv(t) = 0 \quad \text{for all} \;\; t \in \mathbb{R} . \tag{13.7}$$

Suppose also that the vector functions $\mathbf{u}$, $\mathbf{v}$ *satisfy* (13.4), (13.5) *with* F *defined by* (13.3). *Then there exists a solution* $u \in W_{\mathrm{loc}}^{\ell}(\mathbb{R})$ *of* (12.84) *such that* (13.2) *holds.*

Proof. (i) By the definition of $\mathbb{S}_{\mathrm{loc}}(\mathbb{R})$ (see Sect. 12.6), and by Lemma 12.6.1, the functions (13.6) belong to $\mathbb{S}_{\mathrm{loc}}(\mathbb{R})$. Applying the operators P and $I_{\ell} - P$ to both sides of the system (12.88) and using (12.52) we arrive at (13.4) and (13.5).

(ii) Summing up (13.4) and (13.5) we obtain the system (12.88), where $U(t) = \mathbf{u}(t) + \mathbf{v}(t) \in \mathbb{S}_{\mathrm{loc}}(\mathbb{R})$. Since the right-hand side F of (12.88) has the form (13.3), the first $\ell - 1$ equations in (12.88) are

$$D_t U_j(t) = U_{j+1}(t) \quad \text{for} \quad j = 1, \ldots, \ell - 1.$$

This implies $u = U_1 \in W_{\mathrm{loc}}^{\ell}(\mathbb{R})$ and $U_j(t) = D_t^{j-1} u(t)$, $j = 1, \ldots, \ell$. Therefore the last equation in (12.88) becomes (12.84). $\square$

13.3 Solvability of the Infinite-Dimensional Part of the Split System

Our aim in this section is to prove the existence and uniqueness of a solution of the system

$$\big(I_{\ell} D_t + \mathfrak{A} - (I_{\ell} - P)\mathcal{N}(t)\big)\mathcal{U}(t) = \mathcal{F}(t) \quad \text{on} \;\; \mathbb{R} \tag{13.8}$$

with $\mathcal{F} \in \mathbb{Y}_{\mathrm{loc}}(\mathbb{R})$.

Theorem 13.3.1. (*Existence*) *Let*

$$\mathcal{F}(t) = (I_{\ell} - P)\mathrm{col}\,(0, \ldots, 0, f) \tag{13.9}$$

with $f \in L_{2,\mathrm{loc}}(\mathbb{R}; H_0)$, *and let*

$$\int_0^{\infty} e^{k_- \tau} \parallel f \parallel_{L_2(\tau, \tau+1; H_0)} d\tau$$

$$+ \int_{-\infty}^0 e^{k_+ \tau} \parallel f \parallel_{L_2(\tau, \tau+1; H_0)} d\tau < \infty. \tag{13.10}$$

Then the equation (13.8) *has a solution* $\mathcal{U} \in \mathbb{X}_{\mathrm{loc}}(\mathbb{R})$, *which satisfies*

$$\parallel \mathcal{U} \parallel_{\mathbb{X}(t, t+1)} \leq c \Big\{ \int_{-\infty}^t e^{-k_+(t - \tau)} \parallel f \parallel_{L_2(\tau, \tau+1; H_0)} d\tau$$

$$+ \int_t^{\infty} e^{-k_-(t - \tau)} \parallel f \parallel_{L_2(\tau, \tau+1; H_0)} d\tau \Big\}, \tag{13.11}$$

(c *depends only on the coefficients of the pencil* $\mathcal{A}(\lambda)$ *and* $k_{\pm}$, k_0).

Proof. Since the spectrum of $\mathcal{A}(\lambda)$ in the strip $k_- \leq \Im\lambda \leq k_+$ is placed on the line $\Im\lambda = k_0$, $k_0 \in (k_-, k_+)$, the same is true for a strip $s_- \leq \Im\lambda \leq s_+$ for some $s_\pm$ such that $s_- < k_-$ and $s_+ > k_+$.

Let $\mathfrak{L}$ be the inverse operator defined at the end of Sect. 12.7 for the strip $s_- < \Im\lambda < s_+$. By (12.82)

$$\| \mathfrak{L}\mathcal{F} \|_{X(t,t+1)} \leq c \int_{\mathbb{R}} g^{(s)}(t-\tau) \| f \|_{L_2(\tau,\tau+1;H_0)} \, d\tau, \tag{13.12}$$

where

$$g^{(s)}(t) = \begin{cases} (s_+ - s_-)^{-1} e^{-s_+ t} & \text{for } t \geq 0 \\ (s_+ - s_-)^{-1} e^{-s_- t} & \text{for } t < 0 \end{cases} \tag{13.13}$$

is Green's function of the operator

$$(\partial_t + s_+)(-\partial_t - s_-) \quad \text{on} \quad \mathbb{R}$$

(compare with (12.61)).

Formally, the solution $\mathcal{U}$ of the equation (13.8) can be written as the series

$$\sum_{k=0}^{\infty} (\mathfrak{L}(I_\ell - P)\mathcal{N})^k \mathfrak{L}\mathcal{F}. \tag{13.14}$$

We introduce the sequence

$$f_k = \left(\mathcal{N} \, \mathfrak{L}\left((I_\ell - P)\mathcal{N} \, \mathfrak{L}\right)^k \mathcal{F} \right)_\ell, \quad k = 0, 1, \ldots \tag{13.15}$$

Then the series (13.14) can be rewritten as

$$\mathfrak{L}\mathcal{F} + \mathfrak{L}(I_\ell - P)\mathrm{col}(0, \ldots, \sum_{k=0}^{\infty} f_k). \tag{13.16}$$

We show that the series

$$\sum_{k=0}^{\infty} f_k \tag{13.17}$$

converges in $L_{2,\mathrm{loc}}(\mathbb{R}; H_0)$. We have $f_0 = (\mathcal{N} \, \mathfrak{L} \, \mathcal{F})_\ell$ and

$$f_k = \left(\mathcal{N} \, \mathfrak{L}(I_\ell - P)\mathrm{col}\,(0, \ldots 0, f_{k-1}) \right)_\ell, \quad k = 1, \ldots$$

By Corollary 12.9.2 (with k_+ and k_- replaced by s_+ and s_-)

$$\| f_k \|_{L_2(t,t+1;H_0)} \leq c\rho_0 \int_{\mathbb{R}} g^{(s)}(t-\tau) \| f_{k-1} \|_{L_2(\tau,\tau+1;H_0)} \, d\tau,$$

where ρ_0 is the constant in (13.1). Therefore

$$\| f_k \|_{L_2(t,t+1;H_0)}$$
$$\leq (c\rho_0)^{k+1} \int_{\mathbb{R}^{k+1}} g^{(s)}(t-\tau_1)g^{(s)}(\tau_1-\tau_2)\ldots g^{(s)}(\tau_k-\tau)$$
$$\times \| f \|_{L_2(\tau,\tau+1;H_0)} \, d\tau_1 \ldots d\tau_k d\tau, \quad k = 0, 1, \ldots \tag{13.18}$$

Denote by $\mathbf{g}_*(t)$ Green's function of the operator

$$(\partial_t + s_+)(-\partial_t - s_-) - c\rho_0 \qquad (13.19)$$

given by the Neumann series

$$\mathbf{g}_*(t - \tau) = g^{(s)}(t - \tau) + \sum_{k=1}^{\infty}(c\rho_0)^k \int_{\mathbb{R}^k} g^{(s)}(t - \tau_1)g^{(s)}(\tau_1 - \tau_2)$$

$$...g^{(s)}(\tau_k - \tau)d\tau_1\tau_2...\tau_k. \qquad (13.20)$$

From the beginning we can assume that

$$c\rho_0 < \min\left\{(s_+ - k_+)(k_+ - s_-), (s_+ - k_-)(k_- - s_-), (s_+ - s_-)^2/4\right\}, \quad (13.21)$$

where c is the same as in (13.19). Then

$$s_+(c\rho_0) > k_+ \quad \text{and} \quad s_-(c\rho_0) < k_-,$$

where $s_\pm(x)$ are given by (6.17) with $s_\pm$ instead of $k_\pm$.

Applying Proposition 6.3.1 with $k_\pm$ replaced by $s_\pm$ and $\omega = c\rho_0$ we get

$$\mathbf{g}_*(t) = o(e^{-k_\pm t}) \quad \text{as} \quad t \to \pm\infty. \qquad (13.22)$$

By (13.18) and (13.20)

$$\sum_{k=0}^{\infty}||f_k||_{L_2(t,t+1;H_0)} \le c\,\rho_0 \int_{\mathbb{R}} \mathbf{g}_*(t - \tau)\,\|\,f\,\|_{L_2(\tau,\tau+1;H_0)}\,d\tau. \qquad (13.23)$$

Hence the series (13.17) converges in $L_{2,\text{loc}}(\mathbb{R}; H_0)$ to a function Φ. Using (13.23) and (5.15) we obtain

$$\int_{\mathbb{R}} g^{(s)}(-\tau)\,\|\,\Phi\,\|_{L_2(\tau,\tau+1;H_0)}\,d\tau$$

$$\le \int_{\mathbb{R}} \mathbf{g}_*(-\tau)\,\|\,f\,\|_{L_2(\tau,\tau+1;H_0)}\,d\tau < \infty.$$

Therefore the vector function

$$(I_\ell - P)\text{col}(0, ..., \Phi)$$

belongs to the domain of $\mathfrak{L}$. Thus the expression (13.16) is well defined and we denote it by $\mathcal{U}$. The estimates (13.12) and (13.23) imply

$$\|\,\mathcal{U}\,\|_{X(t,t+1)} \le c\left(\int_{\mathbb{R}} g^{(s)}(t - \tau)\,\|\,f\,\|_{L_2(\tau,\tau+1;H_0)}\,d\tau\right.$$

$$\left. +c\rho_0 \int_{\mathbb{R}^2} g^{(s)}(t - \tau_1)\mathbf{g}_*(\tau_1 - \tau)\,\|\,f\,\|_{L_2(\tau,\tau+1;H_0)}\,d\tau_1 d\tau\right).$$

Using (5.15) (with $g = g^{(s)}$, $g_\omega = \mathbf{g}_*$ and $\omega = c\rho_0$) we can rewrite the last inequality as

$$\| \, \mathcal{U} \, \|_{\mathbb{X}(t,t+1)} \leq c \int_{\mathbb{R}} \mathbf{g}_*(t - \tau) \, \| \, f \, \|_{L_2(\tau,\tau+1;H_0)} \, d\tau$$

and by (13.22) we arrive at (13.11).

Clearly, the vector function $\mathcal{U}$ is a solution of (13.8). $\square$

13.4 Uniqueness for the Infinite-Dimensional Part of the Split System

From (13.10) and (13.11) it follows that

$$\| \, \mathcal{U} \, \|_{\mathbb{X}(t,t+1)} = o\!\left(e^{-k_\mp t}\right) \quad \text{as} \quad t \to \pm\infty. \tag{13.24}$$

(Compare with the proof of (3.17)). Now we show that the weaker condition (13.25) describes the uniqueness class for the system (13.8).

Theorem 13.4.1. (Uniqueness) Let $\mathcal{U} \in \mathbb{X}_{\mathrm{loc}}(\mathbb{R})$ satisfy (13.8) with $\mathcal{F} = 0$. If

$$\| \, \mathcal{U} \, \|_{\mathbb{X}(t,t+1)} = O\!\left(e^{-k_\mp t}\right) \quad \text{as} \quad t \to \pm\infty. \tag{13.25}$$

holds then $\mathcal{U} = 0$.

Proof. Put

$$f(t) = \big(\mathcal{N}(t)\mathcal{U}(t)\big)_\ell \tag{13.26}$$

and let $\mathcal{F}$ be given by (13.9). By (12.91), $f \in L_{2,\mathrm{loc}}(\mathbb{R})$ and

$$\| \, f \, \|_{L_2(t,t+1;H_0)} \leq c\,\rho_0 \, \| \, \mathcal{U} \, \|_{\mathbb{X}(t,t+1)} , \tag{13.27}$$

where ρ_0 is the constant in (13.1).

Let $s_\pm$ be the same numbers as in the proof of Theorem 13.3.1. Due to (13.25) and Theorem 12.8.2 we can apply Theorem 12.7.1 to the equation (12.1) with F given by (13.9), (13.26) and $k_\pm$ replaced by $s_\pm$. Using (13.27) we arrive at

$$\| \, \mathcal{U} \, \|_{\mathbb{X}(t,t+1)} \leq v(t) \tag{13.28}$$

with

$$v(t) = c\,\rho_0 \int_{\mathbb{R}} g^{(s)}(t - \tau) \, \| \, \mathcal{U} \, \|_{\mathbb{X}(\tau,\tau+1)} \, d\tau, \tag{13.29}$$

where $g^{(s)}$ is the same Green's function as in the proof of Theorem 13.3.1.

Let $h(t) = \|\mathcal{U}\|_{\mathbb{X}(t,t+1)}$. Then (13.28) can be rewritten as

$$h(t) \leq c\rho_0 \int_{\mathbb{R}} g^{(s)}(t - \tau) h(\tau) d\tau.$$

By (5.30)

$$\limsup_{\tau \to \pm\infty} \frac{\mathbf{g}_*(t-\tau)}{\mathbf{g}_*(-\tau)} \geq \frac{C}{\mathbf{g}_*(-t)}.$$

where C is a positive constant and $\mathbf{g}_*$ is Green's function of (13.19) given by (13.20). Therefore, the relation (13.25) together with (13.22) results in

$$h(t) = o\left(\limsup_{\tau \to \pm\infty} \frac{\mathbf{g}_*(t-\tau)}{\mathbf{g}_*(-\tau)} \right).$$

Applying Lemma 5.4.1(ii), with $g_\omega = \mathbf{g}_*$ and $g = g^{(s)}$, we obtain $h = 0$. $\square$

Denote by $\mathcal{D}_{\mathfrak{M}}$ the class of functions $\mathcal{F}$ expressed in the form (13.9) with $f \in L_{2,\mathrm{loc}}(\mathbb{R}; H_0)$ subject to (13.10). We introduce the operator

$$\mathfrak{M} : \mathcal{D}_{\mathfrak{M}} \to \mathbb{X}_{\mathrm{loc}}(\mathbb{R}), \tag{13.30}$$

which maps $\mathcal{F} = (I_\ell - P)\mathrm{col}\,(0, \dots, 0, f)$ to the solution $\mathcal{U} \in \mathbb{X}_{\mathrm{loc}}(\mathbb{R})$ of (13.8) subject to (13.24) (by Theorem 13.3.1 and 13.4.1 there exists one and only one such a solution). By (13.14)

$$\mathfrak{M}\mathcal{F} = \mathfrak{L} \sum_{k=0}^\infty \mathfrak{L}\Big[(I_\ell - P)\mathcal{N}\,\mathfrak{L}\Big]^k \mathcal{F}, \tag{13.31}$$

where the series on the right converges in $\mathbb{X}_{\mathrm{loc}}(\mathbb{R})$. In Theorem 13.3.1 we have obtained the estimate

$$\| \mathfrak{M}\mathcal{F} \|_{\mathbb{X}(t,t+1)} \leq c\Big\{ \int_{-\infty}^t e^{-k_+(t-\tau)} \| f \|_{L_2(\tau,\tau+1;H_0)} \, d\tau$$
$$+ \int_t^\infty e^{-k_-(t-\tau)} \| f \|_{L_2(\tau,\tau+1;H_0)} \, d\tau \Big\}. \tag{13.32}$$

This will prove useful in Sect. 13.6 and 13.7.

13.5 The Function α

Let $L_\infty(a, b; H_0)$ be the space of measurable bounded vector functions on the interval (a,b) with values in H_0 supplied with the norm

$$\operatorname*{ess\,sup}_{t \in (a,b)} \|U(t)\|_{H_0}.$$

By $L_{\infty,\mathrm{loc}}(\mathbb{R}; H_0)$ we denote the space of vector functions U which belong to $L_\infty(a, b; H_0)$ for arbitrary finite a and b, $a < b$.

We shall also use the notation $L_{\infty,\mathrm{loc}}(a, \infty; H_0)$, where $a > -\infty$, for the space of the vector functions U on (a, ∞) such that $U \in L_\infty(a, b; H_0)$ for all finite $b > a$. In the special case $\mathcal{H}_0 = \mathbb{C}$ we shall write simply $L_\infty(a, b)$, $L_{\infty,\mathrm{loc}}(\mathbb{R})$ and $L_{\infty,\mathrm{loc}}(a, \infty)$.

We introduce the function

$$\alpha(t) = \| (\mathcal{N}P)_\ell \|_{L_\infty(t,t+1;\mathcal{H}_0) \to L_2(t,t+1;\mathcal{H}_0)} . \qquad (13.33)$$

We give some equivalence relations for $\alpha(t)$ in terms of the scalar operator $N(t, D_t)$ and the special solutions $U_{kj}^{(\nu)}$ of $\mathcal{A}(D_t)U = 0$ (see (2.9)).

Consider the set of indices

$$Q = \{(\nu, k, j) : \Im\lambda_\nu = k_0, \ k = 1, \ldots, J_\nu, \ j = 0, \ldots, m_{\nu k} - 1\}. \qquad (13.34)$$

Clearly the number of indices in Q is equal to the total algebraic multiplicity κ of eigenvalues of $\mathcal{A}(\lambda)$ on the line $\Im\lambda = k_0$.

Proposition 13.5.1. *The function α is equivalent both to*

$$\alpha_1(t) = \sum_{(\nu,k,j)\in Q} \left(\int_t^{t+1} \| N(\tau, D_\tau)U_{kj}^{(\nu)}(\tau - t) \|_0^2 \, d\tau \right)^{1/2}, \qquad (13.35)$$

and to

$$\alpha_2(t) = \sum_{(\nu,k,j)\in Q} \left(\int_t^{t+1} \| N(\tau, D_x)U_{kj}^{(\nu)}(x)|_{x=0} \|_0^2 \, d\tau \right)^{1/2}. \qquad (13.36)$$

The equivalence means

$$c_1\alpha_j(t) \le \alpha(t) \le c_2\alpha_j(t) \quad \text{for} \quad t \in \mathbb{R}, \qquad (13.37)$$

where c_1 and c_2 are positive constants and $j = 1, 2$.

Proof. Let $\mathbf{u} \in L_\infty(t, t+1; \mathcal{H}_0)$. Using (12.49) and (12.31) we have

$$(\mathcal{N}\,P\,\mathbf{u})_\ell(\tau) = i \sum_{(\nu,k,j)\in Q} (\mathbf{u}(\tau), \mathcal{V}_{kj}^{(\nu)}(\tau - t))_{(H_0)^\ell} N(\tau, D_\tau)U_{kj}^{(\nu)}(\tau - t).$$

$$(13.38)$$

Hence

$$\| (\mathcal{N}\,P\,\mathbf{u})_\ell \|_{L_2(t,t+1;\mathcal{H}_0)}$$
$$\le c\,\alpha_1(t) \sum_{(\nu,k,j)\in Q} \sup_{\tau\in(t,t+1)} |(\mathbf{u}(\tau), \mathcal{V}_{kj}^{(\nu)}(\tau - t))_{(H_0)^\ell}|.$$

Since $e^{-i\bar{\lambda}_\nu x}\mathcal{V}_{kj}^{(\nu)}(x)$ is a polynomial in x with coefficients in $\mathcal{H}_0$ (see (2.10) and the definition of $\mathcal{V}_{kj}^{(\nu)}$ in Sect. 12.4), we get the right-hand inequality in (13.37) for $j = 1$.

If $\mathbf{u}(\tau) = \mathcal{U}_{kj}^{(\nu)}(\tau - t)$ then by (12.41) and (13.38) we obtain

$$\mathcal{N}\, P\, \mathbf{u}(\tau) = N(\tau, D_\tau) U_{kj}^{(\nu)}(\tau - t),$$

and we arrive at the left inequality in (13.37) for $j = 1$.

We now prove the estimates

$$c_1 \alpha_2(t) \le \alpha_1(t) \le c_2 \alpha_2(t) \quad \text{for} \quad t \in \mathbb{R}, \tag{13.39}$$

which give (13.37) for $j = 2$. Using (2.9) we get

$$N(\tau, D_\tau) U_{k, m_{\nu k} - 1 - h}^{(\nu)}(\tau - t)$$

$$= e^{i\lambda_\nu(\tau - t)} N(\tau, \lambda_\nu + D_\tau) D_\tau^h \Phi_k^{(\nu)}\big(i(\tau - t)\big) \tag{13.40}$$

$$= e^{i\lambda_\nu(\tau - t)} \sum_{s=0}^{m_{\nu k} - 1 - h} \frac{(i(\tau - t))^s}{s!} N(\tau, \lambda_\nu + D_x) D_x^{h+s} \Phi_k^{(\nu)}(i\,x)\Big|_{x=0},$$

which yields

$$N(\tau, D_\tau) U_{k,h}^{(\nu)}(\tau - t)$$

$$= e^{i\lambda_\nu(\tau - t)} \sum_{s=0}^{h} \frac{(i(\tau - t))^s}{s!} N(\tau, D_x) U_{k,h-s}^{(\nu)}(x)\Big|_{x=0}. \tag{13.41}$$

This formula gives the right-hand inequality (13.39).

Analogous calculations lead to

$$N(\tau, D_x) U_{k,h}^{(\nu)}(x)\Big|_{x=0}$$

$$= e^{i\lambda_\nu(t - \tau)} \sum_{s=0}^{h} \frac{(i(t - \tau))^s}{s!} N(\tau, D_\tau) U_{k,h-s}^{(\nu)}(\tau - t), \tag{13.42}$$

which implies the left inequality (13.39). $\square$

In what follows we shall use the notation $\approx$ to denote the equivalence relation explained in Proposition 13.5.1.

We conclude this section with one-sided estimates for α.

Proposition 13.5.2. *The following inequalities hold:*

$$\alpha(t) \ge c \,\big\| \big(N(\cdot, D_x) U_{kj}^{(\nu)}(x)\big|_{x=0}, \, V_{nq}^{(\mu)}(0)\big)_{H_0} \big\|_{L_2(t, t+1)} \tag{13.43}$$

for all (ν, k, j), (μ, n, q) in Q, and

$$\alpha(t) \le c\, \rho(t), \tag{13.44}$$

$$\alpha(t) \le c \int_{t-1}^{t+1} \alpha(\tau) d\tau. \tag{13.45}$$

Proof. The lower estimate (13.43) for α follows from (13.36) and (13.37), while (13.44) is a consequence of (13.35), (13.37) and the definition (12.86) of ρ.

In order to obtain (13.45) we put

$$h(\tau) = \sum_{(\nu,k,j)\in Q} \| N(\tau, D_x)U_{kj}^{(\nu)}(x)\Big|_{x=0} \|_0 \,.$$

We have

$$\int_{t-1}^{t+1} \| h \|_{L_2(x,x+1)}\, dx = \int_{t}^{t+1} \big(\| h \|_{L_2(x,x+1)} + \| h \|_{L_2(x-1,x)} \big) dx.$$

For $x \in (t, t+1)$ the union of the intervals $(x, x+1)$ and $(x-1, x)$ contains $(t, t+1)$. Therefore

$$\| h \|_{L_2(x,x+1)} + \| h \|_{L_2(x-1,x)} \geq c \| h \|_{L_2(t,t+1)}$$

and we arrive at

$$\| h \|_{L_2(t,t+1)} \leq c \int_{t-1}^{t+1} \| h \|_{L_2(x,x+1)}\, dx \qquad (13.46)$$

which, together with (13.36) and (13.37), implies (13.45). $\square$

13.6 Reduction of the Split System to a Finite Dimensional System

Here (and below) we denote by $g = g(t)$ Green's function of the operator

$$\big(\partial_t + k_+\big)\big(-\partial_t - k_-\big) \quad \text{on} \quad \mathbb{R}$$

defined by (12.61).

Lemma 13.6.1. *Let* $\mathbf{u} \in L_{\infty,\text{loc}}(\mathbb{R}; \mathcal{H}_0)$ *and* $P\mathbf{u}(t) = \mathbf{u}(t)$ *for all* $t \in \mathbb{R}$. *We suppose that*

$$\int_{\mathbb{R}} g(-\tau)\alpha(\tau) \| \mathbf{u} \|_{L_\infty(\tau,\tau+1;\mathcal{H}_0)}\, d\tau < \infty. \qquad (13.47)$$

Then the vector function $(I_\ell - P)\mathcal{N}\mathbf{u}$ *belongs to the domain* $\mathcal{D}_{\mathfrak{M}}$ *of the operator* $\mathfrak{M}$ *(see the end of* Sect. 13.4*) and the estimate*

$$\| \mathfrak{M}(I_\ell - P)\mathcal{N}\mathbf{u} \|_{X(t,t+1)}$$

$$\leq c \int_{\mathbb{R}} g(t - \tau)\alpha(\tau) \| u \|_{L_\infty(\tau,\tau+1;\mathcal{H}_0)}\, d\tau \qquad (13.48)$$

holds.

Proof. By definition of $\mathcal{N}$ (see (12.90))

$$\mathcal{N}\mathbf{u} = \text{col}\,(0,\ldots,0,(\mathcal{N}\mathbf{u})_\ell)$$

and by (13.33) the estimate

$$\| (\mathcal{N}\mathbf{u})_\ell \|_{L_2(t,t+1;H_0)} \leq \alpha(t) \| \mathbf{u} \|_{L_\infty(t,t+1;\mathcal{H}_0)} \tag{13.49}$$

holds. Therefore the inclusion $(I_\ell - P)\mathcal{N}\mathbf{u} \in \mathcal{D}_{\mathfrak{M}}$ follows from (13.47). The estimate (13.48) is a consequence of (13.31) and (13.49). $\square$

As a direct corollary of Lemma 13.6.1 and Theorem 13.3.1, we obtain the following assertions about solutions of (13.5).

Theorem 13.6.2. *Let* $\mathbf{u} \in L_{\infty,\text{loc}}(\mathbb{R}; \mathcal{H}_0)$, $P\mathbf{u}(t) = \mathbf{u}(t)$ *for all* $t \in \mathbb{R}$ *and let* $f \in L_{2,\text{loc}}(\mathbb{R}; H_0)$. *Suppose that*

$$\int_{\mathbb{R}} g(-\tau)\Big(\alpha(\tau) \| \mathbf{u} \|_{L_\infty(\tau,\tau+1;\mathcal{H}_0)} + \| f \|_{L_2(\tau,\tau+1;H_0)} \Big) d\tau < \infty. \tag{13.50}$$

Then (13.5), *considered as an equation for* $\mathbf{v}$, *has a solution in* $\mathbb{X}_{\text{loc}}(\mathbb{R})$ *represented by*

$$\mathbf{v} = \mathfrak{M}(I_\ell - P)(\mathcal{N}\mathbf{u} + F), \tag{13.51}$$

where $F = \text{col}\,(0,\ldots 0, f)$, *and the estimate*

$$\| \mathbf{v} \|_{\mathbb{X}(t,t+1)} \leq c \int_{\mathbb{R}} g(t - \tau)\Big(\alpha(\tau) \| \mathbf{u} \|_{L_\infty(\tau,\tau+1;\mathcal{H}_0)}$$
$$+ \| f \|_{L_2(\tau,\tau+1;H_0)} \Big) d\tau \tag{13.52}$$

holds.

By (13.50) and (12.61) the solution $\mathbf{v}$ in the above theorem satisfies

$$\| \mathbf{v} \|_{\mathbb{X}(t,t+1)} = o\big(e^{-k_\mp t}\big) \quad \text{as} \quad t \to \pm\infty. \tag{13.53}$$

(Compare with the proof of (3.17)).

Remark 13.6.3. By Proposition 2.3.5 and by the definition of $\mathbb{S}(t, t + 1)$ (see Sect. 12.6)

$$\| \mathbf{u} \|_{L_\infty(t,t+1;\mathcal{H}_0)} \leq c \| \mathbf{u} \|_{\mathbb{S}(t,t+1)}. \tag{13.54}$$

Therefore Lemma 13.6.1 and Theorem 13.6.2 remain valid if we replace $L_{\infty,\text{loc}}(\mathbb{R}; \mathcal{H}_0)$ by $\mathbb{S}_{\text{loc}}(\mathbb{R})$.

Theorem 13.6.4. *Let* u *and* f *be as* Theorem 13.6.2. *Then a solution* $\mathbf{v} \in \mathbb{X}_{\mathrm{loc}}(\mathbb{R})$ *of* (13.5) *subject to*

$$\| \mathbf{v} \|_{\mathbb{X}(t,t+1)} = O\left(e^{-k_{\mp}t}\right) \quad as \quad t \to \pm\infty. \tag{13.55}$$

is unique.

Proof. The result follows by reference to Theorem 13.4.1. □

Let f and u satisfy (13.50). Substituting (13.51) into (13.4) we get the equation for u:

$$\left(D_t + \mathfrak{A} - P\mathcal{N}(t)\right)\mathbf{u}(t) - (\mathcal{K}\mathbf{u})(t) = \mathbf{f}(t), \tag{13.56}$$

where

$$(\mathcal{K}\mathbf{u})(t) = P\,\mathcal{N}(t)\Big(\mathfrak{M}\big((I_\ell - P)\mathcal{N}\mathbf{u}\big)\Big)(t) \tag{13.57}$$

and

$$\mathbf{f}(t) = P\,F(t) + P\,\mathcal{N}(t)\Big(\mathfrak{M}\big((I_\ell - P)F\big)\Big)(t). \tag{13.58}$$

By Lemma 13.6.1 the operator $\mathcal{K}$ can be defined on

$$\mathcal{D}_{\mathcal{K}} = \Big\{\mathbf{u} \in L_{\infty,\mathrm{loc}}\big(\mathbb{R}; \mathcal{H}_0\big) : P\mathbf{u}(t) = \mathbf{u}(t) \text{ a.e. and (13.47) holds}\Big\}. \tag{13.59}$$

We always suppose that the function $f \in L_{2,\mathrm{loc}}(\mathbb{R}; H_0)$ satisfies (13.10). Then by Theorem 13.3.1 we can apply $\mathfrak{M}$ to the vector function $(I_\ell - P)F$. Hence the second term in the right-hand side of (13.58) is well defined.

Under various complementary assumptions imposed on the operators $\mathcal{A}$ and N, we show in subsequent chapters that the leading term in asymptotics of solutions to (12.84) is described by the equation (13.56). In fact, this equation is a κ-dimensional system of ordinary differential equations perturbed by the nonlocal operator $\mathcal{K}$.

In the next section we obtain a crucial estimate for $\mathcal{K}$ which shows that this operator can be considered as a weak perturbation.

In the following two theorems we describe the connection between solutions of (13.56) and (12.84).

Theorem 13.6.5. *Suppose* $f \in L_{2,\mathrm{loc}}(\mathbb{R}; H_0)$ *is subject to* (13.10). *Let* $u \in W_{\mathrm{loc}}^\ell(\mathbb{R})$ *be a solution of* (12.84) *satisfying*

$$\int_{\mathbb{R}} g(-\tau)\alpha(\tau) \| u \|_{W^\ell(\tau,\tau+1)} \, d\tau < \infty \tag{13.60}$$

and

$$\| u \|_{W^\ell(t,t+1)} = O\left(e^{-k_{\mp}t}\right) \quad as \quad t \to \pm\infty. \tag{13.61}$$

Then the vector functions u *and* v *defined by* (13.6) *have the following properties:*

(i) $\mathbf{u} \in \mathbb{S}_{\mathrm{loc}}(\mathbb{R}) \cap \mathcal{D}_{\mathcal{K}}$ *and* u *satisfies* (13.56);
(ii) v *is expressed by* (13.51).

Proof. By Theorem 13.2.1 **u** and **v** belong to $\mathbb{S}_{\mathrm{loc}}(\mathbb{R})$ and satisfy (13.4) and (13.5). The inequalities (13.10) and (13.60) imply

$$(I_\ell - P)(\mathcal{N}\,\mathbf{u} + F) \in \mathcal{D}_{\mathcal{K}}.$$

From (13.61) it follows that

$$\| \mathbf{v} \|_{\mathbb{S}(t,t+1)} = O\left(e^{-k_\mp t}\right) \quad \text{as} \quad t \to \pm\infty.$$

By using Theorems 13.3.1 and 13.4.1 we derive (13.51) from (13.5) which gives (ii).

Since $\mathbf{u} = P\mathrm{col}(D_t^{j-1}u)_{j=1}^\ell$, where $u \in W_{\mathrm{loc}}^\ell(\mathbb{R})$, we have that $\mathbf{u} \in \mathbb{S}_{\mathrm{loc}}(\mathbb{R})$. Therefore the inclusion $\mathbf{u} \in \mathcal{D}_{\mathcal{K}}$ follows from (13.60) and (13.6). In order to obtain (i), it remains to insert (13.51) into (13.4). $\square$

Theorem 13.6.6. *Let* $f \in L_{2,\mathrm{loc}}(\mathbb{R}; H_0)$ *satisfy* (13.10). *Also let* $\mathbf{u} \in \mathbb{S}_{\mathrm{loc}}(\mathbb{R})$ $\cap \mathcal{D}_{\mathcal{K}}$ *be a solution of* (13.56) *and* $P\mathbf{u}(t) = \mathbf{u}(t)$ *for all* $t \in \mathbb{R}$. *Then there exists a solution* $u \in W_{\mathrm{loc}}^\ell(\mathbb{R})$ *of* (12.84) *such that* (13.2) *holds with* **v** *given by* (13.51).

Proof. Since $\mathbf{u} \in \mathcal{D}_{\mathcal{K}}$, the inclusion

$$(I_\ell - P)(\mathcal{N}\mathbf{u} + F) \in \mathcal{D}_{\mathfrak{M}} \tag{13.62}$$

follows from (13.10). Hence the function (13.51) is well defined and satisfies (13.5). Moreover, by (13.57) and (13.58),

$$\mathcal{K}\mathbf{u} + \mathbf{f} = P\mathcal{N}\mathbf{v} + PF.$$

Inserting this in (13.56) we arrive at (13.4). Reference to Theorem 13.2.1 completes the proof. $\square$

13.7 Estimates for the Operator $\mathcal{K}$ and the Function f

Here our goal is to estimate the operator (13.57) and the function (13.58). Along with $\alpha(t)$ we shall use the function

$$\beta(t) = \| P\,\mathcal{N} \|_{\mathbb{X}(t,t+1) \to L_2(t,t+1;\mathcal{H}_0)}. \tag{13.63}$$

Its importance is explained by the following assertion.

Proposition 13.7.1. (i) *Let* $\mathcal{D}_{\mathcal{K}}$ *be defined by* (13.59). *Then* $\mathcal{K}$ *maps* $\mathcal{D}_{\mathcal{K}}$ *into* $L_{2,\mathrm{loc}}(\mathbb{R}; \mathcal{H}_0)$ *and the following estimate holds:*

$$\| \mathcal{K}\,\mathbf{u} \|_{L_2(t,t+1;\mathcal{H}_0)} \leq c\,\beta(t) \int_{\mathbb{R}} g(t - \tau)\alpha(\tau) \| \mathbf{u} \|_{L_\infty(\tau,\tau+1;\mathcal{H}_0)} d\tau. \tag{13.64}$$

(ii) *Let* $f \in L_{2,\mathrm{loc}}(\mathbb{R})$ *be subject to* (13.10). *Then the function* **f** *given by* (13.58) *satisfies*

$$\|\mathbf{f}\|_{L_2(t,t+1;\mathcal{H}_0)} \le c(\|f\|_{L_2(t,t+1;H_0)}$$
$$+\beta(t) \int_{\mathbb{R}} g(t-\tau)\|f\|_{L_2(\tau,\tau+1;H_0)}d\tau). \qquad (13.65)$$

The constant c in both inequalities depends only on $\mathcal{A}(\lambda)$, $k_\pm$ *and* k_0.

Proof. By the definitions of β and $\mathcal{K}$, the $L_2(t,t+1;\mathcal{H}_0)$–norm of left-hand side in (13.57) is majorized by

$$\beta(t) \| \mathfrak{M}(I_\ell - P)\mathcal{N} \, \mathbf{u} \|_{\mathbb{X}(t,t+1)} . \qquad (13.66)$$

Due to $P\mathbf{u} = \mathbf{u}$ and (13.33),

$$\| (\mathcal{N}\mathbf{u})_\ell \|_{L_2(\tau,\tau+1)} \le \alpha(\tau) \| \mathbf{u} \|_{L_\infty(\tau,\tau+1;\mathcal{H}_0)} . \qquad (13.67)$$

By the definition of $\mathcal{D}_\mathcal{K}$ one can apply the operator $\mathfrak{M}$ to $(I_\ell - P)\mathcal{N} \, \mathbf{u}$. The estimate (13.32) together with (13.66) and (13.67) gives (13.64).

The estimate (13.65) follows from the definition of the function β and from (13.32). $\square$

13.8 Estimates for the Function β

We start with a simple property of β.

Proposition 13.8.1. *The inequality*

$$\beta(t) \le c \int_{t-1}^{t+1} \beta(\tau)d\tau \qquad (13.68)$$

is valid.

Proof. Let

$$\mathcal{U}(\tau) = (I_\ell - P)\mathrm{col} \left(D_\tau^{j-1}v(\tau) \right)_{j=1}^{\ell} \qquad (13.69)$$

be a vector function in $\mathbb{X}(t,t+1)$. We extend v to the interval $(t-1,t+2)$ in such a way that

$$\| v \|_{W^\ell(t-1,t+2)} \le c \| v \|_{W^\ell(t,t+1)},$$

where c does not depend on v and t. We have

$$\| P\mathcal{N}\mathcal{U} \|_{L_2(t,t+1;\mathcal{H}_0)} \le \int_t^{t+1} \| P\mathcal{N}\mathcal{U} \|_{L_2(\tau-1,\tau+1;\mathcal{H}_0)} \, d\tau$$

$$\le c \int_t^{t+1} \left(\| P\mathcal{N}\mathcal{U} \|_{L_2(\tau-1,\tau;\mathcal{H}_0)} + \| P\mathcal{N}\mathcal{U} \|_{L_2(\tau,\tau+1;\mathcal{H}_0)} \right)d\tau$$

$$\le c \int_t^{t+1} \left(\beta(\tau-1) \| \mathcal{U} \|_{\mathbb{X}(\tau-1,\tau)} + \beta(\tau) \| \mathcal{U} \|_{\mathbb{X}(\tau,\tau+1)} \right)d\tau$$

$$\le c \int_{t-1}^{t+1} \beta(\tau)d\tau \| \mathcal{U} \|_{\mathbb{X}(t-1,t+2)} \le c \int_{t-1}^{t+1} \beta(\tau)d\tau \| \mathcal{U} \|_{\mathbb{X}(t,t+1)},$$

which implies (13.68). $\square$

In the following proposition we estimate β by the function ρ.

Proposition 13.8.2. *The estimate*

$$\beta(t) \le c\,\rho(t) \tag{13.70}$$

holds.

Proof. Let $\mathcal{U} \in \mathbb{X}(t, t+1)$ and

$$\|\,\mathcal{U}\,\|_{\mathbb{X}(t,t+1)} = 1.$$

By the definition of $\mathbb{X}(t, t+1)$ (see Sect. 12.6) there exists $v \in W^\ell(t, t+1)$ such that (13.69) holds and the norm of v in $W^\ell(t, t+1)$ is arbitrarily close to 1.

Since by (12.49) and (12.35)

$$(P\mathcal{N}\mathcal{U})(\tau) = i \sum_{(\nu,k,j)\in Q} \left((\mathcal{N}\mathcal{U})_\ell(\tau),\ V_{kj}^{(\nu)}(0) \right)_{\mathcal{H}_0} \mathcal{U}_{kj}^{(\nu)}(0),$$

and since the functions $\mathcal{U}_{kj}^{(\nu)}$ are linearly independent in $\mathcal{H}_0$ we obtain that β is equivalent to

$$\sup_{\|\mathcal{U}\|_{\mathbb{X}(t,t+1)}=1} \sum_{(\nu,k,j)\in Q} \left\| \left((\mathcal{N}\mathcal{U})_\ell,\ V_{kj}^{(\nu)}(0) \right)_{\mathcal{H}_0} \right\|_{L_2(t,t+1)} . \tag{13.71}$$

By (13.69) and (12.49) we obtain

$$(\mathcal{N}\mathcal{U})_\ell(\tau) = N(\tau, D_\tau) v(\tau) \tag{13.72}$$
$$-i \sum_{(\mu,n,q)\in Q} \left(\mathrm{col}\,(D_\tau^{p-1} v(\tau))_{p=1}^\ell,\ V_{nq}^{(\mu)}(0) \right)_{(\mathcal{H}_0)^\ell} N(\tau, D_x) U_{nq}^{(\mu)}(x) \Big|_{x=0}.$$

This formula together with (13.43) and (13.42) gives

$$\left\| \left((\mathcal{N}\mathcal{U})_\ell,\ V_{kj}^{(\nu)}(0) \right)_{\mathcal{H}_0} \right\|_{L_2(t,t+1)}$$
$$\le c\,\rho(t)\left(\|v\|_{W^\ell(t,t+1)} + \sum_{s=1}^\ell \| D_\tau^{s-1} v \|_{L_\infty(t,t+1;H_{\ell-s})} \right).$$

Since β is equivalent to (13.71), the estimate (13.70) follows from the last inequality and Proposition 2.3.5. $\square$

In the next proposition we give an equivalence relation for $\beta(t)$ in terms of $N(t, D_t)$, $U_{kj}^{(\nu)}(0)$ and $V_{kj}^{(\nu)}(0)$. To this end, we introduce the operator

$$\mathcal{B}(\lambda, z) = (\lambda - z)^{-1}\big(\mathcal{A}(\lambda) - \mathcal{A}(z)\big),$$

and put

$$\mathcal{B}^{(s)}(\lambda, z) = \partial_\lambda^s \mathcal{B}(\lambda, z).$$

One can verify directly that

$$\mathcal{B}(\lambda, z) = \sum_{p=1}^{\ell} E_p(\lambda) z^{p-1}, \tag{13.73}$$

where $E_p(\lambda)$ are defined by (12.22).

Proposition 13.8.3. *The relation $\beta \approx \beta_1$ is valid, where*

$$\beta_1(t) = \sup_v \sum_{(\nu,k,j)\in Q} \left(\int_t^{t+1} \left| \left(N(\tau, D_\tau)v(\tau), V_{kj}^{(\nu)}(0) \right)_{H_0} \right. \right.$$

$$- i \sum_{(\nu,n,q)} \sum_{s=0}^{q} \frac{1}{s!} \left(\mathcal{B}^{(s)}(\lambda_\mu, D_\tau)v(\tau), V_{n,q+s}^{(\mu)}(0) \right)_{H_0}$$

$$\times \left. \left(N(\tau, D_x)U_{nq}^{(\mu)}(x) \Big|_{x=0}, V_{kj}^{(\nu)}(0) \right)_{H_0} \right|^2 d\tau \right)^{1/2} \tag{13.74}$$

with the supremum taken over all

$$v \in W^\ell(t, t+1) \quad \text{with} \quad \| v \|_{W^\ell(t,t+1)} = 1. \tag{13.75}$$

Proof. It was shown in the proof of Proposition 13.8.2 that the function β is equivalent to (13.71). By using (12.34) and (12.35) we get

$$\left(\text{col } (D_\tau^{p-1} v(\tau))_{p=1}^\ell, \mathcal{V}_{nq}^{(\mu)}(0) \right)_{H_0^\ell}$$

$$= \sum_{p=1}^{\ell} \left(D_\tau^{p-1} v(\tau), E_p^*(D_x) V_{nq}^{(\mu)}(x) \Big|_{x=0} \right)_{H_0}. \tag{13.76}$$

Due to (12.43),

$$E_p^*(D_x) V_{nq}^{(\mu)}(x) \Big|_{x=0} = \sum_{s=0}^{\ell} \frac{1}{s!} E_p^{*(s)}(\overline{\lambda}_\mu) V_{n,q+s}^{(\mu)}(0).$$

Hence the right-hand side of (13.76) is equal to

$$\sum_{p=1}^{\ell} \sum_{s=0}^{\ell} \frac{1}{s!} \left(E_p^{(s)}(\lambda_\mu) D_\tau^{p-1} v(\tau), V_{n,q+s}^{(\mu)}(0) \right)_{H_0},$$

where

$$E_p^{(s)}(\lambda) = \partial_\lambda^s E_p(\lambda).$$

By (13.73)

$$\left(\mathrm{col}\,(D_\tau^{p-1}v(\tau))_{p=1}^\ell,\ \mathcal{V}_{nq}^{(\mu)}(0)\right)_{(H_0)^\ell}$$

$$=\sum_{s=0}^\ell \frac{1}{s!}\left(\mathcal{B}^{(s)}(\lambda_\mu,D_\tau)v(\tau),\ V_{n,q-s}^{(\mu)}(0)\right)_{H_0}.$$

Using this formula together with (13.72) we transform (13.71) into (13.74). $\square$

We conclude this section with an upper estimate for β; this estimate proves to be rather convenient in applications.

Proposition 13.8.4. *The inequality*

$$\beta(t) \leq c \sum_{(\nu,k,j)\in Q} \sum_{m=1}^\ell \sup_{\tau\in(t,t+1)} \|\, N_m^*(\tau)V_{kj}^{(\nu)}(0)$$

$$-i \sum_{(\mu,n,q)\in Q} \left(N(\tau,D_x)U_{nq}^{(\mu)}(x)\Big|_{x=0},\ V_{kj}^{(\nu)}(0)\right)_{H_0}$$

$$\times \sum_{s=0}^q \frac{1}{s!} E_{\ell-m+1}^{(s)*}(\overline{\lambda}_\mu)V_{n,q+s}^{(\mu)}(0)\,\|_{H_m^*}\,. \tag{13.77}$$

holds.

Proof. Denote by $S_{kj,m}^{(\nu)}(\tau)$ the vector under the sign of the H_m^*-norm in (13.77). Due to (13.73), the function (13.74) can be written as

$$\beta_1(t) = \sup_v \sum_{(\nu,k,j)\in Q} \left(\int_t^{t+1} \Big| \sum_{m=1}^\ell (D_\tau^{\ell-m}v(\tau),\ S_{kj,m}^{(\nu)}(\tau))_{H_0}\Big|^2 d\tau\right)^{1/2},$$

where the supremum is taken over (13.75). This formula along with Proposition 2.3.5 shows that $\beta_1(t)$ does not exceed the right-hand side of (13.77). Reference to Proposition 13.8.3 completes the proof. $\square$

If $\ell = 1$, the equivalence relation for β from Proposition 13.8.3 takes the form

$$\beta(t) \approx \sup_v \sum_{(\nu,k,j)\in Q} \left(\int_t^{t+1} \Big|(N(\tau)v(\tau),\psi_{kj}^{(\nu)})_{H_0}\right.$$

$$\left. - \sum_{(\mu,n,q)\in Q} (v(\tau),\psi_{nq}^{(\mu)})_{H_0}(N(\tau)\phi_{n,m_{\mu n}-1-q}^{(\mu)},\psi_{kj}^{(\nu)})_{H_0}\Big|^2 d\tau\right)^{1/2}$$

where v belongs to the set (13.75) with $\ell = 1$. Hence the estimate

$$\beta(t) \le c \sup_{\tau \in (t,t+1)} \sum_{(\nu,k,j) \in Q} \| N^*(\tau)\psi_{kj}^{(\nu)}$$

$$- \sum_{(\mu,n,q) \in Q} \left(N(\tau)\phi_{n,m_{\mu n}-1-q}, \psi_{kj}^{(\nu)} \right)_{H_0} \psi_{nq}^{(\mu)} \|_{H_1^*}$$

is valid.

13.9 The Finite Dimensional System in the Matrix Form

We fix canonical sets of Jordan chains (12.28) subject to the biorthogonality condition (12.29). Let $\mathcal{U}_{kj}^{(\nu)}$ and $\mathcal{V}_{kj}^{(\nu)}$ be the same functions as in Sect. 12.4. Since the image of the Riesz projector P (see (12.49)) coincides with the linear hull of $\{\mathcal{U}_{k,j}^{(\nu)}(0)\}$ where $(\nu,k,j) \in Q$, the vector function

$$\mathbf{u} = P \operatorname{col}(u, D_t u, \ldots, D_t^{\ell-1} u)$$

can be represented as

$$\mathbf{u}(t) = \sum_{(\nu,k,j) \in Q} h_{kj}^{(\nu)}(t)\mathcal{U}_{kj}(0). \tag{13.78}$$

Here the coefficients $h_{kj}^{(\nu)}$ are in the space $W_{2,\mathrm{loc}}^1(\mathbb{R})$ of scalar functions which, together with their first derivatives, belong to $L_{2,\mathrm{loc}}(\mathbb{R})$. The number of the triples (ν,k,j) in (13.78) is equal to the total algebraic multiplicity of the eigenvalues of $\mathcal{A}(\lambda)$ on the line $\Im\lambda = k_0$.

In order to obtain an equation for the coefficients $h_{kj}^{(\nu)}$ it is convenient to use the following block matrix and block vector notations: The matrix

$$M = \{M_{\nu k}^{\mu n}\}$$

is formed by $m_{\mu n} \times m_{\nu k}$ matrices $M_{\nu k}^{\mu n}$. The indices ν, μ numerate the eigenvalues λ_ν, λ_μ of $\mathcal{A}(\lambda)$ on the line $\Im\lambda = k_0$, whereas k, n numerate the eigenvectors

$$\mathcal{U}_{k0}^{(\nu)}(0), \quad \mathcal{U}_{n0}^{(\mu)}(0)$$

corresponding to these eigenvalues (see Proposition 12.4.2). The numbers $m_{\nu k}$ are the partial multiplicities of λ_ν. In accordance with the definition of the matrix M, we represent an arbitrary vector $h \in \mathbb{C}^\kappa$ in the block form $\{h_k^{(\nu)}\}$, where

$$h_k^{(\nu)} = \operatorname{col}\left(h_{kj}^{(\nu)}\right)_{j=0}^{m_{\nu k}-1}.$$

We introduce the matrices

$$\Lambda = \operatorname{diag}(\lambda_\nu I_{\nu k}), \quad J = \operatorname{diag}(J_{\nu k}), \tag{13.79}$$

where $I_{\nu k}$ is the identity matrix of size $m_{\nu k}$ and $J_{\nu k}$ is the square matrix of the same size, whose elements on the first diagonal above the leading diagonal are equal to 1 and zero otherwise. Let $\mathcal{R}$ be the block matrix

$$\{\mathcal{R}_{\mu n}^{\nu k}\},$$

where $\mathcal{R}_{\mu n}^{\nu k}$ is the rectangular matrix of size $m_{\nu k} \times m_{\mu n}$ with the elements

$$i\Big(\mathcal{N}(t)\mathcal{U}_{kj}^{(\nu)}(0),\, \mathcal{V}_{nq}^{(\mu)}(0)\Big)_{(H_0)^\ell}$$

or equivalently (by (12.31) and (12.35)),

$$\mathcal{R}_{\mu n}^{\nu k} = i\Big(N(t, D_\tau)\mathcal{U}_{kj}^{(\nu)}(\tau)\Big|_{\tau=0},\, \mathcal{V}_{nq}^{(\mu)}(0)\Big)_{H_0}. \tag{13.80}$$

Inserting (13.78) in (13.56), then multiplying by $\mathcal{V}_{nq}^{(\mu)}(0)$ in $(H_0)^\ell$ and using (12.41) for $\tau = 0$, we obtain the system

$$-iD_t h_{nq}^{(\mu)}(t) + \sum_{(\nu,k,j)\in Q} h_{kj}^{(\nu)}\Big(((\mathfrak{A} - P\mathcal{N}(t))\mathcal{U}_{kj}^{(\nu)}(0),\, \mathcal{V}_{nq}^{(\mu)}(0))_{(H_0)^\ell}$$
$$-((\mathcal{K}h_{kj}^{(\nu)}\mathcal{U}_{kj}^{(\nu)}(0))(t),\, \mathcal{V}_{nq}^{(\mu)}(0))_{(H_0)^\ell}\Big) = \big(q(t),\, \mathcal{V}_{nq}^{(\mu)}(0)\big)_{(H_0)^\ell}.$$

By (12.44), (12.45) and (12.41) this system can be written as the matrix equation

$$\Big(D_t - \Lambda - J - \mathcal{R}(t)\Big)h(t) - (Kh)(t) = q(t). \tag{13.81}$$

In this equation, the block components of q are

$$q_n^{(\mu)} = \mathrm{col}\Big(q_{np}^{(\mu)}\Big)_{p=1}^{m_{\mu n}-1} \tag{13.82}$$

with

$$q_{np}^{(\mu)}(t) = i\Big(f(t),\, V_{np}^{(\mu)}(0)\Big)_{H_0}$$
$$+i\Big((\mathcal{N}\,\mathfrak{M}(I_\ell - P)F)_\ell(t),\, V_{np}^{(\mu)}(0)\Big)_{H_0}, \tag{13.83}$$

where F is given by (13.3). The operator K in (13.81) is a matrix form of the operator (13.57), i.e.

$$\Big((K\,h)(t)\Big)_{np}^{(\mu)} = i \sum_{(\nu,k,j)\in Q} \Big(\mathcal{K}_{\tau \to t}(h_{kj}^{(\nu)}(\tau)\mathcal{U}_{kj}^{(\nu)}(0)),\, \mathcal{V}_{np}^{(\mu)}(0)\Big)_{(H_0)^\ell}. \tag{13.84}$$

By using (13.59) we can define K on the set of κ-dimensional vector functions with components in $L_{\infty,\mathrm{loc}}(\mathbb{R})$ subject to

$$\int_{\mathbb{R}} g(-\tau)\alpha(\tau)\, \|\, h\,\|_{(L_\infty(\tau,\tau+1))^\kappa}\, d\tau < \infty. \tag{13.85}$$

It will be denoted by $\mathcal{D}_K$. The only information about K we need is the following estimate, which is equivalent to (13.64):

$$\|Kh\|_{(L_2(t,t+1))^\kappa} \leq c\,\beta(t)\int_{\mathbb{R}} g(t-\tau)\alpha(\tau)\,\|\,h\,\|_{(L_\infty(\tau,\tau+1))^\kappa}\,d\tau. \qquad (13.86)$$

The inequality (13.65) implies

$$\|\,q\,\|_{(L_2(t,t+1))^\kappa} \leq c\Big(\,\|\,f\,\|_{L_2(t,t+1;H_0)}$$
$$+\beta(t)\int_{\mathbb{R}} g(t-\tau)\,\|\,f\,\|_{L_2(\tau,\tau+1;H_0)}\,d\tau\Big). \qquad (13.87)$$

We now reformulate Theorems 13.6.5 and 13.6.6 with respect to the system (13.81).

Theorem 13.9.1. *Suppose that* $f \in L_{2,\mathrm{loc}}(\mathbb{R}; H_0)$ *is subject to (13.10). Let* $u \in W^\ell_{\mathrm{loc}}(\mathbb{R})$ *be a solution of (12.84) satisfying (13.60) and (13.61) and let*

$$h^{(\nu)}_{kj}(t) = i\left(\mathrm{col}\,(D^{s-1}_t u(t))^\ell_{s=1},\ \mathcal{V}^{(\nu)}_{kj}(0)\right)_{(H_0)^\ell}. \qquad (13.88)$$

Then

$$\mathrm{col}\,\big(u(t), D_t u, \dots, D^{\ell-1}_t u(t)\big) = \sum_{(\nu,k,j)\in Q} h^{(\nu)}_{kj}(t)\,\mathcal{U}^{(\nu)}_{kj}(0) + \mathbf{v}(t), \qquad (13.89)$$

where the scalar functions $h^{(\nu)}_{kj}$ *are subject to*

$$\int_{\mathbb{R}} g(-\tau)\alpha(\tau)\,\|\,h^{(\nu)}_{kj}\,\|_{L_\infty(\tau,\tau+1)}\,d\tau < \infty.$$

and satisfy (13.81).
   The remainder $\mathbf{v}$ *in (13.89) admits the estimate*

$$\|\,\mathbf{v}\,\|_{X(t,t+1)} \leq c\int_{\mathbb{R}} g(t-\tau)\Big(\alpha(\tau)\sum_{(\nu,k,j)\in Q}\|\,h^{(\nu)}_{kj}\,\|_{L_\infty(\tau,\tau+1)}$$
$$+\,\|\,f\,\|_{L_2(\tau,\tau+1;H_0)}\,\Big)d\tau\,. \qquad (13.90)$$

The converse statement has the form

Theorem 13.9.2. *Let* $f \in L_{2,\mathrm{loc}}(\mathbb{R}; H_0)$ *satisfy (13.10). Also let* $h \in \big(W^1_{2,\mathrm{loc}}(\mathbb{R})\big)^\kappa \cap \mathcal{D}_K$ *be a solution of (13.81). Then there exists a solution* $u \in W^\ell_{\mathrm{loc}}(\mathbb{R})$ *of (12.84) such that (13.89) holds with*

$$\mathbf{v}(t) = \Big(\mathfrak{M}(I_\ell - P)(\mathcal{N}\sum_{(\nu,k,j)\in Q} h^{(\nu)}_{kj}U^{(\nu)}_{kj}(0) + F)\Big)(t), \qquad (13.91)$$

where F *is given by (13.3).*

13.10 Operator Differential Equations on a Semiaxis

13.10.1 Preliminaries

As in previous sections we suppose that all the eigenvalues of $\mathcal{A}(\lambda)$ in the strip $k_- \leq \Im\lambda \leq k_+$ are concentrated on the line $\Im\lambda = k_0$, $k_0 \in (k_-, k_+)$.

We consider the equation

$$\Big(\mathcal{A}(D_t) - N(t, D_t)\Big)u(t) = f(t) \quad \text{for} \quad t > t_0, \tag{13.92}$$

where $f \in L_{2,\text{loc}}(t_0, \infty; H_0)$. We assume that the function ρ defined by (12.86) satisfies

$$\rho(t) \to 0 \quad \text{as} \quad t \to +\infty. \tag{13.93}$$

The number t_0 in (13.92) is supposed to be sufficiently large. This should guarantee a smallness of ρ on $(t_0, +\infty)$.

It will be convenient to assume that $N(t, D_t) = 0$ and $f(t) = 0$ for $t < t_0$. Clearly, $\alpha(t)$, $\beta(t)$ and $\rho(t)$ vanish for $t \leq t_0 - 1$. We need the following simple property of α, β and ρ.

Lemma 13.10.1. *For $t \in (t_0, t_0 - 1)$ the estimates*

$$\alpha(t) \leq \alpha(t_0), \quad \beta(t) \leq c\beta(t_0) \quad \text{and} \quad \rho(t) \leq c\rho(t_0) \tag{13.94}$$

are valid. Here the constant c depends only on ℓ.

Proof. Let U be an arbitrary vector function in $L_\infty(t, t+1; H_0)$ and let $\mathring{U}$ be the extension of U by zero to the semiaxis $(t+1, \infty)$. By (13.33) and by $\mathcal{N}(\tau) = 0$ for $\tau < t_0$ we have

$$\|(\mathcal{N}P)_\ell U\|_{L_2(t,t+1;H_0)} = \|(\mathcal{N}P)_\ell \mathring{U}\|_{L_2(t_0, t_0+1; H_0)}$$
$$\leq \alpha(t_0)\|\mathring{U}\|_{L_\infty(t_0, t_0+1; H_0)} \leq \alpha(t_0)\|U\|_{L_\infty(t, t+1; H_0)}.$$

Hence, $\alpha(t) \leq \alpha(t_0)$.

The required estimates for β and ρ can be proved similarly if, instead of the extension by zero, one uses the extension

$$W^\ell(t, t+1) \ni u \to u^* \in W^\ell(t_0, t_0 + 1)$$

such that

$$\|u^*\|_{W^\ell(t_0, t_0+1)} \leq c\|u\|_{W^\ell(t, t+1)}.$$

$\square$

This lemma shows, in particular, that ρ satisfies (13.1).

Consider the equation

$$\Big(D_t + \mathfrak{A} - P\,\mathcal{N}(t)\Big)\mathbf{u}(t) - (\mathcal{K}\mathbf{u})(t) = \mathbf{f}(t) \quad \text{for} \quad t > t_0, \tag{13.95}$$

where $\mathbf{f}$ is given by (13.58) with F (or, equivalently, f) extended by zero for $t < t_0$. Using $\mathcal{N}(t) = 0$ for $t < t_0$ we deduce from (13.57) that

$$\mathcal{K} = \chi_0 \mathcal{K} \chi_0 ,$$

where χ_0 is the characteristic function of the semiaxis $t > t_0$. Since $\mathcal{K}\mathbf{u}$ does not depend upon the values of $\mathbf{u}$ on $(-\infty, t_0)$, we can also consider $\mathcal{K}$ as defined on the set $\mathcal{D}_\mathcal{K}(t_0, \infty)$ of functions given only for $t > t_0$ and such that $\mathbf{u} \in L_{\infty,\mathrm{loc}}(t_0, \infty; \mathcal{H}_0)$ and

$$\int_{t_0}^{\infty} e^{k-\tau} \alpha(\tau) \parallel \mathbf{u} \parallel_{L_\infty(\tau,\tau+1;\mathcal{H}_0)} d\tau < \infty. \qquad (13.96)$$

By (13.64) we have

$$\parallel \mathcal{K}\mathbf{u} \parallel_{L_2(t,t+1;\mathcal{H}_0)}$$
$$\leq c\,\beta(t) \int_{t_0-1}^{\infty} g(t-\tau)\alpha(\tau) \parallel \mathbf{u} \parallel_{L_\infty(\tau,\tau+1;\mathcal{H}_0)} d\tau. \qquad (13.97)$$

Here we assume that $\mathbf{u}$ in the right-hand side is extended by zero for $t < t_0$. From (13.65) it follows that

$$\|\mathbf{f}\|_{L_2(t,t+1;\mathcal{H}_0)} \leq c\big(\|f\|_{L_2(t,t+1;H_0)}$$
$$+ \beta(t) \int_{t_0-1}^{\infty} g(t-\tau)\|f\|_{L_2(\tau,\tau+1;H_0)} d\tau\big). \qquad (13.98)$$

13.10.2 An Existence Result

An analog of Theorem 13.6.6 reads as follows

Theorem 13.10.2. *Let $f \in L_{2,\mathrm{loc}}(t_0, \infty; H_0)$ be subject to*

$$\int_{t_0}^{\infty} e^{k-\tau} \parallel f \parallel_{L_2(\tau,\tau+1;H_0)} d\tau < \infty. \qquad (13.99)$$

Also let

$$\mathbf{u} \in \mathbb{S}_{\mathrm{loc}}(t_0, \infty) \cap \mathcal{D}_\mathcal{K}(t_0, \infty)$$

be a solution of (13.95) with $\mathbf{f}$ defined by (13.58) (where $F(t) = 0$ for $t < t_0$) and

$$P\mathbf{u}(t) = \mathbf{u}(t)$$

for $t > t_0$. Then there exists a solution $u \in W^\ell_{\mathrm{loc}}(t_0, \infty)$ of (13.92) such that (13.2) holds for $t > t_0$ with $\mathbf{v}$ given by (13.51), where F is defined by (13.3). Moreover, $\mathbf{v}$ satisfies the estimate

$$\parallel \mathbf{v} \parallel_{\mathbb{S}(t,t+1)} \leq c \int_{t_0-1}^{\infty} g(t-\tau)\big(\alpha(\tau) \parallel \mathbf{u} \parallel_{L_\infty(\tau,\tau+1;\mathcal{H}_0)}$$
$$+ \parallel f \parallel_{L_2(\tau,\tau+1;H_0)} \big) d\tau \qquad (13.100)$$

(In this formula $\mathbf{u}(t) = 0$ and $f(t) = 0$ for $t < t_0$.)

Proof. The inclusion (13.62) follows from (13.96) and (13.99). Therefore, the function $\mathbf{v}$ given by (13.51) is well defined and satisfies (13.5). The estimate (13.100) is a corollary of (13.52) and Lemma 12.6.2. The equation (13.95) implies (13.4) for $t > t_0$. Summing up (13.4) and (13.5) we arrive at (12.88) on the semiaxis $t > t_0$. Since F has the form (13.3) we obtain

$$U = \mathbf{u} + \mathbf{v} = \mathrm{col}(u, D_t u, ..., D_t^{\ell-1} u)$$

with u satisfying (13.92). Since $\mathbf{u}, \ \mathbf{v} \in \mathbb{S}_{\mathrm{loc}}(t_0, \infty)$, it follows that $u \in W_{\mathrm{loc}}^{\ell}(t_0, \infty)$. $\square$

13.10.3 A Representation for Solutions to (13.92)

Let ζ be a smooth function such that $\zeta(t) = 1$ for $t > t_0 + 1$ and $\zeta(t) = 0$ for $t < t_0$. We rewrite (13.92) as

$$\Big(\mathcal{A}(D_t) - N(t, D_t)\Big)(\zeta u)(t) = f_1(t), \quad t \in \mathbb{R}, \tag{13.101}$$

where

$$f_1(t) = (\zeta f)(t) + \big[\mathcal{A}(D_t) - N(t, D_t), \zeta(t)\big]u(t). \tag{13.102}$$

Applying Theorem 13.6.5 to the equation (13.101) we arrive at

Theorem 13.10.3. *Suppose that* $f \in L_{2,\mathrm{loc}}(t_0, \infty; H_0)$ *is subject to* (13.99). *Let* $u \in W_{\mathrm{loc}}^{\ell}(t_0, \infty)$ *be a solution of* (13.92) *satisfying*

$$\int_{t_0}^{\infty} e^{k-\tau} \alpha(\tau) \, \| u \|_{W^{\ell}(\tau, \tau+1)} \, d\tau < \infty \tag{13.103}$$

and

$$\| u \|_{W^{\ell}(t, t+1)} = O\big(e^{-k-t}\big) \quad as \quad t \to +\infty. \tag{13.104}$$

Then the vector functions $\mathbf{u}$ *and* $\mathbf{v}$ *defined by* (13.6) *(with u replaced by ζu) have the following properties:*

(i) $\mathbf{u} \in \mathbb{S}_{\mathrm{loc}}(\mathbb{R}) \cap \mathcal{D}_{\mathcal{K}}$, $\mathbf{u}(t) = 0$ *for* $t < t_0$, *and* $\mathbf{u}$ *satisfies* (13.56), *where* $\mathbf{f}$ *is given by* (13.58) *with*

$$F(t) = \mathrm{col}\,(0, \dots, 0, f_1(t)). \tag{13.105}$$

The function $\mathbf{f}$ *is subject to*

$$\| \mathbf{f} \|_{(L_2(t, t+1; \mathcal{H}_0))^{\kappa}} \le c\Big(\| f \|_{L_2(t, t+1; H_0)} \tag{13.106}$$

$$+ \beta(t) \int_{t_0-1}^{\infty} g(t-\tau) \, \| f \|_{L_2(\tau, \tau+1; H_0)} \, d\tau + e^{-k+t} \, \| u \|_{W^{\ell-1}(t_0, t_0+1)} \Big).$$

(ii) *The vector function* $\mathbf{v}$ *is expressed by* (13.51) *with* F *given by* (13.105). *Moreover,* $\mathbf{v}(t)$ *equals zero for* $t < t_0$ *and satisfies*

$$\| \mathbf{v} \|_{\mathbb{S}(t, t+1)} \le c\Big(\int_{t_0-1}^{\infty} g(t-\tau) \big(\alpha(\tau) \, \| u \|_{(L_{\infty}(\tau, \tau+1; \mathcal{H}_0))^{\kappa}}$$

$$+ \| f \|_{L_2(\tau, \tau+1; H_0)} \big) d\tau + e^{-k+t} \, \| u \|_{W^{\ell-1}(t_0, t_0+1)} \Big). \tag{13.107}$$

13.10.4 The Matrix Form of System (13.95)

Using the notation from Sect. 13.9 we write (13.95) in the matrix form

$$(D_t - \Lambda - J - \mathcal{R}(t))h(t) - (K\ h)(t) = q(t) \quad \text{for} \quad t > t_0. \tag{13.108}$$

The operator K is defined by (13.84) on the set $\mathcal{D}_K(t_0, \infty)$ of vector functions

$$h \in \left(L_{\infty,\mathrm{loc}}(t_0, \infty)\right)^{\kappa}$$

satisfying

$$\int_{t_0}^{\infty} e^{k-\tau} \alpha(\tau) \parallel h \parallel_{(L_{\infty}(\tau,\tau+1))^{\kappa}} d\tau < \infty \tag{13.109}$$

and

$$\parallel Kh \parallel_{(L_2(t,t+1))^{\kappa}} \le c\,\beta(t) \int_{t_0-1}^{\infty} g(t-\tau)\alpha(\tau) \parallel h \parallel_{(L_{\infty}(\tau,\tau+1))^{\kappa}} d\tau. \tag{13.110}$$

(Here we extend h by 0 to the left of t_0.) The right-hand side $q(t)$ is introduced by (13.82) and (13.83), where F is extended by 0 to the semiaxis $t < t_0$. The inequality (13.87) leads to

$$\parallel q \parallel_{(L_2(t,t+1))^{\kappa}} \le c\Big(\parallel f \parallel_{L_2(t,t+1;H_0)} \tag{13.111}$$
$$+\beta(t) \int_{t_0-1}^{\infty} g(t-\tau) \parallel f \parallel_{L_2(\tau,\tau+1;H_0)} d\tau\Big),$$

where $f(t) = 0$ for $t < t_0$.

We shall use the notation $W_{2,\mathrm{loc}}^1(t_0, \infty)$ for the space of scalar functions on (t_0, ∞) which belong, together with their first derivative, to $L_2(t_0, t_1)$ for all finite $t_1 > t_0$.

Obviously Theorems 13.10.2 and 13.10.3 can be rewritten in the following way.

Theorem 13.10.4. *Let $f \in L_{2,\mathrm{loc}}(t_0, \infty; H_0)$ satisfy (13.99). Furthermore, let*

$$h \in \left(W_{2,\mathrm{loc}}^1(t_0, \infty)\right)^{\kappa} \cap \mathcal{D}_K(t_0, \infty)$$

be a solution of (13.108) with q given by (13.82) and (13.83), where $F(t) = \mathrm{col}\,(0, \ldots, 0, f(t))$. Then there exists a solution

$$u \in W_{\mathrm{loc}}^{\ell}(t_0, \infty)$$

of (13.92) such that (13.89) holds for $t > t_0$ with $\mathbf{v}$ determined by (13.91). Hence, q satisfies (13.111) and $\mathbf{v}$ is estimated by

$$\parallel \mathbf{v} \parallel_{\mathbb{S}(t,t+1)} \le c \int_{t_0-1}^{\infty} g(t-\tau)\Big(\alpha(\tau) \parallel h \parallel_{(L_{\infty}(\tau,\tau+1))^{\kappa}}$$
$$+ \parallel f \parallel_{L_2(\tau,\tau+1;H_0)} \Big)d\tau. \tag{13.112}$$

Theorem 13.10.5. *Let all assumptions on f and u from Theorem 13.10.3 be fulfilled. Then the vector functions h and $\mathbf{v}$ defined by (13.88) and (13.6) (where u is replaced by ζu) have the following properties:*
    (i) $h \in \left(W^1_{2,\mathrm{loc}}(\mathbb{R})\right)^\kappa \cap \mathcal{D}_K$, $h(t) = 0$ for $t < t_0$, and h satisfies (13.108), where q is given by (13.82) and (13.83) with F determined by (13.105) and f is replaced by f_1 given by (13.102). The estimate

$$\| \, q \, \|_{(L_2(t,t+1))^\kappa} \leq c\Big(\, \| \, f \, \|_{L_2(t,t+1;H_0)} \tag{13.113}$$

$$+ \beta(t) \int_{t_0-1}^\infty g(t-\tau) \, \| \, f \, \|_{L_2(\tau,\tau+1;H_0)} \, d\tau + e^{-k_+t} \, \| \, u \, \|_{W^{\ell-1}(t_0,t_0+1)} \Big)$$

holds.
    (ii) $\mathbf{v}$ *is expressed by (13.91) with F given by (13.105), $\mathbf{v}(t) = 0$ for $t < t_0$ and*

$$\| \, \mathbf{v} \, \|_{S(t,t+1)} \leq c\Big(\int_{t_0-1}^\infty g(t-\tau)\big(\alpha(\tau) \, \| \, h \, \|_{(L_\infty(\tau,\tau+1))^\kappa}$$

$$+ \, \| \, f \, \|_{L_2(\tau,\tau+1;H_0)} \, \big)d\tau + e^{-k_+t} \, \| \, u \, \|_{W^{\ell-1}(t_0,t_0+1)} \Big). \tag{13.114}$$

13.11 Properties of Zeros
of the Finite Dimensional System

13.11.1 The System on $\mathbb{R}$

Consider the equation

$$\Big(D_t - \Lambda - J - \mathcal{R}(t)\Big)h(t) - (Kh)(t) = 0 \quad \text{on} \quad \mathbb{R}. \tag{13.115}$$

Here we use the same notation as in Sect. 13.9.

Theorem 13.11.1. *Equation (13.115) has κ linearly independent solutions*

$$h^{(1)}, \ldots, h^{(\kappa)} \in \left(W^1_{2,\mathrm{loc}}(\mathbb{R})\right)^\kappa \tag{13.116}$$

with the following properties:
    (a) *For every $\varepsilon > 0$, the inequality holds:*

$$\| \, h^{(n)} \, \|_{(W^1_2(t,t+1))^\kappa} \leq c_\varepsilon e^{-(k_0 \mp \varepsilon)t} \quad \text{for} \quad t \gtrless 0. \tag{13.117}$$

    (b) *Let $h \in \left(W^1_{2,\mathrm{loc}}(\mathbb{R})\right)^\kappa$ be a non-trivial solution of (13.115) subject to (13.85) and*

$$\| \, h \, \|_{(W^1_2(t,t+1))^\kappa} \leq c \, e^{-k_\mp t} \quad \text{for} \quad t \gtrless 0. \tag{13.118}$$

Then

$$h = \sum_{j=1}^{\kappa} c_j \, h^{(j)}, \tag{13.119}$$

where $c_j, \ldots, c_\kappa$ are constants, and

$$\limsup_{t \to \pm\infty} e^{(k_0 \pm \varepsilon)t} \, \| \, h \, \|_{(W_2^1(t,t+1))^\kappa} > 0 \tag{13.120}$$

for all $\varepsilon > 0$.

Proof. Let $u_1, \ldots, u_\kappa$ be a basis in $\mathfrak{X}(L)$ (see Proposition 9.2.2, where $k_-^{(1)} < k_-$, $k_+^{(1)} = k_-^{(2)} = k_0$ and $k_+^{(2)} > k_+$). Put

$$\mathbf{u_n}(t) = P \, \mathrm{col} \, \big(u_n(t), \ldots, D_t^{\ell-1} u_n(t) \big), n = 1, \ldots, \kappa.$$

Then

$$\mathbf{u_n}(t) = \sum_{(\nu,k,j) \in Q} \big(h^{(n)} \big)_{kj}^{(\nu)}(t) \, \mathcal{U}_{kj}^{(\nu)}(0), \tag{13.121}$$

where $h^{(n)}$ is given by

$$\big(h^{(n)} \big)_{kj}^{(\nu)}(t) = i \Big(\, \mathrm{col} \, (D_t^{s-1} u_n(t))_{s=1}^{\ell}, \, \mathcal{V}_{kj}^{(\nu)}(0) \Big)_{(H_0)^\ell} \tag{13.122}$$

Clearly, $h^{(n)} \in (W_{2,\mathrm{loc}}^1(\mathbb{R}))^\kappa$. Furthermore, $h^{(n)}$ satisfies (13.115) by Theorem 13.9.1. The estimate (13.117) follows from (9.5).

Let us prove the linear independence of $\{h^{(n)}\}$, or equivalently, the linear independence of $\{\mathbf{u_n}\}$. Suppose that

$$\sum_{j=1}^{\kappa} c_j \mathbf{u}_j = 0$$

and put $u = c_1 u_1 + \ldots + c_\kappa u_\kappa$. Then by Theorem 13.2.1 the vector function

$$\mathbf{v}(t) = (I - P) \, \mathrm{col} \, \big(u(t), \ldots, D^{\ell-1} u(t) \big)$$

satisfies (13.8) with $F = 0$. Moreover, $\mathbf{v}$ is subject to (13.53) due to the estimate (9.5) for u_n. Using Theorem 13.6.4 we get $\mathbf{v} = 0$ and hence $u = 0$. Since $u_1, \ldots, u_\kappa$ are linearly independent, we have $c_1 = \ldots = c_\kappa = 0$.

It remains to prove statement (b). By Theorem 13.9.2 there exists a solution $u \in W_{\mathrm{loc}}^\ell(\mathbb{R})$ of (12.84) with $f = 0$ such that (13.89) holds. From (13.118), (13.85) and (13.90), it follows that u satisfies (9.4). Applying Proposition 9.2.2 and Corollary 9.2.4 we obtain that

$$u(t) = \sum_{n=1}^{\kappa} c_n \, u_n(t), \tag{13.123}$$

and that u is subject to (9.12). The representation (13.123) implies (13.119). The lower estimate (9.12) combined with (13.89) together with (13.90) yields

$$\| h \|_{(W_2^1(t,t+1))^\kappa} + c \int_{\mathbb{R}} g(t-\tau)\alpha(\tau) \| h \|_{(W_2^1(t,t+1))^\kappa} \, d\tau$$

$$\geq c\, e^{-(k_0\pm\varepsilon)t}, \ c > 0, \tag{13.124}$$

for large positive $\pm t$. By assuming that (13.120) fails, we can use the boundedness of α to arrive at a contradiction with (13.124). $\square$

13.11.2 The System on a Semiaxis

Consider the system

$$\bigl(D_t - \Lambda - J - \mathcal{R}(t)\bigr)h(t) - (K\,h)(t) = 0 \quad \text{for} \ \ t > t_0. \tag{13.125}$$

Theorem 13.11.2. *The equation* (13.125) *has κ linear independent solutions*

$$h^{(1)}, \dots, h^{(\kappa)} \in \left(W_{2,\mathrm{loc}}^1(t_0,\infty) \right)^\kappa \tag{13.126}$$

with the following properties.
   (a) *For every $\varepsilon > 0$*

$$\| h^{(n)} \|_{(W_2^1(t,t+1))^\kappa} \leq c_\varepsilon e^{-(k_0-\varepsilon)t} \quad \text{for} \ \ t > t_0. \tag{13.127}$$

   (b) *Let $c_1, \dots, c_\kappa$ be constants such that*

$$\left\| \sum_{n=1}^{\kappa} c_n h^{(n)} \right\|_{(W_2^1(t,t+1))^\kappa} \leq c\, e^{-(k_0+\varepsilon)t} \quad \text{for} \ \ t > t_0 \tag{13.128}$$

with some $\varepsilon > 0$. Then $c_n = 0$, $n = 1, \dots, \kappa$
   (c) *Let*

$$h \in \left(W_{2,\mathrm{loc}}^1(t_0,\infty) \right)^\kappa$$

be a solution of (13.125) *satisfying* (13.109) *and*

$$\| h \|_{(W_2^1(t,t+1))^\kappa} \leq c\, e^{-k_- t} \quad \text{for} \ \ t > t_0. \tag{13.129}$$

Then there exist constants $c_1, \dots, c_\kappa$ such that

$$\left\| h - \sum_{n=1}^{\kappa} c_n \, h^{(n)} \right\|_{(W_2^1(t,t+1))^\kappa} \leq c\, e^{-k_+ t} \quad \text{for} \ \ t > t_0. \tag{13.130}$$

Proof. By Corollary 9.3.5 there exists a κ–dimensional space $\mathbf{X}(L)$ of solutions to (13.92) with $f = 0$ whose non-zero elements have the characteristic exponent $-k_0$. This space satisfies the assumptions of Theorem 12.2.3 and, therefore the assertions (i) and (ii) of the same theorem hold. Let $u_1, \dots, u_\kappa$ be a basis in $\mathbf{X}(L)$. Put

$$\mathbf{u}_n(t) = P \, \mathrm{col}\, \bigl(u_n(t), \dots, D_t^{\ell-1} u_n(t)\bigr).$$

We represent $\mathbf{u}_n$ in the form (13.121) for $t > t_0$ with $h^{(n)}$ given by (13.122). It is clear that $h^{(n)}$ satisfies (13.126) and (13.125).

Inequality (13.127) follows from the right-hand estimate in (12.12) and (13.122). To prove (b) we introduce the function

$$u(t) = \sum_{j=1}^{\kappa} c_j\, u_j(t).$$

By Theorem 13.10.3 and by (13.128) the function u admits the estimate

$$\| u \|_{W^\ell(t,t+1)} \le c\, e^{-(k_0+\varepsilon)t}$$

for large t. Due to the left-hand inequality in (12.12) we obtain $c_n = 0$, $n = 1, \ldots, \kappa$.

It remains to prove (c). By Theorem 13.10.4 there exists a solution $u \in W^\ell_{\mathrm{loc}}(t_0, \infty)$ of (13.92) with $f = 0$ such that (13.89), where $\mathbf{v}$ is given by (13.91), holds. Due to (13.129) and (13.112) the function u satisfies (12.13). By Theorem 12.2.3(ii) there exist constants $c_1, c_2, \ldots, c_\kappa$ such that

$$\|u - \sum_{j=1}^{\kappa} c_j\, u_j\|_{W^\ell(t,t+1)} \le C e^{-k_+ t}.$$

This yields (13.130). $\square$

13.12 Estimate for a Neumann Series

Let $k_- < k_+$ and $m_\pm$ be positive integers. We set

$$\mathbf{g}(t) = \begin{cases} e^{-k_+ t}(1 + t)^{m_+ - 1} & \text{if } t \ge 0 \\ e^{-k_- t}(1 + |t|)^{m_- - 1} & \text{if } t < 0. \end{cases}$$

We use the notation

$$Q(\tau, s) = \mathfrak{q}(\tau)\delta(\tau - s) + \mathcal{Q}(\tau, s),$$

where $\mathfrak{q}$ and $\mathcal{Q}$ are non-negative measurable functions on $\mathbb{R}$ and $\mathbb{R}^2$ and δ is the Dirac function. Our aim here is to estimate the Neumann series

$$\mathbf{g}_Q(t, \tau) = \sum_{k=0}^{\infty} \int_{\mathbb{R}^{2k}} \mathbf{g}(t - \tau_1) Q(\tau_1, s_1) \mathbf{g}(s_1 - \tau_2) Q(\tau_2, s_2) \ldots$$
$$\times Q(\tau_k, s_k) \mathbf{g}(s_k - \tau) d\tau_1 \ldots d\tau_k ds_1 \ldots ds_k. \qquad (13.131)$$

These estimates will be used in what follows. We begin with three lemmas.

Lemma 13.12.1. *Let*

$$q_0 = \int \int_{\tau \geq s} e^{k_+(\tau-s)}(1+\tau-s)^{m_+-1}Q(\tau,s)d\tau ds$$

$$+ \int \int_{\tau < s} e^{k_-(\tau-s)}(1+s-\tau)^{m_--1}Q(\tau,s)d\tau ds < \infty. \qquad (13.132)$$

Then

$$\left(\int_{-\infty}^t \int_{-\infty}^x + \int_t^\infty \int_x^\infty\right) \mathbf{g}(t-\tau)Q(\tau,s)\mathbf{g}(s-x)d\tau ds$$

$$\leq c q_0 \mathbf{g}(t-x), \qquad (13.133)$$

where c depends only on $k_\pm$ and $m_\pm$.

Proof. Consider the integral $\int\limits_{-\infty}^t \int\limits_{-\infty}^x$ in (13.133). Let $t \geq x$. One can check directly that for $\tau \leq t$, $s \leq x$,

$$\mathbf{g}(t-\tau)\mathbf{g}(s-x)/\mathbf{g}(t-x)$$
$$= e^{(k_+-k_-)(s-x)+k_+(\tau-s)}(1+t-\tau)^{m_+-1}(1+x-s)^{m_--1}(1+t-x)^{1-m_+}$$
$$\leq c\, e^{k_+(\tau-s)}(1+|\tau-s|)^{m_+-1},$$

and we arrive at (13.133).

If $t < x$ we have for $\tau \leq t$, $s \leq x$

$$\mathbf{g}(t-\tau)\mathbf{g}(s-x)/\mathbf{g}(t-x)$$
$$= e^{(k_+-k_-)(\tau-t)+k_-(\tau-s)}(1+t-\tau)^{m_+-1}(1+x-s)^{m_--1}(1+x-t)^{1-m_-}$$
$$\leq c\, e^{k_-(\tau-s)}(1+|\tau-s|)^{m_--1},$$

and we obtain (13.133) again.

The other integral in (13.133) is estimated analogously. □

Lemma 13.12.2. *The following estimates hold:*

$$\int_t^\infty \int_{-\infty}^x \mathbf{g}(t-\tau)Q(\tau,s)\mathbf{g}(s-x)d\tau ds \leq c\, q_0\mathbf{g}(t-x) \qquad (13.134)$$

for $t \geq x$, and

$$\int_{-\infty}^t \int_x^\infty \mathbf{g}(t-\tau)Q(\tau,s)\mathbf{g}(s-x)d\tau ds \leq c\, q_0\mathbf{g}(t-x) \qquad (13.135)$$

for $t \leq x$. Here q_0 is given by (13.132).

Proof. If $\tau \geq t \geq x \geq s$ then

$$g(t-\tau)g(s-x)/g(t-x) = e^{(k_+ - k_-)(t-\tau+s-x)+k_+(\tau-s)}$$
$$\times (1+\tau-t)^{m_- - 1}(1+x-s)^{m_- - 1}(1+t-x)^{1-m_+}$$
$$\leq c\, e^{k_+(\tau-s)},$$

and we arrive at (13.134).

The inequality (13.135) is proved analogously. $\square$

Lemma 13.12.3. *Let*

$$q_1 = \int \int_{\tau \geq s} e^{k_+(\tau-s)}(1+|s|)^{m_+ - 1}(1+\tau-s)^{m_+ - 1}Q(\tau,s)d\tau ds$$

$$+ \int \int_{\tau < s} e^{k_-(\tau-s)}(1+|\tau|)^{m_- - 1}(1+s-\tau)^{m_- - 1}Q(\tau,s)d\tau ds. \quad (13.136)$$

Then for $t, x \geq 0$,

$$\left(\int_{-\infty}^{t} \int_{x}^{\infty} + \int_{t}^{\infty} \int_{-\infty}^{x} \right) g(t-\tau)Q(\tau,s)g(s-x)d\tau ds$$

$$\leq c\, q_1 g(t-x), \quad (13.137)$$

where c depends only on $k_\pm, m_\pm$.

Proof. Consider the integral $\int\limits_{-\infty}^{t} \int\limits_{x}^{\infty}$ in (13.137). If $t \leq x$ then (13.137) follows from (13.135).

Let $t > x$. Then for $\tau \leq t$ and $s \geq x$

$$g(t-\tau)g(s-x)/g(t-x)$$
$$= e^{k_+(\tau-s)}(1+t-\tau)^{m_+ - 1}(1+s-x)^{m_+ - 1}(1+t-x)^{1-m_+}.$$

One can easily show that

$$(1+t-\tau)(1+s-x)(1+t-x)^{-1} \leq (1+|\tau-s|)(1+s)$$

and we obtain (13.137). The integral $\int\limits_{t}^{\infty} \int\limits_{-\infty}^{x}$ is considered analogously. $\square$

The main result is contained in the following

Theorem 13.12.4. *Let q and Q be non-negative measurable functions on $\mathbb{R}$ and $\mathbb{R}^2$ and let $q(\tau) = 0$ and $Q(\tau,s) = 0$ if $\tau < 0$ or $s < 0$. Then there exists a positive constant c_0 depending on $k_\pm, m_\pm$ such that if $q_1 \leq c_0$ (with q_1 given by (13.136)), then the series (13.131) is convergent for every $t, \tau \geq 0$. Moreover,*

$$g_Q(t,\tau) \leq c\, g(t-\tau) \quad \text{for} \quad t,\tau \geq 0,$$

where c depends only on $k_\pm, m_\pm$.

Proof. Since $q_1 \geq q_0$, we can use Lemmas 13.12.1–13.12.3 to obtain

$$\int_0^\infty \int_0^\infty \mathbf{g}(t - \tau)Q(\tau, s)\mathbf{g}(s - x)d\tau ds \leq c_1 q \, \mathbf{g}(t - x).$$

Hence, for $t, \tau \geq 0$

$$\mathbf{g}_Q(t, \tau) \leq \sum_{k=0}^\infty (c_1 q)^k \, \mathbf{g}(t - \tau).$$

This completes the proof. $\square$

14. Power-Exponential Asymptotics

14.1 Introduction

In order to give an idea of the results in the present chapter, we consider the ordinary second order differential equation

$$D_t^2 u + \rho(t)u = 0, \quad t > t_0,$$

where $\rho(t) > 0$. It is well known and easily checked (see Hartman (1964), Ch. XI, Sect. 9) that the condition

$$\int_{t_0}^{\infty} \rho(t)t\,dt < \infty$$

is equivalent to the existence of two solutions with asymptotics

$$u_1(t) = t(1 + o(1)) \quad \text{and} \quad u_2(t) = 1 + o(1) \quad \text{as } t \to +\infty.$$

In order that the stronger asymptotic formula

$$u_1(t) = t + o(1)$$

be valid, it is both necessary and sufficient that

$$\int_{t_0}^{\infty} \rho(t)t^2\,dt < \infty.$$

In this chapter we obtain results of a similar nature for both the homogeneous equation

$$\big(\mathcal{A}(D_t) - N(t, D_t)\big)u(t) = 0 \quad \text{for} \quad t > t_0, \tag{14.1}$$

and the nonhomogeneous equation

$$\big(\mathcal{A}(D_t) - N(t, D_t)\big)u(t) = f(t) \quad \text{for} \quad t > t_0, \tag{14.2}$$

when t_0 is sufficiently large. We give conditions on $N(t, D_t)$ ensuring the asymptotic equivalence of solutions to (14.1), (14.2) and special solutions of $\mathcal{A}(D_t)u = 0$.

We make the same assumptions on the spectrum of $\mathcal{A}(\lambda)$ as in Ch. 13; we also use the same notation: $k_\pm$, k_0, m_0, κ. Again we suppose that $\rho(t)$ vanishes at $+\infty$.

A sufficient condition for all the results of Sections 14.3 and 14.4 is

$$\int_{t_0}^{\infty} t^p \chi(t) dt < \infty \tag{14.3}$$

with a non-negative integer p, and χ introduced by

$$\chi(t) = \gamma(t) + t^{m_0-1}\beta(t) \int_{t_0}^{\infty} e^{-a|t-\tau|}\alpha(\tau)d\tau. \tag{14.4}$$

Here $a > 0$, the functions α, β are defined in Sections 13.5 and 13.7 and

$$\gamma(t) = \sum_{(\nu,k,j)\in Q} \sum_{(\mu,n,q)\in Q} t^{j+m_{\mu n}-q-m_{\nu k}}$$

$$\times \left(\int_t^{t+1} \left| \left(N(\tau, D_s) U_{kj}^{(\nu)}(s) \Big|_{s=0}, V_{nq}^{(\mu)}(0) \right)_{H_0} \right|^2 d\tau \right)^{1/2}, \tag{14.5}$$

where Q is the set of indices given by (13.34).

More restrictive, but easier to check, conditions are obtained from (14.3) if we replace χ by one of the following three functions

$$t^{m_0-1}\alpha(t), \ \gamma(t) + t^{m_0-1}\rho^2(t), \ t^{m_0-1}\rho(t)$$

(see Corollary 14.4.2).

In Sect.14.3 we show that (14.3) implies the existence of κ linearly independent solutions $u_{kj}^{(\nu)}(t)$ which are asymptotically equivalent to the solutions $U_{k,j}^{(\nu)}(t)$ of $\mathcal{A}(D_t)u = 0$. This results in certain power-exponential asymptotics for any solution of (14.1) subject to

$$\| u \|_{W^\ell(t,t+1)} = O(e^{-k_- t}) \quad \text{as} \quad t \to +\infty. \tag{14.6}$$

This estimate is obtained in Theorem 14.4.1. For example, in the case $p \geq m_0 - 1$ under the conditions $\alpha(t) = o(t^{-p})$ and (14.3), we prove the existence of constants $c_{kj}^{(\nu)}$ such that

$$\left\| u - \sum_{(\nu,k,j)\in Q} c_{kj}^{(\nu)} U_{kj}^{(\nu)} \right\|_{W^\ell(t,t+1)} = o(e^{-k_0 t} t^{m_0-1-p}) \tag{14.7}$$

(see Corollary 14.4.2(ii)). This means that under the perturbation N, the form of the asymptotics of solutions to $\mathcal{A}(D_t)u = 0$ is preserved modulo $o(e^{-k_0 t})$.

Even the roughest of the above mentioned sufficient conditions,

$$\int_{t_0}^{\infty} t^{m_0+p-1}\rho(t) dt < \infty, \tag{14.8}$$

is precise for scalar ordinary differential operators (see Remark 14.4.4). However (14.3) is better than (14.8) because the function χ may vanish identically when $\rho > 0$.

In Sections 14.5-14.7 we study the asymptotics of the same form for solutions of the nonhomogeneous equation (14.2) subject to (14.6). We assume that the right-hand side satisfies

$$\int_{t_0}^{\infty} e^{k_0\tau}\tau^{m_0-1} \| f \|_{L_2(\tau,\tau+1;H_0)} \, d\tau < \infty. \qquad (14.9)$$

As in the case of the homogeneous equation (14.1), our requirements on the operator $N(t, D_t)$ concern only the functions α, β and γ. From Theorem 14.7.1 and Remark 14.7.2 we obtain, for example, that the condition

$$\int_{t_0}^{\infty} \tau^{2(m_0-1)}\alpha(\tau)d\tau < \infty$$

implies (14.7) with $p = m_0 - 1$ (clearly, α can be replaced by ρ).

In Sect.14.8 we suppose that f is subject to

$$\int_{t_0}^{\infty} e^{2k_0\tau}\tau^{2(m_0+\sigma)} \| f \|_{L_2(\tau,\tau+1;H_0)}^2 \, d\tau < \infty, \qquad (14.10)$$

$\sigma > -1/2$. We are interested in conditions on $N(t, D_t)$, which ensure the asymptotics (14.7) with $p = m_0 - 1$ and with the remainder term

$$v(t) = u(t) - \sum_{(\nu,k,j)\in Q} c_{kj}^{(\nu)} U_{kj}^{(\nu)}(t)$$

satisfying

$$\int_{t_0}^{\infty} e^{2k_0\tau}\tau^{2\sigma} \| v \|_{W^\ell(\tau,\tau+1)}^2 \, d\tau < \infty.$$

We show, in particular, that this is guaranteed by

$$\int_{t_0}^{\infty} \tau^{2(\sigma+2m_0-1)}\alpha^2(\tau)d\tau < \infty. \qquad (14.11)$$

In Remark 14.8.3 we prove that this requirement is precise in a particular sense.

Sect. 14.9 contains some applications of the above results to the asymptotic behavior of solutions to arbitrary order elliptic equations with variable coefficients in a neighborhood of a point. We give Dini type conditions on the coefficients ensuring that the form of asymptotics is the same as in the case of constant coefficients.

14.2 Special Solutions of the Finite Dimensional System

14.2.1 The Functions α, β, γ and χ

As in Sect.13.10, we suppose that the operator N is extended to the semiaxis $t < t_0$ by zero.

Together with α and β we use the functions γ defined by (14.5) and χ given by (14.4), where we put

$$a = \frac{1}{2} \min (k_+ - k_0, \ k_0 - k_-). \tag{14.12}$$

In the special case when the eigenvalues λ_ν such that $\Im \lambda_\nu = k_0$ have no generalized eigenvectors, the form of γ, α and β can be simplified:

$$\gamma(t) = \sum_{\Im \lambda_\nu = \Im \lambda_\mu = k_0} \sum_{k=1}^{J_\nu} \sum_{j=1}^{J_\nu} \Big(\int_t^{t+1} |(N(\tau, \lambda_\nu) \phi_k^{(\nu)}, \psi_j^{(\mu)})|^2 d\tau \Big)^{1/2}, \tag{14.13}$$

where $\phi_k^{(\nu)}$, $\psi_j^{(\mu)}$ are eigenvectors corresponding to the eigenvalues λ_ν, $\overline{\lambda}_\mu$ of the pencils $\mathcal{A}(\lambda)$, $\mathcal{A}^*(\lambda)$;

$$\alpha(t) = \sum_{\Im \lambda_\nu = k_0} \sum_{j=1}^{J_\nu} \| N(\cdot, \lambda_\nu) \phi_j^{(\nu)} \|_{L_2(t, t+1; H_0)}; \tag{14.14}$$

$$\beta(t) = \sup_v \sum_{\Im \lambda_\nu = k_0} \sum_{k=1}^{J_\nu} \Big(\int_t^{t+1} |(N(\tau, D_\tau) v(\tau), \psi_k^{(\nu)})_{H_0} \tag{14.15}$$

$$- \sum_{\Im \lambda_\mu = k_0} \sum_{j=1}^{J_\mu} (\mathcal{B}(\lambda_\mu, D_\tau) v(\tau), \psi_j^{(\mu)})_{H_0} (N(\tau, \lambda_\mu) \phi_j^{(\mu)}, \psi_k^{(\nu)})_{H_0}|^2 d\tau \Big)^{1/2},$$

where $\mathcal{B}(\lambda, z) = (\lambda - z)^{-1}(\mathcal{A}(\lambda) - \mathcal{A}(z))$ and the supremum is taken over the set (13.75).

We note that $\alpha(t)$, $\beta(t)$, $\gamma(t)$ and $\chi(t)$ vanish for $t < t_0 - 1$. Clearly,

$$\gamma(t) \le c\alpha(t). \tag{14.16}$$

Lemma 14.2.1. *The inequality*

$$\chi(t) \le c \int_{t-1}^{t+1} \chi(\tau) d\tau$$

holds.

Proof. Due to (13.68) it suffices to prove that

$$\gamma(t) \leq c \int_{t-1}^{t+1} \gamma(\tau)d\tau.$$

This inequality follows from the definition of γ and from the estimates

$$\| r \|_{L_2(t,t+1)} \leq \int_{t}^{t+1} \| r \|_{L_2(x-1,x+1)} \, dx$$

$$\leq \int_{t-1}^{t} \| r \|_{L_2(x,x+1)} \, dx + \int_{t}^{t+1} \| r \|_{L_2(x,x+1)} \, dx,$$

where r is an arbitrary function in $L_2(t-1, t+2)$. $\square$

14.2.2 Lemma on Special Solutions

In the next lemma we construct κ solutions of the finite dimensional system (13.125). We shall use the block numeration introduced in Sect.13.9.

Lemma 14.2.2. *Let p be a non-negative integer and let ε be a positive number depending on $A_1, \ldots, A_\ell$, $k_\pm$, k_0 and p. If*

$$\int_{t_0-1}^{\infty} t^p \chi(t)dt < \varepsilon \tag{14.17}$$

then there exist solutions

$$h(t) = H_{n,s}^{(\mu)}(t), \ (\mu, n, s) \in Q,$$

of system (13.125) *such that*

$$H_{ns}^{(\mu)}(t) = e^{i\lambda_\mu t}\left(\Pi_{ns}^{(\mu)}(t) + \Theta_{ns}^{(\mu)}(t)\right), \tag{14.18}$$

where
(i) $\Pi_{ns}^{(\mu)}(t) = \left\{\left(\Pi_{ns}^{(\mu)}\right)_{kj}^{(\nu)}(t)\right\}$ is the block vector with the components

$$\left(\Pi_{ns}^{(\mu)}\right)_{kj}^{(\nu)}(t) = \begin{cases} (it)^{s-j}/(s-j)! & \text{if } (\nu, k) = (\mu, n) \text{ and } j \leq s \\ 0 & \text{otherwise}; \end{cases}$$

(ii) the remainder term $\Theta_{ns}^{(\mu)}$ satisfies

$$\| \Theta_{n,s}^{(\mu)} \|_{\left(W_2^1(t,t+1)\right)^\kappa}$$
$$\leq c\, t^s \left(\int_{t}^{\infty} \left(\frac{\tau}{t}\right)^p \chi(\tau)d\tau + \int_{t_0-1}^{t} \left(\frac{\tau}{t}\right)^{p+1} \chi(\tau)d\tau \right) \tag{14.19}$$

if $p < s$ and

$$\| \Theta_{n,s}^{(\mu)} \|_{\left(W_2^1(t,t+1)\right)^\kappa} \leq c\, t^s \int_{t-1}^{\infty} \left(\frac{\tau}{t}\right)^p \chi(\tau)d\tau \tag{14.20}$$

if $p \geq s$.

Proof. Let J be the block diagonal matrix diag $(J_{\nu k})$ introduced in Sect.13.9. Clearly, $\exp(itJ) = M(it)$, where

$$M(it) = \text{diag}(M_{\nu k}(it)) \tag{14.21}$$

with

$$M_{\nu k}(z) = \begin{pmatrix} 1 & z & \frac{z^2}{2} & \cdots & \frac{z^{m_{\nu k}-1}}{(m_{\nu k}-1)!} \\ 0 & 1 & z & \cdots & \frac{z^{m_{\nu k}-2}}{(m_{\nu k}-2)!} \\ \vdots & \vdots & \vdots & \vdots & \vdots \\ 0 & 0 & 0 & \cdots & 1 \end{pmatrix}.$$

We set

$$\triangle(z) = \text{diag}(\triangle_{\nu k}(z)) \, ,$$

where

$$\triangle_{\nu k}(z) = \text{diag}(1, z, \ldots, z^{m_{\nu k}-1}) \, .$$

Then

$$M(z) = \triangle(z)^{-1} M(1) \triangle(z). \tag{14.22}$$

Substituting

$$h(t) = M(it)\triangle(it)^{-1} Z(t)$$

into (13.125) and using (14.22) we obtain

$$\left(D_t - \Lambda - \triangle'(it)\triangle(it)^{-1}\right) Z(t) - (TZ)(t) = 0, \tag{14.23}$$

where

$$(TZ)(t) = M(1)^{-1}\triangle(it)\mathcal{R}(t)\triangle(it)^{-1}M(1)Z(t)$$
$$+ M(1)^{-1}\triangle(it)K_{\tau \to t}\left(\triangle(i\tau)^{-1}M(1)Z(\tau)\right)(t).$$

By (14.5) and (13.110) we arrive at

$$\| TZ \|_{L_2(t,t+1)} \leq c \left(\gamma(t) \| Z \|_{(L_\infty(t,t+1))^\kappa} \right.$$
$$\left. + t^{m_0-1}\beta(t) \int_{t_0-1}^\infty g(t-\tau)\alpha(\tau) \| Z \|_{(L_\infty(\tau,\tau+1))^\kappa} \, d\tau \right), \tag{14.24}$$

where g is Green's function defined by (12.61). Here the vector function Z is extended by zero for $t < t_0$.

We are looking for a solution of (14.23) in the form

$$Z(t) = e^{i\lambda_\mu t}(it)^s \left(Y_{ns}^{(\mu)} + \mathcal{Y}_{ns}^{(\mu)}(t) \right),$$

where $Y_{ns}^{(\mu)}$ is the vector with components

$$\left(Y_{ns}^{(\mu)} \right)_{kj}^{(\nu)} = \delta_\nu^\mu \, \delta_k^n \, \delta_j^s.$$

The remainder

$$\mathcal{Y}(t) = \mathcal{Y}_{ns}^{(\mu)}(t)$$

satisfies

$$\left(D_t - \Lambda - \Delta'(it)\Delta(it)^{-1} + (\lambda_\mu + \frac{s}{it})I\right)\mathcal{Y}(t)$$
$$- \left(\Gamma(Y_{ns}^{(\nu)} + \mathcal{Y})\right)(t) = 0, \tag{14.25}$$

where $I = \mathrm{diag}(I_{\nu k})$ is the identity matrix and the operator Γ is defined by

$$(\Gamma X)(t) = e^{-i\lambda_\mu t}(it)^{-s}T_{\tau \to t}\left(e^{i\lambda_\mu \tau}(i\tau)^s X(\tau)\right)(t)$$

with $X \in (L_{\infty,\mathrm{loc}}(t_0, \infty))^\kappa$. Using (14.24) we obtain

$$\| \Gamma X \|_{(L_2(t,t+1))^\kappa} \le c \left(\gamma(t) \| X \|_{(L_\infty(t,t+1))^\kappa}\right.$$
$$\left. + t^{m_0-1}\beta(t)\int_{t_0-1}^\infty g(t-\tau)\alpha(\tau)e^{k_0(t-\tau)}\left(\frac{\tau}{t}\right)^s \| X \|_{(L_\infty(\tau,\tau+1))^\kappa} \, d\tau\right).$$

Here and in what follows X is extended by zero for $t < t_0$. By (12.61),

$$g(t - \tau) \le c\, e^{-k_0(t-\tau)-2a|t-\tau|} \tag{14.26}$$

with a given by (14.12). Hence, using the inequality

$$\left(\frac{\tau}{t}\right)^s \le c_{s,\varepsilon}e^{\varepsilon|t-\tau|}, \quad t,\tau \ge 1, \tag{14.27}$$

we arrive at

$$\| \Gamma X \|_{(L_2(t,t+1))^\kappa} \le c \left(\gamma(t) \| X \|_{(L_\infty(t,t+1))^\kappa}\right.$$
$$\left. + t^{m_0-1}\beta(t)\int_{t_0-1}^\infty e^{-a|t-\tau|}\alpha(\tau) \| X \|_{(L_\infty(\tau,\tau+1))^\kappa} \, d\tau\right). \tag{14.28}$$

To obtain an integral equation for $\mathcal{Y}$ we construct Green's matrix function $\mathcal{G}$ of the diagonal system

$$\left(D_t - \Lambda - \Delta'(it)\Delta(it)^{-1} + (\lambda_\mu + \frac{s}{it})I\right)\mathcal{G}(t,\tau) = \delta(t-\tau)I.$$

Put $\mathcal{G}(t,\tau) = \mathrm{diag}(\mathcal{G}_{\nu k}(t,\tau))$, where $\mathcal{G}_{\nu k}(t,\tau)$ is the diagonal matrix of size $m_{\nu k}$ whose elements $\mathcal{G}_{\nu k,q}(t,\tau), q = 0, \ldots, m_{\nu k} - 1$, on the diagonal are determined in the following way: if $q < s - p$ then

$$\mathcal{G}_{\nu k,q}(t,\tau) = \begin{cases} ie^{i(\lambda_\nu - \lambda_\mu)(t-\tau)}(t/\tau)^{q-s} & \text{for } t \ge \tau, \\ 0 & \text{for } t < \tau, \end{cases}$$

and if $q \ge s - p$ then

$$\mathcal{G}_{\nu k,q}(t,\tau) = \begin{cases} 0 & \text{for } t > \tau, \\ -ie^{i(\lambda_\nu - \lambda_\mu)(t-\tau)}(t/\tau)^{q-s} & \text{for } t \le \tau. \end{cases}$$

This Green's matrix function satisfies

$$\| \mathcal{G}(t,\tau) \|_{\mathbb{C}^\kappa} \le \begin{cases} c\,(\tau/t)^{p+1} & \text{if } t \ge \tau, \\ c\,(\tau/t)^p & \text{if } t \le \tau, \end{cases} \tag{14.29}$$

for $p < s$ and

$$\| \mathcal{G}(t,\tau) \|_{\mathbb{C}^\kappa} \le \begin{cases} 0 & \text{if } t \ge \tau, \\ c\,(\tau/t)^p & \text{if } t \le \tau, \end{cases} \tag{14.30}$$

for $p \ge s$.

We rewrite (14.25) as the integral equation

$$\mathcal{Y}(t) = (\mathbb{Q}\mathcal{Y})(t) + (\mathbb{Q}Y_{ks}^{(\mu)})(t), \quad t > t_0, \tag{14.31}$$

where the operator $\mathbb{Q}$ is defined by

$$(\mathbb{Q}X)(t) = \int_{t_0}^{\infty} \mathcal{G}(t,\tau)(\Gamma X)(\tau)d\tau.$$

To prove the existence of a solution to (14.31) in $L_\infty(t_0,\infty)$ it is sufficient to show that the norm of the operator

$$\mathbb{Q} : L_\infty(t_0,\infty) \to L_\infty(t_0,\infty) \tag{14.32}$$

is less than 1.

We consider only the case $p < s$, since the opposite case can be treated analogously. By (14.29) we get

$$\| (\mathbb{Q}X)(t) \|_{\mathbb{C}^\kappa} \le c \left\{ \int_t^\infty \left(\frac{\tau}{t}\right)^p \| \Gamma X \|_{(L_2(\tau,\tau+1))^\kappa} d\tau \right. $$
$$\left. + \int_{t_0-1}^t \left(\frac{\tau}{t}\right)^{p+1} \| \Gamma X \|_{(L_2(\tau,\tau+1))^\kappa} d\tau \right\}.$$

Hence, using (14.28), we obtain

$$\| (\mathbb{Q}X)(t) \|_{\mathbb{C}^\kappa} \tag{14.33}$$
$$\le c \left(\int_t^\infty \left(\frac{\tau}{t}\right)^p \chi(\tau)d\tau + \int_{t_0-1}^t \left(\frac{\tau}{t}\right)^{p+1} \chi(\tau)d\tau \right) \| X \|_{(L_\infty(t_0,\infty))^\kappa} \ .$$

This implies

$$\| \mathbb{Q} \|_{(L_\infty(t_0,\infty))^\kappa \to (L_\infty(t_0,\infty))^\kappa} \le c \int_{t_0-1}^\infty \tau^p \chi(\tau)d\tau.$$

Due to (14.17) the norm of $\mathbb{Q}$ is small. Therefore the equation (14.31) has a bounded solution $\mathcal{Y}$ and we arrive at (14.18) with

$$\Theta_{ns}^{(\mu)}(t) = \triangle(it)^{-1}M(1)(it)^s\mathcal{Y}(t). \tag{14.34}$$

From (14.31) and (14.33) we conclude that

$$\| \mathcal{Y}(t) \|_{\mathbb{C}^{\kappa}} \le c \left(\int_t^{\infty} \left(\frac{\tau}{t}\right)^p \chi(\tau)d\tau + \int_{t_0-1}^t \left(\frac{\tau}{t}\right)^{p+1} \chi(\tau)d\tau \right). \tag{14.35}$$

We estimate the first derivative of $\mathcal{Y}$ by using the equation (14.25):

$$\| \mathcal{Y}' \|_{(L_2(t,t+1))^{\kappa}} \le c \left(\| \mathcal{Y} \|_{(L_2(t,t+1))^{\kappa}} \right.$$

$$\left. + \| \Gamma(Y_{ks}^{(\mu)} + \mathcal{Y}) \|_{(L_2(t,t+1))^{\kappa}} \right).$$

Applying (14.28) and (14.35), and taking into account the boundedness of $Y_{ks}^{(\mu)}$ we have

$$\| \mathcal{Y}' \|_{(L_2(t,t+1))^{\kappa}} \le c \left(\int_t^{\infty} \left(\frac{\tau}{t}\right)^p \chi(\tau)d\tau \right.$$

$$\left. + \int_{t_0-1}^t \left(\frac{\tau}{t}\right)^{p+1} \chi(\tau)d\tau + \chi(t) \right). \tag{14.36}$$

Due to Lemma 14.2.1, the last term in (14.36) can be removed. Combining (14.35) and (14.36) with (14.34) we obtain (14.19). $\square$

14.3 Special Solutions of (14.1)

14.3.1 Construction of Solutions with Prescribed Asymptotics

Lemma 14.3.1. *Let p be a non-negative integer such that*

$$\int_{t_0}^{\infty} \tau^p \chi(\tau)d\tau < \infty. \tag{14.37}$$

Then for sufficiently large t_0 the equation (14.1) has κ solutions

$$u_{ns}^{(\mu)} \in W_{\mathrm{loc}}^{\ell}(t_0, \infty), \quad (\mu, n, s) \in Q .$$

These solutions have the asymptotics:

$$\mathrm{col}\left(D_t^{j-1} u_{ns}^{(\mu)}(t)\right)_{j=1}^{\ell} = \mathrm{col}\left(D_t^{j-1} U_{ns}^{(\mu)}(t)\right)_{j=1}^{\ell} + W_{ns}^{(\mu)}(t), \tag{14.38}$$

where the remainder $W_{ns}^{(\mu)}$ belongs to $\mathbb{S}_{\mathrm{loc}}(t_0, \infty)$ and satisfies

$$\| W_{ns}^{(\mu)} \|_{\mathbb{S}(t,t+1)} \le c\, e^{-k_0 t} t^s \left(\int_t^{\infty} \left(\frac{\tau}{t}\right)^p \chi(\tau)d\tau \right. \tag{14.39}$$

$$\left. + \int_{t_0-1}^t \left(\frac{\tau}{t}\right)^{p+1} \chi(\tau)d\tau + \int_{t_0-1}^{\infty} e^{-a|t-\tau|} \alpha(\tau)d\tau \right)$$

if $p < s$ and

$$\| \mathcal{W}_{ns}^{(\mu)} \|_{\mathbb{S}(t,t+1)} \leq c \, e^{-k_0 t} t^s \left(\int_t^\infty \left(\frac{\tau}{t} \right)^p \chi(\tau) d\tau \right.$$

$$\left. + \int_{t_0-1}^\infty e^{-a|t-\tau|} \alpha(\tau) d\tau \right) \qquad (14.40)$$

if $p \geq s$.

Proof. Let

$$\mathcal{U}(t) = e^{i\lambda_\mu t} \sum_{(\nu,k,j)\in Q} \left(\Pi_{ns}^{(\mu)} \right)_{kj}^{(\nu)}(t) \mathcal{U}_{kj}^{(\nu)}(0),$$

where $\Pi_{ns}^{(\mu)}$ are the same vector polynomials as in Lemma 14.2.2. We will transform the right-hand side of the last inequality. By the definition of $\Pi_{ns}^{(\mu)}$ we have

$$\mathcal{U}(t) = e^{i\lambda_\mu t} \sum_{q=0}^s \frac{(it)^{s-q}}{(s-q)!} \mathcal{U}_{nq}^{(\mu)}(0). \qquad (14.41)$$

The formula (12.42) together with (12.30) and (12.31) gives

$$\mathcal{U}_{nq}^{(\mu)}(t) = (D_t - \lambda_\nu)^{m_{\mu n}-1-q} \mathcal{U}_{n,m_{\mu n}-1}^{(\mu)}(t). \qquad (14.42)$$

This, along with Taylor's formula for the vector polynomial

$$e^{-i\lambda_\mu t} \mathcal{U}_{n,m_{\mu n}-1}^{(\mu)}(t),$$

implies

$$\mathcal{U}_{n,m_{\mu n}-1}^{(\mu)}(t) = e^{i\lambda_\mu t} \sum_{q=0}^{m_{\mu n}-1} \frac{(it)^q}{q!} \mathcal{U}_{n,m_{\mu n}-1-q}^{(\mu)}(0). \qquad (14.43)$$

Now applying $(D_t - \lambda_\mu)^{m_{\mu n}-1-s}$ to both sides of (14.43) and using (14.42) we obtain

$$\mathcal{U}_{ns}^{(\mu)}(t) = e^{i\lambda_\mu t} \sum_{q=0}^s \frac{(it)^q}{q!} \mathcal{U}_{n,s-q}^{(\mu)}(0).$$

Comparing the last formula with (14.41) we find that

$$\mathcal{U}(t) = \mathcal{U}_{ns}^{(\mu)}(t). \qquad (14.44)$$

By Theorem 13.10.4 and Lemma 14.2.2 there exists a solution

$$u_{ns}^{(\mu)} \in W_{\text{loc}}^\ell(t_0, \infty)$$

of (14.1) such that

$$\text{col}\left(D_t^{j-1} u_{ns}^{(\mu)}(t) \right)_{j=1}^\ell = \mathcal{U}_{ns}^{(\mu)}(t) + \mathcal{W}_{ns}^{(\mu)}(t), \qquad (14.45)$$

where

$$\mathcal{W}_{ns}^{(\mu)}(t) = e^{i\lambda_\mu t} R_{ns}^{(\mu)}(t) + \mathfrak{M}(I_\ell - P)\mathcal{N}\mathbf{u} \qquad (14.46)$$

with $\mathfrak{M}$ defined by (13.30) and (13.31). Here

$$R_{ns}^{(\mu)}(t) = \sum_{(\nu,k,j)\in Q} (\Theta_{ns}^{(\mu)})_{kj}^{(\nu)}(t)\mathcal{U}_{kj}^{(\nu)}(0)$$

and

$$\mathbf{u}(t) = \sum_{(\nu,k,j)\in Q} (H_{ns}^{(\mu)})_{kj}^{\nu}(t)\mathcal{U}_{kj}^{(\nu)}(0)$$

with $H_{ns}^{(\mu)}$ given by (14.18). By (12.30) and (12.31) the first terms in the right-hand sides of (14.38) and (14.45) coincide.

Now we estimate $\mathcal{W}_{ns}^{(\mu)}(t)$. Due to (13.32), (13.33) and (14.26) the $\mathbb{S}(t, t+1)$-norm of the second term in the right-hand side of (14.46) does not exceed

$$c\, e^{-k_0 t} t^s \int_{t_0-1}^{\infty} e^{-2a|t-\tau|}\alpha(\tau)\,\| \Pi_{ns}^{(\mu)} + \Theta_{ns}^{(\mu)} \|_{(L_\infty(\tau,\tau+1))^\kappa}\,d\tau. \qquad (14.47)$$

(Here $\Pi_{ns}^{(\mu)}$ and $\Theta_{ns}^{(\mu)}$ are extended by 0 for $t < t_0$.) Using the definition of $\Pi_{n,s}^{(\mu)}$ (see Lemma 14.2.2 (i)) and the estimates (14.19), (14.20) for $\Theta_{n,s}^{(\mu)}$ together with (14.27) we estimate (14.47) by

$$c\, e^{-k_0 t} t^s \int_{t_0-1}^{\infty} e^{-a|t-\tau|}\alpha(\tau)\,d\tau.$$

This, together with (14.19) and (14.20), implies (14.39) and (14.40) with $\int_{t-1}^{\infty}$ instead of $\int_t^{\infty}$. Due to (14.16) the integral

$$\int_{t-1}^{t} \left(\frac{\tau}{t}\right)^p \chi(\tau)\,d\tau$$

is majorized by the last integral in (14.40). $\square$

14.3.2 Main Result

We now can obtain an asymptotic representation for solutions of (14.1) in the space $W_{\text{loc}}^\ell(t_0, \infty)$.

Theorem 14.3.2. *Let p be a non-negative integer such that (14.37) holds. Then for sufficiently large t_0 the equation (14.1) has κ solutions*

$$u_{ns}^{(\mu)} \in W_{\text{loc}}^\ell(t_0, \infty), \quad (\mu, n, s) \in Q\,.$$

These have the asymptotics:

$$u_{ns}^{(\mu)}(t) = U_{ns}^{(\mu)}(t) + w_{ns}^{(\mu)}(t), \qquad (14.48)$$

where the remainder $w_{ns}^{(\mu)}(t)$ belongs to $W_{\text{loc}}^\ell(t_0, \infty)$ and satisfies the estimate

$$\| w_{ns}^{(\mu)} \|_{W^\ell(t,t+1)} \le c\, e^{-k_0 t} t^s \left(\int_t^\infty \left(\frac{\tau}{t}\right)^p \chi(\tau)\, d\tau \right.$$

$$\left. + \int_{t_0-1}^\infty \left(\frac{\tau}{t}\right)^{p+1} \chi(\tau)\, d\tau + \int_{t_0-1}^\infty e^{-a|t-\tau|} \alpha(\tau)\, d\tau \right) \qquad (14.49)$$

if p < s, and the estimate

$$\| w_{ns}^{(\mu)} \|_{W^\ell(t,t+1)} \le c\, e^{-k_0 t} t^s \left(\int_t^\infty \left(\frac{\tau}{t}\right)^p \chi(\tau)\, d\tau \right.$$

$$\left. + \int_{t_0-1}^\infty e^{-a|t-\tau|} \alpha(\tau)\, d\tau \right) \qquad (14.50)$$

if p ≥ s.

Proof. Since the vector function in the left-hand side of (14.38) and the first vector function on the right of (14.38) satisfy the equation

$$(I_\ell D_t + \mathfrak{A} - \mathcal{N}(t))\mathcal{U}(t) = 0 \quad \text{on} \quad t > t_0,$$

the same is valid for $\mathcal{W}_{ns}^{(\mu)}(t)$. Therefore

$$\mathcal{W}_{ns}^{(\mu)}(t) = \text{col} \left(D_t^{j-1} w_{ns}^{(\mu)}(t) \right)_{j=1}^\ell.$$

The representation (14.48) follows from (14.38); due to the definition of the space $\mathbb{S}(t, t+1)$,

$$\| \mathcal{W}_{ns}^{(\mu)} \|_{\mathbb{S}(t,t+1)} = \| w_{ns}^{(\mu)} \|_{W^\ell(t,t+1)}.$$

Hence the required estimates for $w_{ns}^{(\mu)}$ follow from (14.39) and (14.40). $\square$

Remark 14.3.3. Since the right-hand sides of (14.49) and (14.50) are

$$o\!\left(e^{-k_0 t} t^{s-p}\right),$$

the remainder term in Theorem 14.3.2 satisfies

$$\| w_{ns}^{(\nu)} \|_{W^\ell(t,t+1)} = o\!\left(e^{-k_0 t} t^{s-p}\right) \quad \text{as} \quad t \to +\infty. \qquad (14.51)$$

Therefore

$$\| u_{ns}^{(\mu)} \|_{W^\ell(t,t+1)} \le c\, e^{-k_0 t} t^s. \qquad (14.52)$$

In the next example we mention a special case in which the estimate of the remainder term in Theorem 14.3.2 becomes simpler.

Example 14.3.4. Let

$$\alpha(t) + t\chi(t) \le \text{const } t^{-b}, \quad b > 0, \tag{14.53}$$

for large positive t. Then the remainder $w_{ns}^{(\mu)}$ in (14.48) satisfies

$$\| w_{ns}^{(\mu)} \|_{W^\ell(t,t+1)} \le \begin{cases} c\, e^{-k_0 t} t^{s-b} \log t & \text{if } b \text{ is integer and } b \le s \\ c\, e^{-k_0 t} t^{s-b} & \text{for other values of } b. \end{cases}$$

Now we show that the zeros $u_{kj}^{(\nu)}$ are linearly independent modulo $o(e^{-k_0 t})$.

Proposition 14.3.5. *The relation*

$$\Big\| \sum_{(\nu,k,j)\in Q} c_{kj}^{(\nu)} u_{kj}^{(\nu)} \Big\|_{W^\ell(t,t+1)} = o(e^{-k_0 t}) \tag{14.54}$$

(where $c_{kj}^{(\nu)}$ are constants), implies

$$c_{kj}^{(\nu)} = 0, \quad (\nu,k,j) \in Q.$$

Proof. Let Q_s be the set of $(\nu,k,j) \in Q$ such that $j = s$. From (14.54) and (14.51) it follows that

$$\Big\| \sum_{(\nu,k,j)\in Q_{m_0-1}} c_{kj}^{(\nu)} U_{kj}^{(\nu)} \Big\|_{W^\ell(t,t+1)} = o(e^{-k_0 t} t^{m_0-1}).$$

Since the degree of the polynomials $e^{-i\lambda_\nu t} U_{kj}^{(\nu)}(t)$ in the last sum is $m_0 - 1$ and their leading coefficients are linearly independent (for fixed ν), we obtain

$$c_{kj}^{(\nu)} = 0 \quad \text{for} \quad (\nu,k,j) \in Q_{m_0-1}.$$

By repeating this argument we show that

$$c_{kj}^{(\nu)} = 0 \quad \text{for} \quad Q_s, \ s = m_0 - 2, \ldots, 0.$$

The proof is complete. $\square$

Remark 14.3.6. By Proposition 14.3.5 and Theorem 14.3.2 the linear hull of $u_{kj}^{(\nu)}$, $(\nu,k,j) \in Q$, can play the role of space $\mathbf{X}(L)$ in Theorem 12.2.3.

14.4 Asymptotics of Arbitrary Solutions of (14.1)

14.4.1 Main Result

In this section we obtain estimates for the difference between an arbitrary solution u of the homogeneous equation (14.1), satisfying the growth condition (14.55), and a linear combination of special solutions $U_{kj}^{(\nu)}$. These estimates show that the principal terms in the asymptotics of u at $+\infty$ have the same form as in the case of the non-perturbed equation $\mathcal{A}(D_t)u = 0$.

Theorem 14.4.1. *Let p be a non-negative integer such that (14.37) holds and let*

$$u \in W_{\text{loc}}^{\ell}(t_0, \infty)$$

be a solution of (14.1) subject to

$$\| u \|_{W^{\ell}(t,t+1)} = O\big(e^{-k-t}\big) \quad \text{as} \quad t \to +\infty. \tag{14.55}$$

Then there exist constants $c_{kj}^{(\nu)}$, $(\nu, k, j) \in Q$, such that

$$u(t) = \sum_{(\nu,k,j) \in Q} c_{kj}^{(\nu)} U_{kj}^{(\nu)}(t) + v(t), \tag{14.56}$$

where $v \in W_{\text{loc}}^{\ell}(t_0, \infty)$ and

$$\| v \|_{W^{\ell}(t,t+1)} \le Ce^{-k_0 t} t^{m_0-1} \Big(\int_t^{\infty} \Big(\frac{\tau}{t}\Big)^p \chi(\tau) d\tau$$
$$+ \int_{t_0-1}^{t} \Big(\frac{\tau}{t}\Big)^{p+1} \chi(\tau) d\tau + \int_{t_0-1}^{\infty} e^{-a|t-\tau|} \alpha(\tau) d\tau \Big) + Ce^{-k+t} \tag{14.57}$$

if $p \le m_0 - 1$ and

$$\| v \|_{W^{\ell}(t,t+1)} \le Ce^{-k_0 t} t^{m_0-1} \Big(\int_t^{\infty} \Big(\frac{\tau}{t}\Big)^p \chi(\tau) d\tau$$
$$+ \int_{t_0-1}^{\infty} e^{-a|t-\tau|} \alpha(\tau) d\tau \Big) + Ce^{-k+t} \tag{14.58}$$

if $p \ge m_0 - 1$. The constants $c_{kj}^{(\nu)}$ and C satisfy

$$\sum_{(\nu,k,j) \in Q} |c_{kj}^{(\nu)}| + C \le c \| u \|_{W^{\ell}(t_0,t_0+1)}, \tag{14.59}$$

where c is independent of u.

Proof. Due to Remark 14.3.6 and Theorem 12.2.3, there exist constants $c_{kj}^{(\nu)}$ such that

$$u(t) = \sum_{(\nu,k,j)\in Q} c_{kj}^{(\nu)} u_{kj}^{(\nu)}(t) + w(t),$$

where

$$\| w \|_{W^\ell(t,t+1)} \leq C_1 e^{-k+t} \quad \text{for} \quad t > t_0$$

and

$$\sum_{(\nu,k,j)\in Q} |c_{kj}^{(\nu)}| + |C_1| \leq c \, \| u \|_{W^\ell(t_0,t_0+1)} \,.$$

Using the asymptotic representation (14.48) for $u_{kj}^{(\nu)}$ and the estimates (14.49) and (14.50) we complete the proof. $\square$

14.4.2 Refinement of the Asymptotics

Since the right-hand sides in (14.57) and (14.58) satisfy the rough estimate

$$\| v \|_{W^\ell(t,t+1)} = o\left(e^{-k_0 t} t^{m_0-1}\right),$$

it follows that the sum in (14.56) contains the main terms in the asymptotics of u. In the next assertion we formulate stronger restrictions to the perturbation operator, which ensure a better estimate for v.

Corollary 14.4.2. *The remainder v in (14.56) satisfies*

$$\| v \|_{W^\ell(t,t+1)} = o\left(e^{-k_0 t} t^{m_0-1-p}\right) \quad \text{as} \quad t \to +\infty \tag{14.60}$$

if one of the following conditions is valid:

(i)
$$\alpha(t) = o\left(t^{-p}\right) \quad \text{as } t \to +\infty, \tag{14.61}$$

and
$$\int_{t_0}^{\infty} t^p \chi(t) dt < \infty; \tag{14.62}$$

(ii)
$$\int_{t_0}^{\infty} t^{m_0+p-1} \alpha(t) dt < \infty; \tag{14.63}$$

(iii)
$$\int_{t_0}^{\infty} \left(t^{m_0+p-1} + t^{2p}\right) \rho^2(t) dt < \infty \tag{14.64}$$

and
$$\int_{t_0}^{\infty} t^p \gamma(t) dt < \infty. \tag{14.65}$$

Each of the conditions (i), (ii), (iii) implies (14.37).

Proof. (i) From (14.61) we obtain

$$\int_{t_0}^{\infty} e^{-a|t-\tau|}\alpha(\tau)d\tau = o(t^{-p}) \quad \text{as} \quad t \to +\infty,$$

which together with (14.62) implies (14.37). Hence the relation (14.60) follows from (14.57) and (14.58).

(ii) By (13.43)

$$\int_{t_0}^{\infty} t^p \gamma(t)dt \le c \int_{t_0}^{\infty} t^{m_0+p-1}\alpha(t)dt.$$

Since β is bounded we have

$$\int_{t_0}^{\infty} t^{m_0+p-1}\beta(t) \int_{t_0}^{\infty} e^{-a|t-\tau|}\alpha(\tau)d\tau dt \le c \int_{t_0}^{\infty} t^{m_0+p-1}\alpha(t)dt.$$

Hence (14.63) leads to (14.37). Moreover, the inequality (13.45) together with (14.63) gives

$$\alpha(t) = o(t^{1-m_0-p}).$$

Hence

$$\int_{t_0}^{\infty} e^{-a|t-\tau|}\alpha(\tau)d\tau = o(t^{1-m_0-p}),$$

and the result follows from (14.57) and (14.58).

(iii) Due to (13.44) and (13.70) the estimate (14.37) is a consequence of (14.64) and (14.65). By (5.43) the inequality (14.64) implies

$$\rho(t) = o(t^{-p})$$

and therefore α satisfies (14.61). The result follows from (14.57) and (14.58). □

If $p \ge m_0 - 1$ and $\alpha(t) = o(t^{-p})$ then by Corollary 14.4.2 (i), the theorem explicitly describes the polynomial exponential terms in the asymptotics of u. However, in the case $p < m_0 - 1$, the terms $U_{kj}^{(\nu)}(t)$ with $j < m_0 - 1 - p$ are weaker at infinity than the estimate (14.57) for the remainder term v. To avoid this disadvantage it makes sense to formulate the following consequence of the theorem.

Corollary 14.4.3. *Let p be an integer, $0 < p < m_0 - 1$, and let (14.61) and (14.62) hold. Let $u \in W_{loc}^{\ell}(t_0, \infty)$ be a solution of (14.1) subject to (14.55). Then there exist constants $c_{kj}^{(\nu)}$ such that*

$$\left\| u - \sum_{\substack{(\nu,k,j)\in Q, \\ j \ge m_0-p-1}} c_{kj}^{(\nu)} U_{kj}^{(\nu)} \right\|_{W^{\ell}(t,t+1)} = o(e^{-k_0 t} t^{m_0-1-p}). \tag{14.66}$$

Proof. We rewrite (14.56) as

$$u(t) - \sum_{\substack{(\nu,k,j)\in Q,\\ j\geq m_0-1-p}} c_{kj}^{(\nu)} U_{kj}^{(\nu)}(t) = \sum_{\substack{(\nu,k,j)\in Q,\\ j<m_0-1-p}} c_{kj}^{(\nu)} U_{kj}^{(\nu)}(t) + v(t). \qquad (14.67)$$

Since $e^{-k_0 t} U_{k,j}^{(\nu)}$ are polynomials of degree j, the $W^\ell(t,t+1)$ norm of the sum in the right-hand side of (14.67) can be estimated by

$$c\, e^{-k_0 t} t^{m_0-2-p}.$$

This together with (14.60) leads to (14.66). $\square$

Remark 14.4.4. By (13.44) and Corollary 14.4.2 the asymptotics

$$\|u - \sum_{(\nu,k,j)\in Q} c_{kj}^{(\nu)} U_{kj}^{(\nu)}\|_{W^\ell(t,t+1)} = o(e^{-k_0 t} t^{m_0-1-p})$$

is valid if

$$\int_{t_0}^{\infty} \tau^{m_0+p-1} \rho(\tau) d\tau < \infty. \qquad (14.68)$$

We note that the exponent $m_0 + p - 1$ is best possible. In fact, the following result is obtained in Kozlov, Maz'ya (1997), Theorem 7.4.1:

Let σ, p be integers such that $0 \leq p \leq \sigma \leq m_0 - 1$. Then the condition

$$\int_{t_0}^{\infty} \tau^{m_0+p-1} \rho(\tau) a\tau < \infty$$

is both necessary and sufficient for the existence of a solution to the ordinary differential equation

$$(\partial_t + k_0)^{m_0} (-\partial_t - k_-)^{m_-} w(t) - \rho(t) w(t) = 0 \quad \text{for } t > t_0$$

with the asymptotics

$$w(t) = e^{-k_0 t} t^\sigma (1 + o(t^{-p})) \quad \text{as } t \to \infty.$$

Here k_0, k_- are real numbers, $k_0 > k_-$, and m_0, m_- are positive integers, and ρ is a non-negative measurable function bounded by a small constant for $t > t_0$.

We give an example where the remainder term in the asymptotics of u can be written quite explicitly.

Example 14.4.5. Let (14.53) hold. Then by the estimates (14.66) and (14.58),

$$\| u - \sum_{\substack{(\nu,k,j)\in Q\\ j>m_0-1-b}} c_{kj}^{(\nu)} U_{kj}(\nu) \|_{W^\ell(t,t+1)}$$

$$\leq \begin{cases} c\, e^{-k_0 t} t^{m_0-1-b} \log t & \text{if } b \text{ is integer and } b \leq m_0 - 1, \\ c\, e^{-k_0 t} t^{m_0-1-b} & \text{for other values of } b. \end{cases}$$

14.5 Nonhomogeneous Finite Dimensional System

In the next lemma we construct a special solution of the nonhomogeneous system (13.108). We shall make use of the function

$$g^{(a)}(t) = \begin{cases} e^{-(k_0+a)t} & \text{for } t \geq 0 \\ e^{-k_0 t}(1+|t|)^{m_0-1} & \text{for } t < 0 \end{cases} \tag{14.69}$$

with a given by (14.12).

The following estimates (where g is given by (12.61)) can be checked directly:

$$\int_{\mathbb{R}} g^{(a)}(t-\tau)g(\tau-x)d\tau \leq c\, g^{(a)}(t-x) \tag{14.70}$$

and

$$\int_{\mathbb{R}} g(\tau-x)g^{(a)}(x-t)dx \leq c\, g^{(a)}(\tau-t). \tag{14.71}$$

Lemma 14.5.1. *Let*

$$\int_{t_0-1}^{\infty} \tau^{m_0-1} \| \mathcal{R} \|_{(L_2(\tau,\tau+1))^{\kappa \times \kappa}} d\tau$$

$$+ \int_{t_0-1}^{\infty} \int_{t_0-1}^{\infty} \tau^{m_0-1} \beta(\tau) e^{-a|\tau-x|} \alpha(x) d\tau dx < \varepsilon, \tag{14.72}$$

where $\mathcal{R}$ is the matrix function in (13.108) and ε is a small number. Also let $q \in \left(L_{2,\text{loc}}(t_0,\infty)\right)^{\kappa}$ and

$$\int_{t_0}^{\infty} e^{k_0\tau} \tau^{m_0-1} \| q \|_{(L_2(\tau,\tau+1))^{\kappa}} d\tau < \infty. \tag{14.73}$$

Then the equation (13.108) has a solution $h \in \left(W_{2,\text{loc}}^1(t_0,\infty)\right)^{\kappa}$, which satisfies

$$\| h \|_{(W_2^1(t,t+1))^{\kappa}} \leq c \int_{t_0-1}^{\infty} g^{(a)}(t-\tau) \| q \|_{(L_2(\tau,\tau+1))^{\kappa}} d\tau. \tag{14.74}$$

Proof. We put $\mathcal{G}(t) = 0$ for $t \geq 0$ and

$$\mathcal{G}(t) = -i\, e^{i(\Lambda+J)t} \quad \text{for} \quad t < 0.$$

Then

$$\left(D_t - \Lambda - J\right)\mathcal{G}(t) = \delta(t)$$

and we note that (13.108) follows from the integral equation on $\mathbb{R}$:

$$h(t) = \int_{\mathbb{R}} \mathcal{G}(t-\tau)\left(\mathcal{R}(\tau)h(\tau) + (Kh)(\tau) + q(\tau)\right)d\tau,$$

where $q(t) = 0$ for $t < t_0$. (In fact, the integration here is restricted to $t < t_0$ since $\mathcal{N}(t) = 0$ for $t < t_0$.) We solve this equation by an iterative procedure. We take $h_0 = 0$ and

$$h_{k+1}(t) = \int_{\mathbb{R}} \mathcal{G}(t - \tau)\big(\mathcal{R}(\tau)h_k(\tau) + (Kh_k)(\tau) + q(\tau)\big)d\tau$$

for $k = 0, 1, \ldots$

In order to obtain the convergence of $\{h_k\}$ in $(L_{\infty,\text{loc}}(t_0, \infty))^\kappa$ we introduce the sequence

$$y_k(t) = \| h_k - h_{k-1} \|_{(L_2(t,t+1))^\kappa}, \; k = 1, 2, \ldots$$

Since $\exp(iJt)$ is a polynomial matrix of degree $m_0 - 1$ (see the proof of Lemma 14.2.2) it follows that

$$\| \mathcal{G} \|_{(L_2(t,t+1))^\kappa} \le c\, g^{(a)}(t).$$

By (13.110) we have

$$y_{k+1}(t) \le c \int_{\mathbb{R}} g^{(a)}(t - \tau)\Big(\| \mathcal{R} \|_{(L_2(\tau,\tau+1))^{\kappa \times \kappa}}\, y_k(\tau)$$

$$+\beta(\tau) \int_{\mathbb{R}} g(\tau - s)\alpha(s)y_k(s)ds\Big)d\tau,$$

where g is the function (12.61).

From Theorem 13.12.4 with $\mathbf{g} = g^{(a)}$ and

$$Q(\tau, s) = \| \mathcal{R} \|_{(L_2(\tau,\tau+1))^{\kappa \times \kappa}}\, \delta(\tau - s) + \beta(\tau)g(\tau - s)\alpha(s),$$

it follows that the function

$$\sum_{k=0}^{\infty} \int_{\mathbb{R}^{2k}} g^{(a)}(t - \tau_1)Q(\tau_1, s_1)g^{(a)}(s_1 - \tau_2)Q(\tau_2, s_2)$$

$$\times Q(\tau_k, s_k)g^{(a)}(s_k - \tau)d\tau_1 \ldots d\tau_k ds_1 \ldots ds_k$$

does not exceed $c\, g^{(a)}(t - \tau)$ for $t, \tau \ge 0$ if the number q_1 defined by (13.136) is sufficiently small. In our case, the smallness of q_1 follows from (14.72). Hence

$$\sum_{k=1}^{\infty} y_k(t) \le c \int_{\mathbb{R}} g^{(a)}(t - \tau) \| q \|_{(L_2(\tau,\tau+1))^\kappa}\, d\tau.$$

Therefore $\{h_k\}$ converges in $(L_{\infty,\text{loc}}(t_0, \infty))^\kappa$ and its limit h satisfies

$$\| h \|_{(L_2(t,t+1))^\kappa} \le c \int_{\mathbb{R}} g^{(a)}(t - \tau) \| q \|_{(L_2(\tau,\tau+1))^\kappa}\, d\tau. \tag{14.75}$$

Using (13.108), (13.110) and (13.43) we have

$$\| h' \|_{(L_2(t,t+1))^\kappa} \le c \left(\| h \|_{(L_\infty(t,t+1))^\kappa} + \| q \|_{(L_2(t,t+1))^\kappa} \right.$$

$$\left. + \beta(t) \int_{\mathbb{R}} g(t - \tau)\alpha(\tau) \| h \|_{(L_\infty(\tau,\tau+1))^\kappa} \, d\tau \right). \tag{14.76}$$

Using (14.75), (14.71) and boundedness of α and β the last term in the brackets in (14.76) can be estimated by

$$\le c \int_{\mathbb{R}} g^{(a)}(t - \tau) \| q \|_{(L_2(\tau,\tau+1))^\kappa} \, d\tau.$$

Now the inequality

$$\| q \|_{L_2(t,t+1)} \le c \int_{t-1}^{t+1} \| q \|_{(L_2(\tau,\tau+1))^\kappa} \, d\tau$$

together with (14.75) and (14.76) leads to (14.74). $\square$

Corollary 14.5.2. *Let* (14.72) *hold and let* $f \in L_{2,\mathrm{loc}}(t_0, \infty; H_0)$ *satisfy* (14.9). *Then the equation* (13.108) *has a solution* $h \in \left(W^1_{2,\mathrm{loc}}(t_0, \infty) \right)^\kappa$ *such that*

$$\| h \|_{(W^1_2(t,t+1))^\kappa} \le c \int_{t_0-1}^{\infty} g^{(a)}(t - \tau) \| f \|_{L_2(\tau,\tau+1;H_0)} \, d\tau \tag{14.77}$$

($g^{(a)}$ *is given by* (14.69)).

Proof. Using (14.70) and (14.74) together with (13.111) we arrive at (14.77). $\square$

14.6 A Special Solution of the Nonhomogeneous System (14.2)

The following theorem gives a solution of the nonhomogeneous equation (14.2). This will be used in Sect.14.7 in order to construct the asymptotics of an arbitrary solution to (14.2).

Theorem 14.6.1. *Suppose that the operator* $N(t, D_t)$ *satisfies*

$$\int_{t_0}^{\infty} t^{m_0-1} \| \mathcal{R} \|_{(L_2(t,t+1))^{\kappa \times \kappa}} \, dt < \infty, \tag{14.78}$$

and

$$\int_{t_0}^{\infty} \int_{t_0}^{\infty} \tau^{m_0-1} \beta(\tau) e^{-a|\tau-x|} \alpha(x) dx d\tau < \infty, \tag{14.79}$$

and that $f \in L_{2,\mathrm{loc}}(t_0, \infty; H_0)$ *is subject to* (14.9).

If t_0 *is sufficiently large that the equation* (14.2) *has a solution* $u \in W^\ell_{\mathrm{loc}}(t_0, \infty)$ *such that*

$$\| u \|_{W^{\ell}(t,t+1)} \le c \int_{t_0-1}^{\infty} g^{(a)}(t-\tau) \, \| f \|_{L_2(\tau,\tau+1;H_0)} \, d\tau, \qquad (14.80)$$

where $g^{(a)}$ is given by (14.69) and $f(t) = 0$ for $t < t_0$.

Proof. Using Theorem 13.10.4 along with Lemma 14.5.1 we obtain the existence of a solution $u \in W_{\text{loc}}^{\ell}(t_0, \infty)$ of (14.2) such that

$$\text{col}\big(u(t), \ldots, D_t^{\ell-1}u(t)\big) = \sum_{(\nu,k,j)\in Q} h_{kj}^{(\nu)}(t)\mathcal{U}_{kj}^{(\nu)}(0) + \mathbf{v}(t),$$

where the vector function

$$h = \{h_{kj}^{(\nu)}\}_{(\nu,k,j)\in Q} \in \big(W_{2,\text{loc}}^1(t_0, \infty)\big)^{\kappa}$$

is the solution of (13.108) from Lemma 14.5.1. By Theorem 13.10.4

$$\| \mathbf{v} \|_{\mathbb{S}(t,t+1)} \le c \int_{t_0-1}^{\infty} g(t-\tau)\Big(\alpha(\tau) \, \| h \|_{(W_2^1(\tau,\tau+1))^{\kappa}}$$
$$+ \| f \|_{L_2(\tau,\tau+1;H_0)} \Big)d\tau, \qquad (14.81)$$

where both h and f equal zero for $t < t_0$. Since u satisfies (12.84) on (t_0, ∞) with F given by (13.3) it follows that

$$\| u \|_{W^{\ell}(t,t+1)} \le c \big(\| h \|_{(W_2^1(t,t+1))^{\kappa}} + \| \mathbf{v} \|_{\mathbb{S}(t,t+1)} \big).$$

Therefore

$$\| u \|_{W^{\ell}(t,t+1)} \le c \big(\| h \|_{(W_2^1(t,t+1))^{\kappa}}$$
$$+ \int_{t_0-1}^{\infty} g(t-\tau)\Big(\alpha(\tau) \, \| h \|_{(W_2^1(\tau,\tau+1))^{\kappa}} + \| f \|_{L_2(\tau,\tau+1;H_0)} \Big)\Big)d\tau.$$

This together with (14.77) and (14.71) gives (14.80). $\square$

Corollary 14.6.2. *Let all conditions of Theorem 14.6.1 be fulfilled. Then the solution u in this theorem satisfies*

$$\| u \|_{W^{\ell}(t,t+1)} = o\big(e^{-k_0 t}\big) \quad as \quad t \to +\infty.$$

Proof. Follows from (14.80) and (14.9). $\square$

Remark 14.6.3. The conditions (14.78) and (14.79) can be replaced by a simpler one

$$\int_{t_0}^{\infty} \tau^{m_0-1}\alpha(\tau)d\tau < \infty.$$

This follows from (13.43) and the boundedness of β.

14.7 Asymptotics of Solutions to the Nonhomogeneous System (14.2)

Theorem 14.7.1. *Suppose that*

$$\int_{t_0}^{\infty} \tau^{m_0-1}\big(\chi(\tau)+\parallel \mathcal{R} \parallel_{(L_2(\tau,\tau+1))^\kappa}\big)d\tau < \infty. \tag{14.82}$$

Let $f \in L_{2,\mathrm{loc}}(t_0,\infty;H_0)$ satisfy (14.9) and let $u \in W_{\mathrm{loc}}^\ell(t_0,\infty)$ be a solution of (14.2) subject to (14.6). Then there exist constants $c_{kj}^{(\nu)}$ such that

$$u(t) = \sum_{(\nu,k,j)\in Q} c_{kj}^{(\nu)} U_{kj}^{(\nu)}(t) + w(t) \quad for \quad t > t_0, \tag{14.83}$$

where the remainder w satisfies

$$\parallel w \parallel_{W^\ell(t,t+1)} \le c\Big\{ \Big(\int_{t_0-1}^{\infty} e^{k_0\tau}\tau^{m_0-1} \parallel f \parallel_{L_2(\tau,\tau+1;H_0)} d\tau$$

$$+ \parallel u \parallel_{W^\ell(t_0,t_0+1)}\Big)e^{-k_0 t}\Big(\int_t^{\infty} \tau^{m_0-1}\chi(\tau)d\tau$$

$$+ t^{m_0-1}\int_{t_0-1}^{\infty} e^{-a|t-\tau|}\alpha(\tau)d\tau + e^{(k_0-k_+)t}\Big)$$

$$+ \int_{t_0-1}^{\infty} g^{(a)}(t-\tau) \parallel f \parallel_{L_2(\tau,\tau+1;H_0)} d\tau\Big\}. \tag{14.84}$$

Proof. Denote by u_0 the solution of (14.2) which was constructed in Theorem 14.6.1. Then $u-u_0$ satisfies the homogeneous equation (14.1) and by Corollary 14.6.2 and by (14.6)

$$\parallel u - u_0 \parallel_{W^\ell(t,t+1)} = o\big(e^{-k_- t}\big) \quad as \quad t \to +\infty.$$

Applying Theorem 14.4.1 we obtain

$$u(t) = \sum_{(\nu,k,j)\in Q} c_{kj}^{(\nu)} U_{kj}^{(\nu)}(t) + v(t) + u_0(t),$$

where $c_{kj}^{(\nu)}$ are constants and v satisfies (14.58). By (14.59) the constants $c_{kj}^{(\nu)}$ and the constant C in (14.58) are estimated as

$$\sum_{(\nu,k,j)\in Q} |c_{kj}^{(\nu)}| + C \le c \parallel u - u_0 \parallel_{W^\ell(t_0,t_0+1)}$$

$$\le c\Big(\parallel u \parallel_{W^\ell(t_0,t_0+1)} + \int_{t_0-1}^{\infty} e^{k_0\tau}\tau^{m_0-1} \parallel f \parallel_{L_2(\tau,\tau+1;H_0)} d\tau\Big).$$

The result follows from this inequality and from (14.80) and (14.58). $\square$

Remark 14.7.2. Let

$$\int_{t_0}^{\infty} \tau^{2m_0-2}\alpha(\tau)d\tau < \infty . \tag{14.85}$$

By (13.80), (13.43) and (14.5) we have

$$\|\mathcal{R}\|_{(L_2(\tau,\tau+1))^\kappa} + \tau^{1-m_0}\gamma(\tau) \le c\alpha(\tau).$$

Hence the assumption (14.85) gives (14.82). By Theorem 14.7.1 the representation (14.83) holds with w satisfying (14.84). Standard estimates of the right-hand side (14.84) show that

$$\|w\|_{W^\ell(t,t+1)} = o(e^{-k_0 t})$$

if α is subject to (14.85). Therefore (14.83) is an asymptotic representation of the solution $u(t)$.

Theorem 14.7.1 together with the estimate

$$\chi(t) \le c\, t^{m_0-1}\rho(t),$$

(which is a consequence of (13.44) and (13.70)), implies the following rougher but more simply formulated variant of the theorem.

Corollary 14.7.3. *Let*

$$\int_{t_0}^{\infty} \tau^{2m_0-2}\rho(\tau)d\tau < \infty$$

and let $u \in W^\ell_{\mathrm{loc}}(t_0, \infty)$ be a solution of (14.2) satisfying (14.6). Suppose that (14.9) holds. Then for $t > t_0$ the asymptotic representation (14.83) is valid where

$$\| w \|_{W^\ell(t,t+1)} \le c \Big\{ \Big(\int_{t_0-1}^{\infty} e^{k_0\tau}\tau^{m_0-1} \| f \|_{L_2(\tau,\tau+1;H_0)} \, d\tau$$

$$+ \| u \|_{W^\ell(t_0,t_0+1)} \Big) e^{-k_0 t} \Big(\int_t^{\infty} \tau^{2(m_0-1)}\rho(\tau)d\tau$$

$$+ t^{m_0-1} \int_{t_0-1}^{t} e^{-a(t-\tau)}\rho(\tau)d\tau + e^{(k_0-k_+)t}\Big)$$

$$+ \int_{t_0-1}^{\infty} g^{(a)}(t-\tau) \| f \|_{L_2(\tau,\tau+1;H_0)} \, d\tau \Big\}.$$

14.8 Asymptotics of Solutions in a Weighted Sobolev Space

Here we use Theorem 14.7.1 to obtain an estimate for the remainder term in the asymptotics (14.83) in terms of the norm of a certain weighted Sobolev space.

Theorem 14.8.1. *Let $\sigma < -1/2$ and let*

$$\int_{t_0}^{\infty} \tau^{2(\sigma+m_0)} \left(\chi^2(\tau) + \tau^{-2}\alpha^2(\tau) \right) d\tau < \infty. \tag{14.86}$$

Also let $u \in W_{\mathrm{loc}}^{\ell}(t_0, \infty)$ be a solution of (14.2) subject to (14.6). If (14.10) holds then the remainder w in (14.83) satisfies

$$\int_{t_0}^{\infty} e^{2k_0\tau} \tau^{2\sigma} \sum_{j=0}^{\ell} \| D_\tau^j w(\tau) \|_{\ell-j}^2 \, d\tau \tag{14.87}$$

$$\leq c \Big\{ \int_{t_0-1}^{\infty} e^{2k_0\tau} \tau^{2(m_0+\sigma)} \| f \|_{L_2(\tau,\tau+1;H_0)}^2 \, d\tau + \| u \|_{W^{\ell-1}(t_0,t_0+1)}^2 \Big\}.$$

Proof. The inequality (14.86) together with $\sigma < -1/2$ guarantees (14.82). By Proposition 4.2.1, a), the left-hand side of (14.87) is estimated by

$$\int_{t_0}^{\infty} e^{2k_0\tau} \tau^{2\sigma} \| w \|_{W^{\ell}(\tau,\tau+1)}^2 \, d\tau.$$

Thus (14.87) follows from (14.84). $\square$

Remark 14.8.2. The condition (14.86) follows from one of the assumptions

(i) $\displaystyle \int_{t_0}^{\infty} \tau^{2(\sigma+m_0)} \left(\gamma(t) + t^{m_0-1}\rho^2(t) + t^{-1}\alpha(\tau) \right)^2 d\tau < \infty,$

(ii) $\displaystyle \int_{t_0}^{\infty} \tau^{2(\sigma+2m_0-1)} \alpha^2(\tau) d\tau < \infty. \tag{14.88}$

Proof. (i) By (13.44) and (13.70)

$$\alpha(t) + \beta(t) \leq c\, \rho(t).$$

Hence

$$\int_{t_0}^{\infty} \tau^{2(\sigma+m_0)} \chi^2(\tau) d\tau \leq c \int_{t_0}^{\infty} \tau^{2(\sigma+m_0)} \left(\gamma(\tau) + \tau^{m_0-1}\rho^2(\tau) \right)^2 d\tau.$$

(ii) By (13.43)

$$\chi(\tau) \leq c\, \tau^{m_0-1} \Big(\alpha(\tau) + \int_{t_0}^{\infty} e^{-a|\tau-x|}\alpha(x)dx \Big).$$

Therefore (14.88) implies (14.86). $\square$

Remark 14.8.3. Condition (14.88), and an even rougher sufficient condition

$$\int_{t_0}^{\infty} \tau^{2(\sigma + 2m_0 - 1)} \rho^2(\tau) d\tau < \infty, \tag{14.89}$$

is best possible in some sense. In fact, by assuming that ρ is decreasing for $t \leq t_0$, we can obtain that (14.89) is necessary for (14.87) in the special case

$$L(t, D_t) = (\partial_t + k_+)^{m_+} (-\partial_t - k_0)^{m_0} - \rho(t) I,$$

where $k_+ > k_0$. To check this we assume that the solution u of $Lu = 0$ for $t > t_0$ can be represented in the form

$$u(t) = e^{-k_0 t} t^{m_0 - 1} + v(t),$$

where $v(t) = o(e^{-k_0 t})$ as $t \to +\infty$. By Theorem 7.4.3 from Kozlov, Maz'ya (1997), this implies that

$$\int_{t_0}^{\infty} \tau^{2m_0 - 2} \rho(\tau) d\tau < \infty.$$

The remainder term v satisfies

$$(\partial_t + k_+)^{m_+} (-\partial_t - k_0)^{m_0} v(t) = \rho(t) e^{-k_0 t} (t^{m_0 - 1} + o(1)) \quad \text{for} \quad t > t_0.$$

We extend ρ to $(-\infty, t_0)$ by setting it to 0 there. Then

$$v(t) = e^{-k_+ t} p(t) + \int_{t_0}^{\infty} g(t - \tau) \rho(\tau) e^{-k_0 \tau} (\tau^{m_0 - 1} + o(1)) d\tau,$$

where p is a polynomial of degree $m_+ - 1$ and g is Green's function given by (3.23) and (3.24) with k_-, m_- replaced by k_0, m_0. The integral on the right is greater than

$$c\, e^{-k_0 t} \int_{2t}^{3t} (1 + \tau - t)^{m_0 - 1} \tau^{m_0 - 1} \rho(\tau) d\tau \geq$$
$$\geq c_1 e^{-k_0 t} t^{2m_0 - 1} \rho(3t).$$

Therefore the finiteness of the integral

$$\int_{t_0}^{\infty} e^{2k_0 \tau} \tau^{2\alpha} |v(\tau)|^2 d\tau$$

implies (14.89). $\square$

14.9 Asymptotic Behaviour of Solutions to Elliptic Equations near an Interior Point

Let $P(D_x)$ be the elliptic operator (0.7) with constant coefficients and let $Q(x, D_x)$ be the operator (6.50) with measurable coefficients in the punctured ball $B_{r_0}\setminus\{0\}$. We introduce the function

$$S(r) = \sup_{K_r} \left\{ \sum_{|\alpha|=2m} |q_\alpha(x) - p_\alpha| + \sum_{|\alpha|<2m} |x|^{2m-|\alpha|}|q_\alpha(x)| \right\},$$

where $K_r = \{x \in \mathbb{R}^n : e^{-1}r < |x| < r\}$, and assume that $S(r)$ does not exceed a small positive constant.

We shall formulate three theorems on the asymptotic representations as $x \to 0$ for solutions $u \in W^{2m}_{2,\mathrm{loc}}(B_{r_0}\setminus\{0\})$ of

$$Q(x, D_x)u = 0 \quad \text{on} \quad B_{r_0}\setminus\{0\} \tag{14.90}$$

satisfying

$$\|u\|_{W^{2m}_2(K_r)} = O(r^{k+\delta}) \tag{14.91}$$

for some $\delta > 0$ and integer k. These theorems follow from Corollary 14.4.2 combined with Proposition 1.5.3.

Theorem 14.9.1. *Let $2m < n$ and*

$$\int_0^{r_0} S(r)|\log r|^{\gamma-1}\frac{dr}{r} < \infty, \tag{14.92}$$

where γ is a positive integer.
  (i) *If $k \geq 0$ then*

$$u(x) = \sum_{|\alpha|=k+1} c_\alpha x^\alpha + v(x), \tag{14.93}$$

where $c_\alpha = \mathrm{const}$ and

$$\|v\|_{W^{2m}_2(K_r)} = o\left(r^{k+1}|\log r|^{1-\gamma}\right). \tag{14.94}$$

  (ii) *If $k \leq 2m - n - 1$ then*

$$u(x) = \sum_{|\alpha|=2m-n-k-1} C_\alpha D_x^\alpha G(x) + v(x), \tag{14.95}$$

where $C_\alpha = \mathrm{const}$, G is the positive homogeneous of degree $2m - n$ fundamental solution for the operator $P(D_x)$ and v satisfies (14.94).
  (iii) *If $k = 2m - n$ then*

$$u(x) = \mathrm{const} + v(x), \tag{14.96}$$

where

$$\|v\|_{W^{2m}_2(K_r)} = o\left(|\log r|^{1-\gamma}\right).$$

Remark 14.9.2. The asymptotics (14.96) can be made more precise under the assumption that the operator Q contains no derivatives of order $|\alpha| < s$, i.e.

$$Q(x, D_x) = \sum_{s \le |\alpha| \le 2m} q_\alpha(x) D_x^\alpha.$$

The formula (14.96) can be replaced by

$$u(x) = \sum_{|\alpha| \le s} c_\alpha x^\alpha + v(x),$$

where

$$\|v\|_{W_2^{2m}(K_r)} = o\left(r^s |\log r|^{1-\gamma}\right).$$

In fact, the proof of Theorem 9.5.2 shows that

$$u(x) = \sum_{|\alpha| \le s-1} c_\alpha x^\alpha + w(x),$$

where

$$\|w\|_{W_2^{2m}(K_r)} = O\left(r^{s-1+\delta}\right).$$

Since $Qw = 0$, the result follows by Theorem 14.9.1 (i) applied to w and $k = s - 1$. $\square$

Theorem 14.9.3. *Let n be odd, $2m > n$ and let S be subject to (14.92). Then*

$$u(x) = \sum_{|\alpha|=k+1} c_\alpha x^\alpha + \sum_{|\beta|=2m-n-k-1} C_\beta D_x^\beta G(x) + v(x), \qquad (14.97)$$

where c_α and C_β are constants and v satisfies (14.94). If either $k < -1$ or $k \ge 2m - n$, then the first or the second sum in (14.97) is omitted.

Theorem 14.9.4. *Let n be even and $2m \ge n$.*
    (i) If $k \le -2$ and (14.92) holds then u satisfies (14.95) with v subject to (14.94).
    (ii) If $k \ge 2m - n$ and (14.92) holds then u satisfies (14.93) with v subject to (14.94).
    (iii) Let $-1 \le k \le 2m - n - 1$ and let (14.92) be valid with $p \ge 2$. Then u is represented by (14.97) with

$$\|v\|_{W_2^{2m}(K_r)} = o\left(r^{k+1} |\log r|^{3-\gamma}\right).$$

In order to simplify the statements of Theorems 14.9.1 - 14.9.4 we have restricted ourselves to somewhat rough estimates of the remainder v. Improved estimates, stated in terms of $S(r)$, can be obtained directly from Theorem 14.4.1 and Corollary 14.4.2.

14.10 Comments

When estimating the remainders in the asymptotic formulae of this chapter, we used the L_2-norm and the family of Hilbert spaces $H_0, ..., H_\ell$. However, all our arguments are applicable to the case of the Banach spaces $B_0, ..., B_\ell$ and the L_p-norm, $1 < p < \infty$, considered in Sect. 10.9. One should replace H_0 by B_0 and L_2, W^ℓ by L_p, W_p^ℓ in the corresponding statements.

The preservation of the power-exponential asymptotics was established in Kozlov, Maz'ya (1988) and Kozlov, Maz'ya (1991-1996), report 6, under the condition

$$\int_{t_0}^{\infty} \tau^{m-1} \rho(\tau) d\tau < \infty$$

with $m > m_0$. In Sections 14.3 and 14.4 we refined this result by imposing restrictions on the perturbation $N(t, D_t)$ which were stated only in terms of α, β and γ (see Theorems 14.3.2 and 14.4.1).

Condition (14.89) (which ensures the estimate (14.87) for the remainder term in the power-exponential asymptotics) improves the condition

$$\rho(t) = O(t^{-\delta}) \quad \text{as } t \to +\infty$$

with $\delta > \sigma + 2m_0 - 1/2$ obtained in Bagirov, Kondratiev (1991).

15. The Case of One Simple Eigenvalue on the Line

15.1 Introduction

In the present chapter we suppose that there is one eigenvalue λ_0 on the line $\Im\lambda = k_0$ and that only one eigenfunction φ corresponds to λ_0 (up to multiplication by a constant). We also assume that there are no generalized eigenvectors corresponding to λ_0.

Let φ and ψ be the eigenvectors of the operator pencils $\mathcal{A}(\lambda)$, $\mathcal{A}^*(\lambda)$ corresponding to the eigenvalues λ_0, $\overline{\lambda_0}$ which satisfy

$$\left(\frac{d}{d\lambda}\mathcal{A}(\lambda)\big|_{\lambda=\lambda_0}\varphi, \psi\right)_{H_0} = 1 . \tag{15.1}$$

(This is a special case of (1.20)).

As in the theory of ordinary differential equations, this is the simplest situation for the asymptotic analysis of the equations (14.1) and (14.2). As in Chapters 12–14 we require that $\rho(t) \to 0$ as $t \to +\infty$.

After establishing a variant of the comparison principle for scalar integral inequalities in Sect.15.2, we turn in Sect.15.3 to an upper estimate for solutions of (14.2) satisfying

$$\| u \|_{W^\ell(t,t+1)} \leq c\, e^{-k_- t} \quad \text{for} \quad t > t_0. \tag{15.2}$$

For solutions of (14.1) this estimate takes the form

$$\| u \|_{W^\ell(t,t+1)} \leq c\, \exp\left\{ -\Im \int_{t_0}^t \lambda(\tau)d\tau + c_0 \int_{t_0}^t \gamma(\tau)d\tau \right\},$$

where $c_0 = \text{const} > 0$,

$$\lambda(\tau) = \lambda_0 + \left(N(\tau, \lambda_0)\varphi, \psi \right)_{H_0} \tag{15.3}$$

and

$$\chi(\tau) = \beta(\tau) \int_{t_0-1}^\infty e^{-a|\tau-x|}\alpha(x)dx. \tag{15.4}$$

The functions α and β are the same as previously, but here they are much simpler than in the general case (see (15.15) and (15.16)). We show by an example that the above estimate is precise in a sense.

In Sect. 15.4 and 15.5 we find the asymptotic formula for solutions to (14.1):

$$u(t) \sim c \, \exp \Big(i \int_{t_0}^t \lambda(\tau) d\tau \Big) \varphi, \quad t \to +\infty, \tag{15.5}$$

which is valid when

$$\int_{t_0}^\infty \chi(\tau) d\tau < \infty.$$

According to Sect.15.6, the same asymptotic formula holds for solutions of the nonhomogeneous equation (14.2) if

$$\int_{t_0}^\infty \exp \Big(\Im \int_{t_0}^t \lambda(\tau) d\tau \Big) \, \| \, f \, \|_{L_2(t,t+1;H_0)} \, dt < \infty. \tag{15.6}$$

15.2 A Comparison Principle for Integral Inequalities

The following comparison principle will be used in the sequel.

Lemma 15.2.1. *Let $\varkappa$ be a real number and let $Q = Q(t,\tau)$ be a non-negative measurable function on $(t_0,\infty) \times (t_0,\infty)$ such that*

$$\operatorname*{ess\,sup}_{t \geq t_0} \int_{t_0}^\infty e^{\varkappa(\tau-t)} Q(t,\tau) d\tau < 1. \tag{15.7}$$

Additionally, let f be a non-negative function on $[t_0,\infty)$ which is subject to

$$f(t) \leq c \, e^{\varkappa t}. \tag{15.8}$$

Suppose p_+ and p_- are non-negative functions satisfying

$$p_+(t) \geq \int_{t_0}^\infty Q(t,\tau) p_+(\tau) d\tau + f(t) \tag{15.9}$$

and

$$p_-(t) \leq \int_{t_0}^\infty Q(t,\tau) p_-(\tau) d\tau + f(t). \tag{15.10}$$

If

$$p_+(t) + p_-(t) = O\big(e^{\varkappa t}\big) \tag{15.11}$$

then

$$p_-(t) \leq p_+(t). \tag{15.12}$$

Proof. Let B be the Banach space of measurable functions on (t_0, ∞) supplied with the norm

$$\text{ess sup}_{t > t_0} e^{-\varkappa t} |w(t)|.$$

By (15.7), the norm of the operator

$$w \to \int_{t_0}^{\infty} Q(\cdot, \tau) w(\tau) d\tau$$

in B is less then 1. Hence, by the Banach Theorem, the equation

$$v(t) = \int_{t_0}^{\infty} Q(t, \tau) v(\tau) d\tau + f(\tau)$$

has a single solution in B, and the iterations

$$v_{k+1}(t) = \int_{t_0}^{\infty} Q(t, \tau) v_k(\tau) d\tau + f(t), \quad k = 0, 1, \ldots,$$

converge to v in B whatever be the initial choice of $v_0 \in B$.

Due to (15.11) $p_+, p_- \in B$. Putting $v_0 = p_+$ we see by (15.9) that $v_0 \geq v_1 \geq \ldots \geq v_k \geq \ldots$. Therefore $p_+ \geq v$.

Putting $v_0 = p_-$ we derive from (15.10) that $v_0 \leq v_1 \leq \ldots \leq v_k \leq \ldots$ Hence $p_- \leq v$ and we arrive at (15.12). $\square$

15.3 Estimate for Solutions of the Scalar Equation (13.108)

As mentioned in Sect. 15.1, we suppose that there is only one simple eigenvalue λ_0 in the strip $k_- \leq \Im \lambda \leq k_+$ and that $k_0 = \Im \lambda_0 \in (k_-, k_+)$. In this case, the equation (13.108) takes the form

$$(D_t - \lambda(t)) h(t) - (Kh)(t) = q(t) \quad \text{for} \quad t > t_0, \tag{15.13}$$

where $\lambda(\tau)$ is given (15.3). By (13.110) the operator K satisfies the estimate

$$\| Kh \|_{L_2(t, t+1)} \leq c \, \beta(t) \int_{t_0-1}^{\infty} g(t - \tau) \alpha(\tau) \| h \|_{L_\infty(\tau, \tau+1)} \, d\tau, \tag{15.14}$$

where g is Green's function defined in (12.61) and $h(t) = 0$ for $t < t_0$. The functions α and β are expressed as

$$\alpha(t) = \left(\int_t^{t+1} \| N(\tau, \lambda_0) \varphi \|_0^2 \, d\tau \right)^{1/2} \tag{15.15}$$

and

$$\beta(t) = \sup_{v} \Big(\int_{t}^{t+1} \big| (N(\tau, D_\tau)v(\tau), \psi)_{H_0}$$

$$- (\mathcal{B}(\lambda_0, D_\tau)v(\tau), \psi)_{H_0} (N(\tau, \lambda_0)\varphi, \psi)_{H_0} \big|^2 d\tau \Big)^{1/2}, \qquad (15.16)$$

where the operator $\mathcal{B}$ is defined by (13.73) and the supremum is taken over the set (13.75).

By (13.43), (13.44) and (13.70)

$$\alpha(t) + \beta(t) + \Big(\int_{t}^{t+1} \big| (N(\tau, \lambda_0)\varphi, \psi)_{H_0} \big|^2 d\tau \Big)^{1/2} \le c\,\rho(t). \qquad (15.17)$$

Furthermore, by Proposition 13.8.4

$$\beta(t) \le c \sum_{m=1}^{\ell} \sup_{\tau \in (t, t+1)} \| N_m^*(\tau)\psi$$

$$- (N(\tau, \lambda_0)\varphi, \psi)_{H_0} \sum_{p=0}^{m-1} \overline{\lambda}_0^p A_{m-1-p}^* \psi \|_{H_m^*} .$$

Lemma 15.3.1. *Let* $q \in L_{2,\mathrm{loc}}(t_0, \infty)$ *satisfy*

$$\int_{t_0}^{\infty} e^{\Im \int_{t_0}^{\tau} \lambda(x)dx} \| q \|_{L_2(\tau, \tau+1)} d\tau < \infty. \qquad (15.18)$$

Also let $h \in W_{2,\mathrm{loc}}^1(t_0, \infty)$ *be a solution of (15.13) such that*

$$\| h \|_{W_2^1(t, t+1)} \le c\, e^{-(k_0-a)t} \quad for \quad t > t_0 , \qquad (15.19)$$

where a is given by (14.12). Then

$$\| h \|_{W_2^1(t, t+1)} \le ce^{-\Im \int_{t_0}^{t} \lambda(\tau)d\tau + c_0 \int_{t_0}^{t} \chi(\tau)d\tau}$$

$$\times \Big(\| h \|_{L_\infty(t_0, t_0+1)} + \int_{t_0-1}^{\infty} e^{\Im \int_{t_0}^{\tau} \lambda(x)dx} \| q \|_{L_2(\tau, \tau+1)} d\tau \Big) \quad (15.20)$$

with some constants c and c_0*. Here* χ *is the function defined in (15.4).*

Proof. Put

$$y(t) = e^{-i \int_{t_0}^{t} \lambda(\tau)d\tau} h(t).$$

From (15.13) one derives

$$D_t y(t) = K_{\tau \to t} \Big(e^{i \int_{t}^{\tau} \lambda(x)dx} y(\tau) \Big) + e^{-i \int_{t_0}^{t} \lambda(\tau)d\tau} q(t). \qquad (15.21)$$

Using (15.21) and (15.14) we obtain the following inequality for the function $p_-(t) = \| y \|_{L_\infty(t, t+1)}$:

$$p_-(t) \le z(t) + \int_{t_0-1}^{\infty} Q(t,\tau)p_-(\tau)d\tau \,, \tag{15.22}$$

where

$$Q(t,\tau) = c\,\alpha(\tau)\int_{t_0}^{t+1} \beta(x)g(x-\tau)e^{-\Im\int_\tau^x \lambda(s)ds}dx \tag{15.23}$$

and

$$z(t) = p_-(t_0) + c\int_{t_0-1}^{t+1} e^{\Im\int_{t_0}^\tau \lambda(s)ds}\,\|\,q\,\|_{L_2(\tau,\tau+1)}\,d\tau.$$

Since

$$\int_{t_0-1}^{t_0} Q(t,\tau)p_-(\tau)d\tau \le c\,p_-(t_0),$$

we can rewrite (15.22) as

$$p_-(t) \le cz(t) + \int_{t_0}^{\infty} Q(t,\tau)p_-(\tau)d\tau$$

Due to (15.17) and (13.1) we have

$$\alpha(t) + \beta(t) + \gamma(t) + \left(\int_t^{t+1} \left|(N(\tau,\lambda_0)\varphi,\psi)_{H_0}\right|^2 d\tau\right)^{1/2} \le c\rho_0 \,, \tag{15.24}$$

where ρ_0 is the constant in (13.1). Therefore (15.18) and (15.19) imply that z is bounded on (t_0,∞) and

$$p_-(t) \le c\,e^{\varkappa t} \,,$$

where $t \ge t_0$ and $\varkappa = 3a/2$. From the definition (15.23) of $Q(t,\tau)$ we obtain

$$\sup_{t \ge t_0} \int_{t_0}^{\infty} e^{\varkappa(\tau-t)}Q(t,\tau)d\tau < 1.$$

Hence, to estimate w we can apply Lemma 15.2.1. Let

$$p_+(t) = 2Ce^{c_0\int_{t_0}^t \chi(\tau)d\tau} \,,$$

where c_0 is a positive constant and

$$C = c_*\left(p_-(t_0) + \int_{t_0-1}^{\infty} e^{\Im\int_{t_0}^\tau \lambda(s)ds}\,\|\,q\,\|_{L_2(\tau,\tau+1)}\,d\tau\right).$$

Here c_* is a constant which is chosen in such a way that $C \ge cz(t)$ for $t > t_0$. Our goal is to find the constant c_0 such that (15.9) holds (with $f(t) = C$) for sufficiently small ρ_0.

By (15.23) and (15.24) we get

$$Q(t,\tau)p_+(\tau) \le c\,\alpha(\tau) \int_{t_0}^{t+1} \beta(x)e^{-a|x-\tau|}p_+(x)dx.$$

Hence

$$\int_{t_0}^{\infty} Q(t,\tau)p_+(\tau)d\tau \le 2cC \int_{t_0}^{t+1} \chi(x)e^{c_0 \int_{t_0}^{x} \chi(\tau)d\tau}\,dx$$

$$= \frac{c}{c_0}\big(p_+(t+1) - p_+(0)\big) \le \frac{c}{c_0}\big(e^{c_0 c\varepsilon}p_+(t) - 2C\big).$$

Choosing $c_0 = 2c$ and assuming that

$$\rho_0 \le (2c^2)^{-1}\log 2$$

we get

$$p_+(t) \ge C + \int_{t_0}^{\infty} Q(t,\tau)p_+(\tau)d\tau.$$

By Lemma 15.2.1 we obtain

$$p_-(t) \le p_+(t), \tag{15.25}$$

Using (15.21) and (15.14) together with (15.25) we get

$$\| y' \|_{L_2(t,t+1)} \le ce^{\Im \int_{t_0}^{t} \lambda(\tau)d\tau} \| q \|_{L_2(t,t+1)}$$

$$+ c\Big(\| y \|_{L_\infty(t_0,t_0+1)} + \int_{t_0-1}^{\infty} e^{\Im \int_{t_0}^{\tau} \lambda(s)ds} \| q \|_{L_2(\tau,\tau+1)}\,d\tau \Big)\chi(t)p_+(t);$$

this, along with (15.25), gives the estimate for $\|h'\|_{L_2(t,t+1)}$ by the right-hand side in (15.20). $\square$

Now we prove the main result of this section.

Theorem 15.3.2. *Let* $f \in L_{2,\mathrm{loc}}(t_0,\infty)$ *be subject to (15.6). Also let* $u \in W^{\ell}_{\mathrm{loc}}(t_0,\infty)$ *be a solution of (14.2) (with sufficiently large* t_0*) such that (15.2) is valid. Then the estimate*

$$\| u \|_{W^{\ell}(t,t+1)} \le c\,C \exp\Big(-\Im \int_{t_0}^{t} \lambda(\tau)d\tau + c_0 \int_{t_0}^{t} \chi(\tau)d\tau \Big) \tag{15.26}$$

holds, where c *does not depend on* u *and* f*, whereas* χ *is given by (15.4) and*

$$C = \| u \|_{W^{\ell}(t_0,t_0+1)} + \int_{t_0-1}^{\infty} e^{\Im \int_{t_0}^{\tau} \lambda(x)dx} \| f \|_{L_2(\tau,\tau+1;H_0)}\,d\tau, \tag{15.27}$$

(here $f(t) = 0$ *for* $t < t_0$*).*

Proof. We apply Theorem 12.2.1(ii) to obtain the estimate

$$\| u \|_{W^\ell(t,t+1)} \leq c_\delta e^{-(k_0-\delta)t} \quad \text{for} \quad t > t_0 \tag{15.28}$$

for an arbitrary positive δ. Choosing t_0 sufficiently large we can suppose, by (13.94), that (13.1) is valid.

Let h satisfy (15.13) with q given by (13.83), where the function F is defined by (13.105). From Theorem 13.10.5 we obtain

$$\text{col} \left(D_t^{j-1} u(t) \right)_{j=1}^\ell = h(t) \, \text{col} \left(\lambda_0^{j-1} \varphi \right)_{j=1}^\ell + \mathbf{v}(t), \tag{15.29}$$

where $\mathbf{v}$ is defined in (13.91). By (13.114) the following estimate is valid:

$$\| \mathbf{v} \|_{\mathbb{S}(t,t+1)} \leq c \Big\{ \int_{t_0-1}^\infty g(t-\tau) \Big(\alpha(\tau) \, \| h \|_{L_\infty(\tau,\tau+1)}$$

$$+ \| f \|_{L_2(\tau,\tau+1;H_0)} \Big) d\tau + \| u \|_{W^{\ell-1}(t_0,t_0+1)} \, e^{-k+t} \Big\}. \tag{15.30}$$

From (15.28) and (15.29) it follows that h satisfies (15.19). Using (13.113) and (15.6) we get

$$\int_{t_0-1}^\infty e^{\Im \int_{t_0}^\tau \lambda(x)dx} \, \| q \|_{L_2(\tau,\tau+1)} \, d\tau \leq c \, C.$$

Hence all conditions of Lemma 15.3.1 are fulfilled. The last estimate together with (15.20) gives

$$\| h \|_{W_2^1(t,t+1)} \leq c \, C \exp \Big(- \Im \int_{t_0}^t \lambda(\tau)d\tau + c_0 \int_{t_0}^t \chi(\tau)d\tau \Big).$$

Hence, using (15.30), the function $\| \mathbf{v} \|_{\mathbb{S}(t,t+1)}$ is majorized by the right-hand side of (15.26). An obvious consequence of (15.29) is the inequality

$$\| u \|_{W^\ell(t,t+1)} \leq c \left(\| h \|_{W_2^1(t,t+1)} + \| \mathbf{v} \|_{\mathbb{S}(t,t+1)} \right).$$

The proof is complete. $\square$

We prove an estimate similar to (15.26) which is formulated in terms of the function

$$\Omega(t) = \sup_{\tau>t} \rho(\tau). \tag{15.31}$$

Corollary 15.3.3. *Let the conditions of* Theorem 15.3.2 *be satisfied. Then*

$$\| u \|_{W^\ell(t,t+1)} \leq c \, C \exp \Big\{ - \Im \int_{t_0}^t \lambda(\tau)d\tau + c_1 \int_{t_0}^t \Omega^2(\tau)d\tau \Big\}, \tag{15.32}$$

where C is given by (15.27).

Proof. By (15.17),

$$\chi(\tau) \le c\Omega(\tau) \int_{t_0-1}^{\infty} e^{-a|\tau - x|} \Omega(x) dx.$$

Therefore

$$\int_{t_0-1}^{t} \chi(\tau) d\tau \le c \Big\{ \int_{t_0-1}^{t} \Omega(x) \int_{x}^{t} e^{-a(\tau - x)} \Omega(\tau) d\tau dx$$

$$+ \int_{t_0-1}^{t} \Omega(\tau) \int_{\tau}^{\infty} e^{-a(x-\tau)} \Omega(x) dx d\tau \Big\}.$$

Hence, using the monotonicity of Ω,

$$\int_{t_0-1}^{t} \chi(\tau) d\tau \le c \int_{t_0-1}^{t} \Omega^2(\tau) d\tau.$$

The result follows from (15.26). $\square$

Now we show that the estimates (15.26) and (15.32) are precise in some sense.

Example 15.3.4. Consider the ordinary differential equation

$$u''(t) - (1 + N(t))u(t) = 0 \quad \text{for} \quad t > t_0 , \tag{15.33}$$

where N is a real valued function, $N(t) \to 0$ as $t \to +\infty$. We put

$$\mathcal{A}(D_t) = -D_t^2 - 1 \quad \text{and} \quad N(t, D_t) = N(t).$$

One can easily check that

$$c_1 \rho(t) \le \Big(\int_{t}^{t+1} |N(\tau)|^2 d\tau \Big)^{1/2} \le c_2 \rho(t).$$

The pencil

$$\mathcal{A}(\lambda) = -\lambda^2 - 1$$

acts in $\mathbb{C}$ and has the simple eigenvalues $\lambda_{\pm} = \pm i$. The eigenvectors $\varphi_{\pm}$ are equal to 1. The corresponding eigenvectors $\psi_{\pm}$ of $\mathcal{A}^*(\lambda)$ satisfy the normalization condition

$$1 = \Big(\mathcal{A}'(\lambda_{\pm})\varphi_{\pm}, \psi_{\pm} \Big)_{\mathbb{C}} = -2\lambda_{\pm} \overline{\psi_{\pm}}.$$

Hence $\psi_{\pm} = \mp i/2$. Let u be a bounded solution of (15.33). We shall use Corollary 15.3.3 with $\lambda_0 = \lambda_+ = i$. Then

$$\lambda(\tau) = i + \frac{i}{2} N(\tau)$$

and by (15.32) we have

$$|u(t)| \leq c\exp\left(-t - \frac{1}{2}\int_{t_0}^{t} N(\tau)d\tau + c_1\int_{t_0}^{t}\Omega^2(\tau)d\tau\right), \tag{15.34}$$

where Ω is defined by (15.31) and c, c_1 are positive constants. On the other hand, under the complementary condition

$$\int_{t_0}^{\infty}|N''(\tau)|d\tau < \infty,$$

the Liouville-Green asymptotic formula holds:

$$u(t) \sim c\,\exp\left(-\int_{t_0}^{t}(1 + N(\tau))^{1/2}d\tau\right)$$

(see Hartman (1964), Ch. XI, Sect. 9). Therefore, the following estimate is valid:

$$|u(t)| \geq c\,\exp\left(-t - \frac{1}{2}\int_{t_0}^{t}N(\tau)d\tau + c_2\int_{t_0}^{t}N^2(\tau)d\tau\right).$$

Up to the values of c, c_1, c_2 and the replacement of N^2 by Ω^2, this lower estimate coincides with the upper one (15.34). $\square$

15.4 Zeros with Prescribed Asymptotics at Infinity

We start with a construction of a special solution to the equation

$$D_t h(t) - \lambda(t)h(t) - (Kh)(t) = 0 \quad \text{for} \quad t > t_0. \tag{15.35}$$

Lemma 15.4.1. *Suppose that*

$$\int_{t_0-1}^{\infty}\int_{t_0-1}^{\infty} e^{-a|t-\tau|}\beta(t)\alpha(\tau)d\tau dt \leq \varepsilon, \tag{15.36}$$

where ε is a small constant. Then the equation (15.35) has a solution

$$h(t) = e^{i\int_{t_0}^{t}\lambda(\tau)d\tau}(1 + r(t)), \tag{15.37}$$

where $\lambda(\tau)$ is given by (15.3) and the remainder r satisfies

$$\|r\|_{W_2^1(t,t+1)} \leq c\int_{t-1}^{\infty}\beta(\tau)\int_{t_0-1}^{\infty}e^{-a|\tau-x|}\alpha(x)dx d\tau \tag{15.38}$$

for $t > t_0$.

Proof. Inserting (15.37) into (15.35) we obtain the equation for r:

$$D_t r(t) - (Tr)(t) = (T1)(t) \quad \text{for} \quad t > t_0, \tag{15.39}$$

where

$$(Tr)(t) = K_{\tau \to t}(e^{i \int_t^\tau \lambda(x)dx} r(\tau)).$$

Using (15.17) we get

$$\left| \int_\tau^t (N(x,\lambda_0)\varphi, \psi)_{H_0} dx \right| \leq c \sup_{x > t_0} \rho(x)(1 + |t - \tau|).$$

Hence the smallness of ρ and the estimate (15.14) together imply

$$\| Tr \|_{L_2(t,t+1)} \leq c\,\beta(t) \int_{t_0-1}^\infty e^{-a|t-\tau|}\alpha(\tau) \| r \|_{L_\infty(\tau,\tau+1)} d\tau, \tag{15.40}$$

where r is extended by zero for $t < t_0$. Now we write (15.39) as the integral equation

$$r(t) - (\mathcal{T}r)(t) = (\mathcal{T}1)(t), \tag{15.41}$$

where

$$(\mathcal{T}z)(t) = -i \int_t^\infty (Tz)(\tau)d\tau.$$

To show the solvability of (15.41) in $L_\infty(t_0, \infty)$ we estimate the norm of $\mathcal{T}$ in this space. We have

$$\left| \int_{t+1}^\infty (Tr)(\tau)d\tau \right| \leq \int_t^\infty \int_\tau^{\tau+1} |(Tr)(x)|dx\,d\tau$$

$$\leq \int_t^\infty \| Tr \|_{L_2(\tau,\tau+1)} d\tau$$

$$\leq c \int_t^\infty \beta(\tau) \int_{t_0-1}^\infty e^{-a|\tau-x|}\alpha(x) \| r \|_{L_\infty(x,x+1)} dx\,d\tau. \tag{15.42}$$

Hence

$$\| \mathcal{T} \|_{L_\infty(t_0,\infty) \to L_\infty(t_0,\infty)} \leq c\,\varepsilon,$$

where ε is the constant in (15.36). Therefore, equation (15.41) is solvable in $L_\infty(t_0, \infty)$. Using (15.42), the boundedness of r and (15.41) we derive the inequality

$$|r(t)| \leq c \int_{t-1}^\infty \beta(\tau) \int_{t_0-1}^\infty e^{-a|\tau-x|}\alpha(x)dx\,d\tau,$$

which implies

$$\| r \|_{L_2(t,t+1)} \leq c \int_{t-1}^\infty \beta(\tau) \int_{t_0-1}^\infty e^{-a|\tau-x|}\alpha(x)dx\,d\tau. \tag{15.43}$$

Equation (15.39) together with (15.40) leads to the estimate

$$\| \, r' \, \|_{L_2(t,t+1)} \le c \, \beta(t) \int_{t_0-1}^{\infty} e^{-a|t-\tau|} \alpha(\tau) d\tau. \tag{15.44}$$

By using (15.43) and (15.44) combined with (13.68) we arrive at (15.38). $\square$

Remark 15.4.2. The condition (15.36) follows from

$$\int_{t_0-1}^{\infty} \big(\alpha(t) + \beta(t)\big)^2 dt \le c \, \varepsilon.$$

Indeed, the left-hand side of (15.36) is the bilinear form of a convolution operator with kernel in L_1. A simpler, more stringent requirement is

$$\int_{t_0-1}^{\infty} \rho^2(t) dt \le c \, \varepsilon$$

(see (15.17)).

We now prove the principal result of this section.

Theorem 15.4.3. *Let the strip* $k_- \le \Im\lambda \le k_+$ *contain only one eigenvalue* λ_0 *with* $\Im\lambda_0 = k_0 \in (k_-, k_+)$. *Suppose that* $\rho(t) \to 0$ *as* $t \to +\infty$ *and*

$$\int_{t_0}^{\infty} \int_{t_0}^{\infty} e^{-a|t-\tau|} \beta(t)\alpha(\tau) dt d\tau < \infty. \tag{15.45}$$

Then for sufficiently large t_0 *the equation*

$$\big(\mathcal{A}(D_t) - N(t, D_t)\big) u(t) = 0 \quad \text{for} \quad t > t_0 \tag{15.46}$$

has a solution $U \in W^\ell(t_0, \infty)$ *such that*

$$\text{col}\big(U(t), D_t U(t), \ldots, D_t^{\ell-1} U(t)\big)$$
$$= e^{i \int_{t_0}^{t} \lambda(\tau) d\tau} \big(\text{col}(\lambda_0^{j-1} \varphi)_{j=1}^\ell + v \big), \tag{15.47}$$

where $v \in \mathbb{S}_{\text{loc}}(t_0, \infty)$ *satisfies*

$$\| \, v \, \|_{\mathbb{S}(t,t+1)} \le c \int_{t_0-1}^{\infty} \alpha(\tau) \big(e^{-a|t-\tau|} + \int_t^{\infty} e^{-a|\tau-x|} \beta(x) dx \big) d\tau \tag{15.48}$$

for $t > t_0$. (*The right-hand side in the last inequality is obviously* $o(1)$ *as* $t \to +\infty$.)

Proof. By Lemma 15.4.1 and Theorem 13.10.4 there exists a solution $U \in W_{\text{loc}}^{\ell}(t_0, \infty)$ of (15.46) such that

$$\text{col}\big(U(t), D_t U(t), \dots, D_t^{\ell-1} U(t)\big)$$

$$= e^{i \int_{t_0}^t \lambda(\tau) d\tau} \big\{ (1 + r(t)) \, \text{col}(\lambda_0^{j-1} \varphi)_{j=1}^{\ell} + \mathbf{w}(t) \big\},$$

where r satisfies (15.38) and $\mathbf{w} \in \mathbb{X}_{\text{loc}}(t_0, \infty)$ is subject to

$$\| \mathbf{w} \|_{\mathbb{S}(t, t+1)} \leq c \int_{t_0 - 1}^{\infty} e^{-a|t-\tau|} \alpha(\tau) d\tau. \tag{15.49}$$

Putting

$$v(t) = r(t) \text{col}(\lambda_0^{j-1} \varphi)_{j=1}^{\ell} + \mathbf{w}(t),$$

we arrive at (15.47). The estimates (15.38), (15.49) together with (12.59) yield (15.48). $\square$

15.5 Asymptotics of Solutions of the Homogeneous Higher Order Equation at $+\infty$

Now we are in a position to write an asymptotic formula for solutions of (15.46).

Theorem 15.5.1. *Let* (15.45) *hold and let* $u \in W_{\text{loc}}^{\ell}(t_0, \infty)$ *be a solution of* (15.46) *satisfying* (15.2). *Then*
   (i) *There exists a constant* c_1 *such that*

$$u(t) = c_1 U(t) + w(t), \tag{15.50}$$

where U *is a special solution of* (15.46) *constructed in* Theorem 15.4.3, *and*

$$\| w \|_{W^{\ell}(t, t+1)} \leq c_2 e^{-k_+ t} \quad \text{for} \quad t > t_0. \tag{15.51}$$

The constants c_1 *and* c_2 *admit the estimate*

$$|c_1| + |c_2| \leq c \| u \|_{W^{\ell}(t_0, t_0+1)}, \tag{15.52}$$

where c *does not depend on* u.
   (ii) *The following asymptotic formula is valid:*

$$\text{col}\big(D_t^{j-1} u(t)\big)_{j=1}^{\ell} = e^{i \int_{t_0}^t \lambda(\tau) d\tau} \Big(c_1 \, \text{col}(\lambda_0^{j-1} \varphi)_{j=1}^{\ell} + v(t) \Big), \tag{15.53}$$

where

$$\| v \|_{\mathbb{S}(t, t+1)} \leq c \Big(\int_{t_0-1}^{\infty} \alpha(\tau) \big(e^{-a|t-\tau|} \tag{15.54}$$

$$+ \int_t^{\infty} e^{-a|\tau-x|} \beta(x) dx \big) d\tau + e^{(k_0 - k_+)t} \Big) \| u \|_{W^{\ell}(t_0, t_0+1)} .$$

Proof. (i) From (15.47) it follows that

$$c_1 e^{-(k_0+\varepsilon)t} \leq \| U \|_{W^{\ell}(t,t+1)} \leq c_2 e^{-(k_0-\varepsilon)t}$$

with an arbitrary positive ε. Since $\kappa = 1$, the function $U(t)$ generates the space $\mathbf{X}(L)$ satisfying the assumptions of Theorem 12.2.3. By using this theorem we obtain (15.50)–(15.52).

(ii) The result follows from (i) and (15.47), (15.48). $\square$

Remark 15.5.2. By (15.17), α and β in (15.54) can be replaced by ρ. For example, let

$$\rho(t) \leq c\, t^{-\nu}, \quad \nu > 1/2.$$

Then the asymptotic formula (15.53) with

$$\| v \|_{S(t,t+1)} \leq c \left(t^{-\nu} + t^{1-2\nu} \right)$$

is valid.

Example 15.5.3. We apply Theorem 15.5.1 to the case $\ell = 1$; that is, to the equation

$$\left(D_t + A - N(t) \right) u(t) = 0 \quad \text{for} \quad t > t_0. \tag{15.55}$$

We assume that the operator $A + \lambda I : H_1 \rightarrow H_0$ is Fredholm for all λ and that the estimate

$$\| w \|_1 + |\tau| \| w \|_0 \leq c \| (A + i\tau I)w \|_0$$

holds for real τ with sufficiently large $|\tau|$. Also, let λ_0 be a single eigenvalue of the pencil $A + \lambda I$ on the line $\Im \lambda = k_0$ (which contains all the spectrum of $\mathcal{A}(\lambda)$ in the strip $k_- \leq \Im \lambda \leq k_+$). We suppose that this eigenvalue is simple and denote the corresponding eigenvector by φ. Let ψ be the eigenvector of $A^* + \lambda I$ corresponding to the eigenvalue $\lambda = \bar{\lambda}_0$, and normalized so that $(\varphi, \psi)_{H_0} = 1$.

We suppose that $N(t)$ is a bounded operator: $H_1 \rightarrow H_0$ and that

$$\| N \|_{W^1(t,t+1) \rightarrow L_2(t,t+1;H_0)} \rightarrow 0 \quad \text{as} \quad t \rightarrow +\infty.$$

The functions α and β admit the equivalence relations

$$\alpha(t) \approx \left(\int_t^{t+1} \| N(\tau)\varphi \|_{H_0}^2 \, d\tau \right)^{1/2} \tag{15.56}$$

and

$$\beta(t) \approx \sup_v \left(\int_t^{t+1} \left| (v(\tau), N^*(\tau)\psi - (N(\tau)\varphi, \psi)_{H_0} \psi)_{H_0} \right|^2 d\tau \right)^{1/2}.$$

The supremum is extended over the set

$$\{v \in W^1(t, t+1) : \| v \|_{W^1(t,t+1)} = 1\}.$$

Clearly,

$$\beta(t) \leq \sup_{\tau \in (t,t+1)} \| N^*(\tau)\psi - (N(\tau)\varphi, \psi)_{H_0} \psi \|_{H_1^*}. \tag{15.57}$$

If $N^*(\tau)\psi \in H_0$ for almost all τ then

$$\beta(t) \leq c \left(\int_t^{t+1} \| N^*(\tau)\psi - (N(\tau)\varphi, \psi)_{H_0} \psi \|_0^2 \, d\tau \right)^{1/2}. \tag{15.58}$$

We suppose that the condition (15.45) is valid. Then Theorem 15.5.1 can be applied to (15.55) and the asymptotic formula (15.53) takes the form

$$u(t) = e^{i\lambda_0 t + i \int_{t_0}^t (N(\tau)\varphi, \psi)_{H_0} \, d\tau} (c\varphi + v(t)),$$

where $c = \text{const}$ and v admits the estimate

$$\| v \|_{W^1(t,t+1)} \leq c \left(\int_t^{\infty} \int_{t_0-1}^{\infty} e^{-a|x-\tau|} \alpha(\tau) d\tau \beta(x) dx \right.$$

$$\left. + \int_{t_0-1}^{\infty} e^{-a|t-\tau|} \alpha(\tau) d\tau + e^{(k_0 - k_+)t} \right). \tag{15.59}$$

Inserting the above upper estimates for α and β into the right-hand side of (15.59) we can characterize the value of the remainder v in more explicit terms.

Example 15.5.4. Let us consider the equation

$$\left(D_t^2 + 2ci D_t + A - N(t) \right) u(t) = 0 \quad \text{for} \quad t > t_0, \tag{15.60}$$

where $c \in \mathbb{R}$ and A is a non-negative selfadjoint operator in H_0 with discrete spectrum $\{\mu_k\}_{k \geq 0}$, $\mu_{k+1} > \mu_k$. We denote by H_2 and H_1 the domains of $I + A$ and $(I + A)^{1/2}$ respectively. Then the operator

$$\mathcal{A}(\lambda) = (\lambda^2 + 2ci\lambda) I + A : H_2 \to H_0$$

is Fredholm for all $\lambda \in \mathbb{C}$ and

$$\| w \|_2 + |\tau| \| w \|_1 + \tau^2 \| w \|_0 \leq c \| \mathcal{A}(\tau)w \|_0$$

for real τ with sufficiently large $|\tau|$.

Let μ_0 be a simple eigenvalue of A with an eigenvector $\varphi, \| \varphi \|_0 = 1$. If $\mu_0 = 0$ we assume that $c \neq 0$. Then there is a single eigenvalue

$$\lambda_{\pm} = i\left(-c \pm (c^2 + \mu_0)^{1/2}\right)$$

of $\mathcal{A}(\lambda)$ on the line $\Im\lambda = -c \pm (c^2 + \mu_0)^{1/2}$. Clearly, $\lambda_{\pm}$ is a simple eigenvalue of $\mathcal{A}(\lambda)$ and φ is its eigenvector. The pencil $\mathcal{A}^*(\lambda)$ has the eigenvector

$$\psi_{\pm} = \mp\frac{1}{2i}(c^2 + \mu_0)^{-1/2}\varphi$$

corresponding to the eigenvalue $-\lambda_{\pm}$ and

$$\left(\frac{d}{d\lambda}\mathcal{A}(\lambda)\varphi, \psi_{\pm}\right)_{H_0}\Big|_{\lambda=\lambda_{\pm}} = 1.$$

The strip

$$-c - (c^2 + \mu_1)^{1/2} < \Im\lambda < -c + (c^2 + \mu_1)^{1/2}$$

contains only the eigenvalues $\lambda_{\pm}$ of the pencil $\mathcal{A}(\lambda)$.

Let $N(t) : H_2 \to H_0$ satisfy

$$\| N \|_{W^2(t,t+1)\to L_2(t,t+1;H_0)} \to 0 \quad \text{as} \quad t \to +\infty. \tag{15.61}$$

The required information on α is contained in (15.56). By (15.16),

$$\beta(t) \approx \sup_{v}\left(\int_t^{t+1}\Big|(N(\tau)v(\tau), \psi_{\pm})_{H_0}\right.$$
$$\left.-\big((\lambda_{\pm} + D_{\tau} + 2ci)v(\tau), \psi_{\pm}\big)_{H_0}(N(\tau)\varphi, \psi_{\pm})_{H_0}\Big|^2 d\tau\right)^{1/2},$$

where the supremum is extended over the set

$$\{v \in W^2(t, t+1) : \| v \|_{W^2(t,t+1)} = 1\}.$$

Hence

$$\beta(t) \leq C \sup_{\tau\in(t,t+1)} \| N^*(\tau)\varphi \|_{H_2^*}. \tag{15.62}$$

If we assume a priori that $N^*(\tau)\varphi \in H_0$, then

$$\beta(t) \leq C\left(\int_t^{t+1} \| N^*(\tau)\varphi \|_{H_0}^2 \, d\tau\right)^{1/2}. \tag{15.63}$$

Now, sufficient conditions for the convergence of the integral (15.45) follow immediately from (15.56), (15.62) and (15.63).

In particular, if $N(t)$ is a selfadjoint operator in H_0, then $\beta(t) \leq C \, \alpha(t)$ by (15.63), and (15.45) follows from

$$\int_{t_0}^{\infty} \int_{t_0}^{\infty} e^{-a|t-\tau|}\alpha(t)\alpha(\tau)dtd\tau < \infty.$$

By using (13.45) one can show that this condition is equivalent to $\alpha \in L_2(t_0, \infty)$.

Under the above conditions, both Theorem 15.4.3 and Theorem 15.5.1 can be applied to equation (15.60), and the asymptotic formula (15.53) is then written as

$$\left(u(t), D_t u(t)\right) = \exp\left(i\lambda_{\pm}t \pm \frac{1}{2}(c^2 + \mu)^{1/2} \int_{t_0}^t \left(N(\tau)\varphi, \varphi\right)_{H_0} d\tau\right)$$
$$\times \left(c_1 \operatorname{col}\left(\varphi, \lambda_{\pm}\varphi\right) + v(t)\right), \tag{15.64}$$

where c_1 is a constant and v is estimated by (15.54). In the case of " $-$ " in this formula, we deal with a solution subject to

$$\|u\|_{W^2(t, t+1)} = O(e^{-kt}) \quad \text{as } t \to +\infty, \tag{15.65}$$

where $k > -c - (c^2 + \mu_1)^{1/2}$. If "+", is taken in (15.64), then the solution is assumed to satisfy (15.65) with $k > -c - (c^2 + \mu_0)^{1/2}$.

Example 15.5.5. The previous example can be applied directly to the asymptotics of solutions of the Schrödinger equation

$$\Delta u(x) + \frac{p(x)}{|x|^2} u(x) = 0 \tag{15.66}$$

in a neighborhood of a point $O \in \mathbb{R}^n$, $n > 2$. We assume that $p(x) = o(1)$ as $x \to O$ and that

$$\int_{|x|<\delta} |p(x)|^2 \frac{dx}{|x|^n} < \infty. \tag{15.67}$$

By using the spherical coordinates (r, θ) and setting $t = \log r^{-1}$ we write (15.66) as

$$D_t^2 u + i(n - 2)D_t u - \delta_\theta u - p(e^{-t}\theta)u = 0 \quad \text{for } t > t_0.$$

Thus, we are dealing here with a special case of the operator considered in Ex. 15.5.4 with

$$c = \frac{n - 2}{2}, \quad A = -\delta_\theta$$

and with the multiplication by $p(e^{-t}\theta)$ as the perturbation operator $N(t)$. Clearly, μ_0, $\mu_1 = n - 1$ and

$$\lambda_+ = 0, \quad \lambda_- = (2 - n)i.$$

We put

$$\varphi = \left(\operatorname{mes}_{n-1}(S^{n-1})\right)^{-1/2}$$

as a normed eigenfunction of $-\delta_\theta$ corresponding to μ_0. Then

$$\psi_{\pm} = \pm\frac{n - 2}{4i}.$$

Condition (15.61) holds because $p(e^{-t}\theta) = o(1)$ as $t \to +\infty$. By (15.56) and (15.63), the functions α and β admit the estimate

$$\alpha(t) + \beta(t) \le \left(\int_t^{t+1} \int_{S^{n-1}} |p(e^{-\tau}\theta)|^2 d\tau d\theta \right)^{1/2},$$

which, together with (15.67) implies (15.45). Hence we deduce the following result from Ex. 15.5.4.

Let $u \in W^2_{2,\text{loc}}(B_{r_0} \setminus \{O\})$ be a solution of (15.66) in $B_{r_0} \setminus \{O\}$.

(i) If

$$\left(r^{-n} \int_{K_r} |u|^2 dx \right)^{1/2} \le C r^{2-n+\varepsilon}$$

with some positive ε then

$$u(x) = \exp \left(\frac{1}{(n-2)\text{mes}_{n-1}(S^{n-1})} \int_{|x|<|y|<r_0} p(y) \frac{dy}{|y|^n} \right) \left(\Lambda + v(x) \right),$$

where $\Lambda = \text{Const}$ and

$$r^{-n} \int_{K_r} |v|^2 dx \to 0 \quad \text{as } r \to 0. \tag{15.68}$$

(ii) If

$$\left(r^{-n} \int_{K_r} |u| dx \right)^{1/2} \le C r^{1-n+\varepsilon}$$

with some positive ε then

$$u(x) = |x|^{2-n} \exp \left(\frac{-1}{(n-2)\text{mes}_{n-1}(S^{n-1})} \int_{|x|<|y|<r_0} p(y) \frac{dy}{|y|^n} \right) \left(\Lambda + v(x) \right),$$

where $\Lambda = \text{Const}$ and v satisfies (15.68).

15.6 Asymptotics of Solutions to the Nonhomogeneous Equation

We now turn to the nonhomogeneous equation (14.2). Our aim is to give a condition on the function f such that the asymptotic form of solutions to (14.2) coincides with that of solutions to the homogeneous equation (14.1). The main tool is the following auxiliary assertion:

Lemma 15.6.1. *Let* (15.36) *hold and let the right-hand side q in* (15.13) *be subject to* (15.18). *Then there exists a solution* $h \in W_{2,\mathrm{loc}}^1(t_0, \infty)$ *of* (15.13) *such that*

$$\| h \|_{W_2^1(t,t+1)} \le c \int_{t_0-1}^{\infty} g_*(t-\tau) e^{\Im \int_t^{\tau} \lambda(x)dx} \| q \|_{L_2(\tau,\tau+1)} \, d\tau, \qquad (15.69)$$

where

$$g_*(t) = \begin{cases} e^{-at/3} & \text{for } t > 0 \\ 1 & \text{for } t \le 0. \end{cases}$$

Proof. Putting

$$h(t) = \exp\left(i \int_{t_0}^t \lambda(\tau)d\tau\right) y(t) \qquad (15.70)$$

(where $\lambda(\cdot)$ is given by (15.3)), we obtain

$$D_t y(t) - (Ty)(t) = \mathfrak{q}(t), \quad \text{for } t > t_0 \qquad (15.71)$$

with

$$(Ty)(t) = K_{\tau \to t}\left(e^{i \int_\tau^t \lambda(x)dx} y(\tau)\right)$$

and

$$\mathfrak{q}(t) = e^{-i \int_{t_0}^t \lambda(\tau)d\tau} q(t). \qquad (15.72)$$

Using (15.14) and (15.24) we get

$$\| Ty \|_{L_2(t,t+1)} \le c\, \beta(t) \int_{t_0-1}^{\infty} e^{-(2a-c\rho_0)|t-\tau|} \alpha(\tau) \| \mathring{y} \|_{L_\infty(\tau,\tau+1)} \, d\tau. \qquad (15.73)$$

where $\mathring{y}$ is the extension of y by zero for $t < t_0$. Now we note that (15.71) (and hence (15.13)) follows from the integral equation on $\mathbb{R}$:

$$y(t) = -i \int_t^{\infty} (Ty)(\tau)d\tau - i \int_t^{\infty} \mathfrak{q}(\tau)d\tau, \qquad (15.74)$$

where $\mathfrak{q}(\tau) = 0$ for $\tau < t_0$.

We construct a solution of (15.74) by means of the following iterative procedure: let $y_0 = 0$ and for $k = 0, 1, \dots$ set

$$y_{k+1}(t) = -i \int_t^{\infty} (Ty_k)(\tau)d\tau - i \int_t^{\infty} \mathfrak{q}(\tau)d\tau. \qquad (15.75)$$

Put

$$z_k(t) = \| y_k - y_{k-1} \|_{L_\infty(t,t+1)} .$$

Using (15.73), we derive from (15.75)

$$z_k(t) \le c \int_{t-1}^{\infty} \beta(\tau) \int_{t_0-1}^{\infty} e^{-(2a-c\rho_0)|\tau-x|} \alpha(x) z_{k-1}(x)dxd\tau. \qquad (15.76)$$

In order to apply Theorem 13.12.4, we rewrite (15.76) as

$$z_k(t) \le \int_{\mathbb{R}} \int_{\mathbb{R}} g_*(t - \tau) Q(\tau, x) z_{k-1}(x) dx d\tau,$$

where

$$Q(\tau, x) = c\, \beta(\tau) e^{-(2a - c\rho_0)|\tau - x|} \alpha(x).$$

The assumption (15.36) implies smallness of q_1 given by (13.136), where $k_+ = a/3$, $k_- = 0$, and $m_+ = m_- = 1$. By Theorem 13.12.4 with $\mathbf{g} = g_*$, the series

$$\sum_{k=0}^{\infty} \int_{\mathbb{R}^{2k}} g_*(t - \tau_1) Q(\tau_1, s_1) g_*(s_1 - \tau) \ldots Q(\tau_k, s_k)$$

$$\times g_*(s_k - \tau) d\tau_1 \ldots d\tau_k ds_1 \ldots ds_k$$

converges. Hence, by the same theorem

$$\sum_{k=0}^{\infty} z_k(t) \le c \int_{\mathbb{R}} g_*(t - \tau) \| \mathbf{q} \|_{L_2(\tau, \tau+1)} \, d\tau.$$

This implies the convergence of $\{y_k\}$ in $L_{\infty,\mathrm{loc}}(t_0, \infty)$, and also that the limit function y satisfies

$$\| y \|_{L_\infty(t, t+1)} \le c \int_{\mathbb{R}} g_*(t - \tau) \| \mathbf{q} \|_{L_2(\tau, \tau+1)} \, d\tau. \tag{15.77}$$

Due to (15.70) and (15.72) this estimate gives

$$\|h\|_{L_\infty(t, t+1)} \le c \int_{t_0 - 1}^{\infty} g_*(t - \tau) e^{\Im \int_t^\tau \lambda(x) dx} \| \mathbf{q} \|_{L_2(\tau, \tau+1)} \, d\tau. \tag{15.78}$$

Using (15.13), (15.14) together with (15.78) we arrive at the estimate for $\|h'\|_{L_2(t, t+1)}$ by the right-hand side of (15.78). The proof is complete. $\square$

If the right-hand side q in (15.13) has the form (13.83), (13.82) we can obtain an estimate for the solution h constructed in Lemma 15.6.1 which is formulated in terms of f.

Corollary 15.6.2. *Let $f \in L_{2,\mathrm{loc}}(t_0, \infty)$ satisfy*

$$\int_{t_0}^{\infty} e^{\Im \int_{t_0}^t \lambda(\tau) d\tau} \| f \|_{L_2(t, t+1; H_0)} \, dt < \infty, \tag{15.79}$$

and let q be given by (13.83), (13.82). Then equation (15.13) has a solution $h \in W_{2,\mathrm{loc}}^1(t_0, \infty)$ such that

$$\| h \|_{W_2^1(t, t+1)} \le c \int_{t_0 - 1}^{\infty} g_*(t - \tau) e^{\Im \int_t^\tau \lambda(x) dx} \| f \|_{L_2(\tau, \tau+1; H_0)} \, d\tau. \tag{15.80}$$

Proof. One can verify directly that (15.79) together with (13.111) gives (15.18). Hence, by Lemma 15.6.1, there exists a solution h of (15.13) which satisfies (15.69). Using (13.111), one can show that the right-hand side of (15.69) is estimated by the right-hand side of (15.80). $\square$

The following result is a corollary of Theorem 13.10.4, Lemma 15.6.1 and Corollary 15.6.2.

Lemma 15.6.3. *Let* (15.45) *hold and let* $f \in L_{2,\mathrm{loc}}(t_0, \infty)$ *be extended by zero for* $t < t_0$ *and subject to* (15.79). *Then there exists a solution* $u \in W_{\mathrm{loc}}^{\ell}(t_0, \infty)$ *of* (14.2) *such that*

$$\| u \|_{W^{\ell}(t,t+1)} \le c \int_{t_0-1}^{\infty} g_*(t-\tau) e^{\Im \int_t^{\tau} \lambda(x)dx} \| f \|_{L_2(\tau,\tau+1;H_0)} \, d\tau \quad (15.81)$$

for $t > t_0$.

Proof. By Theorem 13.10.4 there exists a solution $u \in W_{\mathrm{loc}}^{\ell}(t_0, \infty)$ of (14.2) such that

$$\mathrm{col}\big(D_t^{j-1}u(t)\big)_{j=1}^{\ell} = h(t)\,\mathrm{col}\big(\lambda_0^{j-1}\varphi\big) + \mathbf{v}(t) \quad (15.82)$$

(where h is a solution of (15.13) constructed in Lemma 15.6.1). By (13.112), the vector function $\mathbf{v}$ admits the estimate

$$\| \mathbf{v} \|_{S(t,t+1)}$$
$$\le c \int_{t_0-1}^{\infty} g(t-\tau)\big(\alpha(\tau) \| h \|_{L_{\infty}(\tau,\tau+1)} + \| f \|_{L_2(\tau,\tau+1;H_0)}\big) d\tau (15.83)$$

where h is supposed to be zero for $t < t_0$. Due to (15.82) we have

$$\| u \|_{W^{\ell}(t,t+1)} \le c \left(\| h \|_{W_2^1(t,t+1)} + \| \mathbf{v} \|_{S(t,t+1)} \right).$$

The result follows from (15.80) and (15.83). $\square$

Remark 15.6.4. The estimate (15.81) and the assumption (15.79) together imply

$$\| u \|_{W^{\ell}(t,t+1)} = o\big(e^{-\Im \int_{t_0}^{t} \lambda(x)dx}\big). \quad (15.84)$$

We prove the main result of this section.

Theorem 15.6.5. *Let* (15.45) *hold and let* t_0 *be sufficiently large. Suppose that* $f \in L_{2,\mathrm{loc}}(t_0, \infty; H_0)$ *satisfies* (15.79), *and let* $u \in W^{\ell}(t_0, \infty)$ *be a solution of* (14.2) *which is subject to*

$$\| u \|_{W^{\ell}(t,t+1)} = o\big(e^{-k_-t}\big) \quad as \quad t \to +\infty. \quad (15.85)$$

Then

$$\mathrm{col}\big(u(t), D_t u(t), \ldots, D_t^{\ell-1} u(t)\big)$$

$$= e^{i \int_{t_0}^t \lambda(\tau) d\tau} \big(c \, \mathrm{col}(\lambda_0^{j-1} \varphi)_{j=1}^\ell + w(t)\big), \qquad (15.86)$$

where the remainder w satisfies

$$\| w \|_{S(t, t+1)} = o(1) \quad as \quad t \to +\infty. \qquad (15.87)$$

Proof. By Lemma 15.6.3, there exists a solution $u_0 \in W^\ell(t, t+1)$ of (14.2) such that (15.84) holds. By Theorem 15.5.1(ii)

$$\mathrm{col}\Big(D_t^{j-1}(u - u_0)\Big)_{j=1}^\ell = e^{i \int_{t_0}^t \lambda(\tau) d\tau} \Big(c \, \mathrm{col}\Big(\lambda_0^{j-1} \varphi\Big)_{j=1}^\ell + v\Big),$$

where v satisfies (15.54). Therefore (15.86) is valid with

$$w = e^{-i \int_{t_0}^t \lambda(\tau) d\tau} \mathrm{col}\Big(D_t^{j-1}(u_0)\Big)_{j=1}^\ell + v.$$

Relation (15.87) follows from (15.84) for u_0 and from (15.54). $\square$

Remark 15.6.6. Condition (15.79) is essential for the coincidence of the asymptotic forms of solutions to (14.1) and (14.2). For a larger class of f one can obtain the following more general asymptotic representation for solutions of (14.2) subject to (15.85):

$$\mathrm{col}\big(D_t^{j-1} u(t)\big) \sim y(t) \, \mathrm{col}\big(\lambda_0^{j-1} \varphi\big)_{j=1}^\ell, \qquad (15.88)$$

where y is a solution of the ordinary differential equation

$$D_t y(t) - \lambda(t) y(t) = -i\big(f(t), \psi\big)_{H_0}, \quad t > t_0.$$

We do not wish to go into details since the estimate of the remainder term in this asymptotic expression, and sufficient conditions on f to guarantee the asymptotics are both rather cumbersome to write down.

We conclude this section with a corollary of Theorem 15.5.1 of the Phragmén-Lindelöf type.

Corollary 15.6.7. *Let*

$$\int_{t_0}^\infty \rho^2(t) dt < \infty.$$

Suppose that the strip $k_1 < \Im\lambda < k_2$ is free of eigenvalues of $\mathcal{A}(\lambda)$ and that each line $\Im\lambda = k_j$, $j = 1, 2$, contains one simple eigenvalue λ_j. Let φ_j and ψ_j be eigenvectors of $\mathcal{A}(\lambda)$ and $\mathcal{A}^*(\lambda)$, respectively, corresponding to the eigenvalues λ_j and $\overline{\lambda}_j$, and subject to

$$\left(\frac{d}{d\lambda}\mathcal{A}(\lambda)\big|_{\lambda=\lambda_j}\varphi_j,\psi_j\right)_{H_0}=1.$$

If u is a solution of (14.1) *satisfying*

$$\|u\|_{W^\ell(t,t+1)}=o\left(\exp\left(-\Im\int_{t_0}^t\lambda_1(\tau)d\tau\right)\right)\quad as\quad t\to+\infty$$

then

$$\|u\|_{W^\ell(t,t+1)}\le c\,\exp\left(-\Im\int_{t_0}^t\lambda_2(\tau)d\tau\right). \tag{15.89}$$

Proof. First we note that

$$\|u\|_{W^\ell(t,t+1)}\le c\,\|\,\mathrm{col}(D^{j-1}u)_{j=1}^\ell\,\|_{S(t,t+1)}.$$

Hence, using Theorem 15.5.1 with $k_0=k_1$ and $k_\pm=k_1\pm\varepsilon$ (ε is positive and sufficiently small), we arrive at the estimate

$$\|u\|_{W^\ell(t,t+1)}\le c\,e^{-(k_1+\varepsilon)t}\quad for\quad t>t_0.$$

By applying Theorem 15.5.1 once more with $k_0=k_2$, $k_-=k_1+\varepsilon$, $k_+=k_2+\varepsilon$ we obtain (15.89). $\square$

15.7 Comments

The asymptotic formula

$$u(t)\sim\mathrm{const}\exp\left(i\int_{t_0}^t\Lambda(\tau)d\tau\right)\Phi(t)\quad t\to+\infty$$

(similar to (15.5)) was obtained by Evgrafov (1960) for the equation

$$D_tu-A(t)u=0,$$

where $\Lambda(t)$ is the eigenvalue of the operator $A(t)$ tending to a simple eigenvalue λ_0 of the operator $A(+\infty)$ as $t\to+\infty$ and $\Phi(t)$ is the corresponding eigenvector of $A(t)$. In Evgrafov (1960) it is assumed that the resolvent $R(\lambda)$ of the operator $A(+\infty)$ is compact and has at most a finite number of poles outside a double angular sector containing the imaginary axis. Outside this sector, $\|R(\lambda)\|\le C(1+|\lambda|)^{-1}$ for large $|\lambda|$.

An extension of this result to higher order operator differential equations

$$L(t,D_t)u=0\quad\text{for }t>t_0 \tag{15.90}$$

was given in Maz'ya, Plamenevskii (1972), where the limit operator pencil $L(+\infty,\lambda)$ satisfies conditions I, II from Sect. 2.2.1. The restrictions to the coefficients $A_j(t)$ in that paper involve the derivatives up to order $2\ell-j$,

$1 \leq j \leq \ell$, whereas there are no differentiability assumptions in our Theorems 15.4.3 and 15.5.1. Here we have used a different reduction of (15.90) to a first order system, as well as a different spectral splitting.

Theorem 15.4.3 is an analog of the Hartman-Wintner theorem for solutions of the system of ordinary differential equations of the first order (see Eastham (1989), Sect. 1.5).

16. Several Simple Eigenvalues on the Line

16.1 Introduction

Here we consider the case when there are eigenvalues $\lambda_1, \ldots, \lambda_\kappa$ of the operator pencil $\mathcal{A}(\lambda)$ on the line $\Im\lambda = k_0$, $k_0 \in (k_-, k_+)$ and no other eigenvalues in the strip $k_- \leq \Im\lambda \leq k_+$. We suppose that only one eigenvector φ_j (up to the multiplication by a constant) corresponds to each λ_j and that there are no generalized eigenvectors. We assume that $\rho(t) \to 0$ when $t \to +\infty$, where ρ is the function defined by (12.86).

Let ψ_j be the eigenvector of $\mathcal{A}^*(\lambda)$ corresponding to the eigenvalue $\overline{\lambda}_j$ and satisfying

$$\left(\frac{d}{d\lambda}\mathcal{A}(\lambda)\big|_{\lambda=\lambda_j}\varphi_j, \psi_j\right)_{H_0} = 1. \tag{16.1}$$

The results of this chapter are based upon Lemma 16.2.2, where special solutions of the finite dimensional system (13.125) with prescribed asymptotics at infinity are constructed. This leads to the existence of solutions $U_1, \ldots, U_\kappa$ to (14.1) which have the asymptotic form

$$U_k(t) = \exp\left(i\lambda_k t + i\int_{t_0}^t \left(N(\tau, \lambda_k)\varphi_k, \psi_k\right)_{H_0} d\tau\right)\left(\varphi_k + o(1)\right). \tag{16.2}$$

(Theorem 16.3.1). Hence, using Theorem 12.2.3 we can obtain an asymptotic representation for an arbitrary solution of (14.1) (Theorem 16.3.3).

The case of several eigenvalues on the line $\Im\lambda = k_0$ has been already studied in Sect. 11.4, 11.5. There we even admitted that the geometric multiplicities of the eigenvalues could be greater than 1 and constructed a complete asymptotic expansion. In the present chapter we deal with simple eigenvalues and obtain only a principal term (16.2) of the asymptotics. However, our restrictions on the perturbation are much weaker (see Theorem 16.3.1). We note that (16.2) agrees with the principal term in the asymptotics (11.22) if the coefficients of $N(t, D_t)$ are represented as asymptotic power series in t^{-1}.

16.2 Special Solutions of the Finite Dimensional System

16.2.1 Functions α and β

The finite dimensional system (13.125) takes the following simpler form

$$\big(D_t - \Lambda - \mathcal{R}(t)\big)h(t) - (Kh)(t) = 0, \qquad (16.3)$$

where

$$\Lambda = \mathrm{diag}(\lambda_1, \ldots, \lambda_\kappa)$$

and

$$\mathcal{R} = \{\mathcal{R}_{jk}\}_{j,k=1}^{\kappa}$$

with

$$\mathcal{R}_{jk}(t) = \big(N(t, \lambda_k)\varphi_k, \psi_j\big)_{H_0}.$$

The function α (introduced in Sect.13.5) satisfies the equivalence relation

$$\alpha(t) \approx \sum_{j=1}^{\kappa} \Big(\int_t^{t+1} \| N(\tau, \lambda_j)\varphi_j \|_0^2 \, d\tau \Big)^{1/2} \qquad (16.4)$$

(see Proposition 13.5.1). Since there are no generalized eigenvectors, (13.74) can be written as

$$\beta_1(t) = \sup_v \sum_{j=1}^{\kappa} \Big(\int_t^{t+1} \big| \big(N(\tau, D_\tau)v(\tau), \psi_j\big)_{H_0} \\ - \sum_{k=1}^{\kappa} \big(\mathcal{B}(\lambda_k, D_\tau)v(\tau), \psi_k\big)_{H_0} \big(N(\tau, \lambda_k)\varphi_k, \psi_j\big)_{H_0} \big|^2 d\tau \Big)^{1/2}, (16.5)$$

where $\mathcal{B}$ is defined in (13.73). By Proposition 13.8.3, $\beta \approx \beta_1$. Both functions α and β are majorized by $c\rho$ (see (13.44) and (13.70)). From Proposition 13.8.4 it follows that

$$\beta(t) \leq c \sum_{m=1}^{\ell} \sum_{j=1}^{\kappa} \sup_{\tau \in (t,t+1)} \| N_m^*(\tau)\psi_j \\ - \sum_{k=1}^{\kappa} \big(N(\tau, \lambda_k)\varphi_k, \psi_j\big) E_{\ell-m+1}^*(\overline{\lambda}_k)\psi_k \|_{H_m^*}. \qquad (16.6)$$

16.2.2 Dichotomy Condition

Let

$$\mu_j(t) = \big(N(t, \lambda_j)\varphi_j, \psi_j\big)_{H_0}, \quad j = 1, \ldots, \kappa. \qquad (16.7)$$

We suppose that the following *Dichotomy Condition* is fulfilled; this is similar to the condition in the classical Levinson theorem (see, for example, Eastham (1989), Ch. I).

For each pair of integers j, k in $[1, \kappa]$ $(j \neq k)$ and for all τ and t such that $t_0 \leq \tau \leq t < \infty$, we have either

$$\int_\tau^t \Im\{\mu_j(s) - \mu_k(s)\}ds \leq b_1 \qquad (16.8)$$

or

$$\int_\tau^t \Im\{\mu_j(s) - \mu_k(s)\}ds \geq b_2, \tag{16.9}$$

where b_1 and b_2 are constants.

Lemma 16.2.1. *Let* j, k *be integers in* $[1, n]$ *and* $j \neq k$.
  (i) *If* (16.8) *holds and* (16.9) *fails then*

$$\int_{t_0}^t \Im\{\mu_j(s) - \mu_k(s)\}ds \to -\infty \quad as \quad t \to +\infty.$$

  (ii) *If* (16.9) *holds and* (16.8) *fails then*

$$\int_{t_0}^t \Im\{\mu_j(s) - \mu_k(s)\}ds \to +\infty \quad as \quad t \to +\infty.$$

Proof. It suffices to prove (i) since (ii) reduces to (i) by changing indices $(j, k) \to (k, j)$.

We prove (i) by contradiction. Suppose that (16.8) is valid and that for some sequence $\{t_k\}$, such that $t_k \to \infty$ as $k \to \infty$, we have

$$\int_{t_0}^{t_k} \Im\{\mu_j(s) - \mu_k(s)\}ds \geq c.$$

Using this inequality and (16.8) for $t = t_k$ and $\tau \leq t_k$ we obtain that

$$\int_{t_0}^\tau \Im\{\mu_j(s) - \mu_k(s)\}ds \geq c - b_1,$$

Together with (16.8), this implies (16.9) with $b_2 = c - 2b_1$. $\square$

16.2.3 Special Solutions of the Finite Dimensional System (16.3)

We now introduce some new notation. By $R_1(t)$ we denote the $\kappa \times \kappa$-matrix with elements

$$(R_1(t))_{jk} = \begin{cases} 0 & \text{if } j = k, \\ e^{i(\lambda_k - \lambda_j)t}\left(N(t, \lambda_k)\varphi_k, \psi_j\right)_{H_0} & \text{if } j \neq k. \end{cases} \tag{16.10}$$

Furthermore, we put

$$\chi(t) = \beta(t) \int_{t_0-1}^\infty e^{-a|t-\tau|}\alpha(\tau)d\tau, \tag{16.11}$$

where a is defined by (16.4) and β is given by the right-hand side of (16.5). Since the functions α and β are majorized by $c\rho$ we have

$$\chi(t) \leq c\,\rho(t) \int_{t_0-1}^\infty e^{-a|t-\tau|}\rho(\tau)d\tau.$$

Therefore

$$\chi(t) \to 0 \quad as \quad t \to \infty. \tag{16.12}$$

Lemma 16.2.2. *Suppose that the Dichotomy Condition is valid and that the following assumptions are satisfied:*

(i) *The integral*

$$\int_{t_0}^{\infty} R_1(\tau)d\tau$$

converges and the inequality

$$\left\| \int_t^{\infty} R_1(\tau)d\tau \right\| \le 1/2$$

holds for $t \ge t_0$.

(ii) *The functions $\chi(t)$ and*

$$p(t) = \| p_1 \|_{L_2(t,t+1)}, \tag{16.13}$$

where

$$p_1(t) \tag{16.14}$$
$$= \left\| \int_t^{\infty} R_1(\tau)d\tau\mu(t) - \mu(t)\int_t^{\infty} R_1(\tau)d\tau - R_1(t)\int_t^{\infty} R_1(\tau)d\tau \right\|,$$

with

$$\mu(t) = \mathrm{diag}\big(\mu_1(t), \ldots, \mu_\kappa(t)\big),$$

belong to $L_1(t_0, \infty)$ and their norms are sufficiently small.

Then there exists a collection of solutions h_j, $j = 1, \ldots, \kappa$, of (16.3) which have the asymptotic form

$$h_j(t) = e^{i\lambda_j t + i\int_{t_0}^t \mu_j(\tau)d\tau}\big(e_j + r_j(t)\big), \tag{16.15}$$

where

$$e_j = \mathrm{col}(0, \ldots, 1, \ldots 0), \tag{16.16}$$

with 1 at the jth place, and where $r_j \in \big(W_{2,\mathrm{loc}}^1(t_0, \infty)\big)^\kappa$ with

$$\| r_j \|_{\big(W_2^1(t,t+1)\big)^\kappa} = o(1) \quad as \quad t \to +\infty. \tag{16.17}$$

16.2.4 Proof of Lemma 16.2.2

Transformations of the system (16.3). Put

$$Q(t) = -i\int_t^{\infty} R_1(\tau)d\tau.$$

Let us look for a solution $h = h_j$ of the form

$$h(t) = e^{i\Lambda t}\big(I + Q(t)\big)X(t), \tag{16.18}$$

where X is an unknown vector. From (16.3) it follows that for $t > t_0$

$$D_t X(t) - \mu(t)X(t) + R_2(t)X(t) - (SX)(t) = 0, \qquad (16.19)$$

where

$$R_2(t) = \big(I + Q(t)\big)^{-1}\big\{Q(t)\mu(t) - \mu(t)Q(t) - R_1(t)Q(t)\big\}$$

and

$$(SX)(t) = \big(I + Q(t)\big)^{-1} e^{-i\Lambda t} K_{\tau \to t}\Big(e^{i\Lambda \tau}\big(I + Q(\tau)\big)X(\tau)\Big)(t).$$

By assumption (i), the matrix $\big(I + Q(t)\big)^{-1}$ is invertible and

$$\big\| \big(I + Q(t)\big)^{-1} \big\| \le 2 \quad \text{for} \quad t \ge t_0.$$

From (16.14) it follows that

$$\| R_2(t) \| \le 2p_1(t). \qquad (16.20)$$

Using (13.110) we obtain

$$\| SX \|_{(L_2(t,t+1))^\kappa}$$
$$\le c\,\beta(t) \int_{t_0-1}^{\infty} g(t-\tau)e^{k_0(t-\tau)}\alpha(\tau) \| X \|_{(L_\infty(\tau,\tau+1))^\kappa}\, d\tau, \quad (16.21)$$

where g is given by (12.61) and $X(t)$ is extended by zero for $t < t_0$.

We want to find a solution of (16.19) in the form

$$X(t) = e^{i \int_{t_0}^{t} \mu_j(\tau)d\tau}\big(e_j + Y(t)\big). \qquad (16.22)$$

The vector Y satisfies

$$D_t Y(t) - \mu^{(j)}(t)Y(t) + R_2(t)\big(e_j + Y(t)\big)$$
$$-S_{\tau \to t}\Big(e^{-i\int_\tau^t \mu_j(x)dx}\big(e_j + Y(\tau)\big)\Big) = 0, \qquad (16.23)$$

where

$$\mu^{(j)}(t) = \operatorname{diag}\big((\mu_1 - \mu_j)(t), \ldots, (\mu_\kappa - \mu_j)(t)\big).$$

By Proposition 13.5.2 and the smallness of ρ

$$\int_{t}^{t+1} |\mu_j(\tau)|^2 d\tau \le \varepsilon,$$

where ε is a small positive number. Therefore

$$\Big| \int_\tau^t \mu_j(x)dx \Big| \le \Big| \int_\tau^t \int_{x-1}^{x+1} |\mu_j(y)|dy dx \Big| \le (2\varepsilon)^{1/2}|t - \tau|. \qquad (16.24)$$

Using (16.21) and (16.24) we show that for an arbitrary $Z \in (L_{\infty,\mathrm{loc}}(t_0, \infty))^\kappa$, extended by zero for $t < t_0$:

$$\left(\int_{\tau}^{\tau+1} \left| S_{x \to y} \left(e^{i \int_{y}^{x} \mu_j(s)ds} Z(x) \right) \right|^2 dy \right)^{1/2}$$

$$\leq c \, \beta(\tau) \int_{t_0-1}^{\infty} e^{-a|x-\tau|} \alpha(x) \, \| \, Z \, \|_{(L_\infty(x,x+1))^\kappa} \, dx \qquad (16.25)$$

with a given by (14.12).

Existence of Y. We construct a bounded Green's function $\Phi_k = \Phi_k(t, \tau)$ for the equation $D_t - (\mu_k - \mu_j)(t)$, i.e. a solution of the equation

$$D_t \Phi_k(t, \tau) - (\mu_k - \mu_j)(t) \Phi_k(t, \tau) = \delta(t - \tau)$$

for $t, \tau > t_0$. Due to the Dichotomy Condition and Lemma 16.2.1 there are two possible cases:

(a) inequality (16.8) holds and (16.9) fails;

(b) inequality (16.9) holds.

In case (a) we put

$$\Phi_k(t, \tau) = \begin{cases} i \, e^{i \int_{\tau}^{t} (\mu_k - \mu_j)(x)dx} & \text{for } t > \tau \\ 0 & \text{for } t \leq \tau. \end{cases}$$

In case (b) we use Green's function

$$\Phi_k(t, \tau) = \begin{cases} 0 & \text{for } t \geq \tau \\ -i \, e^{i \int_{\tau}^{t} (\mu_k - \mu_j)(x)dx} & \text{for } t < \tau. \end{cases}$$

Now we introduce the matrix function

$$\Phi(t, \tau) = \text{diag} \left(\Phi_1(t, \tau), \dots, \Phi_\kappa(t, \tau) \right)$$

and write (16.23) as the integral equation

$$Y(t) = (\mathbb{M} Y)(t) + (\mathbb{M} e_j)(t), \qquad (16.26)$$

where $\mathbb{M}$ is defined by

$$(\mathbb{M} Z)(t) = \int_{t_0}^{\infty} \Phi(t, \tau) \Big\{ - R_2(\tau) Z(\tau)$$

$$+ S_{x \to \tau} \left(e^{i \int_{\tau}^{x} \mu_j(y)dy} Z(x) \right) \Big\} d\tau. \qquad (16.27)$$

To prove the solvability of (16.26), we estimate the norm of the operator $\mathbb{M}$ in $(L_\infty(t_0, \infty))^\kappa$. Using the boundedness of the function $(t, \tau) \to \Phi(t, \tau)$ and (16.20) we obtain

$$|\mathbb{M}Z(t)| \le c \int_{t_0-1}^{\infty} p(\tau) \parallel Z(\tau) \parallel_{(L_\infty(\tau,\tau+1))^\kappa} d\tau$$

$$+ \int_{t_0-1}^{\infty} \left(\int_{\tau}^{\tau+1} \left| S_{x \to y} \left(e^{i \int_y^x \mu_j(s)ds} Z(x) \right) \right|^2 dy \right)^{1/2} d\tau, \qquad (16.28)$$

where p is given by (16.13). By (16.25) we get

$$\parallel \mathbb{M} \parallel_{(L_\infty(t_0,\infty))^\kappa \to (L_\infty(t_0,\infty))^\kappa} \le c \int_{t_0-1}^{\infty} (p(\tau) + \chi(\tau)) d\tau.$$

Due to assumption (ii), the norm of the operator $\mathbb{M}$ is sufficiently small. Therefore, equation (16.26) has a solution Y in $(L_\infty(t_0,\infty))^\kappa$.

By combining (16.18) and (16.22) we obtain (16.15) with

$$r_j(t) = e^{i(\Lambda - \lambda_j I)t} \left(Y(t) + Q(t)(e_j + Y(t)) \right), \qquad (16.29)$$

where I is the identity matrix.

Estimate for r_j. Let us show that

$$|Y(t)| \to 0 \quad \text{as} \quad t \to \infty. \qquad (16.30)$$

Put

$$\Psi_k(t) = e^{- \int_{t_0}^{t} \Im(\mu_k - \mu_j)(s)ds}.$$

Using the representation (16.26) together with the boundedness of Y and (16.20), we get

$$|Y(t)| \le c \sum_k \Psi_k(t) \int_{t_0}^{t} (\Psi_k(\tau))^{-1} (p_1(\tau) + \chi(\tau)) d\tau$$

$$+ c \int_{t}^{\infty} (p_1(\tau) + \chi(\tau)) d\tau, \qquad (16.31)$$

where the sum is taken over all k satisfying (a). By Lemma 16.2.1 (i), $\Psi_k(t) \to 0$ as $t \to \infty$ and by (16.8), $\Psi_k(t) \le c\, \Psi_k(\tau)$ for $t \ge \tau \ge t_0$. Therefore the right-hand side of (16.31) does not exceed

$$c \Big\{ \sum_k \Psi_k(t) \int_{t_0}^{N} (\Psi_k(\tau))^{-1} (p_1(\tau) + \chi(\tau)) d\tau$$

$$+ \int_{N}^{\infty} (p_1(\tau) + \chi(\tau)) d\tau \Big\}$$

for all $N, t_0 \le N \le t$. This leads to (16.30).

Since (i) implies $Q(t) \to 0$ as $t \to \infty$, we derive from (16.29) and (16.30) that

$$r_j(t) \to 0 \quad \text{as} \quad t \to \infty. \qquad (16.32)$$

Estimate for r'_j. By (16.23), (16.25) and the boundedness of Y we obtain

$$\left(\int_t^{t+1} |Y'(\tau)|^2 d\tau\right)^{1/2} \leq$$

$$c\left\{\left(\int_t^{t+1} (|\mu(\tau)|^2 + |R_2(\tau)|^2)d\tau\right)^{1/2} + \chi(t)\right\}.$$

By (16.20)

$$\left(\int_t^{t+1} \| R_2(\tau) \|^2 d\tau\right)^{1/2} \leq c\, p(t) \leq c \int_{t-1}^{t+1} p(\tau)d\tau.$$

Since $p \in L_1(t_0, \infty)$ the right-hand side of the last inequality tends to 0 as $t \to \infty$. Due to Proposition 13.5.2 and (16.12) the functions

$$\left(\int_t^{t+1} |\mu(\tau)|^2 d\tau\right)^{1/2} \quad \text{and} \quad \chi(t)$$

also vanish at $+\infty$. Therefore

$$\left(\int_t^{t+1} |Y'(\tau)|^2 d\tau\right)^{1/2} \to 0 \quad \text{as} \quad t \to \infty,$$

which together with (16.32) yields

$$||r'_j||_{(L_2(t,t+1))^\kappa} \to 0 \quad \text{as } t \to +\infty. \tag{16.33}$$

The relation (16.17) follows from (16.32) and (16.33). $\square$

16.3 Asymptotic Formulae for Solutions

Theorem 16.3.1. *Let the Dichotomy Condition in Sect. 16.2 be fulfilled. Suppose that the integral*

$$\int_{t_0}^\infty R_1(\tau)d\tau \tag{16.34}$$

(where R_1 is introduced by (16.10)) converges, and that the functions χ and p (which are given by (16.11) and (16.13)) belong to $L_1(t_0, \infty)$. Then for sufficiently large t_0 there exist solutions $U_k \in W^\ell_{\mathrm{loc}}(t_0, \infty)$, $k = 1, \ldots, \kappa$, of (14.1) such that

$$\mathrm{col}\big(D_t^{j-1}U_k(t)\big)_{j=1}^\ell$$

$$= \exp\left(i(\lambda_k t + \int_{t_0}^t \mu_k(\tau)d\tau)\right)\left(\mathrm{col}(\lambda_k^{j-1}\varphi_k)_{j=1}^\ell + w_k(t)\right), \tag{16.35}$$

where

$$\| w_k \|_{S(t,t+1)} = o(1) \quad \text{as} \quad t \to +\infty. \tag{16.36}$$

Proof. Let

$$h_k = (h_{k1}, h_{k2}, ..., h_{k\kappa})$$

be the vector functions introduced in Lemma 16.2.2. By Theorem 13.10.4 there exists a solution $U_k \in W^\ell_{\text{loc}}(t_0, \infty)$ of (14.1) such that

$$\text{col}\big(D_t^{j-1}U_k(t)\big)^\ell_{j=1} = h_{kk}(t)\text{col}(\lambda_k^{j-1}\varphi_k)^\ell_{j=1}$$
$$+ \sum_{q \neq k} h_{kq}(t)\text{col}(\lambda_q^{j-1}\varphi_q)^\ell_{j=1} + \mathbf{v}_k. \quad (16.37)$$

By (16.15) and (16.17)

$$\|h_{kq}\|_{W_2^1(t,t+1)} = o\Big(e^{-k_0 t - \Im \int_{t_0}^t \mu_k(\tau)d\tau}\Big) \quad \text{as } t \to +\infty \quad (16.38)$$

for $q \neq k$. From (13.112), the following estimate can be derived:

$$\| \mathbf{v}_k \|_{\mathbb{S}(t,t+1)} \leq c \int_{t_0-1}^\infty g(t-\tau)\alpha(\tau) \| h_k \|_{(L_\infty(\tau,\tau+1))^\kappa} d\tau.$$

Since $\alpha(\tau) \to 0$ as $\tau \to +\infty$, the relation (16.15) and the last estimate imply

$$\| \mathbf{v}_k \|_{\mathbb{S}(t,t+1)} = o\Big(e^{-k_0 t - \Im \int_{t_0}^t \mu_k(\tau)d\tau}\Big). \quad (16.39)$$

We put

$$w_k = e^{-i(\lambda_k t + \int_{t_0}^t \mu_k(\tau)d\tau)}\Big(\sum_{q \neq k} h_{kq}(t)\text{col}(\lambda_q^{j-1}\varphi_q)^\ell_{j=1} + \mathbf{v}_k\Big) + r_k(t),$$

where r_k is the same as in (16.15). From (16.37) we arrive at (16.35). The relation (16.36) results from (16.39), (16.38) and (16.17). $\square$

The following lemma will be used to obtain an asymptotic formula for an arbitrary solution u of the equation (14.1) subject to

$$\| u \|_{W^\ell(t,t+1)} \leq c\, e^{-k_- t}. \quad (16.40)$$

Lemma 16.3.2. *Let $\nu_1, \ldots, \nu_\kappa$ be different real numbers and let γ_j, r_j, $j = 1, \ldots, \kappa$, be complex valued locally integrable functions on $[t_0, \infty)$ vanishing at $+\infty$. Suppose that for a positive ε*

$$\sum_{j=1}^\kappa c_j \exp\Big(i\nu_j t + \int_{t_0}^t \gamma_j(\tau)d\tau\Big)(1 + r_j(t)) = o\big(e^{-\varepsilon t}\big), \quad (16.41)$$

where $t \geq t_0$ and $c_j = \text{const}$. Then $c_j = 0$ for $j = 1, ..., \kappa$.

Proof. Let $c_\kappa \neq 0$. Without loss of generality we assume that $\nu_\kappa = 0$, $\gamma_\kappa = 0$ and $r_\kappa = 0$. Let $\delta > 0$ satisfy

$$\exp(i\nu_j\delta) \neq 1, \quad j = 1, \ldots, \kappa - 1.$$

Denote the left-hand side of (16.41) by $f(t)$. Then

$$f(t + \delta) - f(t) = o(e^{-\varepsilon t}). \tag{16.42}$$

The difference on the left is equal to

$$\sum_{j=1}^{\kappa-1} c_j \exp\left(i\nu_j t + \int_{t_0}^t \gamma_j(\tau)d\tau\right)$$

$$\times \left(\exp\left(i\nu_j\delta + \int_t^{t+\delta} \gamma_j(\tau)d\tau\right)\left(1 + r_j(t + \delta)\right) - 1 - r_j(t)\right).$$

Hence, by (16.42),

$$\sum_{j=1}^{\kappa-1} c_j \left(\exp(i\nu_j\delta) - 1\right) \exp\left(i\nu_j t + \int_{t_0}^t \gamma_j(\tau)d\tau\right)$$

$$\times \left(1 + s_j(t)\right) = o(e^{-\varepsilon t}),$$

where $s_j(t) \to 0$ as $t \to +\infty$. Thus we obtain a relation similar to (16.41) with κ replaced by $\kappa - 1$. The result follows by induction. $\square$

Theorem 16.3.3. *Let the assumptions of* Theorem 16.3.1 *be fulfilled. Also let* $u \in W^\ell_{\mathrm{loc}}(t_0, \infty)$ *be a solution of* (14.1) *satisfying* (16.40). *Then*
  (i) *there exist constants* $c_0, \ldots, c_\kappa$ *such that*

$$u(t) = \sum_{j=1}^\kappa c_j U_j(t) + v(t), \tag{16.43}$$

where $\{U_j\}$ *are the solutions of* (14.1) *constructed in* Theorem 16.3.1 *and*

$$\| v \|_{W^\ell(t,t+1)} \leq c_0 e^{-k+t}. \tag{16.44}$$

The constants $c_0, \ldots, c_\kappa$ *are subject to the estimate* (9.45).
  (ii) *The solution* u *admits the asymptotics:*

$$\mathrm{col}\left(D_t^{j-1}u(t)\right) = \sum_{k=1}^\kappa c_k \exp\left(i\left(\lambda_k t + \int_{t_0}^t \mu_k(\tau)d\tau\right)\right)$$

$$\times \left(\mathrm{col}(\lambda_k^{j-1}\varphi_k)_{j=1}^\ell + w_k(t)\right) + \mathrm{col}\left(D_t^{j-1}v(t)\right)_{j=1}^\ell,$$

where w_k *are subject to* (16.36) *and* v *satisfies* (16.44).

Proof. (i) From Lemma 16.3.2 and Theorem 16.3.1 we obtain that the linear hull of U_j, $j = 1, \ldots, \kappa$, may serve as the space $\mathbf{X}(L)$ in Theorem 12.2.3). The result follows from Theorem 12.2.3(ii).
  Part (ii) is a consequence of (i) combined with (16.35). $\square$

16.4 A Second Order Equation

Here we apply Theorem 16.3.3 to the equation

$$\left(D_t^2 + A - N(t)\right)u(t) = 0 \quad \text{for} \quad t > t_0. \tag{16.45}$$

We assume that the operator $A + \lambda^2 : H_2 \to H_0$ is Fredholm for all λ and that

$$\| w \|_2 + |\tau| \| w \|_1 + \tau^2 \| w \|_0 \leq c \| (A - \tau^2)w \|_0$$

for real τ with sufficiently large $|\tau|$. Let the operator A be invertible and have only one eigenvalue $-k^2$ on the negative semiaxis. We suppose that this eigenvalue is simple and denote the corresponding eigenvector by φ. We introduce the eigenvector ψ of A^* corresponding to the eigenvalue $-k^2$ and normalized by

$$(\varphi, \psi)_{H_0} = 1.$$

Under these assumptions the operator pencil $\mathcal{A}(\lambda) = \lambda^2 + A$ satisfies our usual conditions (see Sect.2.2.1). It has only two eigenvalues $\lambda_1 = k$, $\lambda_2 = -k$ on the real line and there are no other eigenvalues in some strip $|\Im\lambda| \leq \varepsilon$ for a sufficiently small ε, $\varepsilon > 0$. The eigenvalues λ_1, λ_2 are simple and have the eigenvectors $\varphi_1 = \varphi$ and $\varphi_2 = \varphi$. The eigenvectors of the adjoint operator pencil corresponding to the eigenvalues λ_1, λ_2 are $\psi_1 = (2k)^{-1}\psi$, $\psi_2 = -(2k)^{-1}\psi$, which implies the normalization condition

$$\left(\frac{d}{d\lambda}\mathcal{A}(\lambda)\varphi_j, \psi_j\right)_{H_0}\bigg|_{\lambda=\lambda_j} = 1, \quad j = 1, 2.$$

By $N(t)$ we denote an operator which maps H_2 continuously into H_0 and satisfies (15.61).

The equation (16.3) takes the form

$$D_t h - k \begin{pmatrix} 1 & 0 \\ 0 & -1 \end{pmatrix} h - \frac{1}{2k}(N(t)\varphi, \psi) \begin{pmatrix} 1 & 1 \\ -1 & -1 \end{pmatrix} h$$
$$+ Kh = 0 \quad \text{for} \quad t > t_0.$$

In this expression, K is the non-local operator described in Sect.13.9 which satisfies the estimate (13.110) with

$$g(t) = (2\varepsilon)^{-1}e^{-\varepsilon|t|}.$$

The functions α, β in (13.110) admit the equivalence relations

$$\alpha(t) \approx \left(\int_t^{t+1} \| N(\tau)\varphi \|^2 \, d\tau\right)^{1/2} \tag{16.46}$$

and

$$\beta(t) \approx \sup_v \left(\int_t^{t+1} \left| \left(v(\tau), N^*(\tau)\psi - \left(N(\tau)\varphi, \psi \right)_{H_0} \psi \right)_{H_0} \right|^2 d\tau \right)^{1/2},$$

where supremum is taken over the set

$$\left\{ v \in W^2(t, t+1) : \| v \|_{W^2(t, t+1)} = 1 \right\}.$$

Clearly,

$$\beta(t) \le c \sup_{\tau \in (t, t+1)} \| N^*(\tau)\varphi - \left(N(\tau)\varphi, \psi \right)_{H_0} \psi \|_{H_2^*}. \tag{16.47}$$

If we know beforehand that the vector

$$N^*(\tau)\psi - \left(N(\tau)\varphi, \psi \right)_{H_0} \psi$$

belongs to H_0 for almost all $\tau \in (t, t+1)$ then

$$\beta(t) \le c \left(\int_t^{t+1} \| N^*(\tau)\psi - \left(N(\tau)\varphi, \psi \right)_{H_0} \psi \|_{H_0}^2 d\tau \right)^{1/2}. \tag{16.48}$$

While checking the validity of the condition

$$\int_{t_*}^{\infty} \int_{t_*}^{\infty} \exp\left(-\frac{1}{2}\varepsilon|t - \tau| \right) \alpha(\tau)\beta(t) d\tau dt < \infty \tag{16.49}$$

(see Theorem 16.3.1) it is convenient to use the majorants for α and β obtained from (16.46)-(16.48).

The matrix $R(t)$ is equal to

$$\frac{1}{2k} \left(N(t)\varphi, \psi \right)_{H_0} \begin{pmatrix} 1 & 1 \\ -1 & -1 \end{pmatrix}$$

and hence

$$\mu_1(t) = -\mu_2(t) = \frac{1}{2k} \left(N(t)\varphi, \psi \right)_{H_0}.$$

The general Dichotomy Condition from Sect.16.2 takes the following simpler form:

(D) *There exist constants b_1, b_2 such that for $t \ge \tau \ge t_0$ either*

$$\int_\tau^t \Im\left(N(x)\varphi, \psi \right)_{H_0} dx \le b_1$$

or

$$\int_\tau^t \Im\left(N(x)\varphi, \psi \right)_{H_0} dx \ge b_2.$$

According to (16.10),

$$R_1(t) = \frac{1}{2k} \left(N(t)\varphi, \psi\right)_{H_0} \begin{pmatrix} 0 & e^{2kit} \\ -e^{-2kit} & 0 \end{pmatrix}.$$

Therefore, condition (i) in Lemma 16.2.2 is equivalent to the convergence of the integrals

$$\int_{t_0}^{\infty} \left(N(t)\varphi, \psi\right)_{H_0} e^{\pm 2kti} dt. \tag{16.50}$$

In the special case under consideration, the function p_1 defined by (16.14) is equivalent to

$$p_2(t) = \left| \left(N(t)\varphi, \psi\right)_{H_0} \right| \sum_{\pm} \left| \int_{t}^{\infty} \left(N(\tau)\varphi, \psi\right) e^{\pm 2ki\tau} d\tau \right|.$$

By collecting all this information together we can state Theorems 16.3.1 and 16.3.3 for the equation (16.45).

Theorem 16.4.1. *Let* (16.49) *and the condition* (**D**) *be valid, and let the integrals* (16.50) *converge. Also let the function*

$$t \rightarrow \left(\int_{t}^{t+1} |p_2(\tau)|^2 d\tau \right)^{1/2}$$

be summable on (t_*, ∞). *Then for a sufficiently large* t_0 *there exist two solutions* $u^{(+)}$, $u^{(-)} \in W_{\mathrm{loc}}^2(t_0, \infty)$ *of the equation* (16.45), *which have the following asymptotic form for* $t \rightarrow +\infty$:

$$\mathrm{col}\left(u^{(\pm)}(t), D_t u^{(\pm)}(t)\right)$$
$$= e^{\pm ikt \pm \frac{i}{2k} \int_{t_0}^{t} \left(N(\tau)\varphi, \psi\right)_{H_0} d\tau} \left(\mathrm{col}(\varphi, \pm k\varphi) + w^{(\pm)}(t)\right),$$

where

$$\| D_\tau w_j^{(\pm)} \|_{L_2(t, t+1; H_{2-j})} + \| w_j^{(\pm)} \|_{L_2(t, t+1; H_{3-j})} = o(1) \tag{16.51}$$

for $j = 1, 2$.

Theorem 16.4.2. *Under the conditions of* Theorem 16.4.1, *an arbitrary solution* $u \in W_{\mathrm{loc}}^2(t_0, \infty)$ *of* (16.45) *subject to*

$$\| u \|_{W^2(t, t+1)} = o\left(e^{\varepsilon t}\right) \quad \text{as} \quad t \rightarrow +\infty$$

has the asymptotic form

$$\mathrm{col}\left(u(t), D_t u(t)\right) = \sum_{\pm} c_{\pm} e^{\pm ikt \pm \frac{i}{2k} \int_{t_0}^{t} \left(N(\tau)\varphi, \psi\right)_{H_0} d\tau}$$
$$\times \left(\mathrm{col}(\varphi, \pm k\varphi) + w^{(\pm)}(t)\right) + \mathrm{col}(v(t), D_t v(t)),$$

where $c_{\pm}$ *are constants and the remainder terms* $w^{(\pm)}$ *and* v *satisfy* (16.51) *and*

$$\sum_{s=0}^{2} \| D_t^s v \|_{L_2(t, t+1; H_{2-s})} \leq c \, e^{-\varepsilon t}.$$

16.5 Application to the Schrödinger Equation in a Cylinder

Consider the Neumann problem

$$\begin{cases} \Delta u + (k^2 + p(x,t))u = 0 & \text{for } t > t_0, \ x \in \Omega \\ \partial_\nu u = 0 & \text{for } t > t_0, \ x \in \partial\Omega, \end{cases} \tag{16.52}$$

where Ω is a bounded domain in $\mathbb{R}^n$, k is a positive number and p is a complex-valued measurable function.

By using the abstract example in Sect. 16.4 we can describe the asymptotic form of the solution $u \in W^2_{2,\mathrm{loc}}(\overline{\Omega} \times (t_0, \infty))$ as $t \to +\infty$.

We assume that p is bounded and that

$$\int_{t_0}^\infty \int_\Omega |p(x,t)|^2 \, dx \, dt < \infty. \tag{16.53}$$

Let

$$\overline{p}(t) = \frac{1}{\mathrm{mes}_n \Omega} \int_\Omega p(x,t) \, dx.$$

We suppose that the integrals

$$\int_{t_0}^\infty \overline{p}(\tau) e^{\pm 2ki\tau} \, d\tau$$

are convergent and that the functions

$$\mathcal{P}_\pm(t) = \overline{p}(t) \int_t^\infty \overline{p}(\tau) e^{\pm 2ki\tau} \, d\tau$$

belong to $L_1(t_0, \infty)$. Moreover, let the following Dichotomy Condition be valid:

There exist constants b_1, b_2 such that for $t \geq \tau \geq t_0$ either

$$\int_\tau^t \Im \overline{p}(s) \, ds \leq b_1$$

or

$$\int_\tau^t \Im \overline{p}(s) \, ds \geq b_2.$$

Theorem 16.5.1. *Let the function p satisfy the conditions just stated. Suppose that $k^2 < \lambda_1$ where λ_1 is the first positive eigenvalue of the Neumann problem for the Laplacian in Ω. If a solution u of (16.52) satisfies*

$$|u(x,t)| = O\big(\exp(\sqrt{\lambda_1 - k^2} - \varepsilon)t \big)$$

with $\varepsilon > 0$ then

$$u(x,t) \sim \sum_\pm c_\pm \exp \Big(\pm i \big(kt + \frac{1}{2k} \int_{t_0}^t \overline{p}(\tau) d\tau \big) \Big)$$

for large t.

Proof. We rewrite our assumptions in the notation of Sect. 16.4. We have

$$A = -\Delta_x - k^2,$$

$N(t)$ is the operator of multiplication by $p(x, t)$,

$$H_0 = L_2(\Omega), \quad H_2 = \{u \in W_2^2(\Omega) : \partial_\nu u = 0 \text{ on } \partial\Omega\},$$

$$\phi = 1, \quad \psi = (\operatorname{mes}_n \Omega)^{-1}.$$

By Hölder's inequality and Sobolev's embedding theorem we have

$$\rho(t) \le c\|p\|_{L_s((t,t+1)\times\Omega)} \le c\sup|p|^{(s-1)/s}\|p\|_{L_2((t,t+1)\times\Omega)}^{1/s}$$

for sufficiently large s. Hence, by (16.53)

$$\rho(t) \to 0 \quad \text{as } t \to +\infty.$$

From (16.46) and (16.48) we obtain

$$\alpha(t) + \beta(t) \le C\Big(\int_t^{t+1} \int_\Omega |p(x,\tau)|^2 dx\, d\tau \Big)^{1/2},$$

and the inequality (16.49) follows from (16.53). Analogously, one can verify that all other conditions of Theorem 16.4.1 are satisfied. Thus, reference to Theorem 16.4.1 completes the proof. $\square$

16.6 Comments

Conditions of Lemma 16.2.2 are close in spirit to the conditions of Theorem 3.1 in Harris, Lutz (1975) (see also Eastham (1989), Sect. 1.11), where a finite dimensional system of ordinary differential equations of the first order has been considered.

17. The Case of a Single Multiple Eigenvalue

17.1 Introduction

In this chapter we consider the case when there is only one eigenvalue λ_0 of $\mathcal{A}(\lambda)$ on the line $\Im\lambda = k_0$ and the geometric multiplicity of λ_0 is 1. Let κ be the algebraic multiplicity of λ_0 and let $\varphi_0, \ldots, \varphi_{\kappa-1}$ be a Jordan chain corresponding to λ_0. Let $\psi_0, \ldots, \psi_{\kappa-1}$ be the Jordan chain of $\mathcal{A}^*(\lambda)$ corresponding to the eigenvalue $\overline{\lambda}_0$ and satisfying the biorthogonality condition (1.20) or equivalently,

$$\sum_{s=0}^{n} \sum_{j=s+1}^{s+\kappa} \frac{1}{j!} \left(\mathcal{A}^{(j)}(\lambda_0)\varphi_{s+\kappa-j}, \psi_{n-s} \right)_{H_0} = \delta_n^0. \tag{17.1}$$

We suppose that the function ρ (which characterizes the perturbation operator $N(t, D_t)$ (see (12.86))) tends to zero as $t \to +\infty$.

Our main result is an asymptotic formula for solutions of the equation (14.1); this is obtained by analyzing the finite dimensional system (13.125). In the present case, the matrix $\mathcal{R} = \{\mathcal{R}_{jk}\}_{j,k=1}^{\kappa}$ is given by

$$\mathcal{R}_{jk}(t) = \sum_{q=0}^{k-1} \frac{1}{q!} \left(N^{(q)}(t, \lambda_0)\varphi_{k-1-q}, \psi_{\kappa-j} \right)_{H_0}, \tag{17.2}$$

where $N^{(q)}(t, \lambda) = \partial_\lambda^q N(t, \lambda)$.

We assume that there exists a positive function r on (t_0, ∞) vanishing at ∞ such that the matrix

$$\mathcal{B}(t) = \left\{ r(t)^{k-j-1} \mathcal{R}_{jk}(t) \right\}_{j,k=1}^{\kappa} \tag{17.3}$$

has a finite limit $\mathcal{B}(\infty)$ as $t \to \infty$. (Clearly, the elements of $\mathcal{B}(\infty)$ with $k \geq j + 1$ are equal to zero). The function r must satisfy several conditions; these are described in Sect.17.3. One of them is the existence of a finite limit

$$q = \lim_{t \to \infty} (r' r^{-2})(t). \tag{17.4}$$

We suppose that the eigenvalues of the matrix

$$\mathcal{B}(\infty) + J + iq \ \text{diag} \ (0, 1, \ldots, \kappa - 1) \tag{17.5}$$

have different imaginary parts, where the elements of the matrix J are equal to 1 on the first diagonal above the leading diagonal and to 0 otherwise.

In Theorem 17.4.1 under certain restrictions on $\mathfrak{B}'$, $(r^{-2}r')'$ and the functions α, β used in previous chapters, we construct κ solutions $u_k \in W^\ell(t_0, \infty)$ of equation (14.1) with the characteristic exponent $-k_0$.

Let $\mathfrak{m}_k(t)$ be eigenvalues of the matrix

$$\mathfrak{B}(t) + J + i\frac{r'(t)}{r^2(t)} \operatorname{diag}(0, 1, \ldots, \kappa - 1) \qquad (17.6)$$

and $\operatorname{col}(T_{jk})_{j=1}^\kappa$ is an eigenvector of (17.5) corresponding to the eigenvalue $\mathfrak{m}_k(\infty)$. The eigenvalues $\mathfrak{m}_k(t)$ are represented in the form

$$\mathfrak{m}_k(t) = \mathfrak{m}_k(\infty) + \mathbb{H}_k\Big((r'r^{-2})(t) - q,\ \mathfrak{B}(t) - \mathfrak{B}(\infty)\Big).$$

Here ,

$$\mathbb{R} \times \mathbb{C}^{\kappa \times \kappa} \ni (z, Z) \to \mathbb{H}_k(z, Z)$$

are functions which are holomorphic at the origin, are determined by the matrix $\mathfrak{B}(\infty)$, and have $\mathbb{H}_k(0, 0) = 0$.

As a consequence of Theorem 17.4.1 we obtain that under the complementary assumption $T_{1k} \neq 0$, $k = 1, \ldots, \kappa$, the solutions u_k admit the asymptotic representation

$$u_k(t) \sim \exp\Big(i\big(\lambda_0 t + \int_{t_0}^t r(\tau)\mathfrak{m}_k(\tau)d\tau\big)\Big)(C_k\varphi_0 + o(1)), \qquad (17.7)$$

where $C_k = \operatorname{const}$. Formula (17.7) together with Theorem 12.2.3 leads to an asymptotic representation for all solutions of (14.1) subject to (15.2) (see Theorem 17.4.3).

We conclude the chapter with an example where the asymptotics (17.7) takes an especially simple form.

17.2 Solution of an Auxiliary Matrix Equation

In the construction of an asymptotic formula for the vector function h (see equation (13.125)) in the next section, we will need the following matrix equation:

$$(\mathcal{H} + \mathcal{Z})\Upsilon = \Upsilon(\mathfrak{m} + \Xi) , \qquad (17.8)$$

where Υ and Ξ are square matrices, and $\Xi = \operatorname{diag}(\Xi_1, \ldots, \Xi_\kappa)$. Here

$$\mathcal{H} = \Big\{\mathcal{H}_{jk}\Big\}_{j,k=1}^\kappa$$

is a matrix with different eigenvalues $\mathfrak{m}_1, \ldots, \mathfrak{m}_\kappa$,

$$\mathfrak{m} = \mathrm{diag}(\mathfrak{m}_1, \ldots, \mathfrak{m}_\kappa), \qquad (17.9)$$

and $\mathcal{Z}$ is a $\kappa \times \kappa$-matrix with elements $\mathcal{Z}_{jk}$, which are small in a certain sense.

Solving of (17.8) is equivalent to the problem of constructing the eigenvalues and the eigenvectors of the matrix $\mathcal{H} + \mathcal{Z}$. In fact, the diagonal elements of $\mathfrak{m} + \Xi$ are eigenvalues of $\mathcal{H} + \mathcal{Z}$ and the columns of the solution Υ are eigenvectors of $\mathcal{H} + \mathcal{Z}$. Although it is a classical fact that these eigenvalues and eigenvectors depend analytically on $\mathcal{Z}$ we give a proof for readers convenience.

First, we consider a simpler matrix equation

$$\big(\mathfrak{m} + Z\big)\big(I + Y\big) = \big(I + Y\big)\big(\mathfrak{m} + \Xi\big) \qquad (17.10)$$

with respect to Ξ and Y, where I is the identity $\kappa \times \kappa$-matrix.

Lemma 17.2.1. *Suppose that*

$$|(\mathfrak{m}_i - \mathfrak{m}_n)^{-1}Z_{im}| < (\kappa + 2\sqrt{\kappa - 1})^{-1} \qquad (17.11)$$

for all $i, n, m, i \neq n$. Then the equation (17.10) has a solution (Y, Ξ) such that

(i) *$Y = Y(Z)$ is a matrix $\{Y_{jk}\}_{j,k=0}^{\kappa}$ with zero diagonal elements and $Y_{jk} = Y_{jk}(Z)$ are represented as absolutely convergent power series with respect to the variables*

$$(\mathfrak{m}_i - \mathfrak{m}_n)^{-1}Z_{im}, \quad i \neq n.$$

Moreover, $Y_{jk}(0) = 0$.

(ii) *$\Xi = \mathrm{diag}(\Xi_1, \ldots, \Xi_\kappa)$ and*

$$\Xi_i = Z_{ii} + \sum_{n=1}^{\kappa} Z_{in}Y_{ni}. \qquad (17.12)$$

Proof. We rewrite equation (17.10) as

$$\mathfrak{m}Y - Y\mathfrak{m} = -Z - ZY + \Xi + Y\Xi. \qquad (17.13)$$

Since $\mathfrak{m}, \Xi$ are diagonal matrices and the diagonal elements of Y are zero, we arrive at (17.12).

From (17.13) and (17.12) we obtain the equation

$$(\mathfrak{m}_k - \mathfrak{m}_j)Y_{jk} \qquad (17.14)$$

$$= Z_{jk} + \sum_{n=1}^{\kappa} Z_{jn}Y_{nk} - Y_{jk}Z_{kk} - Y_{jk}\sum_{n=1}^{\kappa} Z_{kn}Y_{nk}.$$

We are looking for a solution Y of (17.14) in the form

$$Y = \sum_{q=1}^{\infty} Y^{(q)}, \qquad (17.15)$$

where the matrix $Y^{(q)}$ has zero diagonal elements and its non-diagonal elements are homogeneous polynomials of degree κ with respect to the variables $(\mathfrak{m}_n - \mathfrak{m}_m)^{-1} Z_{im}$, $n \neq m$. Then equation (17.14) implies

$$Y_{jk}^{(1)} = (\mathfrak{m}_k - \mathfrak{m}_j)^{-1} Z_{jk}, \tag{17.16}$$

$$Y_{jk}^{(2)} = (\mathfrak{m}_k - \mathfrak{m}_j)^{-1} \left(\sum_{n=1}^{\kappa} Z_{jn} Y_{nk}^{(1)} - Y_{jk}^{(1)} Z_{kk} \right) \tag{17.17}$$

and

$$\begin{aligned} Y_{jk}^{(q+1)} = {} & (\mathfrak{m}_k - \mathfrak{m}_j)^{-1} \left(\sum_{n=1}^{\kappa} Z_{jn} Y_{nk}^{(q)} - Y_{jk}^{(q)} Z_{kk} \right. \\ & \left. - \sum_{s=1}^{q-1} Y_{jk}^{(s)} \sum_{n=1}^{\kappa} Z_{kn} Y_{nk}^{(q-s)} \right) \end{aligned} \tag{17.18}$$

for $q = 2, 3, \ldots$.

It remains to show the absolute convergence of the series (17.15) with the variables

$$z_{nmi} = (\mathfrak{m}_i - \mathfrak{m}_n)^{-1} Z_{im}, \quad i \neq n.$$

Let us introduce the $\kappa^2(\kappa - 1)$-dimensional vector $z = (z_{nmi})$, where $i \neq n$. We have

$$Y_{jk}^{(q)} = \sum_{|\alpha|=q} c_{jk\alpha} z^{\alpha},$$

where α is a $\kappa^2(\kappa - 1)$-multiindex. We put

$$a_q = \max_{j,k} \sum_{|\alpha|=q} |c_{jk\alpha} z^{\alpha}|.$$

It is sufficient to verify the convergence of the series

$$\sum_{q=1}^{\infty} a_q.$$

Assuming that

$$|z_{nmi}| \leq \tau \quad \text{for} \quad i \neq n,$$

and using (17.16), (17.17) and (17.18) we get

$$a_1 \leq \tau, \quad a_2 \leq \kappa \tau^2 \tag{17.19}$$

and

$$a_{q+1} \leq \kappa \tau a_q + (\kappa - 1)\tau \sum_{s=1}^{q-1} a_s a_{q-s} \tag{17.20}$$

for $q = 2, 3, \ldots$.

We introduce an auxiliary function f, which solves the equation

$$f(\tau) = \tau + \kappa \tau f(\tau) + (\kappa - 1)\tau f^2(\tau) \tag{17.21}$$

for small τ and vanishes at $\tau = 0$. Since

$$f(\tau) = 2\tau\left(1 - \kappa\tau + \sqrt{(1 - \kappa\tau)^2 - 4(\kappa - 1)\tau^2}\right)^{-1},$$

this function is holomorphic for

$$|\tau| < \left(\kappa + 2\sqrt{\kappa - 1}\right)^{-1}. \tag{17.22}$$

Using (17.21), we can write recurrence relations for the coefficients f_n in

$$f(\tau) = \sum_{n=1}^{\infty} f_n \tau^n. \tag{17.23}$$

We have

$$f_1 = 1, \quad f_2 = \kappa$$

and

$$f_{q+1} = \kappa f_q + (\kappa - 1) \sum_{s=1}^{q-1} f_s f_{q-s} \tag{17.24}$$

where $q = 2, 3, \ldots$

Let us prove that

$$a_q \le f_q \tau^q, \quad q = 1, \ldots \tag{17.25}$$

For $q = 1, 2$ the estimate (17.25) follows from (17.19). Let $q \le 3$. Then (17.20) and (17.24) imply

$$a_q \le \kappa \tau^q f_{q-1} + (\kappa - 1) \sum_{s=1}^{q-1} f_s f_{q-s} \tau^q = f_q \tau^q.$$

Hence (17.25) and the convergence of (17.23) for τ satisfying (17.22) yield the convergence of (17.15). $\square$

We turn to equation (17.8). Let T be a matrix whose columns coincide with eigenvectors of $\mathcal{H}$. The matrix T is invertible and

$$T^{-1}\mathcal{H}T = \mathfrak{m}, \tag{17.26}$$

where $\mathfrak{m}$ is given by (17.9).

Lemma 17.2.2. *Let*

$$\| \mathcal{Z} \| \le c_\kappa \, \| T \| \, \| T^{-1} \| \min_{j \ne k} |\mathfrak{m}_j - \mathfrak{m}_k|, \tag{17.27}$$

where c_κ is a positive constant which depends only on κ. Then there exists a solution

$$\varUpsilon = \varUpsilon(\mathcal{Z}), \quad \varXi = \varXi(\mathcal{Z})$$

of (17.8) satisfying

(i) $\varUpsilon(\mathcal{Z}) = T(I + Y(\mathcal{Z}))$, *where the diagonal elements of Y are zero, the non-diagonal elements are represented as absolutely convergent series with respect to $\mathcal{Z}_{kj}$ and $Y(0) = O$.*

(ii) *The following representation holds:*

$$\varXi_i(\mathcal{Z}) = (T^{-1}\mathcal{Z}T)_{ii} + \sum_{n=1}^{\kappa}(T^{-1}\mathcal{Z}T)_{in}Y_{ni}(\mathcal{Z}).$$

Proof. From (17.26) and (17.8) we derive the following equation with respect to the pair $(Y, \varXi)$:

$$(\varLambda + T^{-1}\mathcal{Z}T)(I + Y) = (I + Y)(\mathfrak{m} + \varXi).$$

Reference to Lemma 17.2.1 completes the proof. $\square$

17.3 Special Solutions of the Finite Dimensional System

17.3.1 Prerequisites

Under the restrictions to the spectrum of $\mathcal{A}(\lambda)$ described in Sect. 17.1 the system (13.125) becomes

$$\left(D_t - \lambda_0 I - J - \mathcal{R}(t)\right)h(t) - (Kh)(t) = 0 \quad \text{for} \quad t > t_0, \tag{17.28}$$

where $h = (h_1, \ldots, h_\kappa)$, I is the $\kappa \times \kappa$ identity matrix, J is the matrix whose elements on the first diagonal above the leading diagonal equal 1 and zero otherwise. The elements of the matrix $\mathcal{R}$ are

$$\mathcal{R}_{jk}(t) = i\Big(N(t, D_\tau)U_{k-1}(\tau)\big|_{\tau=0}, \, V_{j-1}(0)\Big)_{H_0}$$

$$= \sum_{q=0}^{k} \frac{1}{q!}\Big(N^{(q)}(t, \lambda_0)\varphi_{k-1-q}, \, \psi_{\kappa-j}\Big)_{H_0}, \tag{17.29}$$

where

$$U_h(t) = e^{i\lambda_0 t} \sum_{k=0}^{h} \frac{(it)^k}{k!}\varphi_{h-k} \tag{17.30}$$

and

$$V_h(t) = ie^{i\overline{\lambda}_0 t} \sum_{k=0}^{\kappa-1-h} \frac{(it)^k}{k!} \psi_{\kappa-1-h-k} \qquad (17.31)$$

(see Sect. 13.9 and 13.10). The nonlocal operator K satisfies (13.110). By Proposition 13.5.1 the function α satisfies

$$\alpha(t) \approx \sum_{h=0}^{\kappa-1} \Big(\int_t^{t+1} \| \, N(\tau, D_x) U_h(x) \big|_{x=0} \, \|_0^2 \, d\tau \Big)^{1/2}.$$

Together with (17.30) this implies that

$$\alpha(t) \approx \sum_{h=0}^{\kappa-1} \Big(\int_t^{t+1} \| \sum_{s=0}^{h} \frac{1}{s!} N^{(s)}(\tau, \lambda_0) \varphi_{h-s} \, \|_0^2 \, d\tau \Big)^{1/2}. \qquad (17.32)$$

By (17.30), (17.31) and Proposition 13.8.3,

$$\beta(t) \approx \sup_v \sum_{j=0}^{\kappa-1} \Big(\int_t^{t+1} \big| \big(N(\tau, D_\tau) v(\tau), \psi_{\kappa-1-j} \big)_{H_0} \qquad (17.33)$$

$$- \sum_{k=0}^{\kappa-1} \sum_{s=0}^{\kappa-k-1} \frac{1}{s!} \big(\mathcal{B}^{(s)}(\lambda_0, D_\tau) v(\tau), \psi_{\kappa-1-k-s} \big)_{H_0} \mathcal{R}_{j+1,k+1}(\tau) \big|^2 d\tau \Big)^{1/2}$$

where supremum is taken over the set (13.75).

We introduce the following conditions.

Condition A *There exists a positive function r on (t_0, ∞) vanishing at infinity, with absolutely continuous derivative, such that the function $r'r^{-2}$ and the matrix function (17.3) have finite limits q and $\mathcal{B}(\infty)$.*

Condition B *The eigenvalues $\mathfrak{m}_j(\infty)$, $j = 1, \ldots, \kappa$, of the matrix (17.5) have different imaginary parts.*

In the next lemma we describe simple properties of the function r.

Lemma 17.3.1. a) *The following estimate holds:*

$$r(t) \geq c/t \quad \text{for} \quad t > t_0, \qquad (17.34)$$

with a positive constant c.
  b) *There exists a constant c such that*

$$c^{-1} \leq |r(t)/r(\tau)| \leq c \qquad (17.35)$$

for $t, \tau > t_0$, $|t - \tau| \leq 1$.

Proof. a) By (17.4) the function $t \to (r'r^{-2})(t)$ is bounded. Hence

$$\left| \frac{1}{r(t)} - \frac{1}{r(\tau)} \right| = \left| \int_\tau^t (r'r^{-2})(x)dx \right| \le c|t - \tau| \qquad (17.36)$$

and we arrive at (17.34).

b) The relation (17.35) follows from (17.36) and the boundedness of r. $\square$

We introduce the function

$$p(t) = \| (r'r^{-2})' \|_{L_2(t,t+1)} + \| \mathfrak{B}' \|_{(L_2(t,t+1))^{\kappa \times \kappa}}$$
$$+ r(t)^{1-\kappa} \beta(t) \int_{t_0-1}^\infty e^{-a|t-\tau|} \alpha(\tau)d\tau, \qquad (17.37)$$

where $\mathfrak{B}$ is the matrix given by (17.3).

Lemma 17.3.2. *The function p satisfies*

$$p(t) \le c \int_{t-1}^{t+1} p(\tau)d\tau. \qquad (17.38)$$

Proof. The required estimate for the first and second term in the right-hand side of (17.37) is proved in the same way as (13.45).

By (17.35) and by (13.68) the third term in the right-hand side of (17.37) is estimated by

$$c \int_{t-1}^{t+1} r(\tau)^{1-\kappa} \beta(\tau) \int_{t_0-1}^\infty e^{-a|\tau-x|} \alpha(x)dx.$$

$\square$

17.3.2 Lemma on Special Solutions

Now we find κ solutions of the equation (17.28) with prescribed asymptotics at infinity. In the formulation of next lemma we make use of the following notation.

By $\mathfrak{m}_k(\tau)$ we denote the eigenvalues of the matrix (17.6) and by $\mathrm{col}(T_{jk})_{j=1}^\kappa$ an eigenvector of (17.5) corresponding to the eigenvalue $\mathfrak{m}_k(\infty)$.

Let $T = \{T_{jk}\}_{j,k=1}^\kappa$. Since the columns of T are eigenvectors of (17.5), it follows that

$$T^{-1}(\mathfrak{B}(\infty) + J + iqM)T = \mathfrak{m}(\infty),$$

where $\mathfrak{m}(\infty)$ is the matrix $\mathrm{diag}(\mathfrak{m}_1(\infty), \dots, \mathfrak{m}_\kappa(\infty))$. Due to the smallness of $(r'r^{-2})(t) - q$, $\mathfrak{B}(t) - \mathfrak{B}(\infty)$ (see (17.41)) and Lemma 17.2.2, there exist matrices

$$Y = Y(t) \quad \text{and} \quad \varXi = \varXi(t) = \mathrm{diag}\left(\varXi_1(t), \dots, \varXi_\kappa(t)\right)$$

such that

$$\left(\mathfrak{B}(t) + J + i\frac{r'(t)}{r^2(t)}M\right)T(I + Y(t))$$
$$= T(I + Y(t))(\mathfrak{m}(\infty) + \Xi(t)). \tag{17.39}$$

Clearly, $\mathfrak{m}(t) = \mathfrak{m}(\infty) + \Xi(t)$. By Lemma 17.2.2 the norm of the matrix Y is small and its elements are holomorphic with respect to

$$(r'r^{-2})(t) - q, \quad \mathfrak{B}_{jk}(t) - \mathfrak{B}_{jk}(\infty).$$

Hence and by Condition A from Sect. 17.3.1 the matrix function Y vanishes at ∞ and

$$|D_t Y(t)| \le c\left(|(\frac{r'(t)}{r^2(t)})'| + \| D_t \mathfrak{B}(t) \|\right), \tag{17.40}$$

where $\| \cdot \|$ denotes the norm of matrix.

Lemma 17.3.3. *Let the conditions A and B be satisfied and let*

$$\int_{t_0-1}^{\infty} p(\tau)d\tau \le \varepsilon, \tag{17.41}$$

where ε is a small positive constant. Then equation (17.28) has solutions

$$h^{(k)} = \mathrm{col}\big(h_j^{(k)}\big)_{j=1}^{\kappa} \in \big(W_{2,\mathrm{loc}}^1(t_0, \infty)\big)^{\kappa}$$

such that

$$h_j^{(k)}(t) = \exp\left(i(\lambda_0 t + \int_{t_0}^t r(\tau)\mathfrak{m}_k(\tau)d\tau)\right)$$
$$\times r^{j-1}(t)\big((T(I + Y(t)))_{jk} + w_{jk}(t)\big), \quad k = 1, \ldots, \kappa. \tag{17.42}$$

The remainder term w_{kj} in (17.42) satisfies

$$\| w_{kj} \|_{W_2^1(t,t+1)} = o\left(\int_{t-1}^{\infty} p(\tau)d\tau\right)$$
$$+ O\left(\int_{t_0-1}^{\infty} e^{-a_0|\int_{\tau}^t r(x)dx|}p(\tau)d\tau\right), \tag{17.43}$$

where a_0 is a positive constant. (Note that by convergence of the integral in (17.41) and by (17.34), (17.38) the left-hand side in (17.43) tends to zero as $t \to +\infty$.)

Proof. Let us rewrite (17.3) as

$$\mathcal{R}(t) = r(t)Q(t)\mathfrak{B}(t)Q^{-1}(t), \tag{17.44}$$

where

$$Q(t) = \operatorname{diag}(1, r(t), \ldots, r(t)^{\kappa-1}).$$

We introduce a new unknown vector function y by

$$h(t) = e^{i\lambda_0 t}Q(t)y(t). \tag{17.45}$$

Inserting this into (17.28) and using (17.44) we obtain the equation for y:

$$D_t y(t) - r(t)\Big(i\frac{r'(t)}{r^2(t)}M + \mathfrak{B}(t) + J\Big)y(t)$$
$$-(K_1 y)(t) = 0, \tag{17.46}$$

where

$$M = \operatorname{diag}(0, 1, \ldots, \kappa - 1)$$

and

$$(K_1 y)(t) = (Q(t))^{-1}e^{-i\lambda_0 t}K_{\tau \to t}\big(e^{i\lambda_0 \tau}Q(\tau)y(\tau)\big).$$

By (13.110) and (14.26) we have

$$\| K_1 y \|_{(L_2(t,t+1))^\kappa}$$
$$\leq c\, r(t)^{1-\kappa}\beta(t)\int_{t_0-1}^{\infty} e^{-2a|t-\tau|}\alpha(\tau)\, \| y \|_{(L_\infty(\tau,\tau+1))^\kappa}\, d\tau. \tag{17.47}$$

Changing variables in (17.46):

$$y(t) = T\big(I + Y(t)\big)z(t), \tag{17.48}$$

and using (17.39) we obtain

$$D_t z(t) - r(t)\mathfrak{m}(t)z(t) + \big(I + Y(t)\big)^{-1}D_t Y(t)z(t)$$
$$-(K_2 z)(t) = 0, \tag{17.49}$$

where

$$\mathfrak{m}(t) = \operatorname{diag}\big(\mathfrak{m}_1(t), \ldots, \mathfrak{m}_\kappa(t)\big) \tag{17.50}$$

with $\mathfrak{m}_k(t) = \mathfrak{m}_\kappa(\infty) + \Xi_k(t)$ and

$$(K_2 z)(t) = \big(I + Y(t)\big)^{-1}T^{-1}K_{1,\tau \to t}\Big(T\big(I + Y(t)\big)z(\tau)\Big)(t).$$

By (17.47) we have

$$\| K_2 z \|_{(L_2(t,t+1))^\kappa}$$
$$\leq c\, r(t)^{1-\kappa}\beta(t)\int_{t_0-1}^{\infty} e^{-2a|t-\tau|}\alpha(\tau)\, \| z \|_{(L_\infty(\tau,\tau+1))^\kappa}\, d\tau. \tag{17.51}$$

We are looking for a solution of (17.49) in the form

$$z(t) = e^{i \int_{t_0}^t r(\tau) \mathfrak{m}_k(\tau) d\tau} (e_k + w(t)), \tag{17.52}$$

where $e_k = \mathrm{col}(\delta_k^j)_{j=1}^\ell$. Inserting this in (17.49) we obtain the equation for w:

$$D_t w(t) - r(t)\Theta(t)w(t) + \big(I + Y(t)\big)^{-1} D_t Y(t)\big(e_k + w(t)\big)$$
$$- K_3(e_k + w)(t) = 0 \quad \text{for} \quad t > t_0, \tag{17.53}$$

where

$$\Theta(t) = \mathrm{diag}\,(\theta_1(t), \ldots, \theta_\kappa(t))$$

with

$$\theta_j(t) = \mathfrak{m}_j(t) - \mathfrak{m}_k(t)$$

and where

$$K_3(w)(t) = K_{2,\tau \to t}\Big(e^{i \int_\tau^t r(x)\mathfrak{m}_k(x)dx} w(\tau)\Big).$$

Since r vanishes at ∞, using (17.51) we get

$$\| K_3 w \|_{(L_2(t,t+1))^\kappa}$$
$$\leq c\, r(t)^{1-\kappa} \beta(t) \int_{t_0-1}^\infty e^{-a|t-\tau|}\alpha(\tau)\, \| w \|_{(L_\infty(\tau,\tau+1))^\kappa}\, d\tau. \tag{17.54}$$

To derive an integral equation for w we introduce a solution

$$G(t,\tau) = \mathrm{diag}\Big(G_1(t,\tau), \ldots, G_\kappa(t,\tau)\Big)$$

of the system

$$D_t G(t,\tau) - r(t)\Theta(t)G(t,\tau) = \delta(t-\tau)I$$

as follows:
if $\Im\big(\mathfrak{m}_j(\infty) - \mathfrak{m}_k(\infty)\big) > 0$ then

$$G_j(t,\tau) = \begin{cases} i\, e^{i \int_\tau^t r(x)\theta_j(x)dx} & \text{for } t \geq \tau \\ 0 & \text{for } t < \tau, \end{cases}$$

if $\Im\big(\mathfrak{m}_j(\infty) - \mathfrak{m}_k(\infty)\big) < 0$ then

$$G_j(t,\tau) = \begin{cases} 0 & \text{for } t > \tau \\ -i e^{-i \int_t^\tau r(x)\theta_j(x)dx} & \text{for } t \leq \tau \end{cases}$$

and

$$G_k(t,\tau) = \begin{cases} 0 & \text{for } t > \tau \\ -i & \text{for } t \leq \tau. \end{cases} \tag{17.55}$$

Since $\Im\big(\mathfrak{m}_j(\infty) - \mathfrak{m}_k(\infty)\big) \neq 0$ for $j \neq k$ (see Condition B from Sect. 17.3.1) and since $\varXi(t) \to 0$ as $t \to \infty$, it follows that

$$\| G_j(t,\tau) \| \le c\, e^{-a_0 | \int_\tau^t r(x)dx |} \quad \text{for} \quad t,\tau \ge t_0, \tag{17.56}$$

where a_0 is a positive constant and $j \ne k$.

Using Green's function G we can rewrite (17.53) as the integral equation

$$w(t) = \mathcal{J}(e_k + w)(t), \quad t > t_0, \tag{17.57}$$

where

$$(\mathcal{J}w)(t) = \int_{t_0}^\infty G(t,\tau)\Big((K_3 w)(\tau)$$

$$-(I+Y(\tau))^{-1}(D_\tau Y(\tau))w(\tau)\Big)d\tau. \tag{17.58}$$

In order to prove the solvability of (17.57) in $(L_\infty(t_0,\infty))^\kappa$, we estimate the norm of $\mathcal{J}$. Using (17.55) and (17.56) together with (17.54) and (17.40) we have

$$\| (\mathcal{J}w)(t) \| \le c\Big(\int_{t_0-1}^\infty e^{-a_0 | \int_\tau^t r(x)dx |} p(\tau)d\tau$$

$$+ \int_{t-1}^\infty p(\tau)d\tau\Big) \| w \|_{(L_\infty(t_0,\infty))^\kappa} \tag{17.59}$$

with p given by (17.37). Hence, it follows from (17.41) and (17.34) that the norm of the operator

$$\mathcal{J} : \big(L_\infty(t_0,\infty)\big)^\kappa \to \big(L_\infty(t_0,\infty)\big)^\kappa$$

is small and therefore equation (17.57) has a bounded solution. By (17.57) and (17.59) we get

$$\| w(t) \| \le c\,\Big(\int_{t_0-1}^\infty e^{-a_0 | \int_\tau^t r(x)dx |} p(\tau)d\tau + \int_{t-1}^\infty p(\tau)d\tau\Big). \tag{17.60}$$

Using (17.53) and the boundedness of w we obtain

$$\| Dw \|_{(L_2(t,t+1))^\kappa} \le c\big(\|w\|_{(L_2(t,t+1))^\kappa} + p(t)\big). \tag{17.61}$$

By (17.38) and (17.60) the right-hand side of (17.60) majorizes the norm $\| w \|_{(W_2^1(t,t+1))^\kappa}$ which implies, in particular, that this norm is $o(1)$ as $t \to +\infty$. Hence and by (17.57)

$$\| w - \mathcal{J}(e_k) \|_{(W_2^1(t,t+1))^\kappa}$$

$$= O\Big(\int_{t_0-1}^\infty e^{-a_0 | \int_\tau^t r(x)dx |} p(\tau)d\tau\Big) + o\Big(\int_{t-1}^\infty p(\tau)d\tau\Big). \tag{17.62}$$

By taking into account that the diagonal elements of Y are zero (see Lemma 17.2.2) we obtain from (17.58) that $\|\mathcal{J}(e_k)(t)\|_{(W_2^1(t,t+1))^\kappa}$ is dominated by the right-hand side in (17.62). Therefore, the same holds for w.

The asymptotic formula formula (17.42) with

$$w_{jk}(t) = \left(T(I + Y(t))w(t)\right)_j,\tag{17.63}$$

follows from (17.45), (17.48) and (17.52). By the boundedness of Y and (17.40) the norm $\|w_{jk}\|_{(W_2^1(t,t+1))^\kappa}$ has the same majorant as $\|w\|_{(W_2^1(t,t+1))^\kappa}$. The result follows. $\square$

17.4 Asymptotics of Solutions

Theorem 17.4.1. *Let the conditions* A *and* B *from Sect.17.3 be fulfilled and let*

$$\int_{t_0}^{\infty} p(\tau)d\tau < \infty,\tag{17.64}$$

where p *is given by (17.37). Then for sufficiently large* t_0, *equation (14.1) has solutions* $u_k \in W_{\text{loc}}^\ell(t_0, \infty)$, $k = 1, \ldots, \kappa$, *such that*

$$\text{col}\left(D_t^{s-1}u_k(t)\right)_{s=1}^\ell = \exp\left(i(\lambda_0 t + \int_{t_0}^t r(\tau)\mathfrak{m}_k(\tau)d\tau)\right)\tag{17.65}$$

$$\times \left\{ \sum_{j=1}^\kappa r^{j-1}(t)\left((T(I+Y(t)))_{jk}+w_{jk}(t)\right)\text{col}\left(D_\tau^{s-1}U_{j-1}(\tau)\big|_{\tau=0}\right)_{s=1}^\ell + \mathbf{v}_k(t)\right\}$$

(the notations $\mathfrak{m}_k$, T_{jk}, U_j *were explained in the previous section). The functions* w_{jk} *satisfy (17.43). Furthermore,* $P\mathbf{v}_k = 0$, *where* P *is the Riesz projector introduced in Sect. 12.5, and*

$$\|\,\mathbf{v}_k\,\|_{S(t,t+1)} \le c \int_{t_0-1}^{\infty} e^{-a|t-\tau|}\alpha(\tau)d\tau.\tag{17.66}$$

Proof. By Theorem 13.10.4 and Lemma 17.3.3 there exists a solution $u_k \in W_{\text{loc}}^\ell(t_0, \infty)$ of (14.1) represented in the form

$$\text{col}\left(D_t^{s-1}u_k(t)\right)_{s=1}^\ell = \sum_{j=1}^\kappa h_j^{(k)}(t)\,\text{col}\left(D_\tau^{s-1}U_j(\tau)\big|_{\tau=0}\right)_{s=1}^\ell + \mathbf{v}(t),$$

where $P\mathbf{v} = 0$ and

$$\|\,\mathbf{v}\,\|_{S(t,t+1)} \le c \int_{t_0-1}^{\infty} g(t-\tau)\alpha(\tau)\,\|\,h^{(k)}\,\|_{(L_\infty(\tau,\tau+1))^\kappa}\,d\tau.$$

By (17.42)

$$\|\,h^{(k)}\,\|_{(L_\infty(\tau,\tau+1))^\kappa} \le c\exp\left(-\Im(\lambda_0 \tau + \int_{t_0}^\tau r(x)\mathfrak{m}_k(x)dx)\right),$$

It follows from (14.26) and from $r(t) \to 0$ as $t \to +\infty$ that

$$g(t - \tau) \exp \left(\Im(\lambda_0(t - \tau) + \int_\tau^t r(x)\mathfrak{m}_k(x)dx) \right) \leq c\, e^{-a|t-\tau|}$$

and we arrive at (17.65) and (17.66), where

$$\mathbf{v}_k(t) = \exp \left(- i(\lambda_0 t + \int_{t_0}^t r(\tau)\mathfrak{m}_k(\tau)d\tau) \right) \mathbf{v}(t).$$

□

Corollary 17.4.2. *Let the assumptions of* Theorem 17.4.1 *be fulfilled. Then the solutions* $u_1, \ldots, u_\kappa$ *of* (14.1) *span the space* $\mathbf{X}(L)$ *satisfying the conditions of* Theorem 12.2.3.

Proof. The asymptotic formula (17.65) implies

$$\| u_k \|_{W^\ell(t, t+1)} \leq c \exp \left(- k_0 t + a_1 \int_{t_0}^t r(\tau)d\tau \right)$$

for some positive constant a_1. Since $r(t) \to 0$ as $t \to 0$ this inequality gives the estimate (12.10).

Let us prove (12.11). We set

$$\sigma_{jk}(t) = (T(I + Y(t)))_{jk} + w_{jk}(t) .$$

By (17.65) it suffices to show that the inequality

$$\| \sum_{k=1}^\kappa c_k \exp \left(i \int_{t_0}^t r(\tau)\mathfrak{m}_k(\tau)d\tau \right)$$

$$\times \left(\sum_{j=1}^\kappa r^{j-1}(t)\sigma_{jk}(t)\mathrm{col}\big(D_\tau^{s-1}U_{j-1}(\tau)\big|_{\tau=0}\big)_{s=1}^\ell + \mathbf{v}_k(t) \right) \|_{\mathcal{H}_0} \leq ce^{(k_0 - k_+)t}$$

implies $c_k = 0$. Since $P\mathbf{v}_k = O$ and since the vectors

$$\mathrm{col}\big(D_\tau^{s-1}U_{j-1}(\tau)\big|_{\tau=0}\big)_{s=1}^\ell, \ j = 1, \ldots, \kappa,$$

are linear independent, it follows from the last inequality that

$$|\sum_{k=1}^\kappa \sigma_{jk}(t)c_k \exp \left(i \int_{t_0}^t r(\tau)\mathfrak{m}_k(\tau)d\tau \right)| \leq c\, (r(t))^{1-j} e^{(k_0 - k_+)t} \qquad (17.67)$$

for $j = 1, \ldots, \kappa$. By $Y(\infty) = 0$ and by (17.43) we have $w_{jk}(t) \to 0$ as $t \to +\infty$. Therefore the matrix $\{\sigma_{jk}\}$ is invertible and its inverse is bounded. Hence and by (17.67)

$$\sum_{k=1}^\kappa |c_k \exp \left(i \int_{t_0}^t r(\tau)\mathfrak{m}_k(\tau)d\tau \right)| \leq c\, (r(t))^{1-\kappa} e^{(k_0 - k_+)t}.$$

Using (17.34) and $r(\infty) = 0$ we obtain $c_k = 0$. Thus, (12.11) holds. □

Theorem 17.4.3. *Let* Conditions A *and* B *from* Sect.17.3 *be fulfilled and let (17.64) hold with p given by (17.37). Suppose that $u \in W^\ell_{\mathrm{loc}}(t_0, \infty)$ is a solution of (14.1) which satisfies (14.6). Then*

$$u(t) = \sum_{j=1}^{\kappa} c_j u_j(t) + v(t), \qquad (17.68)$$

where c_j are constants and

$$\| v \|_{W^\ell(t,t+1)} \le c_0 e^{-k+t} \quad for \quad t > t_0. \qquad (17.69)$$

The constants $c_0, \ldots c_\kappa$ are estimated as

$$\sum_{j=0}^{\kappa} |c_j| \le c \| u \|_{W^\ell(t_0, t_0+1)} \qquad (17.70)$$

with c independent of u.

Proof. Follows from Theorem 12.2.3 and Corollary 17.4.2. □

17.5 An Example

Here we give an example of the application of Theorems 17.4.1 and 17.4.3. Consider the equation (14.1) and suppose that $\lambda = 0$ is the only eigenvalue of $\mathcal{A}(\lambda)$ on the line $\Im\lambda = 0$. Let dim $\ker A_\ell = 1$, i.e. the geometric multiplicity of $\lambda = 0$ is equal to 1. Denote by φ and ψ the elements of $\ker A_\ell$ and $\ker A_\ell^*$ respectively. We assume that

$$A_k\varphi = 0, \quad A_k^*\psi = 0 \quad for \quad k = 1, \ldots, \ell - 1.$$

and that

$$(\varphi, \psi)_{H_0} = 1 . \qquad (17.71)$$

Then one of the Jordan chains corresponding to the eigenvalue $\lambda = 0$ is defined by

$$\varphi_0 = \varphi, \quad \varphi_1 = \ldots = \varphi_{\ell-1} = 0.$$

By (17.71) this chain is maximal. Similarly, the operator pencil $\mathcal{A}^*(\lambda)$ has the maximal Jordan chain

$$\psi_0 = \psi, \quad \psi_1 = \ldots = \psi_{\ell-1} = 0,$$

which corresponds to the eigenvalue $\lambda = 0$. One can check that (17.71) is equivalent to the biorthogonality condition (1.20).

As elsewhere in this part of the book, we assume that the function ρ defined by (12.86) tends to zero as $t \to +\infty$.

The functions α and β satisfy the relations:

$$\alpha(t) \approx \sum_{s=1}^{\ell} \left(\int_t^{t+1} \| N_s(\tau)\varphi \|_0^2 \, d\tau \right)^{1/2},$$

$$\beta(t) = \sup_v \left(\int_t^{t+1} \left| ((N - \hat{N})(\tau, D_\tau)v(\tau), \psi)_{H_0} \right|^2 d\tau \right)^{1/2},$$

where $\hat{N}$ is the ordinary differential operator given by

$$\hat{N}(\tau, D_\tau) = \sum_{k=1}^{\ell} (N_k(\tau)\varphi, \psi) D_\tau^{\ell-k}$$

and the supremum is taken over the set (13.75).

All the elements $R_{jk}(t)$ of the matrix R with $j < \ell$ are equal to 0 and

$$R_{\ell k}(t) = \left(N_{\ell+1-k}(t)\varphi, \psi \right)_{H_0}.$$

Therefore, $\mathfrak{B}_{jk}(t) = 0$ for $j < \ell$ and

$$\mathfrak{B}_{\ell k}(t) = r(t)^{k-1-\ell} \left(N_{\ell+1-k}(t)\varphi, \psi \right)_{H_0}.$$

The elements $\mathfrak{B}_{jk}(\infty)$ of the limit matrix $\mathfrak{B}(\infty)$ equal 0 when $j < \ell$, and we denote

$$\mathfrak{B}_{\ell k}(\infty) = b_{\ell+1-k}, \quad k = 1, \ldots, \ell.$$

We now describe all the eigenvalues and the corresponding eigenvectors of the matrix (17.5). Let $e_\ell = 1$ and let $e_1, \ldots, e_{\ell-1}$ be the functions determined by

$$e_k(\mu) = (\mu - ikq)e_{k+1}(\mu) - b_{\ell-k}, \quad k = 1, \ldots, \ell-1,$$

where q is defined by (17.4). We introduce matrices

$$\mathcal{E} = \begin{pmatrix} e_1 & e_2 & \cdots & e_{\ell-1} & 1 \\ 1 & 0 & \cdots & 0 & 0 \\ 0 & 1 & \cdots & 0 & 0 \\ \vdots & \vdots & & \vdots & \vdots \\ 0 & 0 & \cdots & 1 & 0 \end{pmatrix}$$

and

$$\mathfrak{J} = \begin{pmatrix} 1 & 0 & \cdots & 0 & 0 \\ -\mu & 1 & \cdots & 0 & 0 \\ 0 & -\mu + iq & \cdots & 0 & 0 \\ \vdots & \vdots & & \vdots & \vdots \\ 0 & 0 & \cdots & -\mu + i(\ell-2)q & 1 \end{pmatrix}.$$

One verifies directly that

$$\mathcal{E}\Big(\mathfrak{B}(\infty) + J - \mu I + iq \, \text{diag}\,(0, 1, \ldots, \ell-1) \Big)$$
$$= \text{diag}(p, 1, \ldots, 1)\mathfrak{J}, \tag{17.72}$$

where

$$p(\mu) = -e_1\mu + b_\ell = b_\ell - \mu \sum_{k=1}^{\ell-1} b_{\ell-k}(iq - \mu)\dots(i(k-1)q - \mu)$$
$$-\mu(iq - \mu)\dots(i(\ell-1)q - \mu).$$

Moreover, the inverse matrix to $\mathfrak{J}$ has elements π_{jk}, $j, k = 1, \dots, \ell$, which are given by: $\pi_{jk} = 0$ if $k > j$, $\pi_{jj} = 1$ and

$$\pi_{jk}(\mu) = \prod_{s=k}^{j-1} (\mu - i(s-1)q), \quad \text{for} \quad k < j.$$

Denoting by $\mathfrak{m}_1, \dots, \mathfrak{m}_\ell$ the roots of $p(\mu) = 0$, the relation (17.72) implies that $\mathfrak{m}_k$ are eigenvalues and

$$\text{col}(\pi_{j1}(\mathfrak{m}_k))_{j=1}^{\ell}, \quad k = 1, \dots, \ell,$$

are eigenvectors of (17.5).

Condition B from Sect.17.3.1 means that the roots $\mathfrak{m}_1, \dots, \mathfrak{m}_\ell$ of the equation $p(\mu) = 0$ have different imaginary parts. The characteristic equation of the matrix (17.6) is

$$b_\ell(t) - \mu \sum_{k=1}^{\ell-1} b_{\ell-k}(t)(iq(t) - \mu)\dots(i(k-1)q(t) - \mu)$$
$$-\mu(iq(t) - \mu)\dots(i(\ell-1)q(t) - \mu) = 0, \tag{17.73}$$

where

$$b_k(t) = r(t)^{-k}(N_k(t)\varphi, \psi)_{H_0}$$

and

$$q(t) = (r'r^{-2})(t).$$

Now the asymptotic representation for the special solutions in Theorem 17.4.1 can be written as

$$D_t^j u_k(t) = \exp\left(i \int_{t_0}^t r(\tau)\mathfrak{m}_k(\tau)d\tau\right)$$
$$\times \left\{ r^j(t)\left(\prod_{s=0}^{j-1} (\mathfrak{m}_k(\infty) - isq) + w_{jk}(t)\right)\varphi + v_{jk}(t) \right\}, \tag{17.74}$$

where the product on the right is equal to 1 in the case $j = 0$ and

$$\| w_{jk} \|_{W_2^1(t,t+1)} = o(1) \quad \text{as} \quad t \to \infty,$$
$$\| v_{jk} \|_{S(t,t+1)} \le c \int_{t_0}^{\infty} e^{-a|t-\tau|}\alpha(\tau)d\tau.$$

Remark 17.5.1. If $\ell = 2$ then equation (17.73) becomes

$$\mu^2 - \mu\big(b_1(t) + iq(t)\big) + b_2(t) = 0.$$

Therefore, by (17.74), we have the principal term in the asymptotics of the solutions $u_1(t)$ and $u_2(t)$:

$$r^{-1/2}(t) \exp\left(\frac{i}{2} \int_{t_0}^{t} \big((N_1(\tau)\varphi, \psi\big)_{H_0} \right. \tag{17.75}$$

$$\left. \pm \left\{ \big((N_1(\tau)\varphi, \psi\big)_{H_0} + i\,\frac{r'(\tau)}{r(\tau)}\right)^2 - 4\big(N_2(\tau)\varphi, \psi\big)_{H_0}\right\}^{1/2}\right)d\tau\Big)\varphi.$$

Condition B from Sect.17.3 means that the value

$$\lim_{t\to\infty} \frac{1}{r^2(t)}\left\{ \big((N_1(t)\varphi, \psi\big)_{H_0} + i\,\frac{r'(t)}{r(t)}\right)^2 - 4\big(N_2(t)\varphi, \psi\big)_{H_0}\right\}$$

does not belong to $[0, \infty)$.

In particular, if $\big(N_1(t)\varphi, \psi\big)_{H_0} = 0$ and

$$\lim_{t\to\infty} \frac{r'(t)}{r^2(t)} = 0, \quad \int_{t_0}^{\infty} \frac{(r'(t))^2}{r^3(t)}\,dt < \infty,$$

we obtain from (17.75) that $u_1(t)$ and $u_2(t)$ are asymptotically equivalent to

$$\big|(N_2(t)\varphi, \psi\big)_{H_0}\big|^{-1/4} \exp\left(\pm \int_{t_0}^{t} \sqrt{\big(N_2(\tau)\varphi, \psi\big)_{H_0}}\,d\tau\right)\varphi,$$

which is a generalization of the Liouville-Green asymptotic formula for solutions of second order ordinary differential equations (see, for example, Hartman (1964), Ch.11, and Eastham (1989), Ch.2.)

17.6 Comments

Asymptotic formulae for solutions of systems of ordinary differential equations of the first order with coefficient matrices of Jordan type can be found in Eastham (1989), Sect. 1.10, where other sources are mentioned. The asymptotic formula (17.7) corresponds to the main term of the asymptotic series (11.53) obtained by Plamenevskii (1972) for perturbations of $\mathcal{A}(D_t)$ having the form of power series in t^{-1}.

A. Holomorphic Operator Functions

A.1 Introduction

In this Appendix we give an exposition of basic facts from the theory of holomorphic operator functions in a pair of Banach spaces (see Gohberg, Goldberg and Kaashoek (1990), Markus (1980), Wendland (1970) and Mennicken, Möller (1984)).

We have chosen the material which is used in the previous chapters and is also of interest in itself.

The solution of ordinary differential equations with constant operator coefficients can be reduced by the Fourier transform to the inversion of an operator which is polynomially dependent on a complex parameter (see Ch. 2). It is therefore important to have detailed information about the resolvent of the polynomial operator function. In particular, the asymptotics of solutions of abstract differential equations with constant coefficients can be described in terms of the Laurent decomposition of the resolvent near the poles (see Sect. 2.8). Such a decomposition was constructed in Keldysh (1951, 1971) and was extended to meromorphic operator functions in Gohberg, Sigal (1971).

Our principal aim here is to give a complete proof of Keldysh's theorem (Theorem A.10.2). As in the book Gohberg, Goldberg and Kaashoek (1990) we consider holomorphic operator functions instead of polynomial pencils; this does not cause additional difficulties.

In Sect.A.2–A.5 we give definitions and simple results of the operator theory and the spectral theory of holomorphic operator functions. Sect.A.6 contains some auxiliary material from linear algebra. We prove the so called local version of the Smith factorization theorem for matrices (see, for example, Gohberg, Lancaster and Rodman (1982), Ch.1,) and show that the Smith form of the holomorphic matrix function is closely connected with its spectral characteristics. Sect.A.7 is devoted to a representation of the resolvent of a holomorphic matrix function near the pole. In Sect. A.8, A.9 we return to the infinite dimensional case and describe properties of the Fredholm holomorphic operator functions and its adjoint. We complete the Appendix by proving two theorems on the structure of the resolvent of the holomorphic operator function near the pole.

A.2 Prerequisites on Fredholm Operators

Let B_1 and B_2 be Banach spaces with norms

$$\| \cdot \|_{B_1} \quad \text{and} \quad \| \cdot \|_{B_2} .$$

By $\mathcal{L}(B_1, B_2)$ we denote the Banach space of all bounded linear operators: $B_1 \to B_2$. We use the notations $\ker A$ and $\mathrm{Im} A$ for the set of zeros and for the range of the operator A.

Let A be the operator in $\mathcal{L}(B_1, B_2)$ and let

$$n(A) = \dim \ker A,$$

$$d(A) = \dim(B_2/\mathrm{Im} A),$$

where $B_2/\mathrm{Im} A$ is the factor space B_2 modulo $\mathrm{Im} A$. If both $n(A)$ and $d(A)$ are finite then we introduce the index of A:

$$\mathrm{ind} A = n(A) - d(A).$$

Definition A.2.1. The operator $A \in \mathcal{L}(B_1, B_2)$ is called Fredholm if $\mathrm{Im} A$ is closed and $n(A) + d(A) < \infty$.

The properties of Fredholm operators, which are of use in the sequel, are collected in the following proposition. Their proofs can be found, for example, in Gohberg, Lancaster and Rodman (1982) and Prössdorf (1978).

Proposition A.2.2. (i) *Let $A \in \mathcal{L}(B_1, B_2)$ be a Fredholm operator. Then there exists a positive ε such that for $T \in \mathcal{L}(B_1, B_2)$ with $\| T \|_{B_1 \to B_2} \leq \varepsilon$ the operator $A + T$ is also Fredholm and $\mathrm{ind} A = \mathrm{ind}(A + T)$.*

(ii) *If $A \in \mathcal{L}(B_1, B_2)$ is Fredholm and $K \in \mathcal{L}(B_1, B_2)$ is compact then $A + K$ is also Fredholm and $\mathrm{ind}(A + K) = \mathrm{ind} A$.*

(iii) *Let B_1, B_2 and B_3 be Banach spaces and let $A_i \in \mathcal{L}(B_i, B_{i+1})$, $i = 1, 2$, be Fredholm operators. Then $A_1 A_2$ is also Fredholm and $\mathrm{ind}(A_1 A_2) = \mathrm{ind} A_1 + \mathrm{ind} A_2$.*

Let B be a Banach space and let B^* be its dual space, i.e. the space of all bounded linear functionals on B. We assume that a sesquilinear form $< \cdot, \cdot >$ is given on $B \times B^*$ which satisfies the following properties:

(i) the form $< \cdot, \cdot >$ is linear with respect to the first argument and antilinear with respect to the second argument;

(ii) for all $b \in B$, $b_* \in B^*$

$$| < b, b_* > | \leq \| b \|_B \| b_* \|_{B^*}$$

(iii) for any $f \in B^*$ there exists a unique $\psi \in B^*$ such that $f(b) = < b, \psi >$ for all $b \in B$.

Let B_j^* be the dual space of B_j and let $< \cdot, \cdot >_j$ be a sesquilinear form on $B_j \times B_j^*$ satisfying (i)-(iii). For the operator $A \in \mathcal{L}(B_1, B_2)$ we define its adjoint $A^* \in \mathcal{L}(B_2^*, B_1^*)$ by

$$< x, A^*\psi >_1 = < Ax, \psi >_2 \tag{A.1}$$

for all $x \in B_1$, $\psi \in B_2^*$.

The following properties of the adjoint operator follow directly from its definition.

(i) The norms of A^* and A are equal.

(ii) If A_1, $A_2 \in \mathcal{L}(B_1, B_2)$ and $z_1, z_2 \in C$ then

$$(z_1 A_1 + z_2 A_2)^* = \overline{z}_1 A_1^* + \overline{z}_2 A_2^*$$

Proposition A.2.3. *If $A \in \mathcal{L}(B_1, B_2)$ is a Fredholm operator then $A^* \in \mathcal{L}(B_2^*, B_1^*)$ is also Fredholm and* ind $A^* = -$ind A.

Proposition A.2.4. *The operator $A \in \mathcal{L}(B_1, B_2)$ is invertible if and only if $A^* \in \mathcal{L}(B_2^*, B_1^*)$ is invertible.*

A.3 Basic Notions of the Spectral Theory of Holomorphic Operator Functions

We use the notation introduced in Sect.A.2.

Definition A.3.1. Let Ω be a domain in the complex plane $\mathbb{C}$. The operator function

$$F : \Omega \to \mathcal{L}(B_1, B_2) \tag{A.2}$$

is called holomorphic on Ω if it can be represented as the sum of a power series

$$F(\lambda) = \sum_{k=0}^{\infty} F_k (\lambda - \mu)^k, \ F_k \in \mathcal{L}(B_1, B_2) ,$$

which is convergent in $\mathcal{L}(B_1, B_2)$ in a neighbourhood of every point $\mu \in \Omega$.

Definition A.3.2. The spectrum of the holomorphic operator function F is the subset of Ω, where the operator $F(\lambda)$ is not invertible. The points in the complement of the spectrum are called regular.

It is clear that the spectrum of F is closed in Ω.

Definition A.3.3. The number $\lambda_0 \in \Omega$ is called an eigenvalue of F if ker $F(\lambda_0) \neq \{0\}$. Eigenvectors are elements of ker $F(\lambda_0) \backslash \{0\}$.

Definition A.3.4. Let λ_0 be an eigenvalue of F. The dimension of $\ker F(\lambda_0)$ is called the geometric multiplicity of λ_0.

Definition A.3.5. Let λ_0 be an eigenvalue of F and let φ_0 be an eigenvector corresponding to λ_0. The elements $\varphi_1, \ldots, \varphi_{m-1}$ in B_1 are called generalized eigenvectors if they satisfy

$$\sum_{j=0}^{n} \frac{1}{j!} F^{(j)}(\lambda_0)\varphi_{n-j} = 0, \quad n = 1, \ldots, m-1, \tag{A.3}$$

where

$$F^{(j)}(\lambda) = \frac{d^j}{d\lambda^j} F(\lambda).$$

It is said that the ordered collection $\varphi_0, \varphi_1, \ldots, \varphi_{m-1}$ is a Jordan chain corresponding to λ_0.

An equivalent definition of a Jordan chain can be derived from the following.

Proposition A.3.6. *Let λ_0 be an eigenvalue of F and let φ_0 be an eigenvector corresponding to λ_0. The vector function*

$$\Phi(\lambda) = \sum_{k=0}^{m-1} \frac{\varphi_k}{(\lambda - \lambda_0)^{m-k}} \tag{A.4}$$

(where $\varphi_k \in B_1$ and $m \geq 1$) satisfies

$$F(\lambda)\Phi(\lambda) = O(1) \quad \text{for small} \quad |\lambda - \lambda_0| \tag{A.5}$$

if and only if $\{\varphi_j\}_{j=0}^{m-1}$ is a Jordan chain.

Proof. The result follows from the identities

$$F(\lambda)\Phi(\lambda) = \sum_{j=0}^{\infty} \frac{1}{j!} F^{(j)}(\lambda_0)(\lambda - \lambda_0)^j \sum_{k=0}^{m-1} \varphi_k(\lambda - \lambda_0)^{k-m}$$

$$= \sum_{n=0}^{m-1} (\lambda - \lambda_0)^{n-m} \sum_{j=0}^{n} \frac{1}{j!} F^{(j)}(\lambda_0)\phi_{n-j} + O(1).$$

□

Remark A.3.7. From Proposition A.3.6 one can easily obtain another definition of a Jordan chain. The vectors $\varphi_0, \ldots, \varphi_{m-1} \in B_1$ form a Jordan chain corresponding to the eigenvalue λ_0 if and only if there exists a vector polynomial $\varphi(\lambda)$ such that

$$\varphi(\lambda_0) = 0, \quad \varphi^{(j)}(\lambda_0) = j!\varphi_j, \quad j = 0, \cdots, m-1,$$

and λ_0 is a zero of the vector function $F(\lambda)\varphi(\lambda)$ of multiplicity at least m.
Clearly the role of this polynomial is played by

$$\varphi(\lambda) = (\lambda - \lambda_0)^m \Phi(\lambda),$$

where Φ is given by (A.4).

Definition A.3.8. (i) Let λ_0 be an eigenvalue of F. By $S(F, \lambda_0)$ we denote the set of all vector functions Φ represented in the form (A.4) which satisfy (A.5).

(ii). The dimension of $S(F, \lambda_0)$ is called the algebraic multiplicity of λ_0.

If φ_0 is an eigenvector of F corresponding to λ_0 then

$$\varphi_0(\lambda - \lambda_0)^{-1} \in S(F, \lambda_0).$$

Therefore the geometric multiplicity does not exceed the algebraic multiplicity.

We mention the following useful property of $S(F, \lambda_0)$. If $\Phi \in S(F, \lambda_0)$ and $p = p(\lambda)$ is a polynomial then

$$SP\big(p(\lambda)\Phi(\lambda)\big) \in S(F, \lambda_0). \tag{A.6}$$

Here, and in the sequel, SP means the singular part of a meromorphic function.

If we define the product of $\Phi(\lambda)$ with a polynomial $p(\lambda)$ as $SP\big(p(\lambda)\Phi(\lambda)\big)$, then (A.6) implies that $S(F, \lambda_0)$ is a vector space over the ring of polynomials.

Definition A.3.9. Let $\Phi \in S(F, \lambda_0)$ be written in the form (A.4) with $\varphi_0 \neq 0$. The number m and the eigenvector φ_0 are called the degree of Φ ($\deg \Phi$) and the leading coefficient of Φ respectively.

If $\Phi = 0$ we put $\deg \Phi = 0$ and say that the leading coefficient is zero.

A.4 Canonical Generating System in $S(F, \lambda_0)$ and Canonical Set of Jordan Chains

We describe the structure of $S(F, \lambda_0)$ in more detail.

Definition A.4.1. Let λ_0 be an eigenvalue of F of finite algebraic multiplicity. The system

$$\{\Phi_j\}_{j=1}^{J}$$

of elements in $S(F, \lambda_0)$ with $\deg \Phi_{k+1} \leq \deg \Phi_k$, $k = 1, \ldots, J - 1$, is called a canonical generating system in $S(F, \lambda_0)$ if

(i) the leading coefficients of $\Phi_1, \ldots, \Phi_J$ are linearly independent,

(ii) for every $\Phi \in S(F, \lambda_0)$ there exist polynomials $p_1(\lambda), \ldots, p_J(\lambda)$ such that

$$\Phi(\lambda) = SP\left(\sum_{j=1}^{J} p_j(\lambda)\Phi_j(\lambda)\right).$$

It is clear that the leading coefficients of $\Phi_1, \ldots, \Phi_J$ from a basis in ker $F(\lambda_0)$. Hence, J is the geometric multiplicity of λ_0.

From Definition A.4.1 it follows that the elements

$$SP(\lambda - \lambda_0)^s \Phi_k(\lambda), \quad s = 0, \ldots, m_k - 1, \quad k = 1, \ldots, J, \tag{A.7}$$

where $m_k = \deg \Phi_k$, form a basis in $S(F, \lambda_0)$ (over C). Hence

$$m_1 + \cdots + m_J = \dim S(F, \lambda_0). \tag{A.8}$$

The following assertion gives another characterization of canonical generating systems.

Proposition A.4.2. *Let λ_0 be an eigenvalue of F of finite algebraic multiplicity and let non-zero elements $\Phi_1, \ldots, \Phi_J$ satisfy* (ii) *of Definition A.4.1. If*

$$m_1 + \cdots + m_J \leq \dim\ S(F, \lambda_0), \tag{A.9}$$

where $m_j = \deg \Phi_j$, then

$$\{\Phi_j\}_{j=1}^{J}$$

is a canonical generating system in $S(F, \lambda_0)$.

Proof. Due to Definition A.4.1 (ii), the elements (A.7) generate $S(F, \lambda_0)$. Hence, by (A.9), the elements (A.7) form a basis in $S(F, \lambda_0)$. Therefore,

$$(\lambda - \lambda_0)^{m_j - 1}\Phi_j, \quad j = 1, \ldots J,$$

are linearly independent, which implies Definition A.4.1 (i). $\square$

Definition A.4.3. Let λ_0 be an eigenvalue of F of finite algebraic multiplicity. A set of Jordan chains

$$\{\varphi_{k,j}\}_{j=0}^{m_k - 1}, \quad k = 1, \ldots, J, \tag{A.10}$$

of the holomorphic operator function F corresponding to the eigenvalue λ_0 is called canonical if $\varphi_{1,0}, \ldots, \varphi_{J,0}$ form a basis in ker $F(\lambda_0)$, $m_1 \geq \ldots \geq m_J$ and (A.8) holds.

The next proposition connects the notations of the canonical generating system and the canonical set of Jordan chains.

Proposition A.4.4. *Let λ_0 be an eigenvalue of F of finite algebraic multiplicity and let J be the geometric multiplicity of λ_0. Also let*

$$\Phi_k(\lambda) = \sum_{j=0}^{m_k-1} \varphi_{k,j}(\lambda - \lambda_0)^{j-m_k}, \ k = 1, \dots, J,$$

be elements of $S(F, \lambda_0)$. The set $\{\Phi_k\}_{k=1}^{J}$ is a canonical generating system in $S(F, \lambda_0)$ if and only if (A.10) is a canonical set of Jordan chains.

Proof. First, let $\Phi_1, \dots, \Phi_J$ be a canonical generating system in $S(F, \lambda_0)$. Then the vectors $\varphi_{1,0}, \dots, \varphi_{J,0}$ form a basis in $\ker F(\lambda_0)$ and (A.8) holds. Hence $\{\varphi_{k,j}\}_{j=0}^{m_k-1}$ is a canonical set of Jordan chains.

Conversely, let the coefficients of Φ_k, $k = 1, \dots, J$, form a canonical set of Jordan chains. Then the leading coefficients of Φ_k form a basis in $\ker F(\lambda_0)$ and the elements (A.7) form a basis in $S(F, \lambda_0)$. By Proposition A.4.2 $\{\Phi_k\}_{k=1}^{J}$ is a canonical generating system. $\square$

Proposition A.4.5. *Let λ_0 be an arbitrary eigenvalue of the holomorphic operator function (A.2) with finite algebraic multiplicity. Then there exists a canonical generating system.*

Proof. Choose an element $\Phi_1 \in S(F, \lambda_0)$ of maximal degree and denote by φ_1 the leading coefficient of Φ_1.

Suppose that $\Phi_1, \dots, \Phi_k$ have been chosen and that $\varphi_1, \dots, \varphi_k$ are their leading coefficients. Take Φ_{k+1} in such a way that: (i) the vectors $\varphi_1, \dots \varphi_{k+1}$ (where φ_{k+1} is the leading coefficient of Φ_{k+1}) are linearly independent; (ii) the element Φ_{k+1} has maximal degree among all elements with property (i).

Since the leading coefficients belong to $\ker F(\lambda_0)$, this process will terminate by the Jth step, where $J = \dim \ker F(\lambda_0)$. Clearly, $\varphi_1, \dots, \varphi_J$ form a basis in $\ker F(\lambda_0)$ and therefore property (i) in Definition A.4.1 holds.

It remains to verify property (ii) in the same definition. Let $\Phi \in S(F; \lambda_0)$ and let φ be the leading coefficient of Φ. Suppose

$$\deg \Phi_j \geq \deg \Phi \quad \text{and} \quad \deg \Phi_{j+1} < \deg \Phi.$$

According to the construction of $\Phi_1, \dots, \Phi_J$, there exist complex coefficients $\alpha_1, \dots, \alpha_j$ such that

$$\varphi = \alpha_1 \varphi_1 + \dots + \alpha_j \varphi_j.$$

Therefore the degree of

$$\Phi(\lambda) - SP\left(\sum_{k=1}^{J} \alpha_k (\lambda - \lambda_0)^{m_k - m} \Phi_k(\lambda) \right),$$

where $m_k = \deg \Phi_k$ and $m = \deg \Phi$, is less than $\deg \Phi$. Continuing this procedure we arrive at the desired representation for Φ. $\square$

Proposition A.4.6. *Let λ_0 be an eigenvalue of the holomorphic operator function (A.2) with finite algebraic multiplicity, let $\{\Phi_j\}_{j=1}^{J}$ be a canonical generating system in $S(F, \lambda_0)$ and $m_j = \deg \Phi_j$. By $\{\Psi_j\}_{j=1}^{J}$ we denote another collection which is ordered in such a way that*

$$\deg \Psi_k \geq \deg \Psi_{k+1}.$$

The collection $\{\Psi_j\}_{j=1}^{J}$ is a canonical generating system if and only if $\deg \Psi_j = m_j$ and there exist polynomials $p_{ik}(\lambda)$ $(i, k = 1, \ldots, J)$ with the following properties:

(α) $p_{ik}(\lambda)$ has λ_0 as zero of order $m_k - m_i$ if $m_k > m_i$;

(β) the matrix $\left(p_{ik}(\lambda_0)\right)_{i,k=1}^{J}$ is invertible;

(γ) the vector functions Ψ_i, $i = 1, \ldots, J$, are represented by

$$\Psi_i(\lambda) = SP\Big(\sum_{k=1}^{J} p_{ik}(\lambda)\Phi_k(\lambda)\Big). \tag{A.11}$$

Proof. Let $n_k = \deg \Psi_k$. By property (ii) in Definition A.4.1 there exist polynomials $p_{ik}(\lambda)$ such that (A.11) holds.

(i) Suppose that $\{\Psi_j\}_{j=1}^{J}$ is a canonical generating system. Comparing the degrees of Ψ_i and Φ_i we get

$$p_{ik}(\lambda) = (\lambda - \lambda_0)^{m_k - n_i} q_{ik}(\lambda), \quad \text{if} \quad m_k \geq n_i \,,$$

where q_{ik} is a polynomial. We set $q_{ik} = 0$ if $m_k < n_i$ and denote by φ_j and ψ_j the leading coefficients of Φ_j and Ψ_j. By equating the leading coefficients in both sides of (A.11) we obtain

$$\psi_i = \sum_{k=1}^{J} q_{ik}(\lambda_0)\varphi_k.$$

Since the systems $\{\psi_j\}$ and $\{\varphi_j\}$ are linearly independent, the matrix

$$\left(q_{ik}(\lambda_0)\right)_{i,k=1}^{J}$$

is invertible. Let us show that $n_i = m_i$. Suppose that $n_i > m_i$ for some i. This implies $q_{s,k}(\lambda_0) = 0$ for $s \leq i$ and for $k \geq i$, which contradicts the invertibility of

$$\left(q_{ik}(\lambda_0)\right)_{i,k=1}^{J}. \tag{A.12}$$

Hence $n_i \leq m_i$, $i = 1, \ldots, J$. Exchanging the roles of $\{\Phi_j\}$ and $\{\Psi_j\}$ we get the opposite inequality: $m_i \leq n_i$. Thus, we have proved that $n_i = m_i$, $i = 1, \ldots, J$.

Since $q_{ik} = 0$ for $m_k < m_i$, the invertibility of (A.12) is equivalent to (β).

(ii) Suppose that (A.11) holds and the polynomials p_{ik} satisfy (α) and (β). By (α)

$$p_{ik}(\lambda) = (\lambda - \lambda_0)^{m_k - m_i} q_{ik}(\lambda) \quad \text{if} \quad m_k \geq m_i$$

and if we put $q_{ik} = 0$ for $m_k < m_i$ then the matrix (A.12) is invertible if and only if (β) holds.

Since

$$\psi_i = \sum_{k=1}^{J} q_{ik}(\lambda_0)\varphi_k,$$

the leading coefficients of Ψ_i are linearly independent.

It remains to verify (ii) in Definition A.4.1. Due to (β) there exist polynomials $g_{ik}(\lambda)$ such that

$$g_{ik}(\lambda)p_{kj}(\lambda) = \delta_{ik} + O\left(|\lambda - \lambda_0|^N\right),$$

where N is sufficiently large.

Let $\Phi \in S(F, \lambda_0)$. Then

$$\Phi(\lambda) = SP \sum_{j=1}^{J} h_j(\lambda)\Phi_j(\lambda)$$

for some polynomials h_j. Using the equality

$$\Phi_j(\lambda) = \sum_{k=1}^{J} g_{jk}(\lambda)\Psi_k(\lambda),$$

we arrive at the desired representation for Φ. $\square$

Definition A.4.7. The degrees $m_1 \geq ... \geq m_J$ of elements in a canonical generating system are called the partial multiplicities of the eigenvalue λ_0.

By Proposition A.4.6, the partial multiplicities do not depend on the choice of the canonical generating system.

A.5 The Local Equivalence of Holomorphic Operator Functions

Let B_j and C_j, $j = 1, 2$, be Banach spaces. Consider two holomorphic operator functions

$$F : U \rightarrow \mathcal{L}(B_1, B_2)$$

$$G : U \rightarrow \mathcal{L}(C_1, C_2)$$

defined in a neighbourhood U of λ_0. These operator functions are called equivalent at λ_0 if there exist holomorphic operator functions

$$M : U \to \mathcal{L}(B_1, C_1)$$

$$N : U \to \mathcal{L}(C_2, B_2)$$

such that $N(\lambda_0)$ and $M(\lambda_0)$ are invertible operators and

$$F(\lambda) = N(\lambda)G(\lambda)M(\lambda) \tag{A.13}$$

if $\lambda \in U$.

This equivalence relation is symmetric and transitive, i.e. $F \sim G \Leftrightarrow G \sim F$ and $F \sim G$, $G \sim H \Rightarrow F \sim H$.

Let $F \sim G$ and let M be the operator function in (A.13). By $\mathcal{M}$ we mean the mapping

$$S(F, \lambda_0) \supset \Phi \to SP(M\Phi). \tag{A.14}$$

Due to (A.13), $\Phi \in S(F, \lambda_0)$ implies that

$$G(\lambda)M(\lambda)\Phi(\lambda) = O(1) \quad \text{as} \quad \lambda \to \lambda_0.$$

Hence

$$\text{Im}\mathcal{M} \subset S(G, \lambda_0).$$

It is clear that $\mathcal{M}$ is a linear operator and

$$\mathcal{M}\Big(SP(p\Phi)\Big) = SP\Big(p\mathcal{M}\Phi\Big) \tag{A.15}$$

for an arbitrary polynomial p. Since $M(\lambda_0)$ is invertible then

$$\deg \mathcal{M}\Phi = \deg \Phi. \tag{A.16}$$

Proposition A.5.1. *Let F and G be holomorphic operator functions which are equivalent at λ_0. Then*

(i) $\dim S(F, \lambda_0)$ *is finite if and only if* $\dim S(G, \lambda_0)$ *is finite.*

(ii) *Let λ_0 be an eigenvalue of F of finite algebraic multiplicity. The vector-functions $\Phi_1, \ldots, \Phi_J \in S(F, \lambda_0)$ form a canonical generating system if and only if $\mathcal{M}\Phi_1, \ldots, \mathcal{M}\Phi_J$ form a canonical generating system in $S(G, \lambda_0)$.*

(iii) *Let λ_0 be an eigenvalue of F of finite algebraic multiplicity. Then the geometric, partial and algebraic multiplicities of the eigenvalue λ_0 of F and G are the same.*

Proof. (i) It suffices to show that $\mathcal{M}$ is an isomorphism. By (A.13)

$$G(\lambda) = N(\lambda)^{-1}F(\lambda)M(\lambda)^{-1}$$

in a neighbourhood of λ_0. Denote by $\mathcal{M}_1$ the mapping

$$S(G, \lambda_0) \supset \Psi \to SP(M^{-1}\Psi).$$

For an arbitrary $\Phi \in S(F, \lambda_0)$ we have

$$\mathcal{M}_1(\mathcal{M}\Phi) = \mathcal{M}_1(SP(M\Phi))$$
$$= SP(M^{-1}SP(M\Phi)) = SP(M^{-1}M\Phi) = \Phi.$$

In the same way, we verify that

$$\mathcal{M}\mathcal{M}_1\Psi = \Psi, \quad \Psi \in S(G, \lambda_0).$$

Thus $\mathcal{M}_1$ is the inverse operator to $\mathcal{M}$ and hence $\mathcal{M}$ is isomorphic.

(ii) Since $\mathcal{M}$ is an isomorphism, (A.15) implies that $\Phi_1, \ldots, \Phi_J$ satisfy (ii) of Definition A.4.1 if and only if the same is valid for $\mathcal{M}\Phi_1, \ldots, \mathcal{M}\Phi_J$. By (A.16), deg $\Phi_k = $ deg $\mathcal{M}\Phi_k$. Reference to Proposition A.4.2 completes the proof.

(iii) We notice that

(1) partial multiplicities of λ_0 are equal to the degrees of the elements of the canonical generating system;

(2) the geometric multiplicity of λ_0 is equal to the number of elements of the canonical generating system;

(3) the algebraic multiplicity of λ_0 is the sum of partial multiplicities of λ_0.

Therefore the result follows from (ii). $\square$

A.6 The Smith Form of a Holomorphic Matrix Function

The present section (as well as Sect.A.7) contains well-known auxiliary facts from linear algebra which will be used in the subsequent analysis of holomorphic operator functions. We consider $(n \times n)$-matrix functions which are holomorphic at a point λ_0, and characterize their structure near λ_0. Here it is shown that any such matrix function is equivalent to a canonical diagonal matrix function (see Theorem A.6.1, which is a local version of the Smith theorem (Gohberg, Goldberg and Kaashoek (1990), Ch.S1, or Gantmaher (1959)).

We conclude the section by showing that the canonical form is closely connected with notions of the spectral theory of holomorphic operator functions introduced in Sect.A.3, A.4.

Let $\mathcal{H}(\lambda_0)$ be the set of all scalar functions which are holomorphic at λ_0. Furthermore, we denote the set of all $n \times n$ matrix functions holomorphic at λ_0 by $\mathcal{H}_{n \times n}(\lambda_0)$. For the matrix case, the notion of equivalence reads as follows:

Two matrix functions $F, G \in \mathcal{H}_{n \times n}(\lambda_0)$ are called equivalent (the notation $F \sim G$) if, in a vicinity of λ_0,

$$F(\lambda) = N(\lambda)G(\lambda)M(\lambda) , \tag{A.17}$$

where $N, M \in \mathcal{H}_{n \times n}(\lambda_0)$ and the matrices $N(\lambda_0)$ and $M(\lambda_0)$ are invertible.

Theorem A.6.1. *Let $F \in \mathcal{H}_{n \times n}(\lambda_0)$. Then F is equivalent to the diagonal matrix*

$$D(\lambda) = \operatorname{diag}\Big(0, \ldots, 0, (\lambda - \lambda_0)^{m_{q+1}}, \ldots, (\lambda - \lambda_0)^{m_J}, 1, \ldots, 1\Big), \qquad (A.18)$$

where $0 \leq q \leq J \leq n$, and m_s are integers satisfying

$$1 \leq m_J \leq \cdots \leq m_{q+1} < \infty.$$

Proof. We introduce three left elementary transformations on F:

(1) the multiplication of a row by $f \in \mathcal{H}(\lambda_0)$ with $f(\lambda_0) \neq 0$;

(2) the addition to some row of another one multiplied by $f \in \mathcal{H}(\lambda_0)$;

(3) the interchange of rows.

One directly verifies that (1)-(3) are equivalent to the multiplication on the left by the following three matrices respectively

$$\operatorname{diag}(1, \cdots, 1, f(\lambda), 1, \cdots, 1),$$

$$(i) \quad \begin{pmatrix} 1 & \cdots & \cdot & \cdots & \cdot & \cdots & 0 \\ \vdots & \ddots & \vdots & \cdots & \vdots & \cdots & \vdots \\ \cdot & \cdots & 1 & \cdots & f(\lambda) & \cdots & \cdot \\ \vdots & \cdots & \vdots & \ddots & \vdots & \cdots & \vdots \\ \vdots & \cdots & \vdots & \cdots & \ddots & \cdots & \vdots \\ \vdots & \cdots & \vdots & \cdots & \vdots & \ddots & \vdots \\ 0 & \cdots & \cdot & \cdots & \cdot & \cdots & 1 \end{pmatrix} \quad (j)$$

and

$$\begin{pmatrix} 1 & \cdots & \cdot & \cdots & \cdot & \cdots & 0 \\ \vdots & \ddots & \vdots & \cdots & \vdots & \cdots & \vdots \\ \cdot & \cdots & 0 & \cdots & 1 & \cdots & \cdot \\ \vdots & \cdots & \vdots & \ddots & \vdots & \cdots & \vdots \\ \cdot & \cdots & 1 & \cdots & 0 & \cdots & \cdot \\ \vdots & \cdots & \vdots & \cdots & \vdots & \ddots & \vdots \\ 0 & \cdots & \cdot & \cdots & \cdot & \cdots & 1 \end{pmatrix} \quad (i) \quad (j).$$

The determinants of these matrices are non-zero at λ_0. Therefore, to the application of any number of the left elementary operation there corresponds multiplication on the left by a matrix in $\mathcal{H}_{n \times n}(\lambda_0)$ which is invertible at λ_0.

The right elementary operations are defined quite analogously; these transform columns instead of rows. A finite number of right elementary transformations can be interpreted as multiplication on the right by a suitable matrix in $\mathcal{H}_{n \times n}(\lambda_0)$ which is invertible at λ_0.

Hence, it suffices to show that F can be reduced to (A.18) by left and right elementary transformations.

We choose a non-zero element of F with the minimum order of zero at λ_0. By interchanging rows and columns we can always mean the element $F_{11}(\lambda)$. Let

$$F_{11}(\lambda) = (\lambda - \lambda_0)^{a_1} G_{11}(\lambda), \quad a_1 \geq 0,$$

where $G_{11}(\lambda_0) \neq 0$. By the choice of F_{11} we have

$$F_{ij}(\lambda) = (\lambda - \lambda_0)^{a_1} G_{ij}(\lambda)$$

with $G_{ij} \in \mathcal{H}(\lambda_0)$. Subtract the first row multiplied by G_{i1}/G_{11} from the ith row. Then subtract the first column multiplied by G_{1i}/G_{11} from the ith column and multiply the first row by $1/G_{11}$. We arrive at the matrix

$$\begin{pmatrix} (\lambda - \lambda_0)^{a_1} & 0 & \cdots & 0 \\ 0 & f_{22}(\lambda) & \cdots & f_{2n}(\lambda) \\ \vdots & \vdots & \ddots & \vdots \\ 0 & f_{n2}(\lambda) & \cdots & f_{nn}(\lambda) \end{pmatrix}.$$

By applying this procedure to $\left(f_{k\ell}(\lambda) \right)_{k,\ell=2}^{n}$ we obtain the matrix

$$\begin{pmatrix} (\lambda - \lambda_0)^{a_1} & 0 & 0 & \cdots & 0 \\ 0 & (\lambda - \lambda_0)^{a_2} & 0 & \cdots & 0 \\ 0 & 0 & h_{33}(\lambda) & \cdots & h_{3n}(\lambda) \\ \vdots & \vdots & \vdots & \ddots & \vdots \\ 0 & 0 & h_{n3}(\lambda) & \cdots & h_{nn}(\lambda) \end{pmatrix}.$$

Proceeding in the same way, we reduce f to the matrix

$$\mathrm{diag}\left((\lambda - \lambda_0)^{a_1}, (\lambda - \lambda_0)^{a_2}, \ldots, (\lambda - \lambda_0)^{a_s}, 0, \ldots, 0 \right).$$

By using the elementary transformations we interchange the diagonal elements, which completes the proof. $\square$

Definition A.6.2. The matrix $D(\lambda)$ is called the Smith form of $F(\lambda)$.

Remark A.6.3. Since the multiplicities of the root λ_0 of the polynomials $\det F(\lambda)$ and $\det D(\lambda)$ are the same, the number q in (A.18) is zero if and only if $\det F(\lambda)$ is identically equal to zero.

Let $F \in \mathcal{H}_{n \times n}(\lambda_0)$ be such that $\det F(\lambda_0) = 0$, i.e. λ_0 is an eigenvalue of F. The notions of geometric, partial and algebraic multiplicities were defined in Sect.A.3. We now show that these multiplicities are easily interpreted in terms of the canonical form (A.18).

Following Sect.A.3, we denote by $S(F, \lambda_0)$ the set of the rational vector functions

$$\Phi(\lambda) = \sum_{k=1}^{m-1} \frac{\Phi_k}{(\lambda - \lambda_0)^{m-k}}, \quad \Phi_k \in C^n \tag{A.19}$$

$(m \geq 1)$ which satisfy

$$F(\lambda)\Phi(\lambda) = O(1) \quad \text{as} \quad \lambda \to \lambda_0. \tag{A.20}$$

Proposition A.6.4. *The algebraic multiplicity of F at λ_0 is equal to the multiplicity of the root λ_0 of $\det F(\lambda)$.*

Proof. Since the algebraic multiplicity of λ_0 is the dimension of $S(F, \lambda_0)$, the algebraic multiplicities for equivalent matrices coincide by Proposition A.5.1 (iii).

Clearly, the multiplicities of the root λ_0 of $\det F(\lambda)$ and $\det G(\lambda)$ are the same for equivalent matrices. Thus, it suffices to prove the proposition for the Smith form $D(\lambda)$ (see (A.18)). If $q > 0$ in (A.18) then $\det D(\lambda) = 0$ and $\dim S(D, \lambda_0) = \infty$. This leads to the assertion.

Let $q = 0$. Then the multiplicity of the root λ_0 of $\det F(\lambda)$ is $m_1 + \ldots + m_J$. The system

$$(\lambda - \lambda_0)^{-j} e_k, \ j = 1, \ldots, m_k, \ k = 1, \ldots, J,$$

where

$$e_k = \text{col}(0, \ldots, 1, \ldots 0) \tag{A.21}$$

with 1 at the kth place, is a basis in $S(F, \lambda_0)$. Hence $\dim S(F, \lambda_0) = m_1 + \ldots + m_J$. The proof is complete. $\square$

Corollary A.6.5. *The algebraic multiplicity of the eigenvalue λ_0 of F is finite if and only if F is equivalent to (A.18) with $q = 0$.*

Proof. Since F is equivalent to (A.18), the multiplicity of the root λ_0 of $\det F(\lambda)$ is finite if and only if $q = 0$ in (A.18). Reference to the above proposition completes the proof. $\square$

Theorem A.6.6. *Let F be a matrix function of finite algebraic multiplicity which is holomorphic at λ_0, and let (A.18) be its Smith form. Then $q = 0$ in (A.18), J is the geometric multiplicity, $m_1, \ldots, m_J$ in the same formula are partial multiplicities and $m_1 + \ldots + m_J$ is the algebraic multiplicity of λ_0.*

Proof. By Proposition A.5.1 and Corollary A.6.5 it suffices to prove the theorem for F which are equal to the matrix (A.18) (with $q = 0$). Since the geometric multiplicity is the dimension of both $\ker F(\lambda_0)$ and $\ker D(\lambda_0)$, it is equal to J.

We verify that the vector-functions

$$(\lambda - \lambda_0)^{-m_k} e_k, \ k = 1, \ldots, J$$

(where e_k is given by (A.21)) form a canonical generating system in $S(F, \lambda_0)$. Indeed, the leading coefficients of this system are linearly independent. Since $S(F, \lambda_0)$ consists of the elements

$$\sum_{k=1}^{J} \sum_{s=1}^{m_k} c_{ks}(\lambda - \lambda_0)^{-s} e_k,$$

property (ii) in Definition A.4.1 is also satisfied. Hence the partial multiplicities are equal to $m_1, \ldots, m_J$, and by (A.7) the algebraic multiplicity is $m_1 + \ldots + m_J$. $\square$

A.7 The Resolvent of a Holomorphic Matrix Function

Here we obtain a representation for the singular part of the operator $F(\lambda)^{-1}$ near the pole λ_0, when $F \in \mathcal{H}_{n \times n}(\lambda_0)$. First, we introduce some notions connected with the adjoint matrix function.

We equip $\mathbb{C}^n$ with the usual scalar product

$$\langle \varphi, \psi \rangle_{\mathbb{C}^n} = \sum_{k=1}^{n} \varphi_k \overline{\psi}_k.$$

The adjoint matrix for $A = (a_{ij})_{i,j=1}^n$ is defined by

$$A^* = \left(a_{ij}^*\right)_{i,j=1}^n, \quad \text{where} \quad a_{ij}^* = \overline{a}_{ji}.$$

If $F \in \mathcal{H}_{n \times n}(\lambda_0)$ then the matrix function

$$F^*(\lambda) = \left(F(\overline{\lambda})\right)^*$$

is holomorphic at $\overline{\lambda}$ and F^* is called adjoint to F.

Clearly, if F is equivalent to G, i.e.

$$F(\lambda) = N(\lambda)G(\lambda)M(\lambda),$$

then F^* is equivalent to G^* and

$$F^*(\lambda) = M^*(\lambda)G^*(\lambda)N^*(\lambda). \tag{A.22}$$

Let λ_0 be an eigenvalue of F of finite algebraic multiplicity. By Theorem A.6.1, Proposition A.6.4 and (A.22), $F^*(\lambda)$ is equivalent to the diagonal matrix

$$D^*(\lambda) = \operatorname{diag}\left((\lambda - \overline{\lambda}_0)^{m_1}, \ldots, (\lambda - \overline{\lambda}_0)^{m_J}, 1, \ldots, 1\right),$$

where J and $m_1, \ldots, m_J$ are the geometric and partial multiplicities of the eigenvalue λ_0 of the matrix function F. Therefore, by Theorem A.6.6 we have:

Proposition A.7.1. *The geometric, partial and algebraic multiplicities of the eigenvalues λ_0 and $\overline{\lambda}_0$ of F and F^* coincide.*

In the next theorem we describe the structure of the resolvent of a matrix function near an eigenvalue.

Theorem A.7.2. *Let $F \in \mathcal{H}_{n \times n}(\lambda_0)$ have an eigenvalue λ_0 of finite algebraic multiplicity and let J and $m_1, \ldots, m_J$ be its geometric and partial multiplicities. Assume that $X_1, \ldots, X_J$ is a canonical generating system in $S(F, \lambda_0)$. Then there exists a unique canonical generating system $Y_1, \ldots, Y_J$ in $S(F^*, \overline{\lambda}_0)$ such that*

$$F(\lambda)^{-1} = \sum_{j=1}^{J} (\lambda - \lambda_0)^{m_j} \langle \cdot, Y_j(\overline{\lambda}) \rangle_{\mathbb{C}^n} X_j(\lambda) + \Gamma(\lambda), \qquad (A.23)$$

where $\Gamma(\lambda)$ is a holomorphic matrix function in a neighbourhood of λ_0.

Proof. First, we note that the leading coefficients of $X_j(\lambda)$ are linearly independent. Therefore, from

$$\sum_{j=1}^{J} (\lambda - \lambda_0)^{m_j} \langle \cdot, \delta_j(\overline{\lambda}) \rangle_{\mathbb{C}^n} X_j(\lambda) \in \mathcal{H}_{n \times n}(\lambda_0)$$

where $\delta_j \in S(F^*, \overline{\lambda}_0)$, it follows that $\delta_j = 0$; this implies the uniqueness of the system $Y_1, \ldots, Y_J$.

In order to prove the existence of $Y_1, \ldots, Y_J$, we start with a matrix function of the special form

$$D(\lambda) = \mathrm{diag}\Big((\lambda - \lambda_0)^{m_1}, \ldots, (\lambda - \lambda_0)^{m_J}, 1, \ldots, 1 \Big). \qquad (A.24)$$

We have

$$D(\lambda)^{-1} = \sum_{k=1}^{J} \langle \cdot, e_k \rangle_{\mathbb{C}^n} e_k (\lambda - \lambda_0)^{-m_k} + \sum_{k=J+1}^{n} \langle \cdot, e_k \rangle_{\mathbb{C}^n} e_k \qquad (A.25)$$

where e_k is given by (A.21).

As was shown in the proof of Theorem A.6.6, the collection of vector functions

$$X_j(\lambda) = (\lambda - \lambda_0)^{-m_j} e_j, \quad j = 1, \ldots, J,$$

forms a canonical generating system in $S(D, \lambda_0)$. Analogously, the set

$$\{ (\lambda - \overline{\lambda}_0)^{-m_j} e_j \}$$

is a canonical generating system in $S(D^*, \overline{\lambda}_0)$. Hence, by (A.25) we can put

$$Y_j(\lambda) = (\lambda - \overline{\lambda}_0)^{-m_j} e_j, \quad j = 1, \ldots, J.$$

Now let $X_1, \ldots, X_J$ be an arbitrary canonical generating system in $S(D, \lambda_0)$. Then by Proposition A.4.6

$$(\lambda - \lambda_0)^{-m_i} e_i = SP\Big(\sum_{k=1}^{J} p_{ik}(\lambda) X_k(\lambda) \Big), \qquad (A.26)$$

where p_{ik} are polynomials satisfying properties (α) and (β) in the same proposition.

By (A.25) and (A.26),

$$D(\lambda)^{-1} = \sum_{j=1}^{J} \langle \cdot, e_j \rangle_{\mathbb{C}^n} \sum_{k=1}^{J} p_{jk}(\lambda) X_k(\lambda) + \Gamma_1(\lambda)$$

$$= \sum_{k=1}^{J} (\lambda - \lambda_0)^{m_k} \langle \cdot, Y_k(\overline{\lambda}) \rangle_{\mathbb{C}^n} X_k(\lambda) + \Gamma_2(\lambda), \qquad (A.27)$$

where

$$Y_k(\lambda) = SP\Big(\sum_{j=1}^{J} (\lambda - \lambda_0)^{m_j - m_k} p_{jk}(\lambda) (\lambda - \lambda_0)^{-m_j} e_j \Big)$$

and Γ_1, Γ_2 are holomorphic functions in a neighbourhood of λ_0. The polynomials $(\lambda - \lambda_0)^{m_j - m_k} p_{jk}(\lambda)$ also satisfy assumptions (α) and (β) in Proposition A.4.6. Therefore, $\{Y_j\}_{j=1}^{J}$ is a canonical generating system in $S(D, \lambda_0)$.

Now we turn to general F. By Theorem A.6.1,

$$D(\lambda) = N(\lambda) F(\lambda) M(\lambda).$$

Hence

$$F(\lambda)^{-1} = M(\lambda) D(\lambda)^{-1} N(\lambda),$$

and by (A.27) we arrive at

$$F(\lambda)^{-1} = \sum_{k=1}^{J} (\lambda - \lambda_0)^{m_k} \langle \cdot, N^*(\overline{\lambda}) Y_k(\overline{\lambda}) \rangle_{\mathbb{C}^n} M(\lambda) X_k(\lambda)$$
$$+ M(\lambda) \Gamma_2(\lambda) N(\lambda).$$

By Proposition A.5.1, the singular parts of $M(\lambda) X_k(\lambda)$, $k = 1, \ldots, J$, and $N^*(\lambda) Y_k(\lambda)$ form a canonical generating system in $S(F, \lambda_0)$ and $S(F^*, \overline{\lambda}_0)$. According to Proposition A.5.1, an arbitrary canonical generating system in $S(F, \lambda_0)$ can be represented as $SP(M X_k)$ where $\{X_k\}_{k=1}^{J}$ is a canonical generating system in $S(G, \lambda_0)$. $\square$

A.8 Fredholm Holomorphic Operator Functions

Definition A.8.1. The holomorphic operator function (A.2) is called Fredholm if

a) the operator

$$F(\lambda) : B_1 \to B_2 \qquad (A.28)$$

is Fredholm for all $\lambda \in \Omega$;

b) Operator (A.28) is invertible for at least one value λ.

From this definition and from Proposition A.2.2(i), we obtain

$$\text{ind } F(\lambda) = 0 \quad \text{for all} \quad \lambda \in \Omega. \tag{A.29}$$

In order to obtain an analog of the representation (A.23) for the Fredholm holomorphic operator function (A.2), we need an auxiliary holomorphic matrix function f; this is constructed as follows:

Fix a point λ_0 from the spectrum of F. By (A.29)

$$\dim \ker F(\lambda_0) = \dim \operatorname{coker} F(\lambda_0). \tag{A.30}$$

Denote this dimensions (which is equal to the geometric multiplicity of λ_0) by J.

Choose $h_1, \ldots, h_J \in B_1^*$ to be linearly independent on $\ker F(\lambda_0)$. In other words,

$$\sum_{j=1}^{J} \alpha_j \langle \varphi, h_j \rangle_1 = 0 \quad \text{for all} \quad \varphi \in \ker F(\lambda_0)$$

implies $\alpha_1 = \cdots = \alpha_J = 0$. Let $y_1, \ldots, y_J \in B_2$ be linearly independent modulo $\operatorname{Im} F(\lambda_0)$. In other words,

$$\sum_{j=1}^{J} \alpha_j y_j \in \operatorname{Im} F(\lambda_0)$$

implies $\alpha_1 = \cdots = \alpha_J$. We can make this choice of $\{h_j\}$ and $\{y_j\}$ due to (A.30). Put

$$K = \sum_{j=1}^{J} \langle \cdot, h_j \rangle_1 y_j.$$

Consider the operator function

$$E(\lambda) = F(\lambda) + K, \tag{A.31}$$

which is holomorphic in a neighbourhood of λ_0. By the choice of $\{h_j\}$ and $\{y_j\}$, the kernel of $E(\lambda_0)$ is trivial. By Proposition A.2.2 (ii), $\operatorname{ind} E(\lambda_0) = 0$. Hence $E(\lambda_0)$ is invertible.

We define the required matrix function $f(\lambda)$ by

$$f(\lambda)\xi = \operatorname{col}\Big(\xi_k - \sum_{j=1}^{J} \langle E(\lambda)^{-1} y_j, h_k \rangle_1 \xi_j\Big)_{k=1}^{J} \tag{A.32}$$

for all $\xi = \operatorname{col}(\xi_1, \ldots, \xi_J) \in \mathbb{C}^J$. Clearly, f is holomorphic at λ_0.

Moreover, $f(\lambda_0) = 0$. Indeed, from (A.31) we get

$$\varphi = \sum_{j=1}^{J} \langle \varphi, h_j \rangle_1 E(\lambda_0)^{-1} y_j$$

for all $\varphi \in \ker F(\lambda_0)$. This, together with the linear independence of $h_1, \ldots, h_J$ on $\ker F(\lambda_0)$, implies that $f(\lambda_0)\xi = 0$ for all $\xi \in \mathbb{C}^J$.

Lemma A.8.2. *Let F be a holomorphic operator function in a neighbour-hood of λ_0 and let $F(\lambda_0)$ be a Fredholm operator of zero index. Also let*

$$B = \{\varphi \in B_1; \langle \varphi, h_j \rangle_1 = 0 \quad for \quad j = 1, \ldots, J\}.$$

Then the holomorphic operator functions $F(\lambda)$ and $\mathcal{F}(\lambda) : B \times \mathbb{C}^J \to B \times \mathbb{C}^J$ defined by

$$B \times \mathbb{C}^J \ni (\varphi, \xi) \overset{\mathcal{F}(\lambda)}{\to} (\varphi, f(\lambda)\xi) \in B \times \mathbb{C}^J \tag{A.33}$$

are equivalent.

Proof. Since $E(\lambda)$ is invertible in a neighbourhood of λ_0, we derive from (A.31) that

$$F(\lambda) = E(\lambda)(I - E(\lambda)^{-1}K).$$

We take the basis $(e_j)_{j=1}^J$ in $\ker F(\lambda_0)$ such that

$$\langle e_j, h_k \rangle_1 = \delta_k^j. \tag{A.34}$$

Consider the operator

$$P = \sum_{j=1}^J \langle \cdot, h_j \rangle_1 e_j.$$

Then, by (A.34), P is a projection and $\operatorname{Im} P = \ker F(\lambda_0)$. Moreover, $\operatorname{Im}(I - P) = B$. By the equality $KP = K$ we get

$$I - E(\lambda)^{-1}K = (I - PE(\lambda)^{-1}K)(I - (I - P)E(\lambda)^{-1}K). \tag{A.35}$$

Since

$$(I - (I - P)E(\lambda)^{-1}K)^{-1} = I + (I - P)E(\lambda)^{-1}P,$$

the operator functions $F(\lambda)$ and

$$I - PE(\lambda)^{-1}K : B_1 \to B_1$$

are equivalent.

We introduce the operator

$$B_1 \ni \varphi \overset{Q}{\to} ((I - P)x, \, (\langle x, h_j \rangle_1)_{j=1}^J) \in B \times \mathbb{C}^J.$$

One directly verifies that the inverse Q^{-1} is given by

$$B \times \mathbb{C}^J \ni (x_1, \xi) \to x_1 + \sum_{j=1}^J \xi_j e_j.$$

By (A.32) we have

$$Q(I - PE(\lambda)^{-1}K)Q^{-1}(x_1, \xi) = (x_1, f(\lambda)\xi).$$

This, together with (A.35), proves the lemma. $\square$

The following assertion follows from Proposition 1.4, the above lemma and the results of Sect. A.6.

Lemma A.8.3. *Let F be a holomorphic operator function in a neighbourhood of λ_0 and let $F(\lambda_0)$ be a Fredholm operator with zero index. Then:*

(i) λ_0 is an isolated point of the spectrum of F if and only if λ_0 is an eigenvalue of f of finite algebraic multiplicity.

(ii) There exists a neighbourhood V of λ_0 such that for every $\lambda \in V$ the invertibility of $F(\lambda)$ is equivalent to the invertibility of the matrix function $f(\lambda)$.

(iii) If λ_0 is an eigenvalue of F of finite algebraic multiplicity then λ_0 is an eigenvalue of f with the same geometric, partial and algebraic multiplicities.

Proposition A.8.4. *Let (A.2) be a Fredholm holomorphic operator function. Then the spectrum of F consists of isolated eigenvalues of finite algebraic multiplicity.*

Proof. Let Λ be the spectrum of F. By definition, $\Omega \backslash \Lambda$ is not empty. Let $\lambda_0 \in \Omega \cap \partial \Lambda$ and f be a holomorphic matrix function (A.32). By Lemma A.8.3(ii), there exists a neighbourhood V of λ_0 such that the invertibility of F is equivalent to the invertibility of f on V. Since V contains a regular point of F, the determinant of f does not vanish identically. By Corollary A.6.5 and by Lemma A.8.3(i), there exists a neighbourhood V_1 such that $V_1 \cap \partial \Lambda = \{\lambda_0\}$ and λ_0 has a finite algebraic multiplicity. Since the boundary of the spectrum of F consists of isolated points, it coincides with Λ. $\square$

A.9 The Adjoint Holomorphic Operator Function

In what follows we consider the holomorphic operator function (A.2). We shall suppose that there are given sesquilinear forms $\langle \cdot, \cdot \rangle_j$ on $B_j \times B_j^*$, $j = 1, 2$, satisfying properties (i)-(iii) from Sect.A.2. Hence the adjoint operator is defined by (A.1).

Definition A.9.1. The adjoint holomorphic operator function

$$F^* : \{\lambda : \overline{\lambda} \in \Omega\} \to \mathcal{L}(B_2^*, B_1^*), \tag{A.36}$$

is defined by

$$F^*(\lambda) = \left(F(\overline{\lambda})\right)^*. \tag{A.37}$$

Clearly, F^* is holomorphic. In this section we assume that F is Fredholm. By Propositions A.2.3 and A.2.4 this implies that F^* is also Fredholm.

In the proof of the next proposition we need the dual of B, as well as the operator function $\mathcal{F}^*(\lambda)$ adjoint to $\mathcal{F}(\lambda)$ (see (A.33)). Due to (A.34),

$$B^* = \{\psi \in B_1^* : \langle e_j, \psi \rangle_1 = 0 \quad \text{for} \quad j = 1, \dots, J\}. \tag{A.38}$$

Put

$$\langle(\varphi,\xi),(\psi,\eta)\rangle_0 = \langle\varphi,\psi\rangle_1 + \langle\xi,\eta\rangle_{\mathbb{C}^J}. \tag{A.39}$$

Then $\langle\cdot,\cdot\rangle_0$ is a duality on $(B\times\mathbb{C}^J)\times(B^*\times\mathbb{C}^J)$ satisfying (i)-(iii) of Sect.A.2. The adjoint holomorphic operator function $\mathcal{F}^*(\lambda)$ is defined with respect to this duality.

Proposition A.9.2. *Let F be a Fredholm holomorphic operator function (A.2). Then*

(i) *$\lambda_0 \in \mathbb{C}$ is an eigenvalue of F if and only if $\overline{\lambda}_0$ is an eigenvalue of F^*;*

(ii) *the geometric, partial and algebraic multiplicities of λ_0 and $\overline{\lambda}_0$ coincide.*

Proof. By Lemma A.8.2,

$$F(\lambda) = N(\lambda)\mathcal{F}(\lambda)M(\lambda) \tag{A.40}$$

in a neighbourhood of λ_0, where

$$M(\lambda):\ B_1 \to B\times\mathbb{C}^J, N(\lambda):\ B\times\mathbb{C}^J \to B_2$$

are holomorphic at λ_0 and such that $M(\lambda_0)$ and $N(\lambda_0)$ are invertible. Formula (A.40) implies

$$F^*(\lambda) = M^*(\lambda)\mathcal{F}^*(\lambda)N^*(\lambda),$$

where $\mathcal{F}^*(\lambda)$ is the holomorphic operator function:

$$B^*\times\mathbb{C}^J \ni (\varphi,\xi) \overset{\mathcal{F}^*(\lambda)}{\to} (\varphi, f^*(\lambda)\xi) \in B^*\times\mathbb{C}^J.$$

Using Proposition A.5.1 we get that $\overline{\lambda}_0$ is the eigenvalue of F^* and f^* simultaneously, and that its geometric, partial and algebraic multiplicities are the same for both operator functions. Reference to Lemma A.8.3(iii) and Proposition A.7.1 completes the proof. $\square$

The next assertion on zeros of $F^*(\overline{\lambda}_0)$ will be used in Sect.A.10.

Lemma A.9.3. *Let λ_0 be an eigenvalue of F and let $\psi \in \ker F^*(\overline{\lambda}_0)$. If*

$$\langle F(\lambda)\Phi(\lambda),\ \psi\rangle_2 \Big|_{\lambda=\lambda_0} = 0 \tag{A.41}$$

for all $\Phi \in S(F,\lambda_0)$ then $\psi = 0$.

Proof. Suppose that (A.41) holds and $\psi \neq 0$. Choose a canonical generating system

$$\{\Phi_j\}_{j=1}^J$$

in $S(F,\lambda_0)$, and take a basis

$$\{\psi_j\}_{j=1}^J$$

in $\ker\ F^*(\overline{\lambda}_0)$ such that $\psi_1 = \psi$. Since $F(\lambda)\Phi_j(\lambda) = O(1)$, there exist constants $\alpha_1,\ldots,\alpha_J$ such that

$$|\alpha_1| + \ldots + |\alpha_J| \neq 0$$

and

$$\langle F(\lambda) \sum_{k=1}^{J} \alpha_k \Phi_k(\lambda), \psi_s \rangle \Big|_{\lambda=\lambda_0} = 0 \tag{A.42}$$

for $s = 2, \ldots, J$. Put

$$\Phi(\lambda) = \sum_{k=1}^{J} \alpha_k \Phi_k(\lambda),$$

and let

$$h = F(\lambda)\Phi(\lambda)\Big|_{\lambda=\lambda_0}.$$

Due to (A.41) and (A.42),

$$\langle h, \psi_s \rangle_2 = 0 \quad \text{for} \quad s = 1, \ldots, J.$$

Hence there exists at least one solution φ of

$$F(\lambda_0)\varphi = -h.$$

By the definition of h,

$$(\lambda - \lambda_0)^{-1}(\varphi + \Phi(\lambda)) \in S(F, \lambda_0). \tag{A.43}$$

We show that the last inclusion is impossible. Let m denote the degree of the operator function on the left-hand side of (A.43). The leading coefficient can be written as

$$\sum_{m_k=m-1} \alpha_k \varphi_k,$$

where φ_k is the leading coefficient of Φ_k and $m_k = \deg \Phi_k$. By (A.43),

$$(\lambda - \lambda_0)^{-1}(\varphi + \Phi(\lambda)) = \sum_{j=1}^{J} p_j(\lambda)\Phi_j(\lambda), \tag{A.44}$$

where $p_j(\lambda)$ are polynomials. The principal singularity of (A.44) is

$$(\lambda - \lambda_0)^{-m} \sum_{m_j \geq m} c_j \varphi_j,$$

where $c_j = const$. However, the equality

$$\sum_{m_k=m-1} \alpha_k \varphi_k = \sum_{m_j \geq m} c_j \varphi_j$$

is impossible because of the linear independence of $\{\varphi_k\}$. This contradiction shows that $\psi = 0$. $\square$

A.10 The Structure of $F(\lambda)^{-1}$ Near the Pole

We state the main result of this chapter.

Theorem A.10.1. *Let F be a Fredholm holomorphic operator function in a neighbourhood of $\lambda_0 \in \mathbb{C}$. Let λ_0 be an eigenvalue of F and let J and $m_1, \ldots, m_J$ be its geometric and partial multiplicities. Suppose that the vector functions $\Phi_1, \ldots, \Phi_J$ form a canonical generating system in $S(F, \lambda_0)$. Then*

(i) *There exists a unique canonical generating system*

$$\{\Psi_j\}_{j=1}^J \quad in \quad S(F^*, \overline{\lambda}_0)$$

such that in a neighbourhood of λ_0:

$$F(\lambda)^{-1} = \sum_{j=1}^J (\lambda - \lambda_0)^{m_j} \langle \,\cdot\,, \Psi_j(\overline{\lambda}) \rangle_2 \Phi_j(\lambda) + \Gamma(\lambda) , \tag{A.45}$$

where Γ is holomorphic in a neighborhood of λ_0.

(ii) *The collection $\{\Psi_j\}_{j=1}^J$ in (i) satisfies*

$$\langle F(\lambda)\Phi_k(\lambda), \, \Psi_j(\overline{\lambda}) \rangle_2 = \delta_k^j (\lambda - \lambda_0)^{-m_j} + O(1) \quad as \quad \lambda \to \lambda_0 , \tag{A.46}$$

where $\delta_k^j = 0$ if $j \neq k$ and $\delta_k^k = 1$.

(iii) *If $\{\Psi_j\}_{j=1}^J$ is a collection in $S(F^*, \overline{\lambda}_0)$ subject to (A.46), then it is the canonical generating system satisfying (i).*

Proof. (i) *(Existence)* By Theorem A.7.2 there exist canonical generating systems

$$\{X_j^{(0)}\}_{j=1}^J \quad and \quad \{Y_j^{(0)}\}_{j=1}^J$$

in $S(f, \lambda_0)$ and $S(f^*, \overline{\lambda}_0)$ respectively, such that

$$f(\lambda)^{-1} = \sum_{j=1}^J (\lambda - \lambda_0)^{m_j} \langle \cdot, Y_j^{(0)}(\overline{\lambda}) \rangle_{\mathbb{C}^n} X_j^{(0)}(\lambda) + \Gamma_1(\lambda), \tag{A.47}$$

where Γ_1 is a holomorphic matrix function in a neighbourhood of λ_0.

By definition of $\mathcal{F}(\lambda)$ (see (A.33)), the vector functions

$$X_j = \left(0, X_j^{(0)}\right), \quad j = 1, \ldots, J,$$

form a canonical generating system in $S(\mathcal{F}, \lambda_0)$ and

$$Y_j = \left(0, Y_j^{(0)}\right), \quad j = 1, \ldots, J,$$

form a canonical generating system in $S(\mathcal{F}^*, \overline{\lambda}_0)$. Formula (A.47) implies

$$\mathcal{F}(\lambda)^{-1} = \sum_{j=1}^{J} (\lambda - \lambda_0)^{m_j} \langle \cdot, Y_j(\overline{\lambda}) \rangle_0 X_j(\lambda) + \Gamma_2(\lambda) \qquad (A.48)$$

with the duality $\langle \cdot, \cdot \rangle_0$ introduced before Proposition A.9.2. The function Γ_2 is holomorphic in a neighbourhood of λ_0.

Furthermore, by Lemma A.8.2,

$$F(\lambda) = N(\lambda)\mathcal{F}(\lambda)M(\lambda),$$

where $N(\lambda_0)$, $M(\lambda_0)$ are invertible. Hence, using (A.48), we arrive at

$$F(\lambda)^{-1} = \sum_{j=1}^{J} (\lambda - \lambda_0)^{m_j} \langle \cdot, \left(N^*(\overline{\lambda})\right)^{-1} Y_j(\overline{\lambda}) \rangle_2 \left(M(\lambda)\right)^{-1} X_j(\lambda)$$
$$+ \left(M(\lambda)\right)^{-1} \Gamma_2(\lambda) \left(N(\lambda)\right)^{-1}. \qquad (A.49)$$

Applying Proposition A.5.1 we obtain that the functions

$$\Phi_j^{(0)}(\lambda) = SP\left(\left(M(\lambda)\right)^{-1} X_j(\lambda)\right), \ \ j, \dots, J,$$

form a canonical generating system in $S(F, \lambda_0)$, and that

$$\Psi_j^{(0)} = SP\left(\left(N^*(\lambda)\right)^{-1} Y_j(\lambda)\right), \ \ j = 1, \dots, J,$$

form a canonical generating system in $S(F^*, \overline{\lambda}_0)$. Using these notations we can write (A.49) as

$$F(\lambda)^{-1} = \sum_{j=1}^{J} (\lambda - \lambda_0)^{m_j} \langle \cdot, \Psi_j^{(0)}(\overline{\lambda}) \rangle_2 \Phi_j^{(0)}(\lambda) + \Gamma_3(\lambda), \qquad (A.50)$$

where Γ_3 is holomorphic at λ_0.

Now let

$$\{\Phi_j\}_{j=1}^{J}$$

be an arbitrary canonical basis in $S(F, \lambda_0)$. By Proposition A.4.6 there exist polynomials p_{jk} satisfying the properties (α) and (β) of that proposition and subject to

$$\Phi_j^{(0)}(\lambda) = SP \sum_{k=1}^{J} p_{jk}(\lambda) \Phi_j(\lambda).$$

Hence

$$SP \ F(\lambda)^{-1} = SP \sum_{j=1}^{J} \sum_{k=1}^{J} (\lambda - \lambda_0)^{m_j} \langle \cdot, \Psi_j^{(0)}(\overline{\lambda}) \rangle_2 p_{jk}(\lambda) \Phi_j(\lambda)$$

$$= SP \sum^{J} (\lambda - \lambda_0)^{m_k} \langle \cdot, \Psi_k(\overline{\lambda}) \rangle_2 \Phi_k(\lambda)$$

where

$$\Psi_k(\lambda) = SP \sum_{j=1}^{J} (\lambda - \bar{\lambda}_0)^{m_j - m_k} \overline{p_{jk}(\bar{\lambda})} \Psi_j^{(0)}(\lambda).$$

By conditions (α), (β) in Proposition A.4.6, the vector functions

$$P_{kj}(\lambda) = (\lambda - \bar{\lambda}_0)^{m_j - m_k} \overline{p_{jk}(\bar{\lambda})}$$

satisfy

(α') $P_{kj}(\lambda)$ is polynomial and has $\bar{\lambda}_0$ as zero of order $m_j - m_k$ if $m_j > m_k$,

(β') the matrix $\left(P_{kj}(\bar{\lambda}_0)\right)_{k,j=1}^{J}$ is invertible.

Hence, by Proposition A.4.6, $\{\Psi_j\}_{j=1}^{J}$ is a canonical generating system in $S(F^*, \bar{\lambda}_0)$.

(i) (*Uniqueness*) Let

$$\{\Psi_j\}_{j=1}^{J} \quad \text{and} \quad \{\Psi_j^{(0)}\}_{j=1}^{J}$$

be two canonical generating systems satisfying (A.45) with the same $\{\Phi_j\}_{j=1}^{J}$ but with possibly different Γ. Then

$$SP \sum_{j=1}^{J} (\lambda - \lambda_0)^{m_j} \langle x, \Psi_j(\bar{\lambda}) - \Psi_j^{(0)}(\bar{\lambda}) \rangle_2 \Phi_j(\lambda) = 0$$

for all $x \in B_1$. Since the elements

$$\varphi_j = (\lambda - \lambda_0)^{m_j} \Phi_j(\lambda) \Big|_{\lambda = \lambda_0}, \quad j = 1, \ldots, J,$$

are linearly independent we get $\Psi_j = \Psi_j^{(0)}$.

(ii) Formula (A.45) implies

$$g = \sum_{j=1}^{J} (\lambda - \lambda_0)^{m_j} \langle g, \Psi_j(\bar{\lambda}) \rangle_2 F(\lambda) \Phi_j(\lambda) + F(\lambda) \Gamma(\lambda) g. \qquad (A.51)$$

Since $F^*(\lambda)\Psi_k(\lambda) = O(1)$ as $\lambda \to \bar{\lambda}_0$, we get from (A.51):

$$\sum_{j=1}^{J} (\lambda - \lambda_0)^{m_j} \langle g, \Psi_j(\bar{\lambda}) \rangle_2 \langle F(\lambda) \Phi_j(\lambda), \Psi_k(\bar{\lambda}) \rangle_2$$

$$= \langle g, \Psi_k(\bar{\lambda}) \rangle_2 + O(1) \quad \text{as} \quad \lambda \to \lambda_0$$

for arbitrary $g \in B_2$. Hence

$$(\lambda - \lambda_0)^{m_j} \langle F(\lambda) \Phi_j(\lambda), \Psi_k(\bar{\lambda}) \rangle_2 = \delta_j^k + O(|\lambda - \lambda_0|^{m_j}),$$

and we arrive at (A.46).

(iii) Since the canonical generating systems in (A.45) satisfy (A.46), it suffices to prove that the collection

$$\{\Psi_j\}_{j=1}^J$$

subject to (A.46) is unique. The uniqueness directly follows from Lemma A.9.3. $\square$

We restate Theorem A.10.1 in terms of Jordan chains of the operator function F.

Theorem A.10.2. *Let F be a Fredholm holomorphic operator function in a neighbourhood of $\lambda_0 \in \mathbb{C}$. Let λ_0 be an eigenvalue of F and let J and $m_1, \ldots, m_J$ be its geometric and partial multiplicities. Suppose that*

$$\{\varphi_{k,s}\}, \ s = 0, \ldots, m_k - 1, \ k = 1, \ldots, J,$$

is a canonical system of Jordan chains of F corresponding to λ_0. Then

(i) There exists a unique canonical system of Jordan chains

$$\{\psi_{k,s}\}, \ s = 0, \ldots, m_k - 1, \ k = 1, \ldots, J,$$

of F^* corresponding to $\overline{\lambda}_0$ such that in a neighbourhood of λ_0

$$F(\lambda)^{-1} = \sum_{k=1}^{J} \sum_{h=0}^{m_k-1} \frac{P_{k,h}}{(\lambda - \lambda_0)^{m_k-h}} + \Gamma(\lambda), \tag{A.52}$$

where

$$P_{k,h} = \sum_{s=0}^{h} \langle \cdot, \psi_{k,s} \rangle_2 \, \varphi_{k,h-s} \tag{A.53}$$

and Γ is holomorphic in a neighbourhood of λ_0.

(ii) The canonical system $\{\psi_{k,s}\}$ in (i) satisfies

$$\sum_{\substack{\sigma,h,s\geq 0 \\ \sigma+h+s=m_k+n}} \frac{1}{\sigma!} \langle F^{(\sigma)}(\lambda_0)\varphi_{k,h}, \psi_{j,s} \rangle_2 = \delta_k^j \delta_n^0 \tag{A.54}$$

for $n = 0, \ldots, m_j - 1$. Here

$$\varphi_{k,h} = 0, \ \psi_{j,s} = 0 \quad for \quad h \geq m_k, \ s \geq m_j. \tag{A.55}$$

(iii) If

$$\psi_{j,0}, \ldots, \psi_{j,\nu_j-1}, \ j = 1, \ldots, J, \tag{A.56}$$

is a collection of Jordan chains of F^* corresponding to $\overline{\lambda}_0$ which is subject to (A.54), then $\nu_k = m_k$ and the collection (A.56) is the canonical system satisfying (i).

Proof. (i) Put

$$\Phi_k(\lambda) = \sum_{s=1}^{m_k-1} \frac{\varphi_{k,s}}{(\lambda - \lambda_0)^{m_k-s}} \qquad (A.57)$$

and

$$\Psi_k(\lambda) = \sum_{s=1}^{m_k-1} \frac{\psi_{k,s}}{(\lambda - \overline{\lambda}_0)^{m_k-s}}. \qquad (A.58)$$

Then due to Proposition A.4.4, representation (A.52) is equivalent to (A.45).

(ii) Relations (A.54) follow from (A.46).

(iii) By applying Theorem A.10.1(iii) to

$$\{\Phi_k\}_{k=1}^J \quad \text{and} \quad \{\Psi_k\}_{k=1}^J,$$

where

$$\Psi_k(\lambda) = \sum_{s=1}^{\nu_k-1} \frac{\psi_{k,s}}{(\lambda - \overline{\lambda}_0)^{\nu_k-s}},$$

and by equating the coefficients of the same power of $(\lambda - \lambda_0)$ in both sides of (A.46), we obtain the desired assertion. $\square$

The next remarks contain modified forms of the biorthogonality relations (A.45) and (A.54).

Remark A.10.3. Since $F(\lambda)\Phi_k(\lambda) = O(1)$ near λ_0, we can use (A.57) to see that

$$\sum_{\sigma+s=q} \frac{1}{\sigma!} F^{(\sigma)}(\lambda_0)\varphi_{k,s} = 0$$

for $q = 0, \ldots, m_k - 1$. Hence, we rewrite (A.54) as

$$\sum_{q=0}^{n} \sum_{\sigma=0}^{m_k+n-q} \frac{1}{\sigma!} \langle F^{(\sigma)}(\lambda_0)\varphi_{k,m_k+n-q-\sigma}, \psi_{j,q} \rangle_2 = \delta_k^j \, \delta_n^0. \qquad (A.59)$$

Putting $s = n - q$ and using (A.55) we get

$$\sum_{s=0}^{n} \sum_{\sigma=s+1}^{m_k+s} \frac{1}{\sigma!} \langle F^{(\sigma)}(\lambda_0)\varphi_{k,m_k+s-\sigma}, \psi_{j,n-s} \rangle_2 = \delta_k^j \, \delta_n^0. \qquad (A.60)$$

Remark A.10.4. Since $F^*(\lambda)\Psi_j(\lambda) = O(1)$ as $\lambda \to \lambda_0$, we can use (A.58) to see that

$$\sum_{\sigma+s=q} \frac{1}{\sigma!} \left(F^{(\sigma)}(\lambda_0) \right)^* \psi_{j,s} = 0$$

for $q = 0, \ldots, m_j - 1$. Hence, we rewrite (A.54) as

$$\sum_{h=0}^{n} \sum_{\sigma=0}^{m_k+n-h} \frac{1}{\sigma!} \langle F^{(\sigma)}(\lambda_0)\varphi_{k,h}, \ \psi_{j,m_k+n-h-\sigma}\rangle_2 = \delta_k^j \, \delta_n^0 \ .$$

This can be rewritten as

$$\sum_{s=0}^{n} \sum_{\sigma=s+1}^{m_k+s} \frac{1}{\sigma!} \langle F^{(\sigma)}(\lambda_0)\varphi_{k,n-s}, \psi_{j,m_k+s-\sigma}\rangle_2 = \delta_k^j \, \delta_n^0 \tag{A.61}$$

after setting $s = n - h$ and using (A.55).

References

Agmon, S., Douglis, A. and Nirenberg, L. (1959, 1964): Estimates near the boundary for solutions of elliptic partial equations satisfying general boundary conditions I, II. *Comm. Pure Appl. Math.* **12**(1959) 623-729, **17**(1964) 35-92.

Agmon, S., Nirenberg, L. (1963): Properties of solutions of ordinary differential equations in Banach space, *Comm. Pure Appl. Math.* **16**, 121-239.

Agranovich, M.S., Vishik, M.I. (1964): Elliptic problems with parameter and parabolic problems of general type, *Usp. Mat. Nauk* **19**:3, 53-161 (Russian).

Bagirov, L.A., Kondratiev, V.A. (1991): On the asymptotics of solutions to differential equations in the Hilbert space, *Mat. Sb.* **182**:4, 508-525 (Russian).

Birkhoff, G.D. (1908): On the asymptotic character of solutions of certain linear differential equations containing a parameter, *Trans. Amer. Math. Soc.* **9**, 219-231.

Dalec'kiǐ, Ju.L., Krein, M.G. (1974): *Stability of Solutions of Differential Equations in Banach Space*, Amer. Math. Soc., Providence, Rhode Island.

Dauge, M. (1996): Strongly elliptic problems near cuspidal points and edges, *Partial Differential Equations and Functional Analysis, in Memory of Pierre Grisvard,* Birkhäuser, Boston-Basel-Berlin, 93-110.

Davies, E. B. (1980):*One-Parameter Semigroups*, London Math. Soc. Monographs **15**, Academic Press, Inc. , London-New York.

Demidovich, B. P. (1967): *Lectures on the mathematical theory of stability*, Nauka, Moscow (Russian).

Eastham, M.S.P. (1989): *The Asymptotic Solution of Linear Differential Systems. Applications of the Levinson Theorem*, Clarendon Press, Oxford.

Evgrafov, M.A. (1960): Structure of solutions of exponential growth for some operator equations, *Trudy Mat. Inst. Steklov* **60**, 145–180 (Russian).

Fattorini, H.O. (1984): *The Cauchy Problem*, Encyclopedia of Mathematics and its Applications **18**, Addison-Wesley Publishing Co., Reading, Mass.

Fedorjuk, M.V. (1969): Asymptotic methods in the theory of ordinary linear differential equation, *Mat. Sb.* **79**, 477–516. English translation: *Math. USSR Sb.* **8**:4, 451-491.

Gantmacher, F. R. (1959): *The Theory of Matrices*, Vols. 1, 2., Chelsea Publishing Co., New York.

Gohberg, I.C., Krein, M.G. (1969): *Introduction to the Theory of Linear Nonselfadjoint Operators*, Amer. Math. Soc., Providence, Rhode Island.

Gohberg, I.C., Sigal, E.I. (1971): An operator generalization of the logarithmic residue theorem and the theorem of Rouché, *Mat. Sb.* **84**:4, 607–629 (Russian). English translation: *Math. USSR Sb.* **13**, 603–625.

Gohberg, I., Lancaster, P. and Rodman, L. (1982): *Matrix Polynomials*, Academic Press, Inc. , New York-London.

Gohberg, I., Goldberg, S. and Kaashoek, M.A. (1990): *Classes of Linear Operators, Vol. I,* Oper. Theory: Adv. Appl. **49**, Birkhäuser, Basel-Boston-Berlin

Goldstein, J. A. (1993): A survey of semigroups of linear operators and applications. *Semigroups of linear and nonlinear operations and applications*, 9–57, Kluwer Acad. Publ., Dordrecht.

Grisvard, P. (1985): *Elliptic Problems in Nonsmooth Domains*, Monographs and Studies in Mathematics **21**, Pitman, Boston.

Harris, W.A., Lutz, D.A. (1975): Asymptotic integration of adiabatic oscillators, *J. Math.Anal. Appl.* **51**, 76-93.

Hartman, Ph. (1964): *Ordinary Differential Equations*, John Wiley & Sons, Inc., New York-London-Sydney.

Hille, E., Phillips, R. S. (1974): *Functional Analysis and Semi-Groups*, Amer. Math. Soc., Providence, Rhode Island.

Hörmander, L. (1976): *Linear Partial Differential Operators*, Springer, Berlin-Heidelberg-New York.

John, F. (1955): *Plane Waves and Spherical Means Applied to Partial Differential Equations*, Interscience Publishers, New York-London.

Keldysh, M.V. (1951): On the eigenvalues and eigenfunctions of certain classes of non-selfadjoint linear operators, *Dokl. Akad. Nauk SSSR* **77**, 11-14 (Russian).

Keldysh, M.V. (1971): On the completeness of eigenfunctions of some classes of non-selfadjoint linear operators, *Usp. Mat. Nauk* **26**, 15-41 (Russian).

Kondratiev, V.A. (1967): Boundary value problems for elliptic equations in domains with conic or angular points, *Trudy Mosk. Mat. Obshch.* **16**, 209-292. English transaction: in *Moscow Math. Soc.* **10**, 227-313.

Kozlov, V.A. (1989): On singularities of solutions of the Dirichlet problem for elliptic equations in the neighbourhood of corner points, *Algebra i Analiz* **1**:4, 161-177. English translation: *Leningrad Math. J.* **1**:4 (1990) 976-982.

Kozlov, V.A. (1991): On the spectrum of the pencil generated by the Dirichlet problem for an elliptic equation in an angle, *Sib. Mat. J.* **32**:2, 74-87. English translation: *Siberian Math. J.* **32**:2, 238-251.

Kozlov, V.A., Maz'ya, V.G. (1985): The estimates for L_p-averages and asymptotics of solutions of elliptic boundary value problems in a cone I, *Seminar Analysis Operator Equat. and Numer. Anal. 1985/86 Karl-Weierstrass-Institut für Mathematik*, Berlin, 55-92 (Russian).

Kozlov, V.A., Maz'ya, V.G. (1988): Estimates for L_p-averages and asymptotics of solutions of elliptic boundary value problems in a cone II, *Math. Nachr.* **137**, 113-139 (Russian).

Kozlov, V.A., Maz'ya, V.G. (1988): Spectral properties of the operator bundles generated by elliptic boundary value problems in a cone, *Funk. Anal. i Prilozh.* **22**:2, 38-46 (Russian). English translation: *Funct. Anal. Appl.* **22**:2, 114-121.

Kozlov, V.A., Maz'ya, V.G. (1991): On the spectrum of the operator pencil generated by the Dirichlet problem in a cone, *Mat. Sb.* **182**:5, 638-660. English translation: *Math. USSR Sb.* **73**:1 (1992) 27-48.

Kozlov, V.A., Maz'ya, V.G. (1992): Solvability and asymptotic behaviour of solutions of ordinary differential equations with variable operator coefficients, *Journées "Equations aux Dérivées Partielles"*, Saint Jean de Monts, 1er au 5 Juin 1992, V-1–V-12.

Kozlov, V.A., Maz'ya, V.G. (1991-1996): On the asymptotic behaviour of solutions of ordinary differential equations with operator coefficients 1-7, LiTH-MAT-R-91-47 (1991), LiTH-MAT-R-92-18 (1992), LiTH-MAT-R-92-29, (1992), LiTH-MAT-R-92-40 (1992), LiTH-MAT-R-92-53 (1992), LiTH-MAT-R-93-04 (1993), LiTH-MAT-R-96-28 (1996), Linköping University, Sweden.

Kozlov, V.A., Maz'ya, V.G. (1996): Singularities to solutions in mathematical physics problems in non-smooth domains, In *Partial Differential Equations and*

Functional Analysis (in memory of Pierre Grisvard), Birkhäuser, Boston-Basel-Berlin.

Kozlov, V.A., Maz'ya, V.G. (1997): *Theory of a Higher Order Sturm-Liouville Equation*, Lecture Notes in Math. **1659**, Springer, Berlin-Heidelberg.

Kozlov, V.A., Maz'ya, V.G. and Rossmann, J. (1997): *Elliptic Boundary Value Problems in Domains with Point Singularities*, Math. Surveys and Monograhps **52**, Amer. Math. Soc., Providence, Rhode Island.

Krasnosel'skii, M.A., Burd, V.Sh. and Kolesov, Yu.S. (1973): *Nonlinear Almost Periodic Oscillations*, John Wiley & Sons, New York-Toronto..

Krein, S.G. (1971): *Linear Differential Equations in Banach Space*, Translations of Mathematical Monographs **29**, Amer. Math. Soc., Providence, Rhode Island.

Ladyzhenskaya, O.A. (1985): *The Boundary Value Problems of Mathematical Physics*, Applied Mathematical Sciences **49**, Springer, New York-Berlin.

Ladyzhenskaya, O. A., Ural'tseva, N. N. (1968): *Linear and Quasilinear Elliptic Equations*, Academic Press, New York-London.

Lax, P.D. (1957): A Phragmén-Lindelöf theorem in harmonic analysis with application to some questions in the theory of elliptic equations, *Comm. Pure Appl. Math.* **10**, 361-389.

Lefschetz, S. (1969):*Differential Equations: Geometric Theory*, John Wiley & Sons, Inc., New York-London.

Lions, J.-L., Magenes, E. (1972): *Non-Homogeneous Boundary Value Problems and Applications I*, Springer, Berlin-Heidelberg.

Lomov, S.A. (1992): *Introduction to the General Theory of Singular Perturbations*, Amer. Math. Soc., Providence, Rhode Island.

Markus, A.S. (1980): *Introduction to the Spectral Theory of Polynomial Operator Pencils*, Translations of Mathematical Monographs **71**, Amer. Math. Soc., Providence, Rhode Island.

Massera, J. L., Schäffer, J. J. (1966): *Linear Differential Equations and Function Spaces*, Pure and Applied Mathematics **21**, Academic Press, New York-London.

Maz'ya, V.G. (1985): *Sobolev Spaces*, Springer, Berlin-New York.

Maz'ya, V.G., Plamenevskii, B.A. (1972): On the asymptotic behaviour of solutions of differential equations with operator coefficients, *Dokl. Akad. Nauk SSSR Ser. Mat.* **36**, 512-515. English translation: *Soviet Math. Dokl.* **12**, 173-177.

Maz'ya, V.G., Plamenevskii, B.A. (1972): On the asymptotic behaviour of solutions of differential equations in Hilbert space, *Izv. Akad. Nauk SSSR Ser. Mat.* **36**:5. English translation: *Math. USSR Izv.* **6**:5, 1067-1116.

Maz'ya, V.G., Plamenevskii, B.A. (1975): On the coefficients in the asymptotics of solutions of elliptic boundary value problems in a cone, *Zap. Nauch. Sem. LOMI* **52**, 110-127. (Russian).

Maz'ya, V.G., Plamenevskii, B.A. (1977): On the coefficients in the asymptotics of solutions of elliptic boundary value problems in a region with conic points, *Math. Nachr.* **76**, 29-60. English translation: *Amer. Math. Soc. Transl.* **123** (1984) 57-88.

Maz'ya, V.G., Plamenevskii, B.A. (1977): On the asymptotic of the solution of the Dirichlet problem near an isolated singularity of the boundary, *Vestnik Leningrad Univ. Mat. Mekh. Astron.* **13**, 59-66. English translation: *Vestnik Liningrad Univ. Math.* **10** (1982) 295-302.

Maz'ya, V.G., Plamenevskii, B.A. (1978): Estimates in L_p and Hölder classes and the Miranda-Agmon maximum principle for solutions of elliptic boundary value problems in domains with singular points on the boundary, *Math. Nachr.* **81**, 25-82. English translation: *Amer. Math. Soc. Transl.* **123** (1984) 1-56.

Maz'ya, V.G., Plamenevskii, B.A. (1979): On the asymptotics of the fundamental solutions of elliptic boundary value problems in domains with conic points, *Probl. Mat. Anal.* **7**, 100-145. English translation: *Sel. Math. Sov.* **4**:4 (1985) 363-397.

Maz'ya, V., Shaposhnikova, T. (1985): *Theory of Multipliers in Spaces of Differentiable Functions*, Monographs and Studies in Mathematics **23**, Pitman, Boston-London-Melbourne.

Mennicken, R., Möller, M. (1984): Root functions, eigenvectors, associated vectors and the inverse of a holomorphic operator function, *Arch. Math.* **42**, 455-463.

Muckenhoupt, B. (1972): Hardy's inequality with weights, *Studia Math.* **44**:1, 31-38.

Nazarov, S.A., Plamenevskii, B.A. (1994): *Elliptic Problems in Domains with Piecewise Smooth Boundaries*, De Gruyter Expositions in Mathematics **13**, Berlin-New York.

Pazy, A. (1967): Asymptotic expansions of solutions of differential equations in Hilbert space, *Arch. Rat. Mech. Anal.* **24**, 193-218.

Pazy, A. (1974): *Semi-Groups of Linear Operators and Applications to Partial Differential Equations*, Lecture Notes **10**, College Park, Univ. of Maryland.

Reed, M., Simon, B. (1972): *Methods of Modern Mathematical Physics I. Functional Analysis*, Academic Press, New York-San Francisco-London.

Plamenevskii, B.A. (1972): On the existence and asymptotics of solutions of differential equations with unbounded operator coefficients in a Banach space, *Izv. Akad. Nauk SSSR Ser. Mat.* **36**:6 1134-1401; erratum, ibid. **37** (1973), 959. English translation: *Math. USSR Izv.* **6**:6, 1327-1379.

Plamenevskii, B.A. (1973): The asymptotic behavior of the solutions of quasielliptic differential equations with operator coefficients, *Izv. Akad. Nauk SSSR Ser. Mat.* **37**, 1332–1375 (Russian).

Poincaré, H. (1886): Sur les intégrales irrégulières des équations linéaires, *Acta Math.* **8**, 295-344.

Prössdorf, S. (1978): *Some Classes of Singular Equations*, North-Holland Mathematical Library **17**, North-Holland Publishing Co., Amsterdam-New York.

Rudin, W. (1973): *Functional Analysis*, McGraw-Hill, New York.

Stepanov, V.D. (1989): On a weighted inequality of Hardy type for derivatives of higher order, *Trudy Mat. Inst. Steklov* **187**, 178-190. English translation: *Steklov Institute of Math.* **3** (1990) 205-220.

Wasov, W. (1976): *Asymptotic Expansions for Ordinary Differential Equations*, Krieger Publishing, New York.

Wendland, W. (1970): Bemerkungen über die Fredholmschen Sätze, In: *Methoden u. Verfahren d. Mathematischen Physik* B.I. Verlag Mannheim **3**, 141-176.

Index of Notation

Operators

Functions and Functionals

Index

Index of Names

Springer Monographs in Mathematics

Printing: Mercedesdruck, Berlin
Binding: Buchbinderei Lüderitz & Bauer, Berlin